Genetic Mapping and DNA Sequencing:
Principles, Analysis and Applications

Genetic Mapping and DNA Sequencing: Principles, Analysis and Applications

Editor: Ashton Ward

CALLISTO
REFERENCE
www.callistoreference.com

Callisto Reference,
118-35 Queens Blvd., Suite 400,
Forest Hills, NY 11375, USA

Visit us on the World Wide Web at:
www.callistoreference.com

ISBN: 978-1-64116-546-4 (Hardback)

Cataloging-in-Publication Data

Genetic mapping and DNA sequencing : principles, analysis and applications / edited by Ashton Ward.
    p. cm.
Includes bibliographical references and index.
ISBN 978-1-64116-546-4
1. Gene mapping. 2. Nucleotide sequence. 3. Gene mapping--Methodology. 4. Nucleotide sequence--Methodology.
5. Genetics--Technique. I. Ward, Ashton.
QH445.2 .G46 2022
572.863 3--dc23

Table of Contents

Preface...VII

Chapter 1 **Insights into the Genome Sequence of *Chromobacterium amazonense* Isolated from a Tropical Freshwater Lake**..1
Alexandre Bueno Santos, Patrícia Silva Costa, Anderson Oliveira do Carmo,
Gabriel da Rocha Fernandes, Larissa Lopes Silva Scholte, Jeronimo Ruiz,
Evanguedes Kalapothakis, Edmar Chartone-Souza and
Andréa Maria Amaral Nascimento

Chapter 2 **Expanding the miRNA Transcriptome of Human Kidney and Renal Cell Carcinoma**..........10
Adam P. Sage, Brenda C. Minatel, Erin A. Marshall, Victor D. Martinez,
Greg L. Stewart, Katey S. S. Enfield and Wan L. Lam

Chapter 3 **The *β*-Lactamase Gene Profile and a Plasmid-Carrying Multiple Heavy Metal Resistance Genes of *Enterobacter cloacae***...20
Chongyang Wu, Chaoqin Lin, Xinyi Zhu, Hongmao Liu, Wangxiao Zhou, Junwan Lu,
Licheng Zhu, Qiyu Bao, Cong Cheng and Yunliang Hu

Chapter 4 **Genome-Wide Identification and Characterization of Fox Genes in the Honeybee, *Apis cerana*, and Comparative Analysis with Other Bee Fox Genes**........................32
Hongyi Nie, Haiyang Geng, Yan Lin, Shupeng Xu, Zhiguo Li, Yazhou Zhao and
Songkun Su

Chapter 5 **Circular RNAs: Biogenesis, Function, and a Role as Possible Cancer Biomarkers**................44
Luka Bolha, Metka Ravnik-Glavač and Damjan Glavač

Chapter 6 **SCAR Marker for Gender Identification in Date Palm (*Phoenix dactylifera* L.) at the Seedling Stage**..63
Fahad Al-Qurainy, Abdulhafed A. Al-Ameri, Salim Khan, Mohammad Nadeem,
Abdel-Rhman Z. Gaafar and Mohamed Tarroum

Chapter 7 **Transcriptome Analysis of the Thymus in Short-Term Calorie-Restricted Mice using RNA-seq**..69
Zehra Omeroğlu Ulu, Salih Ulu, Soner Dogan, Bilge Guvenc Tuna and
Nehir Ozdemir Ozgenturk

Chapter 8 **Transcriptome Comparison Reveals Distinct Selection Patterns in Domesticated and Wild Agave Species, the Important CAM Plants**..79
Xing Huang, Bo Wang, Jingen Xi, Yajie Zhang, Chunping He, Jinlong Zheng,
Jianming Gao, Helong Chen, Shiqing Zhang, Weihuai Wu,
Yanqiong Liang and Kexian Yi

Chapter 9 **Genome-Wide Identification, Phylogeny, and Expression Analysis of ARF Genes Involved in Vegetative Organs Development in Switchgrass**.......................................91
Jianli Wang, Zhenying Wu, Zhongbao Shen, Zetao Bai, Peng Zhong, Lichao Ma,
Duofeng Pan, Ruibo Zhang, Daoming Li, Hailing Zhang, Chunxiang Fu,
Guiqing Han and Changhong Guo

Chapter 10   **Common DNA Variants Accurately Rank an Individual of Extreme Height** 104
Corinne E. Sexton, Mark T. W. Ebbert, Ryan H. Miller, Meganne Ferrel,
Jo Ann T. Tschanz, Christopher D. Corcoran, Perry G. Ridge and
John S. K. Kauwe

Chapter 11   **Evidence of the Complexity of Gene Expression Analysis in Fish Wild
Populations** 111
Mbaye Tine

Chapter 12   **Whole Genome Sequencing of Greater Amberjack (*Seriola dumerili*) for SNP
Identification on Aligned Scaffolds and Genome Structural Variation Analysis
using Parallel Resequencing** 125
Kazuo Araki, Jun-ya Aokic, Junya Kawase, Kazuhisa Hamada, Akiyuki Ozaki,
Hiroshi Fujimoto, Ikki Yamamoto and Hironori Usuki

Chapter 13   **Comparative Genomics of the First and Complete Genome of "*Actinobacillus
porcitonsillarum*" Supports the Novel Species Hypothesis** 137
Valentina Donà and Vincent Perreten

Chapter 14   **Genome-Wide Expression Profiles of Hemp (*Cannabis sativa* L.) in Response
to Drought Stress** 145
Chunsheng Gao, Chaohua Cheng, Lining Zhao, Yongting Yu, Qing Tang,
Pengfei Xin, Touming Liu, Zhun Yan, Yuan Guo and Gonggu Zang

Chapter 15   **Identification of Wheat Inflorescence Development-Related Genes using a
Comparative Transcriptomics Approach** 158
Lingjie Ma, Sheng-Wei Ma, Qingyan Deng, Yang Yuan, Zhaoyan Wei,
Haiyan Jia and Zhengqiang Ma

Chapter 16   **Relationship of SNP rs2645429 in Farnesyl-Diphosphate Farnesyltransferase 1
Gene Promoter with Susceptibility to Lung Cancer** 171
Mehdi Dehghani, Zahra Samani, Hassan Abidi, Leila Manzouri, Reza Mahmoudi,
Saeed Hosseini Teshnizi and Mohsen Nikseresht

Chapter 17   **Marker-Assisted Introgression of *Saltol* QTL Enhances Seedling Stage Salt
Tolerance in the Rice Variety "Pusa Basmati 1"** 178
Vivek Kumar Singh, Brahma Deo Singh, Amit Kumar, Sadhna Maurya,
Subbaiyan Gopala Krishnan, Kunnummal Kurungara Vinod, Madan Pal Singh,
Ranjith Kumar Ellur, Prolay Kumar Bhowmick and Ashok Kumar Singh

Chapter 18   **Enriching Genomic Resources and Transcriptional Profile Analysis of
Miscanthus sinensis under Drought Stress based on RNA Sequencing** 190
Gang Nie, Linkai Huang, Xiao Ma, Zhongjie Ji, Yajie Zhang, Lu Tang and
Xinquan Zhang

Permissions

List of Contributors

Index

Preface

Proximate DNA sequences on a chromosome are inherited together during sexual reproduction. This tendency of DNA sequences is called genetic linkage. On a chromosome, the closer two genes are, the higher are their chances of being inherited together and lower are the chances of recombination between them. Gene mapping refers to the set of methods that allow the identification of the locus of a gene as well as the distances between genes. It offers evidences regarding the genes that are responsible for the inheritance of disease from parent to child, and also provides clues regarding the location of the genes and the chromosome where they are located. DNA sequencing is the process that determines the nucleic acid sequence or the order of nucleotides in the DNA. It encompasses all methods that can determine the order of the four bases- thymine, guanine, adenine and cytosine. The development of genetic maps and the successful sequencing of the human genome have advanced the frontiers of genetics and medical science. This book aims to shed light on some of the unexplored aspects of genetic mapping and DNA sequencing and the recent researches in these domains. It includes some of the vital pieces of work being conducted across the world, on various topics related to these fields. This book will prove immensely beneficial to geneticists, genetic engineers, students and researchers associated with these areas of study.

Significant researches are present in this book. Intensive efforts have been employed by authors to make this book an outstanding discourse. This book contains the enlightening chapters which have been written on the basis of significant researches done by the experts.

Finally, I would also like to thank all the members involved in this book for being a team and meeting all the deadlines for the submission of their respective works. I would also like to thank my friends and family for being supportive in my efforts.

Editor

Insights into the Genome Sequence of *Chromobacterium amazonense* Isolated from a Tropical Freshwater Lake

Alexandre Bueno Santos,[1] Patrícia Silva Costa,[1] Anderson Oliveira do Carmo,[1] Gabriel da Rocha Fernandes,[2] Larissa Lopes Silva Scholte,[2] Jeronimo Ruiz,[2] Evanguedes Kalapothakis,[1] Edmar Chartone-Souza,[1] and Andréa Maria Amaral Nascimento (iD)[1]

[1]*Departamento de Biologia Geral, Instituto de Ciências Biológicas, Universidade Federal de Minas Gerais, Belo Horizonte, MG, Brazil*
[2]*Centro de Pesquisas René Rachou, FIOCRUZ, Belo Horizonte, MG, Brazil*

Correspondence should be addressed to Andréa Maria Amaral Nascimento; amaral@ufmg.br

Academic Editor: Wilfred van IJcken

Members of the genus *Chromobacterium* have been isolated from geographically diverse ecosystems and exhibit considerable metabolic flexibility, as well as biotechnological and pathogenic properties in some species. This study reports the draft assembly and detailed sequence analysis of *Chromobacterium amazonense* strain 56AF. The de novo-assembled genome is 4,556,707 bp in size and contains 4294 protein-coding and 95 RNA genes, including 88 tRNA, six rRNA, and one tmRNA operon. A repertoire of genes implicated in virulence, for example, hemolysin, hemolytic enterotoxins, colicin V, lytic proteins, and Nudix hydrolases, is present. The genome also contains a collection of genes of biotechnological interest, including esterases, lipase, auxins, chitinases, phytoene synthase and phytoene desaturase, polyhydroxyalkanoates, violacein, plastocyanin/azurin, and detoxifying compounds. Importantly, unlike other *Chromobacterium* species, the 56AF genome contains genes for pore-forming toxin alpha-hemolysin, a type IV secretion system, among others. The analysis of the *C. amazonense* strain 56AF genome reveals the versatility, adaptability, and biotechnological potential of this bacterium. This study provides molecular information that may pave the way for further comparative genomics and functional studies involving *Chromobacterium*-related isolates and improves our understanding of the global genomic diversity of *Chromobacterium* species.

1. Introduction

The genus *Chromobacterium*, belonging to the class Betaproteobacteria and family Chromobacteriaceae (formerly Neisseriaceae) [1], was originally proposed in 1881 by Bergonzini [2]. Until 2007, the only validated species within the genus *Chromobacterium* was *C. violaceum*, the type species of the genus. The current taxonomy of *Chromobacterium* consists of 11 recognized species: *C. violaceum* [2], *C. subtsugae* [3], *C. aquaticum* [4], *C. haemolyticum* [5], *C. piscinae*, *C. pseudoviolaceum* [6], *C. vaccinii* [7], *C. amazonense* [8], *C. rhizoryzae* [9], *C. alkanivorans* [10], and *C. sphagni* [11]. It is evident that, since 2007, there has been a rapid taxonomic expansion of the *Chromobacterium* genus.

Species in this genus have been collected from geographically diverse ecosystems. The wide spread of *Chromobacterium* throughout a variety of environments, such as soil, water, and plant from tropical and subtropical regions, is due to its considerable metabolic flexibility [12–16]. *Chromobacterium* has attracted considerable biotechnological interest for its potential as a biocontrol agent, environmental detoxification, and bioprospecting, in addition to industrial and pharmacological uses. Among its important characteristics, there are the production of chitinase (with fungicide, insecticide, and nematicide activities), polyhydroxyalkanoates (biodegradable plastics), violacein (with anticarcinogenic and antimicrobial activities), and cyanide biogenesis associated with gold recovery production [17]. These attributes were confirmed by the complete sequencing of the first strain of the type species *C. violaceum*, ATCC 12472, and by further genomic studies of *C. vaccinii*, *C. subtsugae*, *C. haemolyticum*, *C. piscinae*, *C. aquaticum*, and *C. pseudoviolaceum*, which showed alternative pathways for energy generation, a large number of open reading frames (ORFs) for transport-related proteins, and systems for stress adaptation, indicating the possible versatility and adaptability of this bacterium. Moreover, in spite of the fact that *Chromobacterium* is mainly considered a free-living microorganism, it is recognized as an important opportunistic pathogen and occasionally leads to lethal infections in mammals [18, 19].

Despite the potential importance of *Chromobacterium* in industry, agriculture, and medicine, few genomic analyses have been performed to gain insights into the biotechnological applications and pathogenicity of *Chromobacterium* species. Here, we present high-quality draft genome sequence of *C. amazonense* strain 56AF, isolated from a tropical freshwater lake. This organism was selected for sequencing due to its high resistance to β-lactam and as part of an ongoing effort to investigate the bacterial diversity on protected area (undisturbed environment). Additionally, a comparative genomic analysis of *Chromobacterium* species sequences available in public databases was performed to gain insight into the core and unique genes. To our knowledge, this is the first reported genome sequencing of a *C. amazonense* isolate.

2. Materials and Methods

2.1. Bacterial Isolate and DNA Extraction. *Chromobacterium* sp. strain 56AF was originally recovered in 2005 from the tropical freshwater Lake Dom Helvécio located in the Rio Doce State Park (Atlantic Rain biome), Minas Gerais, Brazil [12]. This has been a RAMSAR site since 2010 (http://www.ramsar.org) in recognition of its importance for the global conservation of biological diversity. Briefly, *Chromobacterium* sp. strain 56 AF was isolated on 25%-strength nutrient agar (Difco Laboratories, USA) at 25°C. More details were presented in a previous study of our group [12].

For the extraction of genomic DNA (gDNA), strain 56AF was grown in nutrient broth medium (Difco Laboratories, USA) at 25°C with shaking at 150 rpm for 24 h. The gDNA was extracted using the PureLink Genomic DNA kit (Thermo Fisher Scientific, USA) according to the manufacturer's instructions. gDNA quantification was performed with a Qubit® fluorometer (Thermo Fisher Scientific).

2.2. Phylogenetic and Phylogenomic Analyses. After the description of new *Chromobacterium* species, the phylogenetic and taxonomic position of *Chromobacterium* sp. strain 56AF was determined using 16S rRNA gene sequences (1474 bp) from 21 Neisseriaceae and Chromobacteriaceae sequences retrieved from the GenBank database (http://www.ncbi.nlm.nih.gov/). Nucleic acids were aligned using MAFFT V7 [20] with iterative refinement using the G-INS-i strategy. To optimize the dataset for phylogenetic analysis, gap-rich columns were removed from the alignment using TrimAl (Gappyout option)

(version 1.3) [21] available in Phylemon [22]. The best fit model for the multiple sequence alignment was estimated using ProtTest (version 3) [23]. Altogether, 12 different evolutionary models (Blosum62, CpREV, Dayhoff, DCMut, JTT, LG, MtArt, MtMam, MtREV, RtREV, VT, and WAG) were tested with the +I, +G, and +F parameters. The evolutionary model that best fit the data was determined according to the Akaike information criterion (AIC), and support values for each node were estimated using the approximate likelihood-ratio test (aLRT).

Digital DNA-DNA hybridization (dDDH) values between the genome sequences of *Chromobacterium* sp. strain 56AF and the available *Chromobacterium* type strains (GenBank assembly accessions GCA_000711885.1, GCA_000007705.1, GCA_000971335.1, GCA_001676875.1, GCA_001953795.1, GCA_001953775.1, and GCA_001855565.1) were estimated using the online analysis tool Genome-Genome-Distance Calculator 2.1 (GGDC) [24]. The recommended distance formula 2 was taken into account to interpret the results, as it is robust against the use of incomplete genome sequences.

2.3. Genome Sequencing and Assembly. The genome of strain 56AF was prepared using the Nextera DNA Library Preparation Kit (Illumina Inc., USA) and sequenced using a paired-end approach on the Illumina MiSeq platform. The average insert length was ~550 bp. Reads were trimmed and filtered with Trimmomatic software (version 0.32) [25], and sequences with a Phred Q score < 22, length < 35 bp, and ambiguous bases were discarded.

Trimmed reads were assembled de novo with Velvet software (version 1.2.10) [26]. Several assemblies were computed using k-mer values from 31 to 119. Assembly results were compared, and the best assembly (k-mer 95) was selected based on the N50 and the lowest number of contigs. This was further aligned with the Mauve Multiple Genome Alignment program against the complete genome of *C. violaceum* ATCC 12472. Possible misassemblies were examined with QUAST software (version 3.0) [27] (Table S1). The genome coverage was estimated by aligning the raw reads against the assembly with BWA [28] and Samtools [29], revealing a sequence coverage of 16,482x.

2.4. Accession Number. This whole-genome shotgun project has been deposited at DDBJ/ENA/GenBank under the accession number MTBD00000000. The version described in this paper is the first version, MTBD01000000.

2.5. Genome Annotation. ORFs were predicted using the RAST server databases. Additional functional annotations were performed with the SEED database and Kyoto Encyclopedia of Genes and Genomes (KEGG). Additionally, gene discovery was performed with Prodigal software [30]. The identified ORFs were checked against prot2003-2014.fa and cog2003-2014 from the National Center for Biotechnology Information nonredundant sequence database (NCBI-Nr) through a Blastp search for the assignment of Clusters of Orthologous Group (COG) functions. tRNA and rRNA genes were identified using the Aragorn and RNAmmer software programs, respectively. The coding density of the genome was determined with Artemis software (version 16.0.0) [31] based on GenBank files generated by the RAST server. A circular genome map was generated using the BLAST Ring Image Generator (BRIG) program [32]. Phage discovery was conducted using the PHAST tool.

2.6. Comparative Genomic Analysis. The *C. amazonense* strain 56AF genome sequence was compared to the genomes of *C. violaceum* ATCC 12472 [17], *C. haemolyticum* DSM 19808 [33], *C. vaccinii* 21-1 [34], *C. piscinae* ND17 [35], *C. aquaticum* CC-SEYA-1 [36], *C. subtsugae* MWU 2920 [37], *C. pseudoviolaceum* LMG3953 [38], and *C. amazonense* CBMAI 310 (unpublished, GenBank assembly accession: GCA_001855565.1). The predicted proteomes were used with OrthoFinder [39] to identify orthologous genes.

3. Results and Discussion

The phylogenetic position of *Chromobacterium* sp. strain 56AF, previously described by our group [12], was examined using the complete 16S rRNA gene sequences from 21 Neisseriaceae and Chromobacteriaceae sequences retrieved from the GenBank database. The resulting phylogeny shows the evolutionary relationships among these sequences and revealed that *Chromobacterium* sp. 56AF is closely related (94% of identity) to *C. amazonense* CBMAI 310, isolated from water samples from the Rio Negro, in the Amazon, Brazil (Figure 1). The pairwise alignment indicated that the percent identity between these two sequences is over 99.5%. In addition, the phylogenetic tree also demonstrated that the *Chromobacterium* genus forms a well-supported monophyletic clade.

The *C. amazonense* CBMAI 310 genome is 81.6% homologous with that of *Chromobacterium* sp. strain 56AF, whereas the other reference genomes exhibited dDDH values ranging from 23.10 to 42.9% when compared with the *Chromobacterium* sp. strain 56AF genome. Thus, based on the 16S rRNA gene phylogeny and dDDH data, *Chromobacterium* sp. strain 56AF has been renamed as *C. amazonense* strain 56AF.

3.1. General Genomic Features. The *C. amazonense* strain 56AF genome is 4,556,707 bp long and distributed across 141 contigs. A map of the chromosome and the features of the genome is shown in Figure 2 and Table 1, respectively. The N50 is 169,722 bp, and the GC content is 61.95%, consistent with the values reported for *Chromobacterium* species, which range from 62.23% to 64.83%. A total of 85.8% of the final assembly was annotated as coding regions. Out of 4294 predicted coding sequences (CDSs) with an average length of 911 bp, 4205 were protein-coding genes, and 95 were RNA genes, including 88 tRNAs, six rRNAs, and one tmRNA. Among the protein-coding genes, 3298 (76.8%) were matched to functions present in the database, whereas the remaining genes were annotated as hypothetical proteins. The classification of genes into functional categories generated by RAST is shown in Table 2. Prophage sequences identified by PHAST constitute 5.8% of the bacterial

chromosome, including seven intact, two incomplete, and one questionable prophage (Table S2 and Figure S1).

Among the predicted genes present in the genome of *C. amazonense* strain 56AF, we focused on virulence factors, toxins, and secretion systems, which might be of particular importance in terms of pathogenicity, survival, and adaptation to the environment.

3.2. Insights from the Genome Sequence

3.2.1. Virulence Factors. A number of potential *Chromobacterium* virulence factors have been previously identified, including different types of secretion systems and toxins that are likely associated with *C. violaceum* and *C. haemolyticum* fatal infections in humans [17, 34]. Consistent with these reports, the draft genome of *C. amazonense* strain 56AF includes CDSs with significant similarity to virulence factors (Table S3) that were previously described for strains belonging to the *Chromobacterium* genus among others, as discussed in the following sections.

3.2.2. Secretion Systems. There are five well-studied double-membrane-spanning secretion systems (T1SS, T2SS, T3SS, T4SS, and T6SS) present in a wide variety of pathogenic and nonpathogenic gram-negative bacteria. These systems secrete products with diverse functions, such as those involved in adhesion, pathogenicity, adaptation, and survival [40]. Here, we identified a number of particularly interesting proteins that are part of the T1SS, T2SS, T3SS, T4SS, and T6SS machinery.

In gram-negative bacteria, T1SS is composed of three membrane proteins, two in the inner membrane belonging to the ABC transporter and membrane fusion protein (MFP) families and a third porin-like protein in the outer membrane [41], known as LapB, LapC, and TolC, respectively. The products secreted by T1SS are diverse in size and function and frequently associated with nutrient acquisition (lipases and proteases) and virulence, for example, colicin V, which was found in the *C. amazonense* strain 56AF genome.

T2SS consists of a large multiprotein machinery with 12–15 components known as general secretion pathway (Gsp) proteins [40]. Analysis of the *C. amazonense* strain 56AF genome revealed the presence of all of these proteins with the exception of GspC, which may be missing from the genome assembly. An additional T2SS protein, GspN, was identified; however, its function and location are unclear [42].

T3SS is found in a large number of gram-negative pathogens and symbionts [43]. This secretion system is composed of structural, translocator, and effector proteins. Similar to *C. violaceum* ATCC 12472, *C. amazonense* strain 56AF possesses the *Chromobacterium* pathogenicity islands (Cpi) 1, 1a, and 2, whereas *C. haemolyticum* T124 lacks the Cpi-2-like second T3SS. These data suggest the potential virulence of *C. amazonense* strain 56AF.

T4SS is found in both gram-negative and gram-positive bacteria and in some archaea and allows the translocation of DNA and proteins into target cells, playing an important

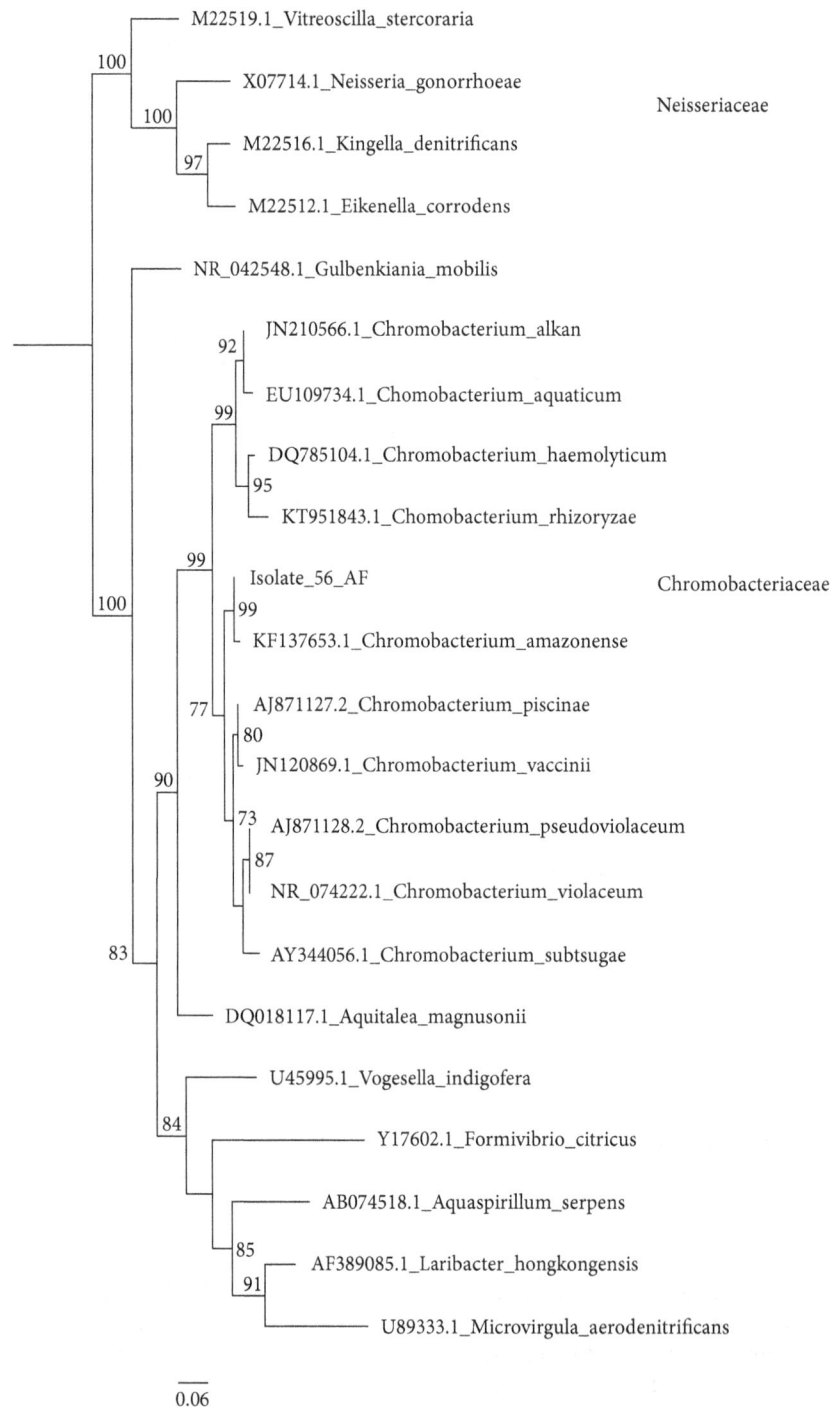

FIGURE 1: Maximum likelihood phylogenetic tree showing the evolutionary position of *C. amazonense* strain 56AF in relation to other *Chromobacterium* species. Neisseriaceae sequences were included as the outgroup. The phylogeny was reconstructed based on 22 partial 16S rRNA gene sequences and 1470 sites. Support values for each node were estimated using the approximate likelihood-ratio test (aLRT). Boxes of different colors highlight distinct families.

role in the pathogenesis of a wide range of bacteria [44]. Among the secretion systems, T4SS is unique for its ability to mediate the conjugation of plasmid DNA [40]. It consists of 12 proteins (VirB1–VirB11 and VirD4), of which only VirB3 and VirB7 were missing from the *C. amazonense* strain 56AF genome assembly. This is the first report of

T4SS in the *Chromobacterium* genus, which is considered to be critical for pathogenicity.

Finally, T6SS is broadly distributed among Proteobacteria. It translocates toxic effector proteins into eukaryotic and prokaryotic cells and has an important role in pathogenesis, bacterial communication, and interactions with the

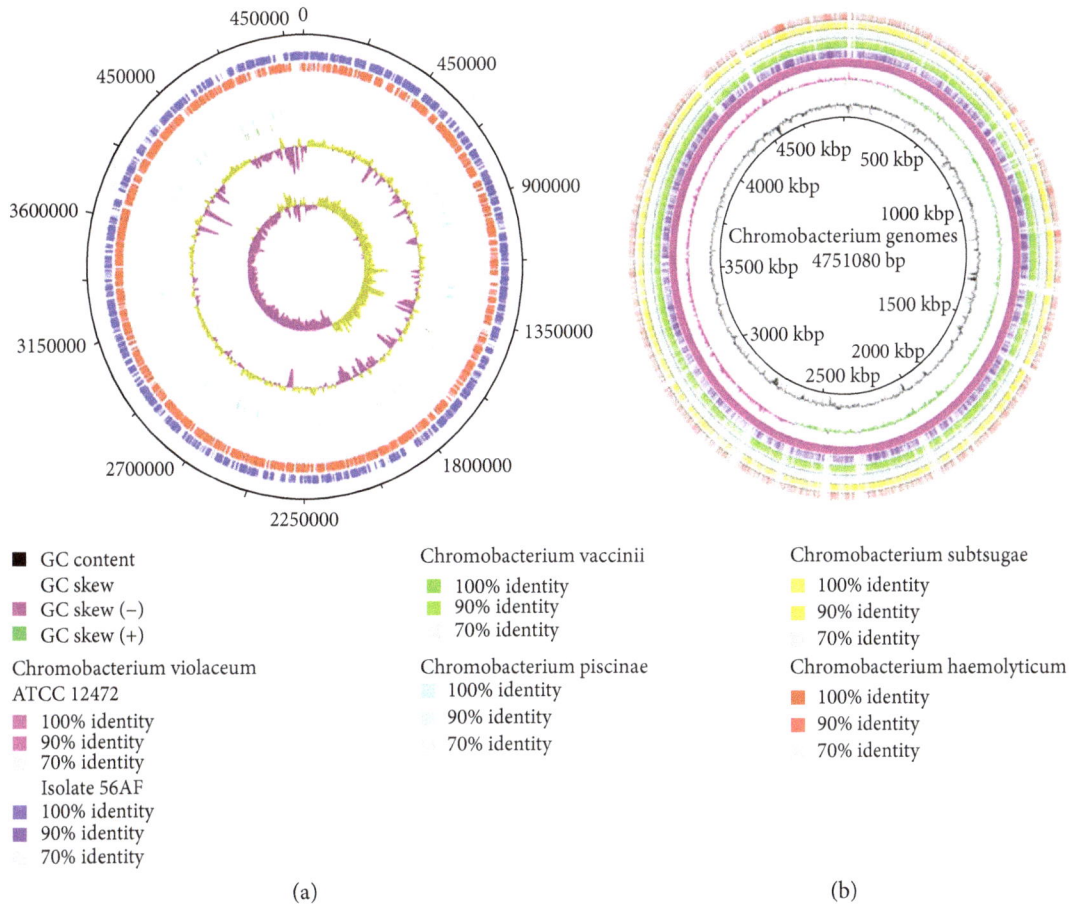

(a)   (b)

FIGURE 2: (a) Circular map of the draft genome of *C. amazonense* strain 56AF. From outer to inner circle: genes on the forward and reverse strands (predicted coding sequences colored), tRNA genes, rRNA genes, GC content, and GC skew. The circular genome map was generated using BRIG. (b) BLAST comparison of the draft genome of *C. amazonense* strain 56AF against the genomes of *Chromobacterium* species. The innermost ring depicts *C. violaceum* ATCC12472, followed by the query sequences of *C. amazonense* strain 56AF, *C. amazonense*, *C. piscinae*, *C. vaccinii*, *C. subtsugae*, *C. haemolyticum*, *C. pseudoviolaceum*, and *C. aquaticum*.

TABLE 1: Comparison of the genomic properties of *Chromobacterium* spp.

Chromobacterium spp.	Size (Mb)	Contig	GC (%)	Gene	rRNA	tRNA	Other RNA	Reference
C. amazonense 56AF	4,56	141	62	4,294	8	89	1	This study
C. violaceum ATCC 12472	4,75	1	64.8	4,438	25	98	5	[32]
C. haemolyticum DSM 19808	5,03	67	62.8	4,595	13	57	4	[33]
C. piscinae ND 17	4,09	223	62.6	3,916	4	81	1	[35]
C. subtsugae MWU 2920	4,67	152	64.9	4,313		77	4	[37]
C. vaccinii MWU 205	4,97	152	64.44	4463	6	74	4	[34]
C. aquaticum CC-SEYA-1	4.77	1,63	64.8	4492		33	4	[36]
C. pseudoviolaceum LMG 3953	4.63	326	64.7	4404		33	5	[12]

environment [45–47]. T6SS consists of a 13-protein core and a set of conserved accessory proteins [48]. In the genome of *C. amazonense* strain 56AF, 7 out of the 13 T6SS-associated proteins (ImpA, ImpB, ImpC, ImpF, ImpG, ImpH, and ImpJ) were found, in addition to the T4SS homolog IcmF. It is also known that T6SS secretes hemolysin and the effector protein VrgG, both of which are present in the *C. amazonense* strain 56AF genome.

3.2.3. Toxins. Toxins can be identified by the presence of a variety of proteins, such as hemolysins, hemolytic entero-toxins, colicin V, lytic proteins, and Nudix hydrolases. These toxins are associated with pathogenic bacteria [49–52] and are present in the genome of *C. amazonense* strain 56AF.

Important virulence factors for host cell invasion were found, for example, CDSs encoding a hydrolase cell-wall associated protein, metalloproteases, lypolitic proteins,

TABLE 2: SEED subsystems distribution of the *Chromobacterium amazonense* strain 56AF genome based on MG-RAST annotation.

Subsystem category	Subsystem feature counts
Sulfur metabolism	32
Phosphorus metabolism	47
Carbohydrates	242
Amino acids and derivatives	417
Fatty acids, lipid, and isoprenoids	103
Metabolism secondary	4
Nitrogen metabolism	13
Metabolism of aromatic compounds	24
Cofactors, vitamins, prosthetic groups, and pigments	242
Nucleosides and nucleotides	90
Dormancy and sporulation	1
Respiration	133
Stress response	134
DNA metabolism	99
Membrane transport	156
Regulation and cell signaling	88
Motility and chemotaxis	176
Cell division and cell cycle	35
Protein metabolism	298
RNA metabolism	135
Iron acquisition and metabolism	34
Phages, prophages, transposable elements and plasmids	65
Miscellaneous	45
Photosynthesis	0
Potassium metabolism	20
Virulence, disease, and defense	88
Cell wall and capsule	166

collagenases, and the proteolytic enzymes KpSS and KpSC. The last two are conserved proteins involved in capsular polysaccharides assembly systems [53], suggesting the potential of *C. amazonense* strain 56AF to synthesize capsular polysaccharides, which can act as important virulence factors in many pathogenic bacteria [53]. Moreover, genes involved in flagellar biosynthesis were identified. Flagella and swimming motility are present in all the genera of the family Chromobacteriaceae [1] and are considered to be important contributors to host colonization for most pathogens.

Unlike other *Chromobacterium* species, the genome of *C. amazonense* strain 56AF contains the gene for pore-forming toxin alpha-hemolysin (*hlyA*), which shares homology with the gene in *Vibrio cholerae*. It should be noted that hemolysin is a well-known virulence factor, and hemolytic activity has been detected in other *Chromobacterium* species, such as *C. violaceum*, *C. haemolyticum*, *C. aquaticum*, and *C. subtsugae* [5, 17, 54]. Thus, some *Chromobacterium* species have the potential to produce different cytolytic toxins that may contribute to their pathogenicity.

A set of genes with potential roles in multiple antibiotic resistance were identified in the *C. amazonense* strain 56AF genome and included CDSs encoding the RND efflux transporter, tripartite multidrug resistance system, Bcr/CflA family drug resistance transporter, multiple antibiotic resistance MarC protein, and resistance to β-lactams. Some efflux pumps not only confer clinically relevant resistance to antibiotics but also have a role in bacterial pathogenicity and may be beneficial for bacterial survival [55]. Regarding β-lactams, *C. amazonense* strain 56AF genome presented CDSs related to β-lactamases, including extended spectrum β-lactamases Ambler class C β-lactamase, penicillinases, metallo-β-lactamase, and metal-dependent hydrolases of the β -lactamase superfamily I PhnP protein. It is worth mentioning that they hydrolyze almost all clinically used β-lactams including carbapenems, thereby representing a therapeutic challenge. These results were supported by antimicrobial susceptibility testing obtained by Lima-Bittencourt et al. [12] for *Chromobacterium* sp. strain 56AF, which found resistance to ampicillin, ampicillin/clavulanic acid, and cefotaxime (third-generation cephalosporin), widely used for the treatment of bacterial infections. Other *Chromobacterium* species also exhibit β-lactamases, such as *C. piscinae* and *C. haemolyticum* [56]. Other antibiotic resistance genes encode resistance to bicyclomycin, fosfomycin, mitomycin, polymyxin (PmrJ, ArnC, and ArnT proteins), and fosmidomycin. Importantly, the genome contains genes related to resistance to several toxic compounds, including tellurite, cobalt, zinc-cadmium (CzcA), and fusaric acid, a mycotoxin produced by several *Fusarium* species that cause diseases in economically important crops [57].

Additionally, the annotation of the *C. amazonense* strain 56AF revealed its potential to produce esterases, lipase, auxins, chitinases, phytoene synthase and phytoene desaturase, polyhydroxyalkanoates, glutathione S-transferases, and violacein, suggesting that this strain might have potential for biotechnological purposes. The genome contains predicted toxin-antitoxin genes, in particular those of *higAB* and *doc/phd*, involved in plasmid stabilization [58], as well as the antitoxin YgiT (MqsA), which helps to mediate the bacterial general stress response [59], although the gene encoding MqsR toxin was absent. Finally, plastocyanin/azurin genes were also found, suggesting that this strain may participate in denitrifying processes.

3.3. Comparison of Eight Chromobacterium Genomes. Based on orthologous groups (OGs) analysis, the eight *Chromobacterium* genomes examined contain a total of 7220 OGs, with a conserved core genome of 2467 OGs present in all species (Figure 3). A large fraction of the flexible genes (65.8%), that is, those present in one or more but not all of the genomes, was mapped, among which 2069 were species-specific and 2684 were shared noncore. These findings are consistent with the great metabolic flexibility of *Chromobacterium*, conferring to their members a possible selective advantage to colonize and to survive in diverse ecosystems. It should be noted that *C. violaceum* contained the largest number of OGs shared with other *Chromobacterium* genomes. Comparisons of *C. amazonense* strain 56AF with the seven

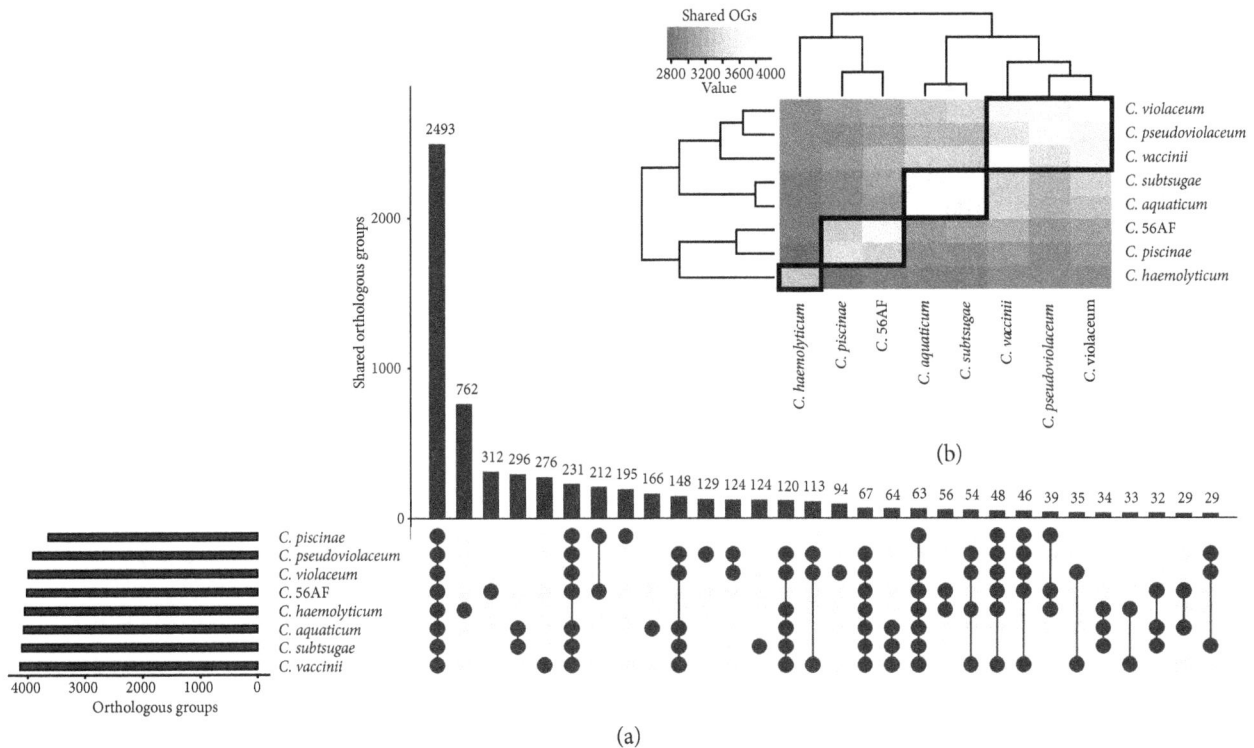

(a)

(b)

FIGURE 3: Distribution of orthologous groups (OGs) among the genomes of *C. amazonense, C. piscinae, C. pseudoviolaceum, C. violaceum, C. amazonense* strain 56AF, *C. haemolyticum, C. aquaticum, C. subtsugae,* and *C. vaccinii*. (a) The bars represent the number of unique or shared OGs among *Chromobacterium* genomes marked with the filled circle. (b) The heat map shows the number of shared OGs between different *Chromobacterium* species. The dendrogram is based on Euclidean distance of the shared OG matrix.

other genomes revealed a subset of OGs (301) that were unique to *C. amazonense* strain 56AF, and these may contribute to the species-specific features of *C. amazonense*. The unique set of *C. amazonense* strain 56AF genes was associated with the COG functional categories carbohydrate transport and metabolism (G); cell cycle control, cell division, and chromosome partitioning (D); amino acid transport and metabolism (E); nucleotide transport and metabolism (F); replication, recombination, and repair (L); defense mechanisms (V); and cell motility (N). Considering that seven out of the eight genomes analyzed here are draft sequences (all except for *C. violaceum*), it is likely that individual genes are missing from these assemblies.

4. Conclusions

C. amazonense strain 56AF is the first strain of the genus known to contain predicted genes associated with five of the six double-membrane-spanning secretion systems (types 1–4 and 6) present in gram-negative bacteria, of which T4SS has not been reported in any other sequenced *Chromobacterium*. Additionally, the presence of several predicted genes encoding potential multidrug resistance and detoxification compounds suggests that this isolate harbors a genomic repertoire capable of dealing with environmental toxins, which may be beneficial to its adaptation and survival. Moreover, *C. amazonense* strain 56AF contains a number of relevant virulence-related genes, suggesting its potential pathogenicity. In summary, our findings confirm

the versatility, adaptability, and biotechnological potential of this bacterium. Moreover, our results provide important insights into the genomics of *Chromobacterium* species, leading to a better understanding of the biology of this genus. Further functional studies will contribute with new information about its pathogenic potential.

Abbreviations

ORFs: Open reading frames
gDNA: Genomic DNA
dDDH: Digital DNA-DNA hybridization
KEGG: Kyoto Encyclopedia of Genes and Genomes
COGs: Clusters of Orthologous Groups
OGs: Orthologous groups
TSS: Type secretion system.

Conflicts of Interest

The authors declare no conflict of interests.

Acknowledgments

This work and publication were supported by the grants from the Fundação de Amparo à Pesquisa do Estado de Minas Gerais (FAPEMIG, APQ 00801/12 and APQ-00463-13), Conselho Nacional de Desenvolvimento Científico e Tecnológico (CNPq), and Coordenação de Aperfeiçoamento de Pessoal de Nível Superior (CAPES).

References

[1] M. Adeolu and R. S. Gupta, "Phylogenomics and molecular signatures for the order *Neisseriales*: proposal for division of the order *Neisseriales* into the emended family *Neisseriaceae* and *Chromobacteriaceae* fam. nov.," *Antonie Van Leeuwenhoek*, vol. 104, no. 1, pp. 1–24, 2013.

[2] C. Bergonzini, "Um nuevo bacterio colorato," *Annuario della Societa dei Naturalisti in Moderna*, vol. 2, pp. 149–158, 1881.

[3] P. A. W. Martin, D. Gundersen-Rindal, M. Blackburn, and J. Buyer, "*Chromobacterium subtsugae* sp. nov., a betaproteobacterium toxic to Colorado potato beetle and other insect pests," *International Journal of Systematic and Evolutionary Microbiology*, vol. 57, no. 5, pp. 993–999, 2007.

[4] C. C. Young, A. B. Arun, W. A. Lai et al., "*Chromobacterium aquaticum* sp. nov., isolated from spring water samples," *International Journal of Systematic and Evolutionary Microbiology*, vol. 58, no. 4, pp. 877–880, 2008.

[5] X. Y. Han, F. S. Han, and J. Segal, "*Chromobacterium haemolyticum* sp. nov., a strongly haemolytic species," *International Journal of Systematic and Evolutionary Microbiology*, vol. 58, no. 6, pp. 1398–1403, 2008.

[6] P. Kämpfer, H. J. Busse, and H. C. Scholz, "*Chromobacterium piscinae* sp. nov. and *Chromobacterium pseudoviolaceum* sp. nov., from environmental samples," *International Journal of Systematic and Evolutionary Microbiology*, vol. 59, no. 10, pp. 2486–2490, 2009.

[7] S. D. Soby, S. R. Gadagkar, C. Contreras, and F. L. Caruso, "*Chromobacterium vaccinii* sp. nov., isolated from native and cultivated cranberry (Vaccinium macrocarpon Ait.) bogs and irrigation ponds," *International Journal of Systematic and Evolutionary Microbiology*, vol. 63, Part 5, pp. 1840–1846, 2013.

[8] C. B. A. Menezes, M. F. Tonin, D. B. A. Corrêa et al., "*Chromobacterium amazonense* sp. nov. isolated from water samples from the Rio Negro, Amazon, Brazil," *Antonie Van Leeuwenhoek*, vol. 107, no. 4, pp. 1057–1063, 2015.

[9] S. Zhou, X. Guo, H. Wang et al., "*Chromobacterium rhizoryzae* sp. nov., isolated from rice roots," *International Journal of Systematic and Evolutionary Microbiology*, vol. 66, no. 10, pp. 3890–3896, 2016.

[10] A. Bajaj, A. Kumar, S. Yadav et al., "Isolation and characterization of a novel Gram-negative bacterium *Chromobacterium alkanivorans* sp. nov., strain IITR-71T degrading halogenated alkanes," *International Journal of Systematic and Evolutionary Microbiology*, vol. 66, no. 12, pp. 5228–5235, 2016.

[11] M. B. Blackburn, R. R. Farrar Jr, M. E. Sparks, D. Kuhar, A. Mitchell, and D. E. Gundersen-Rindal, "*Chromobacterium sphagni* sp. nov., an insecticidal bacterium isolated from Sphagnum bogs," *International Journal of Systematic and Evolutionary Microbiology*, vol. 67, no. 9, pp. 3417–3422, 2017.

[12] C. I. Lima-Bittencourt, S. Astolfi-Filho, E. Chartone-Souza, F. R. Santos, and A. M. A. Nascimento, "Analysis of *Chromobacterium* sp. natural isolates from different Brazilian ecosystems," *BMC Microbiology*, vol. 7, no. 1, p. 58, 2007.

[13] C. I. Lima-Bittencourt, P. S. Costa, F. A. R. Barbosa, E. Chartone-Souza, and A. M. A. Nascimento, "Characterization of a *Chromobacterium haemolyticum* population from a natural tropical lake," *Letters in Applied Microbiology*, vol. 52, no. 6, pp. 642–650, 2011.

[14] E. S. Barreto, A. R. Torres, M. R. Barreto, A. T. R. Vasconcelos, S. Astolfi-Filho, and M. Hungria, "Diversity in antifungal activity of strains of *Chromobacterium violaceum* from the Brazilian Amazon," *Journal of Industrial Microbiology and Biotechnology*, vol. 35, no. 7, pp. 783–790, 2008.

[15] L. T. Dall'Agnol, R. N. Martins, A. C. R. Vallinoto, and K. T. S. Ribeiro, "Diversity of *Chromobacterium violaceum* isolates from aquatic environments of state of Pará, Brazilian Amazon," *Memórias do Instituto Oswaldo Cruz*, vol. 103, no. 7, pp. 678–682, 2008.

[16] K. Ponnusamy, S. Jose, I. Savarimuthu, G. P. Michael, and M. Redenbach, "Genetic diversity study of *Chromobacterium violaceum* isolated from Kolli Hills by amplified ribosomal DNA restriction analysis (ARDRA) and random amplified polymorphic DNA (RAPD)," *Letters in Applied Microbiology*, vol. 53, no. 3, pp. 341–349, 2011.

[17] Brazilian National Genome Project Consortium, A. T. R. de Vasconcelos, D. F. de Almeida et al., "The complete genome sequence of *Chromobacterium violaceum* reveals remarkable and exploitable bacterial adaptability," *Proceedings of the National Academy of Sciences of the United States of America*, vol. 100, no. 20, pp. 11660–11665, 2011.

[18] A. Chattopadhyay, V. Kumar, N. Bhat, and P. L. N. G. Rao, "*Chromobacterium violaceum* infection: a rare but frequently fatal disease," *Journal of Pediatric Surgery*, vol. 37, no. 1, pp. 108–110, 2002.

[19] M. Okada, R. Inokuchi, K. Shinohara et al., "*Chromobacterium haemolyticum*-induced bacteremia in a healthy young man," *BMC Infectious Diseases*, vol. 13, no. 1, p. 406, 2013.

[20] K. Katoh and D. M. Standley, "MAFFT multiple sequence alignment software version 7: improvements in performance and usability," *Molecular Biology and Evolution*, vol. 30, no. 4, pp. 772–780, 2013.

[21] S. Capella-Gutiérrez, J. M. Silla-Martínez, and T. Gabaldón, "trimAl: a tool for automated alignment trimming in large-scale phylogenetic analyses," *Bioinformatics*, vol. 25, no. 15, pp. 1972-1973, 2009.

[22] R. Sánchez, F. Serra, J. Tárraga et al., "Phylemon 2.0: a suite of web-tools for molecular evolution, phylogenetics, phylogenomics and hypotheses testing," *Nucleic Acids Research*, vol. 39, pp. W470–W474, 2011.

[23] D. Darriba, G. L. Taboada, R. Doallo, and D. Posada, "ProtTest 3: fast selection of best-fit models of protein evolution," *Bioinformatics*, vol. 27, no. 8, pp. 1164-1165, 2011.

[24] J. P. Meier-Kolthoff, A. F. Auch, H. P. Klenk, and M. Göker, "Genome sequence-based species delimitation with confidence intervals and improved distance functions," *BMC Bioinformatics*, vol. 14, no. 1, p. 60, 2013.

[25] A. M. Bolger, M. Lohse, and B. Usadel, "Trimmomatic: a flexible trimmer for Illumina sequence data," *Bioinformatics*, vol. 30, no. 15, pp. 2114–2120, 2014.

[26] D. R. Zerbino and E. Birney, "Velvet: algorithms for de novo short read assembly using de Bruijn graphs," *Genome Research*, vol. 18, no. 5, pp. 821–829, 2008.

[27] A. Gurevich, V. Saveliev, N. Vyahhi, and G. Tesler, "QUAST: quality assessment tool for genome assemblies," *Bioinformatics*, vol. 29, no. 8, pp. 1072–1075, 2013.

[28] H. Li and R. Durbin, "Fast and accurate long-read alignment with Burrows-Wheeler transform," *Bioinformatics*, vol. 26, no. 5, pp. 589–595, 2010.

[29] H. Li, B. Handsaker, A. Wysoker et al., "The sequence alignment/map format and SAMtools," *Bioinformatics*, vol. 25, no. 16, pp. 2078-2079, 2009.

[30] D. Hyatt, G.-L. Chen, P. F. LoCascio, M. L. Land, F. W. Larimer, and L. J. Hauser, "Prodigal: prokaryotic gene recognition and translation initiation site identification," *BMC Bioinformatics*, vol. 11, no. 1, pp. 119-130, 2010.

[31] K. Rutherford, J. Parkhill, J. Crook et al., "Artemis: sequence visualization and annotation," *Bioinformatics*, vol. 16, no. 10, pp. 944-945, 2000.

[32] N.-F. Alikhan, N. K. Petty, N. L. Ben Zakour, and S. A. Beatson, "BLAST Ring Image Generator (BRIG): simple prokaryote genome comparisons," *BMC Genomics*, vol. 12, no. 1, p. 402, 2011.

[33] T. Miki and N. Okada, "Draft genome sequence of *Chromobacterium haemolyticum* causing human bacteremia infection in Japan," *Genome Announcements*, vol. 2, no. 6, pp. e01047-e01014, 2014.

[34] K. Vöing, A. Harrison, and S. D. Soby, "Draft genome sequence of *Chromobacterium vaccinii*, a potential biocontrol agent against mosquito (*Aedes aegypti*) larvae," *Genome Announcements*, vol. 3, no. 3, pp. e00477-e00415, 2015.

[35] K.-G. Chan and N. Y. M. Yunos, "Whole-genome sequencing analysis of *Chromobacterium piscinae* strain ND17, a quorum-sensing bacterium," *Genome Announcements*, vol. 4, no. 2, pp. e00081-e00016, 2016.

[36] S. D. Soby, "Draft genome sequence of *Chromobacterium aquaticum* CCSEYA-1, a nonpigmented member of thegenus *Chromobacterium*," *Genome Announcements*, vol. 5, no. 12, pp. e01661-e01616, 2017.

[37] K. Vöing, A. Harrison, and S. D. Soby, "Draft genome sequences of three *Chromobacterium subtsugae* isolates from wild and cultivated cranberry bogs in southeastern Massachusetts," *Genome Announcements*, vol. 3, no. 5, pp. e00998-e00915, 2015.

[38] S. D. Soby, "Draft genome sequence of *Chromobacterium pseudoviolaceum* LMG 3953T, an enigmatic member of the genus *Chromobacterium*," *Genome Announcements*, vol. 5, no. 12, pp. e01632-e01616, 2017.

[39] D. M. Emms and S. Kelly, "OrthoFinder: solving fundamental biases in whole genome comparisons dramatically improves orthogroup inference accuracy," *Genome Biology*, vol. 16, no. 1, p. 157, 2015.

[40] T. R. D. Costa, C. Felisberto-Rodrigues, A. Meir et al., "Secretion systems in Gram-negative bacteria: structural and mechanistic insights," *Nature Reviews Microbiology*, vol. 13, no. 6, pp. 343-359, 2015.

[41] S. Thomas, I. B. Holland, and L. Schmitt, "The type 1 secretion pathway –the hemolysin system and beyond," *Biochimica et Biophysica Acta (BBA) - Molecular Cell Research*, vol. 1843, no. 8, pp. 1629-1641, 2014.

[42] K. V. Korotkov, M. Sandkvist, and W. G. J. Hol, "The type II secretion system: biogenesis, molecular architecture and mechanism," *Nature Reviews Microbiology*, vol. 10, no. 5, pp. 336-351, 2012.

[43] B. Coburn, I. Sekirov, and B. B. Finlay, "Type III secretion systems and disease," *Clinical Microbiology Reviews*, vol. 20, no. 4, pp. 535-549, 2007.

[44] C. E. Alvarez-Martinez and P. J. Christie, "Biological diversity of prokaryotic type IV secretion systems," *Microbiology and Molecular Biology Reviews*, vol. 73, no. 4, pp. 775-808, 2009.

[45] G. Bönemann, A. Pietrosiuk, A. Diemand, H. Zentgraf, and A. Mogk, "Remodelling of VipA/VipB tubules by ClpV-mediated threading is crucial for type VI protein secretion," *EMBO Journal*, vol. 28, no. 4, pp. 315-325, 2009.

[46] B. T. Ho, T. G. Dong, and J. J. Mekalanos, "A view to a kill: the bacterial type VI secretion system," *Cell Host & Microbe*, vol. 15, no. 1, pp. 9-21, 2014.

[47] A. Zoued, Y. R. Brunet, E. Durand et al., "Architecture and assembly of the type VI secretion system," *Biochimica et Biophysica Acta (BBA) - Molecular Cell Research*, vol. 1843, no. 8, pp. 1664-1673, 2014.

[48] F. Boyer, G. Fichant, J. Berthod, Y. Vandenbrouck, and I. Attree, "Dissecting the bacterial type VI secretion system by a genome wide in silico analysis: what can be learned from available microbial genomic resources?," *BMC Genomics*, vol. 10, no. 1, p. 104, 2009.

[49] J. A. Vázquez-Boland, M. Kuhn, P. Berche et al., "Listeria pathogenesis and molecular virulence determinants," *Clinical Microbiology Reviews*, vol. 14, no. 3, pp. 584-640, 2001.

[50] P. H. Edelstein, B. Hu, T. Shinzato, M. A. C. Edelstein, W. Xu, and M. J. Bessman, "*Legionella pneumophila* NudA is a Nudix hydrolase and virulence factor," *Infection and Immunity*, vol. 73, no. 10, pp. 6567-6576, 2005.

[51] E. Ortega, H. Abriouel, R. Lucas, and A. Gálvez, "Multiple roles of *Staphylococcus aureus* enterotoxins: pathogenicity, superantigenic activity, and correlation to antibiotic resistance," *Toxins*, vol. 2, no. 8, pp. 2117-2131, 2010.

[52] N. Jeßberger, R. Dietrich, S. Bock, A. Didier, and E. Märtlbauer, "*Bacillus cereus* enterotoxins act as major virulence factors and exhibit distinct cytotoxicity to different human cell lines," *Toxicon*, vol. 77, pp. 49-57, 2014.

[53] L. M. Willis and C. Whitfield, "KpsC and KpsS are retaining 3-deoxy-d-manno-oct-2-ulosonic acid (Kdo) transferases involved in synthesis of bacterial capsules," *Proceedings of the National Academy of Sciences of the United States of America*, vol. 110, no. 51, pp. 20753-20758, 2013.

[54] P. D. Rekha, C. C. Young, and A. B. Arun, "Identification of *N*-acyl-L-homoserine lactones produced by non-pigmented *Chromobacterium aquaticum* CC-SEYA-1[T] and pigmented *Chromobacterium subtsugae* PRAA4-1[T]," *Biotech*, vol. 1, no. 4, pp. 239-245, 2011.

[55] L. J. V. Piddock, "Multidrug-resistance efflux pumps - not just for resistance," *Nature Reviews Microbiology*, vol. 4, no. 8, pp. 629-636, 2006.

[56] D. D. Gudeta, V. Bortolaia, A. Jayol, L. Poirel, P. Nordmann, and L. Guardabassi, "*Chromobacterium* spp. harbour ambler class A β-lactamases showing high identity with KPC," *Journal of Antimicrobial Chemotherapy*, vol. 71, no. 6, pp. 1493-1496, 2016.

[57] C. W. Bacon, J. K. Porter, W. P. Norred, and J. F. Leslie, "Production of fusaric acid by *Fusarium* species," *Applied and Environmental Microbiology*, vol. 62, no. 11, pp. 4039-4043, 1996.

[58] K. Gerdes, S. K. Christensen, and A. Løbner-Olesen, "Prokaryotic toxin–antitoxin stress response loci," *Nature Reviews Microbiology*, vol. 3, no. 5, pp. 371-382, 2005.

[59] X. Wang, Y. Kim, S. H. Hong et al., "Antitoxin MqsA helps mediate the bacterial general stress response," *Nature Chemical Biology*, vol. 7, no. 6, pp. 359-366, 2011.

Expanding the miRNA Transcriptome of Human Kidney and Renal Cell Carcinoma

Adam P. Sage ⓘ**, Brenda C. Minatel** ⓘ**, Erin A. Marshall** ⓘ**, Victor D. Martinez** ⓘ**,
Greg L. Stewart** ⓘ**, Katey S. S. Enfield** ⓘ**, and Wan L. Lam** ⓘ

Department of Integrative Oncology, British Columbia Cancer Research Centre, Vancouver, BC, Canada

Correspondence should be addressed to Adam P. Sage; asage@bccrc.ca

Academic Editor: Yujing Li

Despite advancements in therapeutic strategies, diagnostic and prognostic molecular markers of kidney cancer remain scarce, particularly in patients who do not harbour well-defined driver mutations. Recent evidence suggests that a large proportion of the human noncoding transcriptome has escaped detection in early genomic explorations. Here, we undertake a large-scale analysis of small RNA-sequencing data from both clear cell renal cell carcinoma (ccRCC) and nonmalignant samples to generate a robust set of miRNAs that remain unannotated in kidney tissues. We find that these novel kidney miRNAs are also expressed in renal cancer cell lines. Moreover, these sequences are differentially expressed between ccRCC and matched nonmalignant tissues, implicating their involvement in ccRCC biology and potential utility as tumour-specific markers of disease. Indeed, we find some of these miRNAs to be significantly associated with patient survival. Finally, target prediction and subsequent pathway analysis reveals that miRNAs previously unannotated in kidney tissues may target genes involved in ccRCC tumourigenesis and disease biology. Taken together, our results represent a new resource for the study of kidney cancer and underscore the need to characterize the unexplored areas of the transcriptome.

1. Introduction

Despite recent advancements in the diagnosis and treatment of kidney cancer, patients are faced with a poor prognosis, especially when diagnosed at a later stage [1]. Kidney cancer is a heterogeneous disease with multiple subtypes, of which clear cell renal cell carcinoma (ccRCC) is the most frequently observed, accounting for 70–75% of cases [2]. While environmental risk factors including hypertension, smoking, obesity, and a history of chronic kidney disease may modulate an individual's susceptibility, ccRCC arises from molecular aberrations that can be both sporadic and inherited [2, 3]. Many of these alterations result from DNA copy number losses, mutations, and hypermethylation events, commonly affecting genes associated with cellular metabolism [4, 5]. The most frequently affected gene is the *von Hippel-Lindau* (VHL) tumour suppressor gene, while other molecular disruptions affecting multiple components of the PI3K-mTOR and AMPK signaling pathways have also been described

[5, 6]. Considering the close association of metabolic reprogramming with ccRCC development and progression, remarkable advances have been made in the treatment of ccRCC patients with the use of antiangiogenic therapies [7]. Despite the increased treatment efficacy of antiangiogenic therapies, patient outcome is impaired by the lack of clinically relevant diagnostic or prognostic markers [8, 9].

The increased availability of next-generation sequencing has led to a dramatic increase in the understanding of noncoding RNAs (ncRNAs). Perhaps, the most well-studied type of ncRNA is microRNAs (miRNAs), short (~22 nt) transcripts that have emerged as critical regulators of gene expression. Since the discovery of *lin-4* in 1993, miRNAs have been found to regulate a multitude of transcripts and their subsequent cellular processes, including proliferation, metabolism, apoptosis, and development [10–12]. Moreover, the relatively long half-life of miRNAs makes them attractive candidates for biomarkers of disease [13]. miRNAs have been observed to be critical to kidney development, physiology, and pathology.

For instance, a cluster of miRNAs (miR-17~92) has been shown to regulate nephrogenesis; both miR-9 and miR-374 are observed to suppress claudin-14 and affect Ca^{2+} readsorption in the ascending limb of Henle; and recent studies have detected aberrant expression of miRNAs in kidney tumours [13–16]. However, despite the mounting evidence of a role for miRNAs in ccRCC, they have yet to be used in kidney cancer diagnostics.

miRNAs have been described to be genus- and tissue-specific; yet, initial characterizations of the human miRNA transcriptome have relied heavily on sequence abundance and conservation, while using relatively low-depth coverage techniques. In light of this, it has been hypothesized that the human genome likely encodes a markedly greater number of miRNAs than currently annotated, which may be able to be identified through a focus on individual tissues and cell lineages [17]. Indeed, recent genome-wide studies have uncovered miRNAs that have previously escaped detection and have observed that these newly detected miRNAs are highly tissue-specific [18]. Additionally, previously uncharacterized miRNAs may in fact represent novel regulators of tissue-specific biology and pathogenesis and may have utility in the clinic as disease markers. Thus, in this study, we use a large-scale analysis of high-throughput sequencing data to probe for novel miRNAs in human kidney tissue. Discovery of these previously unannotated miRNAs provides a new resource to delineate ccRCC pathogenesis.

2. Materials and Methods

2.1. Small RNA-Sequencing and Data Collection. A cohort of clear cell renal cell carcinoma (ccRCC) tumours with paired nonmalignant tissues ($n = 71$), as well as unpaired tumours ($n = 502$), was processed by The Cancer Genome Atlas (TCGA) Research Network (http://cancergenome.nih.gov/). Small RNA-sequencing data were generated on the Illumina HiSeq200 platform and were acquired from the Cancer Genomics Hub (cgHUB) Data Repository (dbgap Project ID: 6208) under the TCGA-KIRC data collection heading. All data analyzed in this experiment are available publically.

2.2. Preprocessing of Small RNA-Sequencing Data. All raw sequence data obtained from TCGA were processed using a previously published custom sequence analysis pipeline designed for small RNA sequence detection [19]. Raw BAM files from TCGA were first converted to FASTQ files of unaligned reads. Unaligned reads were trimmed based on their Phred quality score, which is required to be ≥20. The trimmed reads were then realigned to the current build of the human genome (hg38 annotation) using the Spliced Transcripts Alignment to a Reference (STAR) aligner.

2.3. Detection and Filtering of Novel miRNA Sequences. Through the OASIS online small-RNA-sequencing analysis platform, novel miRNA sequences were predicted using the miRDeep2 algorithm [20]. miRDeep2 takes both relative free energy and the p values associated with random folding to predict species with miRNA-like structure and to generate a miRDeep2 score reflective of the reliability of the prediction.

To confirm the validity of these predictions, stringent manual assessments were performed to generate a robust set of previously unannotated miRNA sequences. Manual filtering was based on (1) an adequate number of sequencing reads covering each locus (≥10 reads); (2) no presence of rRNA/tRNA reads, based on the Rfam database [21]; (3) significant ($p < 0.05$) probability of miRNA-like secondary structure; and (4) removal of duplicate sequences. Standard nucleotide blast (BLASTn) was performed on all remaining predicted novel miRNA sequences using the BLAST+ command line application [22]. This step ensures that any sequences with homology to miRNAs annotated in miRBase v21 are not included in the final list of predicted sequences. Sequences with an expect (E) value of <0.1 were considered previously annotated and discarded from further analyses. The mean and standard deviation of the GC content of all predicted novel miRNA sequences were calculated in order to remove any transcripts with GC content ± 2 standard deviations from the mean. Expression levels of these newly detected sequences were assessed on a per-sample basis using the algorithm featureCounts v1.4.6 [23–25]. Quantification data were normalized according to the weighted trimmed mean of the log expression ratios (trimmed mean of M values, TMM method). Previously unannotated miRNA species were considered to be expressed if the sum of the sequencing reads across all samples was at least 10 reads. Finally, sequences were queried for their presence in five previous studies that have identified novel miRNAs in various tissues [17, 18, 26–28] (Supplemental Table 1).

2.4. Validation of miRNA Expression in Cell Lines. We have performed an in-depth analysis of the small noncoding RNA transcriptome of the National Cancer Institute's cancer cell line panel (NCI-60), which includes eight renal cancer cell lines (A498, CAKI-1, 786-0, TK-10, UO-31, ACHN, RXF393, and SN12C) [28]. Small RNA-sequencing data generated from these renal cancer cell lines were processed as described previously to detect and quantify the expression of the predicted novel miRNA sequences discovered in ccRCC tumours. Cell line characteristics and detailed sequencing information are available in Supplemental Table 2. An expression cutoff (TMM > 0.1) was used to ensure that the sequencing reads were expressed in at least one of the renal cancer cell lines. Predicted miRNA sequences that were expressed in data obtained from both TCGA patient samples and NCI-60 cell lines were considered to be validated and were assigned a unique ID consisting of Knm (kidney novel miRNA sequence), followed by the locus position (i.e., Knm22_2209).

2.5. Differential Expression of Previously Unannotated miRNAs in ccRCC and Nonmalignant Tissue. As performed in the cell line validation step, an expression cutoff of TMM > 0.1 in 10% of samples was used to determine newly detected miRNA loci that had detectable expression in both ccRCC tumour samples and paired nonmalignant tissue ($n = 71$). Fold change values were calculated as the ratio of expression of the newly detected miRNA loci in tumours to their expression in nonmalignant tissue. A Student t-test was performed on the TMM expression values of the newly detected miRNAs in RStudio v3.3.3 to test for statistically significant differences in expression between ccRCC tumours and paired nonmalignant samples. Multiple correction analyses were performed using the Benjamini-Hochberg (BH) correction, to account for the large number of samples and probes being analyzed. Additionally, unsupervised hierarchical clustering analysis using average distance and Pearson correlation metrics of differentially expressed miRNAs was performed to visualize their expression on a per-sample basis (Supplemental Figure 2).

2.5.1. Cell Culture and Real-Time Quantitative PCR (RT-qPCR). The renal cancer cell line TK-10 and the immortalized nonmalignant embryonic kidney cell line HEK-293T were used to further explore the expression of the previously unannotated miRNAs and their deregulation in ccRCC tumours in vitro. TK-10 cells were cultured in RPMI 1640 + 10% FBS, while HEK-293T were cultured in DMEM + 10% FBS. Both cell lines were maintained in an incubator at 37°C and 5% CO_2. Once confluent, cells were harvested for RNA extraction using the Quick-RNA™ MiniPrep Kit (Zymo Research, Catalog number R1055), following manufacturer's guidelines. Custom reverse-transcription primers specific to the mature miRNA sequence were obtained for the novel miRNA candidates Knm3_1968 (GCAGAUUCC CAGAGUGGGACAG) and Knm17_1130 (UGAGGUGGA GGGUUGUGGGA) using the Custom TaqMan® Small RNA Assay Design Tool from Thermo Fisher. The cDNA conversions were performed with the TaqMan MicroRNA Reverse Transcription Kit according to manufacturer's instructions using 2 ng/μL RNA samples for both cell lines. Finally, RT-qPCR analyses using the custom primers generated from Thermo Fisher were performed in triplicate in an Applied Biosystems® 7500 Real-Time PCR System. Relative miRNA expression was calculated via the $2^{-\Delta\Delta Ct}$ method and normalized to the expression of U6 snRNA.

2.6. Survival Analyses of Previously Unannotated miRNAs. Phenotypic information for all ccRCC tumour samples was obtained from GDC (TCGA-KIRC) through UCSC Xena (http://xena.ucsc.edu/). Samples were sorted by high to low miRNA expression, and tertiles were defined. Patients were categorized by vital status and days to death/last follow-up. The Gehan-Breslow-Wilcoxon test was used to assess the significance of the associations between miRNA expression and patient outcome for each miRNA sequence examined. The log-rank test was also considered.

2.7. Protein-Coding Gene Target Prediction and Pathway Enrichment Analysis. To determine the potential target genes of the newly detected kidney miRNAs, they were queried against all human 3′ untranslated region (UTR) sequences, acquired from Ensembl through the BioMart tool (https://www.ensembl.org), using the miRanda v3.3a algorithm [29]. Predicted gene targets of the miRNAs were validated by individually running five separate scrambled sequences through the algorithm. Predicted targets that overlapped between any of the scrambled sequences and the true sequence were discarded. Strict parameters were used in the target prediction analysis, specifically an alignment score of ≥140 and an energy threshold of ≤−20 kcal/mol (Supplemental Table 3). Gene symbols identified by the miRanda algorithm and predicted to be targeted by at least 10% of the previously unannotated miRNAs were submitted to a comprehensive pathway enrichment analysis using pathDIP v.2.5.21.6, which assesses enrichment of the target genes in pathways obtained from 15 distinct public pathway resources (literature-curated (core) pathway memberships) [30]. In this study, we report all pathways enriched with corrected p values ≤ 0.05.

3. Results

3.1. Discovery of Previously Unannotated miRNAs in ccRCC and Normal Kidney Tissue. Small RNA-sequencing data of ccRCC tumours (unpaired ($n = 502$) and tumours with matched nonmalignant tissue ($n = 71$)) were obtained from cgHUB and processed, and data were analyzed using the miRDeep2 prediction algorithm through the OASIS platform [20, 23]. The raw output of this analysis predicted 96 and 280 unique, previously unannotated miRNAs in nonmalignant tissue and ccRCC tumours, respectively. Manual filtering based on read quality, likelihood of miRNA-like secondary structure, and significant folding values, followed by probing the degree of similarity with known miRNAs using the BLASTn algorithm, and removal of sequences with aberrant GC content, resulted in 40 nonmalignant and 143 tumour previously unannotated miRNAs (Figure 1, Supplemental Table 1).

To validate the occurrence of the previously unannotated miRNAs in kidney tissues, we assessed their expression in small RNA-sequencing data generated from the NCI-60 cell line panel [28]. A miRNA was considered to be expressed in the eight renal cell lines included in the panel if it had a normalized expression value of greater than 0.1 in at least one of the renal cancer cell lines. In these cell lines, 26 (65%) novel miRNAs detected in nonmalignant samples and 102 (71%) miRNAs detected in ccRCC tumours were also expressed, strengthening their confidence as true miRNA sequences in human renal tissues and suggesting possible relevance to kidney function and pathology. Thus, we sought to further examine the role that these unannotated and validated miRNAs expressed in kidney tissues may have in ccRCC tumourigenesis and their potential clinical relevance.

3.2. Previously Unannotated miRNAs Are Deregulated in ccRCC Tumours. Recent evidence suggests that widespread disruption of miRNA-coding gene regulatory networks is common in many cancer types. Perturbation of coding gene expression by miRNA-based regulation can be achieved in tumours through loss of a tumour-suppressive miRNA or the gain of an oncogenic miRNA. Thus, we analyzed the expression of these previously unannotated miRNA sequences in tumours and paired nonmalignant tissue to identify potentially novel oncogenic and tumour-suppressive miRNAs in ccRCC.

After filtering for expression in both groups of samples, 59 previously unannotated miRNA loci were considered to

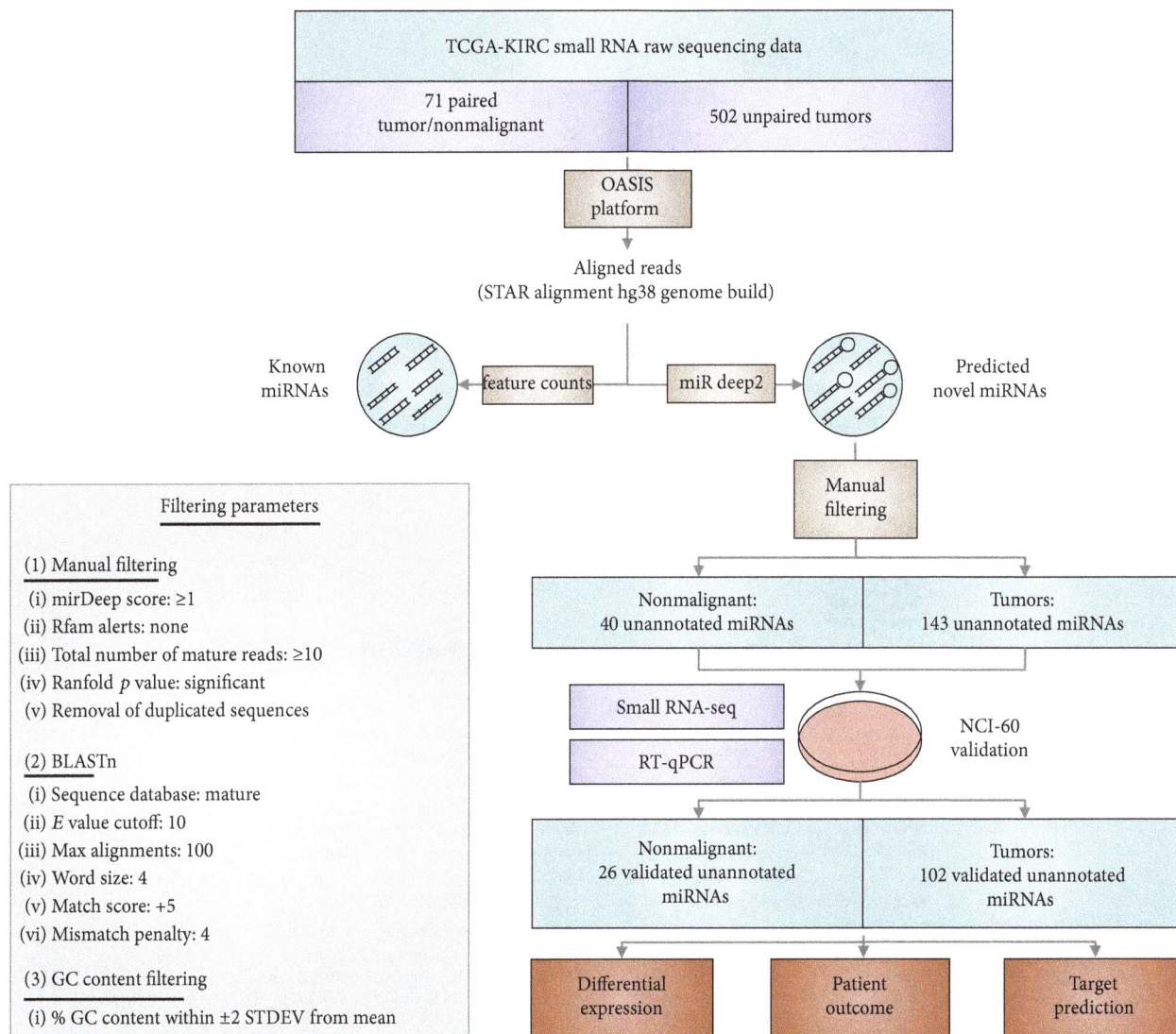

FIGURE 1: Experimental analysis pipeline. Detailed diagram of the analyses used to generate predicted and validated miRNA sequences previously unannotated in kidney tissues, along with subsequent explorations into their biological relevance.

be expressed in both ccRCC tumours and nonmalignant samples. A Student's t-test analysis revealed 30 of these 59 miRNAs to be significantly differentially expressed between the two groups (BH-p < 0.05). An analysis of the average fold change values between paired samples revealed that 14 miRNAs were significantly upregulated in ccRCC tumours, while the other 16 were downregulated (Figure 2). A subset of these miRNAs is of particular interest due to the magnitude of their expression differences between tumour and nonmalignant samples (Figure 2; Supplemental Figure 2). For example, the previously unannotated miRNA Knm22_2209 has an almost complete loss of expression in ccRCC tumours (28-fold downregulated in tumours, BH-p = 5.3×10^{-23}), while previously unannotated miRNAs Knm3_1968 and Knm17_1130 show a 100- and 13-fold increase in expression in ccRCC samples, respectively (BH-p = 1.4×10^{-21}, 2.6×10^{-14}). These findings serve to not only highlight the potential role of currently unannotated miRNAs in kidney cancer but also warrant investigation into their uses as biological markers of cancer onset.

3.3. Patient Outcome Predicted by Previously Unannotated miRNAs. The differential expression of these unannotated miRNAs in paired ccRCC tumour samples suggests their potential roles in kidney cancer and unexplored clinical utility. As such, we sought to examine whether any of the unannotated miRNA sequences deregulated in ccRCC tumours were associated with patient outcome. We examined differences in patient survival in those with high expression of a miRNA to those with low expression, defined by tertiles. Survival analysis was performed using the Gehan-Breslow-Wilcoxon test on phenotype data obtained from UCSC Xena and the expression profiles of the unannotated miRNA loci. Interestingly, two of the significantly differentially expressed unannotated miRNAs in paired ccRCC tumour samples displayed striking associations with patient survival (Figure 3). Again, the unannotated miRNA Knm22_2209 (28-fold downregulated in ccRCC, BH-p = 5.3×10^{-23}) is particularly noteworthy due to the clear correlation between its low expression and poor overall survival (Figure 3(a), p = 0.045).

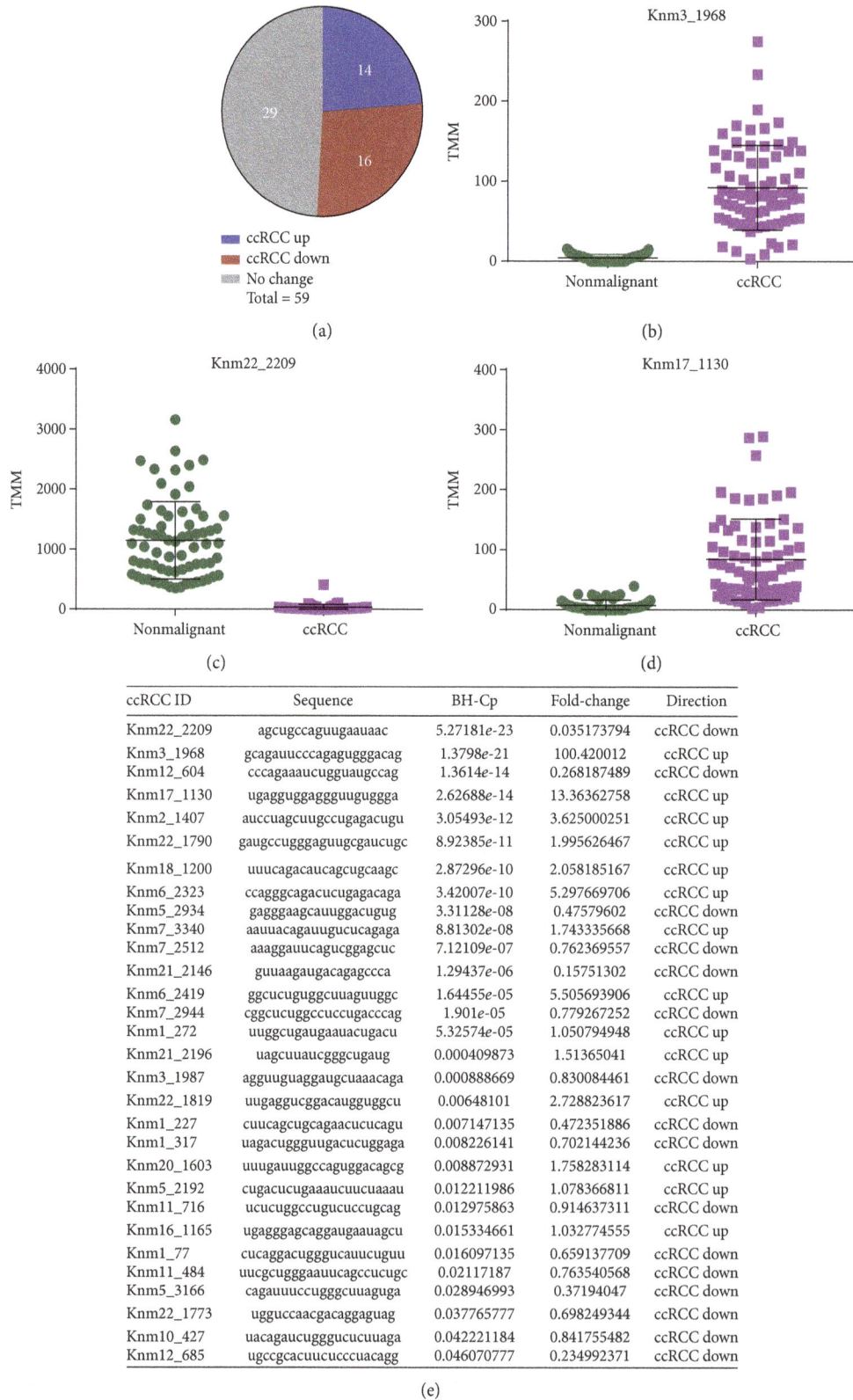

(a)

(b)

(c)

(d)

ccRCC ID	Sequence	BH-Cp	Fold-change	Direction
Knm22_2209	agcugccaguugaauaac	5.27181e-23	0.035173794	ccRCC down
Knm3_1968	gcagauucccagagugggacag	1.3798e-21	100.420012	ccRCC up
Knm12_604	cccagaaaucugguaugccag	1.3614e-14	0.268187489	ccRCC down
Knm17_1130	ugagguggagggvuuguggga	2.62688e-14	13.36362758	ccRCC up
Knm2_1407	auccuagcuugccugagacugu	3.05493e-12	3.625000251	ccRCC up
Knm22_1790	gaugccugggaguugcgaucugc	8.92385e-11	1.995626467	ccRCC up
Knm18_1200	uuucagacaucagcugcaagc	2.87296e-10	2.058185167	ccRCC up
Knm6_2323	ccagggcagacucugagacaga	3.42007e-10	5.297669706	ccRCC up
Knm5_2934	gagggaagcauuggacugug	3.31128e-08	0.47579602	ccRCC down
Knm7_3340	aauuacagaauugcucucagaga	8.81302e-08	1.743335668	ccRCC up
Knm7_2512	aaaggauucagucggagcuc	7.12109e-07	0.762369557	ccRCC down
Knm21_2146	guuaagaugacagagccca	1.29437e-06	0.15751302	ccRCC down
Knm6_2419	ggcucuguggcuuaguuggc	1.64455e-05	5.505693906	ccRCC up
Knm7_2944	cggcucuggccuccugacccag	1.901e-05	0.779267252	ccRCC down
Knm1_272	uuggcugaugaauacugacu	5.32574e-05	1.050794948	ccRCC up
Knm21_2196	uagcuuaucgggcugaug	0.000409873	1.51365041	ccRCC up
Knm3_1987	agguuguaggaugcuaaacaga	0.000888669	0.830084461	ccRCC down
Knm22_1819	uugaggucggacaugguggcu	0.00648101	2.728823617	ccRCC up
Knm1_227	cuucagcugcagaacucucagu	0.007147135	0.472351886	ccRCC down
Knm1_317	uagacuggguugacucuggaga	0.008226141	0.702144236	ccRCC down
Knm20_1603	uuugauuggccaguggacagcg	0.008872931	1.758283114	ccRCC up
Knm5_2192	cugacucugaaaucuucuaaau	0.012211986	1.078366811	ccRCC up
Knm11_716	ucucuggccugucuccugcag	0.012975863	0.914637311	ccRCC down
Knm16_1165	ugagggagcaggaugaauagcu	0.015334661	1.032774555	ccRCC up
Knm1_77	cucaggacugggucauucuguu	0.016097135	0.659137709	ccRCC down
Knm11_484	uucgcugggaauucagccucugc	0.02117187	0.763540568	ccRCC down
Knm5_3166	cagauuuccugggcuuaguga	0.028946993	0.37194047	ccRCC down
Knm22_1773	ugguccaacgacaggaguag	0.037765777	0.698249344	ccRCC down
Knm10_427	uacagaucugggucucuuaga	0.042221184	0.841755482	ccRCC down
Knm12_685	ugccgcacuucucccuacagg	0.046070777	0.234992371	ccRCC down

(e)

Figure 2: Differential expression of previously unannotated miRNAs between nonmalignant and ccRCC samples. (a) Pie chart representing the proportion of miRNAs that are significantly differentially expressed between the samples. (b) Normalized expression of the previously unannotated miRNA Knm3_1968 in individual nonmalignant (green) and ccRCC samples (purple). (c) Normalized expression of Knm22_2209 in individual nonmalignant (green) and ccRCC samples (purple). (d) Normalized expression of Knm17_1130 in individual nonmalignant (green) and ccRCC samples (purple). (e) Summary of differential expression results for each previously unannotated miRNA. BHC-p represents the corrected p value for the differential expression calculated using Student's t-test and Benjamini-Hochberg multiple correction analysis.

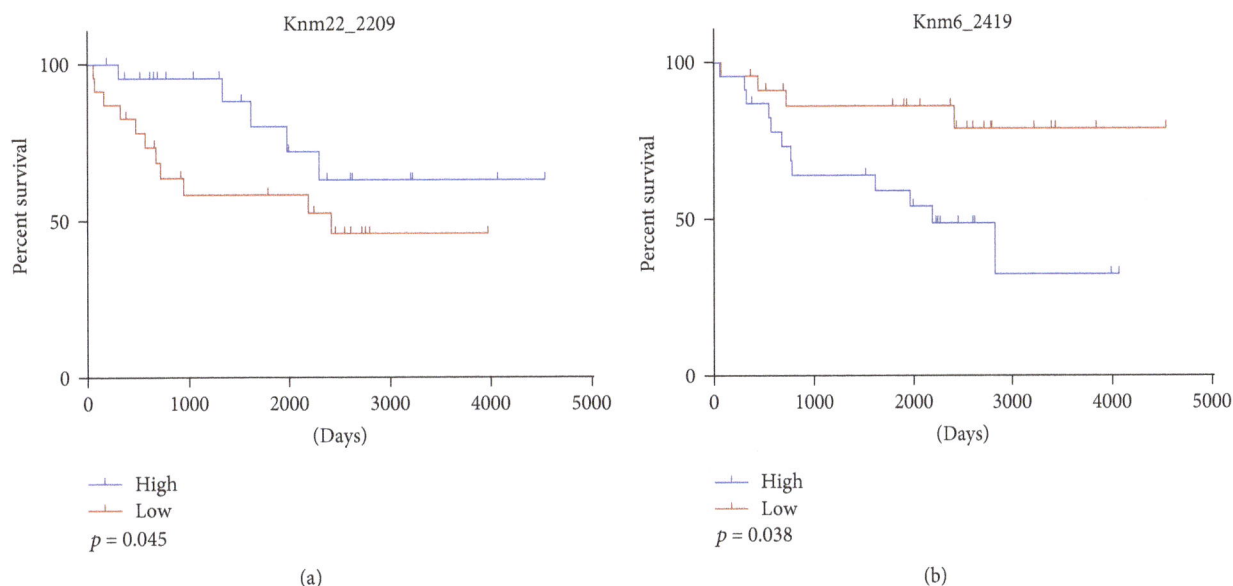

FIGURE 3: ccRCC patient overall survival predicted by previously unannotated miRNAs. Red lines represent patients with expression of the miRNA in the lower tertile of expression, while blue lines represent the upper tertile of expression. Calculation of p values for Knm22_2209 (a) and Knm6_2419 (b) was performed using the Gehan-Breslow-Wilcoxon test.

Alternatively, the Knm6_2419 miRNA is significantly upregulated in ccRCC tumours (FC = 5.5, BH-p = 1.6 × 10^{-5}), and its high expression is significantly associated with a worsened overall patient survival (Figure 3(b), p = 0.038). Taken together, these results emphasize the clinical potential of previously undetected miRNAs.

3.4. Genes and Pathways Targeted by Newly Detected miRNAs. In order to gain deeper insight into the biological relevance of the newly detected miRNAs, target prediction analysis was performed using the miRanda v3.3a algorithm on the set of 30 miRNAs significantly differentially expressed in ccRCC. This algorithm reports potential gene targets for the analyzed sequences based on sequence complementarity and the thermodynamic stability of the complementary RNA duplexes [29]. Predicted target genes with an alignment score ≥ 140 and an energy threshold of ≤ −20 kcal/mol were considered as potential gene targets of our newly detected miRNAs in ccRCC. To generate a broad list of possible gene targets and their associated pathways that may be affected by the regulatory action of the 30 miRNAs, we examined genes that were targeted by at least 10% of these miRNAs (Supplementary Table 3), which were subsequently analyzed for pathway enrichment using pathDIP [30].

Pathway enrichment analysis revealed 63 significantly enriched core pathways (BH-p ≤ 0.05). Interestingly, many of the enriched pathways are associated with cellular response to extracellular stimuli, organ development, and pathways indirectly associated with cellular metabolism, including the axon guidance pathway (110 gene targets), MAPK signaling cascade (58 gene targets), signaling mediated by FGFR and EGFR (70 and 69 gene targets, resp.), as well as the VEGF pathway (65 gene targets), and insulin-mediated signaling (63 gene targets) (Figure 4) [31, 32]. Considering that kidney cancers have been recognized to associate with altered hypoxic signaling, as well as cellular metabolism [4, 33], our results suggest that these previously unannotated miRNAs may play key roles in the regulation of ccRCC development and progression.

3.5. In Vitro Validation of Previously Unannotated miRNAs Deregulated in ccRCC. In order to experimentally confirm the existence of these transcripts and their consequent deregulation in tumour tissues, we performed RT-qPCR using custom primers specific to Knm3_1968 and Knm17_1130, previously unannotated miRNAs strongly overexpressed in ccRCC samples. We used the TK-10 cell line, which is representative of the ccRCC subtype of renal cancer, as well as HEK-293T, which is an immortalized nonmalignant embryonic kidney cell line commonly used as a control cell line to represent nonmalignant kidney tissue. RT-qPCR results were consistent with expression data from RNA-sequencing, wherein both Knm3_1968 and Knm17_1130 were found to be overexpressed in TK-10 cells relative to HEK-293T cells (average RQ = 8.56; 16.54, resp.; Supplemental Figure 3).

4. Discussion

In this study, we discovered miRNAs previously unannotated in kidney tissues using the miRDeep2 algorithm which represents an increase of 11.5% from the current number of kidney-related miRNAs annotated in miRBase v21 [24]. Collectively, our findings underscore the need to accurately define the landscape of miRNA transcription in human tissues, particularly due to their emerging roles in cell biology and disease.

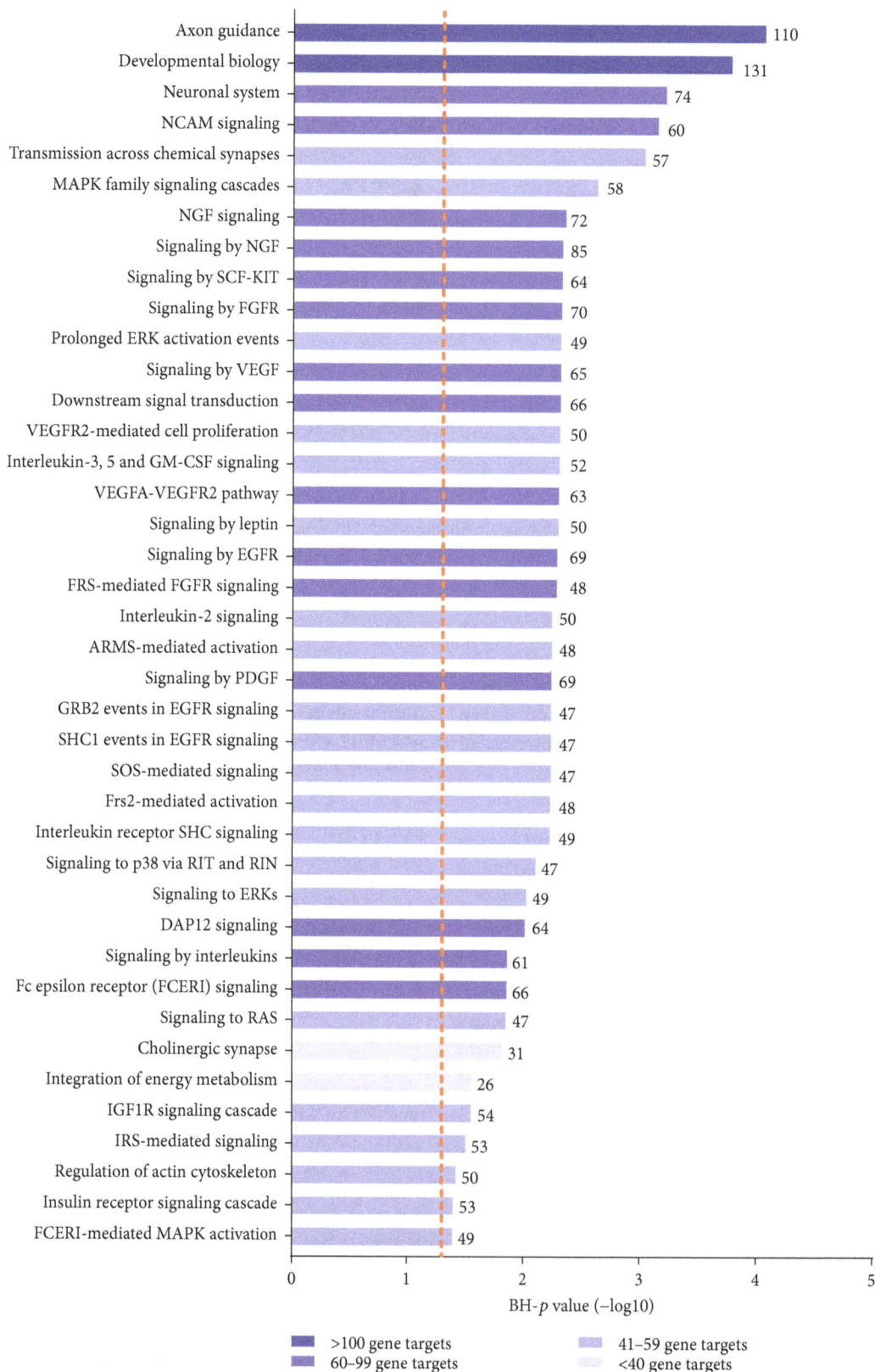

FIGURE 4: Pathways significantly enriched in protein-coding gene targets of previously unannotated miRNAs. Pathways are plotted according to the significance value associated with their enrichment in predicted miRNA gene targets. Individual genes in each pathway are predicted to be targeted by at least 10% of the previously unannotated miRNAs. Bars are coloured by the number of predicted gene targets involved in the pathway, with the total number of gene targets displayed on top of each bar. The orange dashed line represents the significance cutoff of BH-p = 0.05.

The miRDeep2 algorithm integrates artifacts of Dicer-based miRNA processing to generate a score for each predicted miRNA that is reflective of its likelihood of being a true positive [22]. Thus, we further processed the predicted sequences to eliminate sequences that (i) have low true positive detection rates (miRDeep2 score), (ii) share sequencing reads with other ncRNA species (tRNA/rRNA), (iii) do not meet the threshold of significance (ranfold p value), (iv) strongly differ from known miRNAs in G/C content, and (v) are already found in miRBase v21 (BLASTn). Together, these filtering steps enable us to generate a robust set of previously unannotated miRNAs that have low false-discovery rates.

To further validate these miRNAs in kidney tissues, we probed their expression in the NCI-60 panel of cell lines. Although the eight renal cancer cell lines do not reflect the spectrum of ccRCC in patients, they offer the opportunity to assess whether a fraction of the newly identified miRNAs occur in samples independent of TCGA. We found 128 miRNAs newly detected in kidney samples to be expressed in the eight renal cancer cell lines found in the panel. These findings, in tandem with our previous work characterizing the ncRNA transcriptome of the panel [28], serve to confirm the presence of the miRNAs detected in kidney tissues in our analysis and more broadly emphasize the extent of miRNA transcription that has previously escaped detection. Moreover, this highlights the utility of the NCI-60 cell line panel in future explorations of novel miRNA discovery and biology.

After validating their expression in cell lines, we examined the relevance of our panel of miRNAs newly detected in kidney tissues to ccRCC tumourigenesis and disease biology. Strikingly, we found a large proportion of these miRNAs to be significantly differentially expressed between ccRCC tumours and matched nonmalignant tissue. We further observed strongly deregulated miRNA sequences, both up- and downregulated in ccRCC samples. These cases—particularly Knm22_2209, Knm17_1130, and Knm3_1968—suggest that the manipulation of miRNA expression may be a factor in the tumourigenesis of ccRCC tumours. In addition, there are a number of our previously unannotated sequences that were not included in the differential expression analyses as their expression is only detected in either the nonmalignant or ccRCC paired samples. However, the expression of these sequences may be specific to either nonmalignant or cancerous kidney tissue (Supplemental Figure 2), suggesting that these miRNAs may represent exciting candidates for markers of ccRCC development or as therapeutic targets that may display limited off-target effects in normal cells.

The clinical applicability of a subset of the previously unannotated miRNAs is highlighted by our observations that several of the differentially expressed miRNAs are significantly associated with patient outcome. Specifically, the miRNAs Knm22_2209 and Knm6_2419 display significant relationships with the poor outcome of ccRCC patients. While preliminarily, these results highlight the potential impact that novel miRNA discovery can have on both cancer biology and clinical cancer intervention, warranting further characterization of the clinical relevance of previously undetected miRNAs in various pathologies.

The miRNAs discovered in our analyses are observed to target genes critical to both normal and diseased kidney biology. The axon guidance pathway mediates neuronal migration and positioning; in the kidney, this pathway has been shown to play key roles in organ development, in which deletion of critical pathway genes has been shown to lead to after-birth death due to kidney abnormalities [34]. Interestingly, 110 genes in this pathway are each targeted by at least 3 of the previously unannotated miRNAs. Studies have also implicated aberrations in genes from the axon guidance pathway with malignant cell growth and angiogenesis [35]. In fact, kidney cancers have been reported to rely on extensive metabolic reprogramming, in which most of the identified molecular drivers have been shown to participate in pathways related to cellular energy, nutrient metabolism, and oxygen sensing [5, 7]. The newly detected miRNAs in the kidney were found to target several components of the vascular endothelial growth factor (VEGF) pathway, a major regulator of angiogenesis, as well as signaling cascades in response to extracellular stimuli, such as the insulin receptor signaling cascades and FGRF and EGFR signaling.

Specifically, Knm3_1968 is found to target genes such as ACACA, which encodes acetyl-CoA carboxylase alpha, involved in fatty acid synthesis. Additionally, Knm22_2209 targets genes such as RASGRP2, which is associated with the MAPK signaling pathway. Thus, the observation that key pathways in ccRCC tumourigenesis are predicted to be targeted by the miRNAs discovered in our study, particularly axon guidance, metabolic reprogramming, and angiogenesis, emphasizes the potential regulatory role of these transcripts in kidney cancer biology.

Finally, both Knm3_1968 and Knm17_1130 were significantly overexpressed in ccRCC relative to nonmalignant cell lines, as shown by RT-qPCR (Supplemental Figure 3), which confirms the results observed from RNA-sequencing expression data. The expression of our newly detected sequences in cell lines not only confirms their presence in human kidney tissue but is also suggestive of their relevance to ccRCC biology. In light of these observations, further experiments may seek to elucidate the phenotypic consequences of previously unannotated miRNA deregulation.

Taken together, our results suggest that the discovery of previously unannotated miRNAs is an important next step in the exploration of the genomic landscape of tumourigenesis. While performed under the lens of kidney cancer, our findings have implications for cancers of all types, particularly in combination with previous studies that show these types of sequences to have a higher degree of specificity than those that are currently annotated. Although further validation and characterization of these types of sequences are required before these miRNAs can become clinically actionable, our findings highlight the untapped potential of the unexplored areas of the human transcriptome.

5. Conclusions

Here, we present an in-depth and large-scale discovery of previously unannotated miRNA sequences in human kidney samples. We find that not only are these sequences indeed expressed in both patient samples and cell lines but also their expression may be relevant to normal and tumour biology in these tissues. Several of these newly detected miRNAs are both deregulated in ccRCC tumours and associated with poor patient outcome. Moreover, protein-coding genes and subsequent pathways predicted to be targeted by these sequences, such as the VEGF and EGFR pathways, are critical to kidney tumourigenesis. Taken together, our results provide a novel resource for studying kidney cancer biology and underline a need for further identification of miRNAs that have eluded previous detection methods. Through the discovery of previously unannotated sequences, future studies may uncover novel players in tumour biology that may result in better characterization of ccRCC molecular drivers and direct new diagnostic and treatment strategies in the clinic.

Conflicts of Interest

The authors declare that they have no conflict of interest.

Acknowledgments

This work was supported by grants from the Canadian Institutes for Health Research (CIHR FRN-143345). Adam P. Sage, Brenda C. Minatel, Erin A. Marshall, Victor D. Martinez, and Katey S. S. Enfield were supported by scholarships from the University of British Columbia. Adam P. Sage, Erin A. Marshall, and Katey S. S. Enfield were also supported by scholarships from CIHR.

Supplementary Materials

Supplementary 1. Figure 1: genome-wide distribution of previously unannotated miRNA sequences. Purple bars represent $\log_2$ (total read counts) for previously unannotated miRNAs detected in ccRCC samples, while green bars indicate those detected in nonmalignant tissues. The outer circle represents all chromosomes from the human genome.

Supplementary 2. Figure 2: unsupervised hierarchical clustering analysis of differentially expressed previously unannotated miRNA sequences. Both samples and features were clustered using average linkage and Pearson correlation. Standardized expression is represented shown from high (red) to low (green), and samples are coloured according to their malignant status (blue represents nonmalignant and red represents tumour tissue).

Supplementary 3. Figure 3: fold change (RQ) values from RT-qPCR analysis of Knm3_1968 and Knm17_1130 in TK-10 ccRCC cells relative to nonmalignant HEK-293T cells.

Supplementary 4. Table 1: summary of OASIS output and subsequent filtering steps of predicted novel miRNAs detected in kidney tissues. This table lists the raw output of the computational analysis deducing previously undetected miRNAs from RNA-sequencing data of nonmalignant (NM) and clear cell renal cell carcinoma samples (TP) (columns A–E) and the results from further filtering steps (columns F–J) as described in the "Materials and Methods." Table 2: cell line characteristics and next-generation sequencing information for the NCI-60 renal cell lines used to validate the discovery of previously unannotated miRNA sequences. Table 3: predicted target genes of the newly detected miRNAs differentially expressed in clear cell renal cell carcinoma. This table lists the output values of the miRanda target prediction algorithm used to predict the protein-coding genes targeted by previously undetected miRNAs in kidney samples, as described in "Materials and Methods."

References

[1] Canadian Cancer Statistics Advisory Committee, *Canadian Cancer Statistics 2018*, Canadian Cancer Society, Toronto, ON, Canada, 2018.

[2] W. H. Chow, L. M. Dong, and S. S. Devesa, "Epidemiology and risk factors for kidney cancer," *Nature Reviews Urology*, vol. 7, no. 5, pp. 245–257, 2010.

[3] Y. Riazalhosseini and M. Lathrop, "Precision medicine from the renal cancer genome," *Nature Reviews Urology*, vol. 12, no. 11, pp. 655–666, 2016.

[4] W. M. Linehan, R. Srinivasan, and L. S. Schmidt, "The genetic basis of kidney cancer: a metabolic disease," *Nature Reviews Urology*, vol. 7, no. 5, pp. 277–285, 2010.

[5] The Cancer Genome Atlas Research Network, "Comprehensive molecular characterization of clear cell renal cell carcinoma," *Nature*, vol. 499, no. 7456, pp. 43–49, 2013.

[6] I. J. Frew and H. Moch, "A clearer view of the molecular complexity of clear cell renal cell carcinoma," *Annual Review of Pathology: Mechanisms of Disease*, vol. 10, no. 1, pp. 263–289, 2015.

[7] H. I. Wettersten, O. A. Aboud, P. N. Lara, and R. H. Weiss, "Metabolic reprogramming in clear cell renal cell carcinoma," *Nature Reviews Nephrology*, vol. 13, no. 7, pp. 410–419, 2017.

[8] R. Fisher, M. Gore, and J. Larkin, "Current and future systemic treatments for renal cell carcinoma," *Seminars in Cancer Biology*, vol. 23, no. 1, pp. 38–45, 2013.

[9] P. H. Tan, L. Cheng, N. Rioux-Leclercq et al., "Renal tumors: diagnostic and prognostic biomarkers," *The American Journal of Surgical Pathology*, vol. 37, no. 10, pp. 1518–1531, 2013.

[10] K. S. S. Enfield, L. A. Pikor, V. D. Martinez, and W. L. Lam, "Mechanistic roles of noncoding RNAs in lung cancer biology and their clinical implications," *Genetics Research International*, vol. 2012, Article ID 737416, 16 pages, 2012.

[11] R. C. Lee, R. L. Feinbaum, and V. Ambros, "The C. elegans heterochronic gene lin-4 encodes small RNAs with antisense complementarity to lin-14," *Cell*, vol. 75, no. 5, pp. 843–854, 1993.

[12] F. Calore, F. Lovat, and M. Garofalo, "Non-coding RNAs and cancer," *International Journal of Molecular Sciences*, vol. 14, no. 8, pp. 17085–17110, 2013.

[13] P. Trionfini, A. Benigni, and G. Remuzzi, "MicroRNAs in kidney physiology and disease," *Nature Reviews Nephrology*, vol. 11, no. 1, pp. 23–33, 2015.

[14] A. K. Marrone, D. B. Stolz, S. I. Bastacky, D. Kostka, A. J. Bodnar, and J. Ho, "*MicroRNA-17~92* is required for nephrogenesis and renal function," *Journal of the American Society of Nephrology*, vol. 25, no. 7, pp. 1440–1452, 2014.

[15] Y. Gong, V. Renigunta, N. Himmerkus et al., "Claudin-14 regulates renal Ca^{++} transport in response to CaSR signalling via a novel microRNA pathway," *The EMBO Journal*, vol. 31, no. 8, pp. 1999–2012, 2012.

[16] Q. Ma, Z. Peng, L. Wang et al., "miR-19a correlates with poor prognosis of clear cell renal cell carcinoma patients via promoting cell proliferation and suppressing PTEN/SMAD4 expression," *International Journal of Oncology*, vol. 49, no. 6, pp. 2589–2599, 2016.

[17] E. Londin, P. Loher, A. G. Telonis et al., "Analysis of 13 cell types reveals evidence for the expression of numerous novel primate- and tissue-specific microRNAs," *Proceedings of the National Academy of Sciences of the United States of America*, vol. 112, no. 10, pp. E1106–E1115, 2015.

[18] C. Backes, B. Meder, M. Hart et al., "Prioritizing and selecting likely novel miRNAs from NGS data," *Nucleic Acids Research*, vol. 44, no. 6, article e53, 2016.

[19] V. D. Martinez, E. A. Vucic, K. L. Thu et al., "Unique somatic and malignant expression patterns implicate PIWI-interacting RNAs in cancer-type specific biology," *Scientific Reports*, vol. 5, no. 1, article 10423, 2015.

[20] J. An, J. Lai, M. L. Lehman, and C. C. Nelson, "miRDeep* an integrated application tool for miRNA identification from RNA sequencing data," *Nucleic Acids Research*, vol. 41, no. 2, pp. 727–737, 2013.

[21] P. P. Gardner, J. Daub, J. G. Tate et al., "Rfam: updates to the RNA families database," *Nucleic Acids Research*, vol. 37, no. - Database, pp. D136–D140, 2009.

[22] C. Camacho, G. Coulouris, V. Avagyan et al., "BLAST+: architecture and applications," *BMC Bioinformatics*, vol. 10, no. 1, p. 421, 2009.

[23] V. Capece, J. C. Garcia Vizcaino, R. Vidal et al., "Oasis: online analysis of small RNA deep sequencing data," *Bioinformatics*, vol. 31, no. 13, pp. 2205–2207, 2015.

[24] A. Kozomara and S. Griffiths-Jones, "miRBase: annotating high confidence microRNAs using deep sequencing data," *Nucleic Acids Research*, vol. 42, no. D1, pp. D68–D73, 2014.

[25] Y. Liao, G. K. Smyth, and W. Shi, "featureCounts: an efficient general purpose program for assigning sequence reads to genomic features," *Bioinformatics*, vol. 30, no. 7, pp. 923–930, 2014.

[26] M. N. McCall, M.-S. Kim, M. Adil et al., "Toward the human cellular microRNAome," *Genome Research*, vol. 27, no. 10, pp. 1769–1781, 2017.

[27] C. Wake, A. Labadorf, A. Dumitriu et al., "Novel microRNA discovery using small RNA sequencing in post-mortem human brain," *BMC Genomics*, vol. 17, no. 1, p. 776, 2016.

[28] E. A. Marshall, A. P. Sage, K. W. Ng et al., "Small non-coding RNA transcriptome of the NCI-60 cell line panel," *Scientific Data*, vol. 4, article 170157, 2017.

[29] A. J. Enright, B. John, U. Gaul, T. Tuschl, C. Sander, and D. S. Marks, "MicroRNA targets in Drosophila," *Genome Biology*, vol. 5, no. 1, article R1, 2003.

[30] S. Rahmati, M. Abovsky, C. Pastrello, and I. Jurisica, "pathDIP: an annotated resource for known and predicted human gene-pathway associations and pathway enrichment analysis," *Nucleic Acids Research*, vol. 45, no. D1, pp. D419–D426, 2017.

[31] W. Zhang and H. T. Liu, "MAPK signal pathways in the regulation of cell proliferation in mammalian cells," *Cell Research*, vol. 12, no. 1, pp. 9–18, 2002.

[32] L. R. Harskamp, R. T. Gansevoort, H. van Goor, and E. Meijer, "The epidermal growth factor receptor pathway in chronic kidney diseases," *Nature Reviews Nephrology*, vol. 12, no. 8, pp. 496–506, 2016.

[33] C. H. Lee and R. J. Motzer, "Kidney cancer in 2016: the evolution of anti-angiogenic therapy for kidney cancer," *Nature Reviews Nephrology*, vol. 13, no. 2, pp. 69-70, 2017.

[34] U. Grieshammer, le Ma, A. S. Plump, F. Wang, M. Tessier-Lavigne, and G. R. Martin, "SLIT2-mediated ROBO2 signaling restricts kidney induction to a single site," *Developmental Cell*, vol. 6, no. 5, pp. 709–717, 2004.

[35] P. Mehlen, C. Delloye-Bourgeois, and A. Chedotal, "Novel roles for slits and netrins: axon guidance cues as anticancer targets?," *Nature Reviews Cancer*, vol. 11, no. 3, pp. 188–197, 2011.

The β-Lactamase Gene Profile and a Plasmid-Carrying Multiple Heavy Metal Resistance Genes of *Enterobacter cloacae*

Chongyang Wu,[1,2] Chaoqin Lin,[1,2] Xinyi Zhu,[1,2] Hongmao Liu,[1,2] Wangxiao Zhou,[2] Junwan Lu,[3] Licheng Zhu,[3] Qiyu Bao◉,[2] Cong Cheng◉,[3] and Yunliang Hu◉[1,2]

[1]*The Second Affiliated Hospital and Yuying Children's Hospital of Wenzhou Medical University, Wenzhou, Zhejiang 325000, China*
[2]*School of Laboratory Medicine and Life Sciences/Institute of Biomedical Informatics, Wenzhou Medical University, Wenzhou 325035, China*
[3]*College of Medicine and Health, Lishui University, Lishui 323000, China*

Correspondence should be addressed to Cong Cheng; 113246570@qq.com and Yunliang Hu; hyl@wmu.edu.cn

Academic Editor: João Paulo Gomes

In this work, by high-throughput sequencing, antibiotic resistance genes, including class A (bla_{CTX-M}, bla_Z, bla_{TEM}, bla_{VEB}, bla_{KLUC}, and bla_{SFO}), class C (bla_{SHV}, bla_{DHA}, bla_{MIR}, $bla_{AZECL-29}$, and bla_{ACT}), and class D (bla_{OXA}) β-lactamase genes, were identified among the pooled genomic DNA from 212 clinical *Enterobacter cloacae* isolates. Six bla_{MIR}-positive *E. cloacae* strains were identified, and pulsed-field gel electrophoresis (PFGE) showed that these strains were not clonally related. The complete genome of the bla_{MIR}-positive strain (Y546) consisted of both a chromosome (4.78 Mb) and a large plasmid pY546 (208.74 kb). The extended-spectrum β-lactamases (ESBLs) (bla_{SHV-12} and $bla_{CTX-M-9a}$) and AmpC (bla_{MIR}) were encoded on the chromosome, and the pY546 plasmid contained several clusters of genes conferring resistance to metals, such as copper (*pco*), arsenic (*ars*), tellurite (*ter*), and tetrathionate (*ttr*), and genes encoding many divalent cation transporter proteins. The comparative genomic analyses of the whole plasmid sequence and of the heavy metal resistance gene-encoding regions revealed that the plasmid sequences of *Klebsiella pneumoniae* (such as pKPN-332, pKPN-3967, and pKPN-262) shared the highest similarity with those of pY546. It may be concluded that a variety of β-lactamase genes present in *E. cloacae* which confer resistance to β-lactam antibiotics and the emergence of plasmids carrying heavy metal resistance genes in clinical isolates are alarming and need further surveillance.

1. Introduction

Bacteria of the *Enterobacter cloacae complex* (*ECC*), which comprises six species, namely, *E. cloacae*, *E. asburiae*, *E. hormaechei*, *E. kobei*, *E. ludwigii*, and *E. nimipressuralis* [1], are widely distributed in nature. As pathogens, *ECC* species are highly adapted to the environment and are able to contaminate hospital medical devices. Currently, *E. cloacae* and *E. hormaechei* are most frequently isolated from human clinical specimens, and *E. cloacae* is among the *Enterobacter* sp. that have most commonly caused nosocomial infections in the last decade [2]. Furthermore, *E. cloacae* has assumed clinical importance and has emerged as a major human pathogen; it accounts for up to 5% of hospital-acquired bacteremia cases, 5% of nosocomial pneumonia cases, 4% of nosocomial

urinary tract infections, and 10% of postsurgical peritonitis cases [3].

Owing to the low-level but inducible expression of a chromosomal *ampC* gene encoding the AmpC β-lactamase, *E. cloacae* is intrinsically resistant to ampicillin, amoxicillin-clavulanate, and first-generation cephalosporins [4]. Generally, the resistance of *E. cloacae* to third-generation cephalosporins is caused by its overproduction of the AmpC β-lactamases when the production of this cephalosporinase is inducible in the presence of strong β-lactam antibiotics (cefoxitin and imipenem); thus, treatment with third-generation cephalosporins may promote the development of AmpC-overproducing mutants. AmpC-producing organisms become resistant to almost all β-lactam antibiotics, with the exception of cefepime, cefpirome, and carbapenems.

Most chromosomal *ampC* genes are inducible in the presence of certain agents such as cefoxitin and imipenem. Inducible AmpC expression is regulated by AmpR in the presence of two other gene products, namely, AmpD and AmpG. The regulation of AmpC production has been historically understood to require three proteins: AmpG, a plasma membrane-bound permease; AmpD, a cytosolic peptidoglycan-recycling amidase; and AmpR, the transcriptional regulator of AmpC. Derepression has been associated previously with structural defects within the *ampD* gene or with decreased *ampD* expression. Derepression represents the inability of AmpR to keep AmpC expression at constitutively low wild-type levels [5]. As a result, the AmpC enzyme confers resistance to third-generation cephalosporins and is not inhibited by common β-lactamase inhibitors. However, fourth-generation cephalosporins still retain activity against most *Enterobacteriaceae* strains.

In addition to therapeutic antibiotic agents, a large number of other chemical substances with antibacterial activities, such as heavy metals and detergents, are used in human health care and agricultural practices. Recently, concerns have been raised regarding coselection for antibiotic resistance among bacteria exposed to disinfectants and heavy metals (particularly copper, zinc, and mercury) used in some livestock species as growth promoters and therapeutic agents [6]. *Enterobacteriaceae* (including *E. coli*, *K. pneumoniae*, and *E. cloacae*) are highly adept at acquiring resistance genes to all disinfectants, heavy metals, and antibiotics through horizontal gene transfer between different bacteria within the environment; such genes include extended-spectrum beta-lactamases (ESBLs), copper and arsenic resistance systems (the *pco and ars* operons), and enzymes that hydrolyze cephalosporins (AmpC enzymes) [7, 8]. Many gram-negative organisms (such as *E. coli*, *E. cloacae*, and *K. pneumoniae*) encode broad-substrate efflux pumps [6, 7, 9], and a variety of multidrug pumps that have activity against disinfectants are similarly encoded by some gram-positive organisms, including *Staphylococcus aureus* [9, 10]. Besides the efflux pumps, other mechanisms such as chemisorption facilitating cadmium to bind to the bacterial surface also played a role in the heavy metal (cadmium) resistance [11]. Alternatively, in both gram-positive and gram-negative bacteria, mechanisms of acquired resistance to disinfectants may be associated with efflux pump-encoding genes introduced on mobile genetic elements or, in gram-negative bacteria, with mutations causing the constitutive overexpression of efflux pumps. Compared with antibiotics and disinfectants, heavy metals (copper, arsenic, zinc, and mercury) are very persistent in the environment and may accumulate in soil, water, and sediments from agricultural practices as well as from other sources such as aquacultural and industrial effluents [12]. Like the mechanisms of resistance to disinfectants, efflux pumps can expel heavy metal ions; such pumps include elements of the *czc* system, which encodes a pump for zinc, cobalt, and cadmium, and *pcoA*, which is an element of a copper extrusion system (W. [13]).

In this study, we identified β-lactamase genes in 212 clinical *E. cloacae* isolates and sequenced a *bla*$_{MIR}$ gene-carrying strain. Molecular analyses were performed to analyze the function of the resistance genes, and a comparative genomics analysis was conducted to elucidate the potential horizontal gene transfer patterns of genes related to both antibiotic and heavy metal resistance between bacteria of different species or genera. Our analysis revealed the distinct structure of a large plasmid carrying multiple clusters of heavy metal resistance genes that, to our knowledge, have not been described previously in *E. cloacae*.

2. Materials and Methods

2.1. Bacterial Strain Collection, Genomic DNA Extraction, and High-Throughput Sequencing. A total of 212 nonduplicate clinically significant *E. cloacae* strains were isolated from the First Affiliated Hospital of Wenzhou Medical University (Zhejiang, China) between 2008 and 2012. These isolates were identified by a VITEK 60 microbial autoanalyzer (bio-Mérieux, Lyon, France). Bacteria and plasmids used in this study are listed in Table 1. Among the isolates, 31, 36, 43, 32, and 70 strains were isolated in the years 2008, 2009, 2010, 2011, and 2012, respectively. All strains were resistant to a minimum of one or two antibiotics. For pooled genomic DNA sequencing, each clinical strain was incubated in 5 mL of Luria-Bertani (LB) broth at 37°C for approximately 16 h to obtain a concentration equivalent to an optimum optical density ($OD_{600} = 1.5 \pm 0.2$). The cultures were pooled, and genomic DNA was extracted from 100 mL of the mixed bacteria using an AxyPrep Bacterial Genomic DNA Miniprep Kit (Axygen Scientific, Union City, CA, USA). The pooled genomic DNA was sequenced with a HiSeq 2500 DNA sequencer at Annoroad Gene Technology Co. Ltd. (Beijing, China). Reads derived from the HiSeq 2500 sequencing were initially assembled de novo with SOAPdenovo software to obtain contigs of the genome sequences. Glimmer software (http://ccb.jhu.edu/software/glimmer) was used to predict protein-coding genes with potential open reading frames (ORF) > 150 bp in length. BLASTX (https://blast.ncbi.nlm.nih.gov) was used to annotate the predicted protein-coding genes against the nonredundant protein database with an e-value threshold of $1E-5$.

2.2. Collection of Reference Sequences for Resistance Genes and Mapping of Sequencing Reads to the Reference Genes. The nucleotide sequences of all β-lactamase genes, including those encoding for Ambler class A, B, C, and D β-lactamases, were obtained from the GenBank nucleotide database, and the high-throughput sequencing reads were mapped to the reference sequences of the β-lactamase genes as previously described [14]. The relative abundance (sequencing depth) of a certain gene was calculated as the cumulative nucleotide length of the mapped short reads on the gene divided by the gene size.

2.3. Screening of β-Lactamase Gene-Positive Strains and Cloning and Phylogenetic Analysis of bla$_{MIR}$ *Genes.* The *E. cloacae* strains were screened by PCR amplification for the presence of β-lactamase genes as previously described [15–18]. The primers used for the cloning of complete ORFs contained a pair of flanking restriction endonuclease

TABLE 1: Bacteria and plasmids used in this study.

Strain	Relevant characteristic(s)	Reference or source
BL21	*E. coli* BL21 was used as a host for the cloned bla_{MIR} gene	[14]
ATCC25922	*E. coli* ATCC25922 is FDA clinical isolate	[14]
CG34	bla_{MIR}-producing isolate of *E. cloacae* CG34	This study
CG76	bla_{MIR}-producing isolate of *E. cloacae* CG76	This study
CG85	bla_{MIR}-producing isolate of *E. cloacae* CG85	This study
Y546	bla_{MIR}-producing isolate of *E. cloacae* Y546	This study
Y482	bla_{MIR}-producing isolate of *E. cloacae* Y482	This study
Y490	bla_{MIR}-producing isolate of *E. cloacae* Y490	This study
pET28a/BL21	BL21 carrying expression vector pET28a, km^{r}	This study
pET28a-$bla_{\text{MIR-CG34}}$/BL21	BL21 carrying recombinant plasmid pET28a-$bla_{\text{MIR-CG34}}$	This study
pET28a-$bla_{\text{MIR-Y490}}$/BL21	BL21 carrying recombinant plasmid pET28a-$bla_{\text{MIR-Y490}}$	This study
pET28a-$bla_{\text{MIR-CG76}}$/BL21	BL21 carrying recombinant plasmid pET28a-$bla_{\text{MIR-CG76}}$	This study
pET28a-$bla_{\text{MIR-Y546}}$/BL21	BL21 carrying recombinant plasmid pET28a-$bla_{\text{MIR-Y546}}$	This study
pET28a-$bla_{\text{MIR-CG85}}$/BL21	BL21 carrying recombinant plasmid pET28a-$bla_{\text{MIR-CG85}}$	This study
pET28a-$bla_{\text{MIR-Y482}}$/BL21	BL21 carrying recombinant plasmid pET28a-$bla_{\text{MIR-Y482}}$	This study
pET28a-$bla_{\text{CTX-M-9a-Y546}}$/BL21	BL21 carrying recombinant plasmid pET28a-$bla_{\text{CTX-M-9a-Y546}}$	This study
pET28a-$bla_{\text{SHV-12-Y546}}$/BL21	BL21 carrying recombinant plasmid pET28a-bla$_{\text{SHV-12-Y546}}$	This study

km: kanamycin; ^{r}: resistant.

adapters (*BamH*I for the forward primers and *Hind*III for the reverse primers) and were designed using the Primer Premier 5.0 software package; the primers are shown in Table 2. Genomic DNA was extracted from each of the 212 clinical *E. cloacae* isolates using the AxyPrep Bacterial Genomic DNA Miniprep kit (Axygen Scientific, Union City, CA, USA) and was used as the template for PCR amplification. Positive amplification products were verified by sequencing with an ABI 3730 automated sequencer (Shanghai Sunny Biotechnology Co. Ltd., Shanghai, China), and the sequencing results were compared with Basic Local Alignment Search Tool (BLAST) algorithms (https://blast.ncbi.nlm.nih.gov/blast.cgi). The amplicons of the *bla* ORFs were digested with the corresponding restriction endonucleases and ligated into the pET-28a vector with a T4 DNA ligase cloning kit (Takara Bio Inc., Dalian, China). The recombinant plasmid was transformed into competent *E. coli* BL21 cells using the calcium chloride method and cultured on LB agar plates supplemented with ampicillin (200 μg/mL), and the cloned ORFs were confirmed by sequencing. For the phylogenetic analysis, all bla_{MIR} gene sequences were collected from the NCBI nucleotide database using bla_{MIR} as the key search term. A phylogenetic tree of the MIR amino acid sequences from both the database and this work was reconstructed by the maximum likelihood method, and the resulting trees were analyzed with bootstrap values of 100 replicates using MEGA 6.0 (https://www.megasoftware.net/).

2.4. *Antimicrobial Susceptibility Testing.* Antimicrobial susceptibility testing was conducted for all tested antibiotics by the agar dilution method, and the minimum inhibitory concentrations (MICs) were interpreted based on the Clinical and Laboratory Standards Institute (CLSI) breakpoint criteria (CLSI, 2018) (https://clsi.org/standards/products/packages/m02-m07-m100-package/). Strain ATCC 25922 was used as the quality control strain. The 14 antibiotics (or antibiotics combination) used in this work were cephamycins (cefoxitin and cefminox), semisynthetic broad-spectrum penicillins (ampicillin and piperacillin), a first-generation cephalosporin (cefazolin), third-generation cephalosporins (ceftazidime, cefoperazone, cefotaxime, and ceftriaxone), a fourth-generation cephalosporin (cefoselis), a monobactam (aztreonam), an aminoglycoside (kanamycin), and combinations of antibiotics with β-lactamase inhibitors (piperacillin/tazobactam and imipenem/cilastatin sodium hydrate) (Table 3).

2.5. *Whole Genome Sequencing (WGS) of Y546 and Comparative Genomics Analysis.* The *E. cloacae* Y546 genomic DNA was extracted with the AxyPrep Bacterial Genomic DNA Miniprep Kit (Axygen Scientific, Union City, CA, USA) and sequenced with Illumina HiSeq 2500 and Pacific Biosciences sequencers at Annoroad Gene Technology Co. Ltd. (Beijing, China). Reads derived from HiSeq 2500 sequencing were initially assembled de novo with SOAPdenovo software to obtain genome sequence contigs. Reads of approximately 10-20 kb in length from the Pacific Biosciences sequencing were mapped onto the primary assembly for contig scaffolding. Gaps were filled either by remapping the short reads from the HiSeq 2500 sequencing or by sequencing the gap PCR product. Potential ORFs were predicted and annotated using Glimmer3 (http://www.cbcb.umd.edu/software/glimmer) and BASys [19], respectively. GC view software was used to construct the basic genomic features. Annotations were revised using UniProt (http://www.uniprot.org/) and BLAST (https://blast.ncbi.nlm.nih.gov/blast.cgi). Plasmid replicons and plasmid incompatibility

TABLE 2: Primers used in the study for the detection of β-lactamase-encoding genes.

Primer	Target(s)	Sequence (5'-3')	Annealing temperature (°C)	Amplicon size (bp)	Reference
veb-sf	bla_{VEB}	GATTGCTTTAGCCGTTTTGTC	50	452	[15]
veb-sr		ATCGGTTACTTCCTGTTGTTGTTTC			
z-sf	bla_Z	ACAGTTCACATGCCAAAGAGT	50	479	[16]
z-sr		CTTACCGAAAGCAGCAGGTG			
mir-sf	bla_{MIR}	GCCGCACCGATGTCCGAAAAA	50	545	[18]
mir-sr		GGTTTAAAGACCCGCGTCGTCATGG			
azecl-29-sf	$bla_{AZECL-29}$	GTCTTTACGCTAACACCAGCATCGG	50	381	[17]
azecl-29-sr		TCAGCATTTCCCAGCCCAATC			
veb-ff	bla_{VEB}	_CGGGATCC_ATGAAAATCGTAAAAAGGATATTAT	50	918	This study
veb-fr		_CCCAAGCT_TTATTTATTCAAATAGTAATTCCACG			
z-ff	bla_Z	_CGGGATCC_ATGAAAAAGTTAATATTTTTAATTG	50	864	This study
z-fr		_CCCAAGCT_TTAAAATTCCTTCATTACACTCTTG			
mir-ff	bla_{MIR}	_CGGGATCC_ATGATGACAAAATCCCTAAGCTGTG	66	1164	This study
mir-fr		_CCCAAGCT_TTACTGCAGCGCGTCGAGGATACGG			
azecl-29-ff	$bla_{AZECL-29}$	_CGGAATTC_ATGATGAAAAAAAACCTAAGCTGTG	60	1164	This study
azecl-29-fr		_CCCAAGCT_TTACTGCAGCGCGTCGAGGATACG			
ctx-m-9a-ff	$bla_{CTX-M-9a}$	_CGGGATCC_ATGGTGACAAAGAGAGTGCAACGGA	60	876	This study
ctx-m-9a-fr		_CCAAGCTT_TTACAGCCCTTCGGCGATGATTCTC			
shv-12-ff	bla_{SHV-12}	_CGGGATCC_GTGGTTATGCGTTATATTCGCCTGT	60	867	This study
shv-12-fr		_CCAAGCTT_TTAGCGTTGCCAGTGCTCGATCAGC			

Underlined sequences denote restriction endonuclease sites. sf: forward screening primer; sr: reverse screening primer; ff: forward full-length primer; fr: reverse full-length primer.

TABLE 3: The MIC values of antibacterial drugs for the strains (μg/mL).

Strains	AMP	KAN	CAZ	CTX	CPZ	CMN	CEF	CFZ	FOX	CRO	ATM	PRL	PTZ	IMC
ATCC25922	4	2	0.25	0.0625	0.125	1	0.125	2	4	0.0313	0.125	2	4	0.25
BL21	0.5	4	0.0625	0.0156	0.0156	1	0.0313	2	2	0.0156	0.0156	0.5	0.5	0.5
pET-28a/BL21*	0.5	32	0.0625	0.0156	0.0156	1	0.0313	2	2	0.0156	0.0156	0.5	0.5	0.5
CG34	256	0.5	0.125	0.125	0.0625	1024	0.0625	512	1024	0.0313	0.25	2	2	4
Y490	256	2	0.5	0.25	1	>1024	0.125	512	1024	0.5	0.0313	2	4	2
CG76	>1024	8	128	128	64	1024	16	>1024	1024	512	32	512	32	8
Y546	512	2	2	1	2	>1024	0.25	1024	1024	2	1	8	8	4
CG85	>1024	4	1	1	1	>1204	0.25	1024	>1024	1	0.25	4	4	4
Y482	64	2	0.25	0.125	0.25	1024	256	512	0.0625	0	0.125	2	2	0.5
pET-28a-bla_{MIR} (CG34)/BL21*	128	64	4	4	1	32	0.125	256	32	4	1	8	8	1
pET-28a-bla_{MIR} (Y490)/BL21*	128	64	4	4	0.5	32	0.0625	256	32	4	1	8	4	1
pET-28a-bla_{MIR} (CG76)/BL21*	128	64	4	4	1	32	0.0313	256	32	4	1	8	4	1
pET-28a-bla_{MIR} (Y546)/BL21*	128	64	4	4	1	32	0.0625	256	32	4	1	8	8	1
pET-28a-bla_{MIR} (CG85)/BL21*	128	64	4	4	1	32	0.125	256	32	4	1	8	8	1
pET-28a-bla_{MIR} (Y482)/BL21*	128	64	4	4	1	32	0.125	256	32	4	1	8	8	1
pET-28a-$bla_{CTX-M-9a}$/BL21*	8	64	0.25	4	1	4	0.03	64	4	0.03	0.03	2	1	1
pET-28a-bla_{SHV-12}/BL21*	8	64	2	4	1	4	0.25	64	4	0.03	0.03	32	2	2

*IPTG was added to a final concentration of 1 mmol/L to standard Mueller-Hinton (M-H) agar plates. All MIC determinations were performed by agar dilution assays. FOX: cefoxitin; CAZ: ceftazidime; CTX: cefotaxime; AMP: ampicillin; KAN: kanamycin; CPZ: cefoperazone; CMN: cefminox; CEF: cefoselis; CFZ: cefazolin; CRO: ceftriaxone; ATM: aztreonam; PRL: piperacillin; PTZ: piperacillin/tazobactam; IMC: imipenem/cilastatin sodium hydrate.

groups were predicted using Plasmid Finder (https://cge.cbs.dtu.dk//services/PlasmidFinder/). Furthermore, the multilocus sequence typing (MLST) database for *E. cloacae* (https://pubmLst.org/ecloacae/) was used to determine the sequence type of *E. cloacae* Y546.

The plasmid and chromosomal genomic sequences used in this study for the whole genome comparative analysis were downloaded from the NCBI database (http://www.ncbi.nlm.nih.gov). The plasmid and chromosome sequences were selected based on the comparison of the whole genome sequence (pY546) against the sequences of plasmids and chromosomes available in the NCBI database; a cutoff value (max score) of approximately 8700 was defined. For the comparative analysis of the heavy metal gene cluster regions on the pY546 plasmid, the sequences containing corresponding gene clusters with sequence identities of ≥80% with respect to those encoded on pY546 were obtained from the NCBI nucleotide database by BLASTn algorithms. Multiple sequence alignments were performed using MAFFT [20]. Comparisons of the nucleotide sequences were made using BLASTn. Insertion sequences were predicted using ISfinder [21]. Orthologous groups of genes from plasmids or chromosomes were identified using BLASTp and Inparanoid [22]. Additional bioinformatics software was written using Python (https://www.python.org/) and Biopython [23].

2.6. Pulsed-Field Gel Electrophoresis (PFGE). The clonal relatedness of MIR-producing *E. cloacae* isolates was assessed by *Xba*I (Takara Bio Inc., China) PFGE. Briefly, genomic DNA fragments were resolved on a 1% agarose (SeaKem Gold Agarose, Lonza) gel at 14°C, and electrophoresis was conducted at 6 V/cm using a CHEF PFGE instrument (Bio-Rad, USA) at a pulse time gradient of 2.25-55.5 s and a total run time of 18 h. *Salmonella enterica* serovar H9812 was used as a control. Cluster analysis was performed using an unweighted pair-group method with arithmetic (UPGMA) means. Isolates were allocated into genetic similarity clusters using a similarity cutoff value of 80% [24].

2.7. Nucleotide Sequence Accession Numbers. The sequences of the chromosome and the plasmid of Y546 and the bla_{MIR} genes in this work have been submitted to NCBI GenBank with accession numbers of CP032916 (Y546), CP032915 (pY546), MK033024 ($bla_{MIR-CG34}$), MK033023 ($bla_{MIR-Y490}$), MK033025 ($bla_{MIR-CG76}$), MK033026 ($bla_{MIR-Y546}$), MK033021 ($bla_{MIR-CG85}$), and MK033022 ($bla_{MIR-Y482}$), respectively.

3. Results

3.1. Mapping and Screening Results for the β-Lactamase Genes in the Sequenced Bacteria. A total of 75 β-lactamase gene sequences were collected from the database (Supplementary Table S1). The pooled genomic DNA sequences of the 212 isolated strains generated approximately 34.1 gigabases. All reads ranged from 100 to 110 nucleotides in length. The mapping of the sequencing reads onto the reference sequences yielded the identification of resistance genes; in addition, the quantity of mapped reads on a specific reference

could suggest the relative abundance of the reads in the sequenced samples. This analysis showed that the samples contained a total of 12 hits related to β-lactamase resistance genes. The most abundant gene was bla_{TEM}, which had a sequence depth of 466.15 (Supplementary Table S2). The other genotypes with a greater abundance were bla_{SHV}, bla_{DHA}, and bla_{CTX-M} (especially the $bla_{CTX-M-9}$ and $bla_{CTX-M-1}$ subtypes), while the genotypes with a lower abundance were bla_Z, bla_{VEB}, bla_{KLUC}, bla_{MIR}, bla_{SFO}, bla_{AZECL}, bla_{OXA}, and bla_{ACT}. The screening results for the bla_{VEB}, bla_Z, bla_{AZECL}, and bla_{MIR} genotypes revealed that among the 212 strains, only 0.47% (1/212; Y412), 0.94% (2/212; CG3 and CG4), 0.94% (2/212; Y411 and CG90), and 2.83% (6/212; CG34, Y490, CG76, Y546, CG85, and Y482) carried bla_{VEB}, bla_Z, bla_{AZECL}, and bla_{MIR}, respectively.

3.2. Cloning and Functional Detection of the Resistance Genes. Fourteen antimicrobial agents were used to detect the MIC levels of the bla_{MIR}-positive wild-type strains and the recombinant strains expressing the cloned bla_{MIR} genes (pET28a-bla_{MIR}/BL21). The MICs of the 14 antimicrobial agents against these strains are shown in Table 3. The MICs for all 6 bla_{MIR}-positive wild-type strains (CG34, Y490, CG76, Y546, CG85, and Y482) demonstrated that they were resistant to 4 commonly used broad-spectrum beta-lactam antibiotics, including ampicillin, a first-generation cephalosporin (cefazolin), and cephamycins (cefmenoxime and cefoxitin), and *E. cloacae* CG76 displayed higher resistance levels than the other strains to all antibiotics tested. Like the host wild-type strains, the 6 recombinant strains expressing the cloned bla_{MIR} genes (pET28a-bla_{MIR}/BL21) were resistant to ampicillin, cefazolin, cefmenoxime, and cefoxitin. In addition, the recombinants with two other extended-spectrum β-lactamase (ESBL) genes, namely, bla_{SHV-12} and $bla_{CTX-M-9a}$, encoded on the Y546 chromosome displayed higher hydrolytic activity against these four β-lactam antibiotics.

3.3. Clonal Relatedness of the bla_{MIR}-Positive Strains and Genotypes and the Phylogenetic Tree Analysis of the bla_{MIR} Genes. All 6 bla_{MIR}-positive strains (Y490, Y482 CG34, CG76, CG85, and Y546) had distinct *Xba*I PFGE patterns (Figure 1), indicating that the prevalence of bla_{MIR}-positive isolates was caused by disseminated gene transfer. Sequencing results showed that the bla_{MIR} ORFs from strains CG85 and Y482 belonged to the bla_{MIR-17} genotype, while the ORFs from strains CG34, CG76, Y490, and Y546 matched bla_{MIR-5}, bla_{MIR-21}, bla_{MIR-3}, and bla_{MIR-20}, respectively. These ORFs had 99% amino acid similarity to their respective reference sequences. To further analyze the evolutionary relationship of the 6 bla_{MIR} genes identified in this work with other bla_{MIR} genes, we performed a multiple sequence alignment on a total of 30 bla_{MIR} variants including the 6 bla_{MIR} genes identified in this study. The multiple-sequence alignment identified the Pro380Leu variant in $bla_{MIR-Y546}$, $bla_{MIR-Y482}$, and $bla_{MIR-Y490}$ and the Ala381Gln variant in $bla_{MIR-CG76}$, $bla_{MIR-CG85}$, $bla_{MIR-Y546}$, $bla_{MIR-Y482}$, and $bla_{MIR-Y490}$. The Asn206His variant was identified only in $bla_{MIR-CG34}$ (Figure 2). The phylogenetic analysis (Figure 3) showed that with the exception of 2 sequences

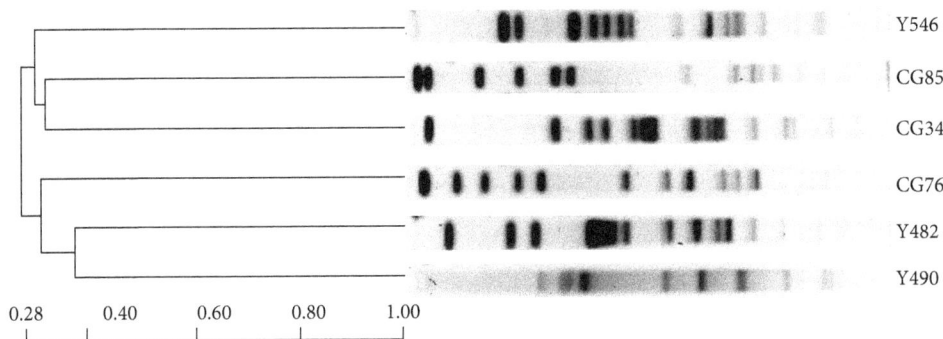

FIGURE 1: Pulsed-field gel electrophoresis (PFGE) analysis of the 6 bla_{MIR}-positive *E. cloacae* isolates. PFGE result showed that all 6 bla_{MIR}-positive isolates had a totally distinct PFGE pattern.

```
                *10         * *  20          30           40           50           60           70           80
CG34    1   MMTKSLSCAL  LLSVASSAFA  APMSEKQLAE  VVERTVTPLM  NAQAIPGMAV  AVIYQGQPHY  FTFGKADVAA  NKPVTPQTLF
CG75    1   MMTKSLSCAL  LLSVTSSAFA  APMSEKQLAE  VVERTVTPLM  NAQAIPGMAV  AVIYQGQPHY  FTFGKADVAA  NKPVTPQTLF
CG85    1   MMTKSLSCAL  LLSVASSAFA  APMSEKQLAE  VVERTVTPLM  NAQAIPGMAV  AVIYQGQPHY  FTFGKADVAA  NKPVTPQTLF
Y482    1   MMTKSLSCAL  LLSVASSAFA  APMSEKQLAE  VVERTVTPLM  NAQAIPGMAV  AVIYQGQPHY  FTFGKADVAA  NKPVTPQTLF
Y490    1   MMTKSLSC TL LLSVASSAFA  APMSEKQLAE  VVERTVTPLM  NAQAIPGMAV  AVIYQGQPHY  FTFGKADVAA  NKPVTPQTLF
Y546    1   MMTKSLSCAL  LLSVTSSAFA  APMSEKQLAE  VVERTVTPLM  NAQAIPGMAV  AVIYQGQPHY  FTFGKADVAA  NKPVTPQTLF
                90          100         *  110         120          130          140          150          160
CG34    81  ELGSISKTFT  GVLGGDAIAR  GEIALGDPVA  KYWPELTGKQ  WQGIRMLDLA  TYTAGGLPLQ  VPDEVTDTAS  LLRFYQNWQP
CG75    81  ELGSISKTFT  GVLGGDAIAR  GETALGDPVA  KYWPELTGKQ  WQGIRMLDLA  TYTAGGLPLQ  VPDEVTDTAS  LLRFYQNWQP
CG85    81  ELGSISKTFT  GVLGGDAIAR  GEIALGDPVA  KYWPELTGKQ  WQGIRMLDLA  TYTAGGLPLQ  VPDEVTDTAS  LLRFYQNWQP
Y482    81  ELGSISKTFT  GVLGGDAIAR  GEIALGDPVA  KYWPELTGKQ  WQGIRMLDLA  TYTAGGLPLQ  VPDEVTDTAS  LLRFYQNWQP
Y490    81  ELGSISKTFT  GVLGGDAIAR  GEIALGDPVA  KYWPELTGKQ  WQGIRMLDLA  TYTAGGLPLQ  VPDEVTDTAS  LLRFYQNWQP
Y546    81  ELGSISKTFT  GVLGGDAIAR  GEIALGDPVA  KYWPELTGKQ  WQGIRMLDLA  TYTAGGLPLQ  VPDEVTDTAS  LLRFYQNWQP
                170         180         190          200         *  210         220          230          240
CG34    161 QWKPGTTRLY  ANASIGLFGA  LAVKPSGMSY  EQAMTTRVFK  PLKLDNTWIN  VPKAEEAHYA  WGYREGKAVH  VSPGMLDAEA
CG75    161 QWKPGTTRLY  ANASIGLFGA  LAVKPSGMSY  EQAMTTRVFK  PLKLDHTWIN  VPKAEEAHYA  WGYREGKAVH  VSPGMLDAEA
CG85    161 QWKPGTTRLY  ANASIGLFGA  LAVKPSGMSY  EQAMTTRVFK  PLKLDHTWIN  VPKAEEAHYA  WGYREGKAVH  VSPGMLDAEA
Y482    161 QWKPGTTRLY  ANASIGLFGA  LAVKPSGMSY  EQAMTTRVFK  PLKLDHTWIN  VPKAEEAHYA  WGYREGKAVH  VSPGMLDAEA
Y490    161 QWKPGTTRLY  ANASIGLFGA  LAVKPSGMSY  EQAMTTRVFK  PLKLDHTWIN  VPKAEEAHYA  WGYREGKAVH  VSPGMLDAEA
Y546    161 QWKPGTTRLY  ANASIGLFGA  LAVKPSGMSY  EQAMTTRVFK  PLKLDHTWIN  VPKAEEAHYA  WGYREGKAVH  VSPGMLDAEA
                250         *  260        *  270         280          290         *            310          320
CG34    241 YGVKTNVKDM  ASW LIANMKP DSLQAPSLKQ  GIALAQSRYW  RVGAMYQGLG  WEMLNWPVDA  KTVVGGSDNK  VALAPLPVAE
CG75    241 YGVKTNVKDM  ASWVIANMKP  DSLQAPSLKQ  GIALAQSRYW  RVGAMYQGLG  WEMLNWPVDA  KTVVGGSDNK  VALAPLPVAE
CG85    241 YGVKTNVKDM  ASWVIANMKP  DSLQAPSLKQ  GIALAQSRYW  RVGAMYQGLG  WEMLNWPVD V KTVVGGSDNK  VALAPLPVAE
Y482    241 YGVKTNVKDM  ASWVIANMKP  DSLQAPSLKQ  GIALAQSRYW  RVGAMYQGLG  WEMLNWPVD V KTVVGGSDNK  VALAPLPVAE
Y490    241 YGVKTNVKDM  ASW LIANMKP DSLHAPSLKQ  GIALAQSRYW  RVGAMYQGLG  WEMLNWPVDA  KTVVGGSDNK  VALAPLPVAE
Y546    241 YGVKTNVKDM  ASW LIANMKP DSLQAPSLKQ  GIALAQSRYW  RVGAMYQGLG  WEMLNWPVDA  KTVVGGSDNK  VALAPLPVAE
                330         340         350          360          370          380
CG34    321 VNPPAPPVKA  SWVHKTGSTG  GFGSYVAFIP  EKQLGIVMLA  NKSYPNPARV  EAAYRILDAL  Q
CG75    321 VNPPAPPVKA  SWVHKTGSTG  GFGSYVAFIP  EKQLGIVMLA  NKSYPNPARV  EAAYRILDAL  Q
CG85    321 VNPPAPPVKA  SWVHKTGSTG  GFGSYVAFIP  EKQLGIVMLA  NKSYPNPARV  EAAYRILDAL  Q
Y482    321 VNPPAPPVKA  SWVHKTGSTG  GFGSYVAFIP  EKQLGIVMLA  NKSYPNPARV  EAAYRILDAL  Q
Y490    321 VNPPAPPVKA  SWVHKTGSTG  GFGSYVAFIP  EKQLGIVMLA  NKSYPNPARV  EAAYRILDAL  Q
Y546    321 VNPPAPPVKA  SWVHKTGSTG  GFGSYVAFIP  EKQLGIVMLA  NKSYPNPARV  EAAYRILDAL  Q
```

FIGURE 2: Comparison of the MIR amino acid sequences from 6 *E. cloacae* isolates.

from CG85 and Y482 that had the same amino acid sequence and were located in the same branch, the 6 MIR proteins were located in unique branches.

3.4. General Features of the Y546 Genome. The genome of *E. cloacae* strain Y546 consists of a chromosome and a plasmid (pY546); the general features of the Y546 genome are shown in Table 4. The chromosome is 4.78 Mb in length, harbors 4312 ORFs, and has an average GC content of 56.02%. In addition to an *ampC* gene bla_{MIR}, the chromosome also encodes two other extended-spectrum β-lactamase (ESBL) genes, namely, $bla_{CTX-M-9}$ and bla_{SHV-12}. MLST determined that *E. cloacae* strain Y546 contains the *leuS*-90, *rpoB*-20,

gyrB-127, *dnaA*-120, *fusA*-25, and *rplB*-12 alleles and belongs to the sequence type ST466. The pY546 plasmid, an IncHI1B plasmid, is 208,740 bp in length and encodes 232 ORFs (Supplementary Table S3), of which 56.89% (132/232) encode proteins with known functions, and it contains a number of accessory modules identified at different sites within the backbone. pY546 contained an incomplete copper resistance operon (*pcoBCDE/cusRS*, ORF91-96) and several clusters of genes related to resistance to other metals, including arsenic (*arsABCDR*, ORF100-105), tetrathionate (*ttrABCDRS*, ORF142-146) and tellurite (*terCDEF*, ORF153-156). This plasmid also encodes numerous metallic ion metabolism and transfer proteins, such as a potassium transporter (*Kef*,

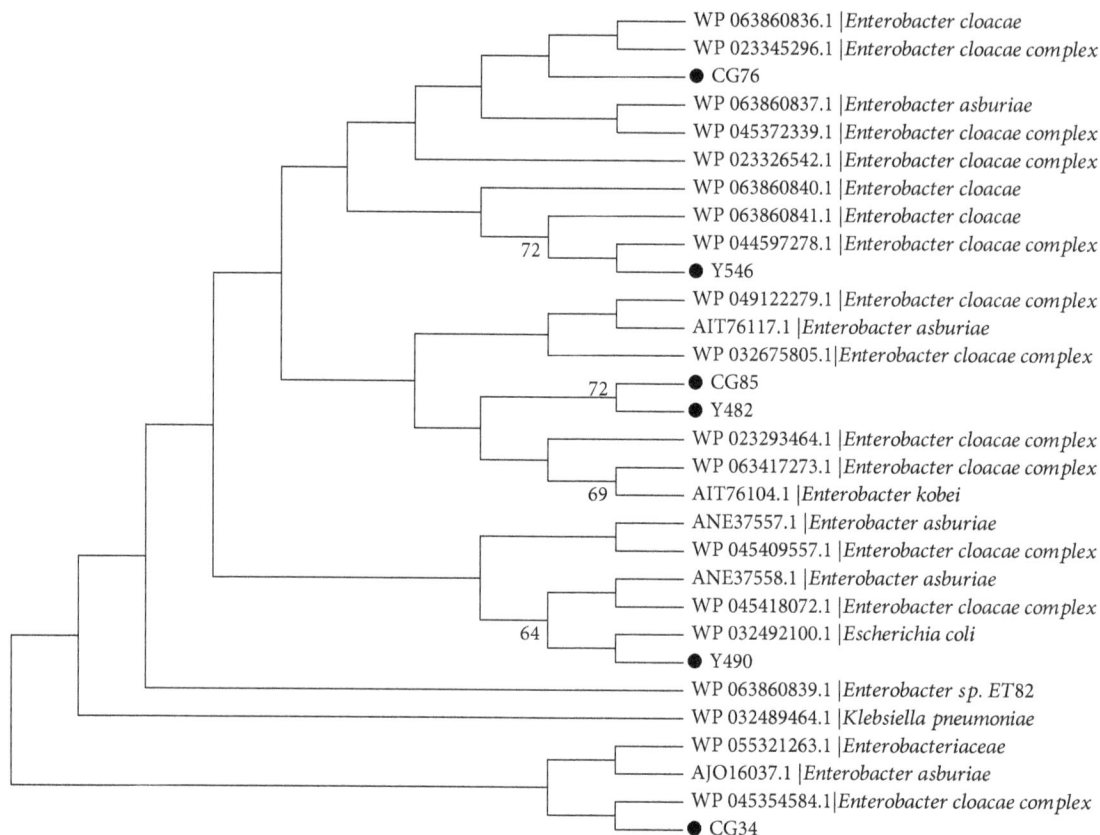

FIGURE 3: Phylogenetic tree of 30 MIR amino acid sequences. The "●" symbols indicate the MIRs of this study.

TABLE 4: General features of the Y546 genome.

	Chromosome	Plasmid
Size (bp)	4,787,919	208,741
GC content (%)	56.02	52.63
ORFs	4312	234
Known proteins	3476 (80.6%)	137 (58.5%)
Hypothetical proteins	836 (19.4%)	97 (41.5%)
Protein coding (%)	88.02	82.80
Average ORF length (bp)	977	738
Average protein length (aa)	324	245
tRNAs	85	0
rRNA operons	(16S-23S-5S) *7 16S-23S-5S-5S	0

ORF109), a fluoride ion transporter (*CrcB*, ORF113), divalent cation transporters (ORF111 and ORF112), and voltage-gated chloride channel proteins (ORF120 and ORF121). In addition, TonB-dependent receptors (TBDRs, ORF12-14), which are involved in the uptake of essential nutrients, are identified in pY546.

3.5. Comparative Genomic Analysis of the pY546 Plasmid Genome. A total of six sequences having greater than 40% nucleotide sequence identity to the pY546 sequence were retrieved from GenBank. Four of these were plasmid sequences, namely, pKPN-332 of *K. pneumoniae* strain KPNIH39 (49%, CP014763.1), pKPN-3967 of *K. pneumoniae* strain KPNIH49 (47%, CP026186.1), plasmid unnamed 1 of *K. pneumoniae* strain KSB2_1B (44%, CP024507.1), and pKPN-262 of *K. pneumoniae* subsp. pneumoniae KPNIH27 (44%, CP007734.1). The other two were chromosome sequences of *E. coli* S43 (41%, CP010237.1) and *E. coli* MEM (40%, CP012378.1) (Figure 4). Comparative genomics analysis showed that pY546 is approximately 80-100 kb smaller than any of the three named plasmids (pKPN-332, pKPN-3967, and pKPN-262). The sequences of 102 genes (43.4%, 102/235) on pY546 showed high similarity (>90%) with those on each of the three named plasmids. These plasmids shared a conserved backbone sequence with pY546; the backbone included the replication initiation gene (*rep*A), stable maintenance genes (*par*AB or *sop*AB), DNA mismatch repair system genes (*mut*S), and so on. On the other hand, all these plasmids possessed their own variable regions, which mainly included heavy metal (*cop*, *ars*, and *ter*) resistance gene clusters and hypothetical genes. The tetrathionate resistance genes (*ttrABCRS*) encoded on pY546 were not present in the three named plasmids. The named plasmids also had multiple copies of common mobile elements, such as transposons and insertion elements (IS). In the two *E. coli* (S43 and MEM) chromosome sequences, the *pco* gene cluster (*pcoABCDRSE*) was detected in MEM but not in S43. Nevertheless, *arsBCD*, which is a part of the arsenic resistance determinants, was present in both S43 and MEM. The

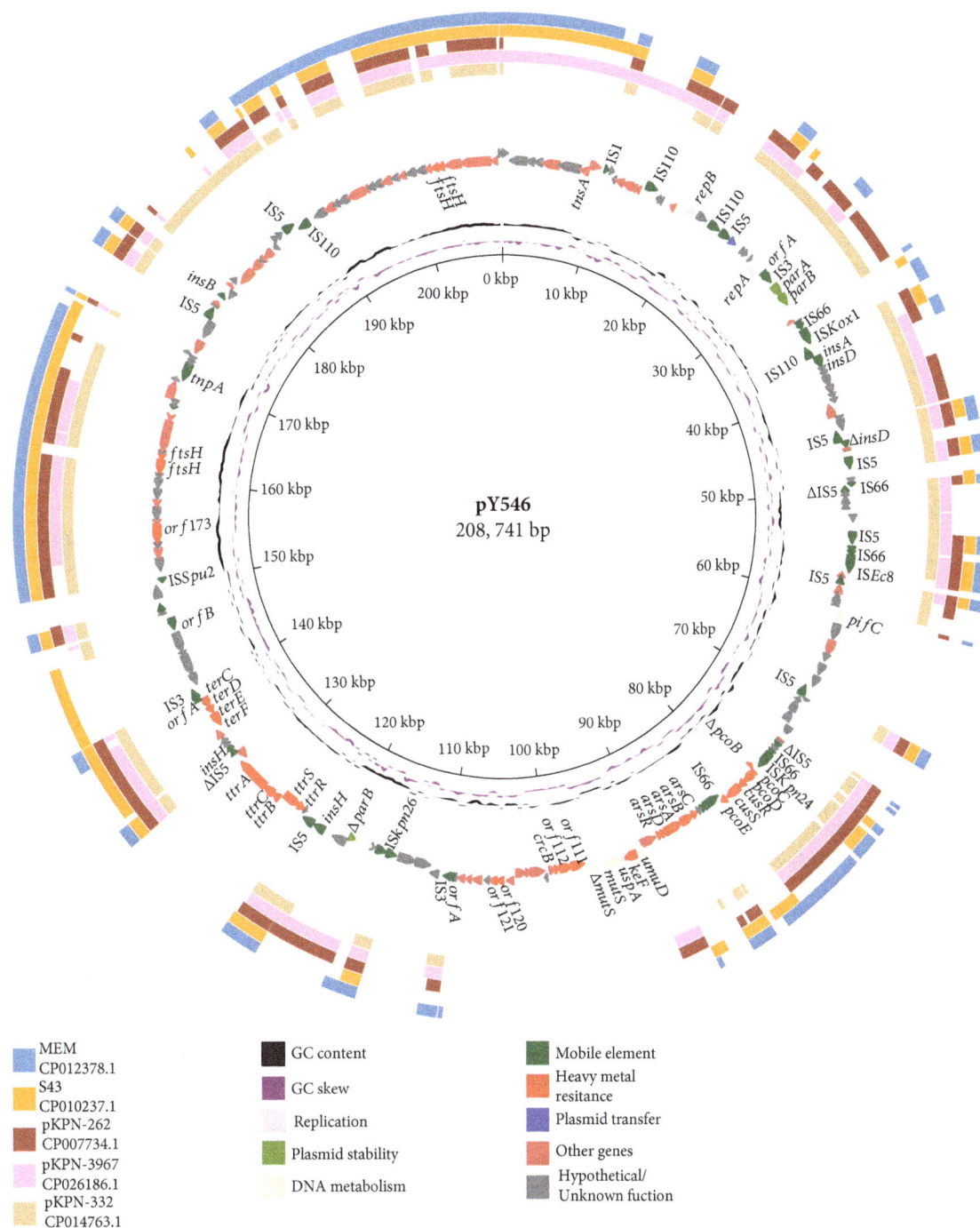

FIGURE 4: Complete sequence of the pY546 plasmid and comparative genomic analysis of the pY546 plasmid sequence with other sequences. The circles (from innermost to outermost) represent (i) the scale in kb; (ii) the cumulative GC skew; (iii) the GC content; (iv) the annotated coding sequences with selected genes indicated according to the gene function: heavy metal resistance genes in red arrows, transposase genes, IS elements in bottle-green arrows, and hypothetical proteins in dark gray arrows; (v) circles (from inside to outside) representing three homologous plasmids (pKPN-332, CP014763.1; pKPN-3967, CP026186.1; and pKPN-262, CP007734.1) and two chromosome fragments (S43, CP010237.1, and MEM, CP012378.1), respectively.

tetrathionate resistance genes were unique to pY546, while the chromosome of S43 carried the *ter* operon, but the MEM chromosome did not (Figure 4). The *ter* operon *terC-DEF* was detected and in the same orientation in the three plasmids pKPN-332, pKPN-3967, and pKPN-262.

3.6. *Comparative Analysis of Copper and Arsenic Resistance Gene Regions on the pY546 Plasmid.* Comparative analysis of an 8.7 kb fragment of pY546 encoding both copper (*pco*) and arsenic (*ars*) operons showed that the 5 sequence fragments with the highest similarity to that of pY546 were 4 fragments from the pKO_JKo3_1 plasmid of *Klebsiella oxytoca* JKo3 (100%, AP014952.1), the pKPN1705-1 plasmid of *Klebsiella quasivariicola* KPN1705 (100%, CP022824.1), the CSK29544_3p plasmid of *Cronobacter sakazakii* ATCC 29544 (100%, CP011050.1), and the pKPN-262 plasmid of *Klebsiella pneumoniae* subsp. *pneumoniae* KPNIH27 (82%, CP007734.1) and 1 fragment from the chromosome of *Escherichia coli* MEM (CP012378.1, 82%) (Figure 5). The fragment sharing the highest sequence identity with that of pY546 (from *E. cloacae*) was in the CSK29544_3p plasmid of *Enterobacter sakazakii*. All sequences except for pY546 contained the complete copper (*pco*) operon structure. pY546, however, contained an incomplete copper (*pco*) operon with a truncated *pcoB* ($\triangle pcopB$) gene and without *pcoA* gene. Four (pY546, CSK29544_3p, pKO_Jko3_1, and pKPN1705-1) sequences contained the complete *ars* operon gene clusters. Moreover, the latter two plasmids (pKO_Jko3_1 and pKPN1705-1) contained an additional functional gene, namely, *arsH*, which encoded an organoarsenical oxidase (NADPH-dependent FMN reductase) [25] and conferred resistance to trivalent forms of organoarsenic compounds. Two of the plasmids (pKO_Jko3_1 and pKPN1705-1) were also identified to contain two copies of *arsA and arsD* with inverted orientations. The CSK29544_3p plasmid contained the same gene arrangement and content as pY546, but pY546 and CSK29544_3p contained fewer genes than the other two plasmids (pKO_Jko3_1 and pKPN1705-1). The pY546 and CSK29544_3p plasmids lacked *arsH* and contained only one copy each of *arsAD*, which was oriented oppositely in these two plasmids. On the other hand, the *arsBCRH* gene cluster was identified in pKPN-262, while the MEM chromosome contained *arsBCR* but lacked *arsH*.

4. Discussion

The production of β-lactamases is the predominant β-lactam resistance mechanism in gram-negative bacteria. The Ambler molecular classification categorizes these β-lactamases into four enzyme classes, namely, A, B, C, and D. Class A, C, and D enzymes all possess an active site serine, whereas class B β-lactamases are metalloenzymes with a Zn^{2+} ion(s) in the active site [26]. In this study, a total of 12 β-lactamase-encoding genes, including 7 class A β-lactamase genes (bla_{SHV}, bla_{CTX-M}, bla_Z, bla_{VEB}, bla_{KLUC}, bla_{SFO}, and bla_{TEM}), 4 class C β-lactamase genes (bla_{MIR}, bla_{DHA}, bla_{ACT}, and $bla_{AZECL-29}$), and 1 class D β-lactamase gene (bla_{OXA}) were identified in 212 *E. cloacae* isolates from a teaching hospital in South China. Over the past years, a variety of metalloenzymes (NDM- and IMP-type) have been found in *E. cloacae* and have contributed to infectious outbreaks in China [27] and Japan [28]. However, we did not detect any genes encoding class B metalloenzymes in these 212 *E. cloacae* isolates.

The class A enzymes are regarded as extended-spectrum β-lactamases (ESBLs) that can hydrolyze extended-spectrum cephalosporins; they are inhibited by clavulanic acid and are spreading widely among *Enterobacteriaceae*. The CTX-M enzymes are replacing the SHV and TEM enzymes as the most prevalent type of ESBL in *Enterobacteriaceae* [29, 30].

Additional clinically relevant types of ESBLs include the VEB, PER, GES, TLA, IBC, SFO-1, BES-1, and BEL-1 types. The SFO-1 β-lactamase was first reported in 1988 in a clinical *E. cloacae* isolate in Japan, and it confers resistance to third-generation cephalosporins (Matsumoto & Inoue, 1999); VEB was reported in China during an outbreak of infection caused by *E. cloacae* [31]. No document has yet reported the identification of the bla_Z gene that encodes class A enzymes in *E. cloacae*, and the bla_Z gene has been found only once in *Staphylococcus aureus* [16]. In this work, however, we isolated two *E. cloacae* strains (CG3 and CG4) that carried the bla_Z gene.

In addition to the bla_{MIR} and bla_{DHA} genes identified in this work, genes encoding AmpC enzymes belonging to Ambler class C and Bush-Jacoby group 1 include bla_{CMY}, bla_{FOX}, bla_{LAT}, bla_{ACT}, bla_{MOX}, bla_{ACC}, and bla_{BIL} and their derivatives (H. [32]). The emergence of AmpC-producing *Enterobacter* spp. has been observed globally in health care-associated settings and in the community [33]. However, unlike most of the AmpC genes, bla_{MIR} has been found only in some strains of several *Enterobacter* spp., mainly in strains of *E. cloacae*. A total of 24 bla_{MIR} nucleotide sequences (between the bla_{MIR-1} and bla_{MIR-21} subtypes) are available in the NCBI nucleotide database; these sequences mainly came from the *ECC*, such as *E. cloacae* and *E. aerogenes*, as well as strains of *K. pneumoniae* and *E. coli*. In this work, 6 MIRs belonged 5 subtypes, including bla_{MIR-3}, bla_{MIR-5}, bla_{MIR-17}, bla_{MIR-21}, and bla_{MIR-20}. Although the bla_{MIR} gene has been primarily identified on bacterial chromosomes, it is also encoded on plasmids. The AmpC β-lactamase bla_{MIR-1} was first described in *K. pneumoniae* plasmids [34]; bla_{MIR-1} confers resistance to penicillins and broad-spectrum cephalosporins, including cefoxitin and ceftibuten, but not to cefepime, cefpirome, meropenem, or imipenem. The resistance features of bla_{MIR} genes in human and animal isolates were different from those of some plasmid-encoded AmpC-type β-lactamase genes, such as bla_{DHA} and bla_{CMY}, and have been reported worldwide to hydrolyze third-generation cephalosporins [35, 36]. The bla_{MIR} gene identified in this work was encoded on the chromosome and showed high sequence identity with other homologous bla_{MIR} genes found in other *Enterobacteriaceae*. Like the other previously reported MIRs, they showed resistance to ampicillin, cefazolin, cefmenoxime, and cefoxitin, but sensitive to fourth-generation cephalosporins (cefoselis) and monobactam (aztreonam).

To adapt to environmental changes, bacteria often harbor genes conferring resistance to toxic metal compounds; these genes include those encoding copper and arsenic ion transportation systems [37]. Copper sulfate is a common feed supplement for pigs, chickens, and calves worldwide. The copper-binding operon system (*PcoBCDE* and *CusR/S*), which is known to transport copper-derived compounds out of the bacterial cell to balance the concentration of copper salts, was elaborated on the pRJ1004 plasmid of *E. coli* isolates from piggeries in which the animals were provided food supplemented with copper sulfate [38]. Despite the identity of arsenic and arsenite compounds as high-toxicity compounds that are neither used in agriculture nor found in either the community or the hospital sector, the presence

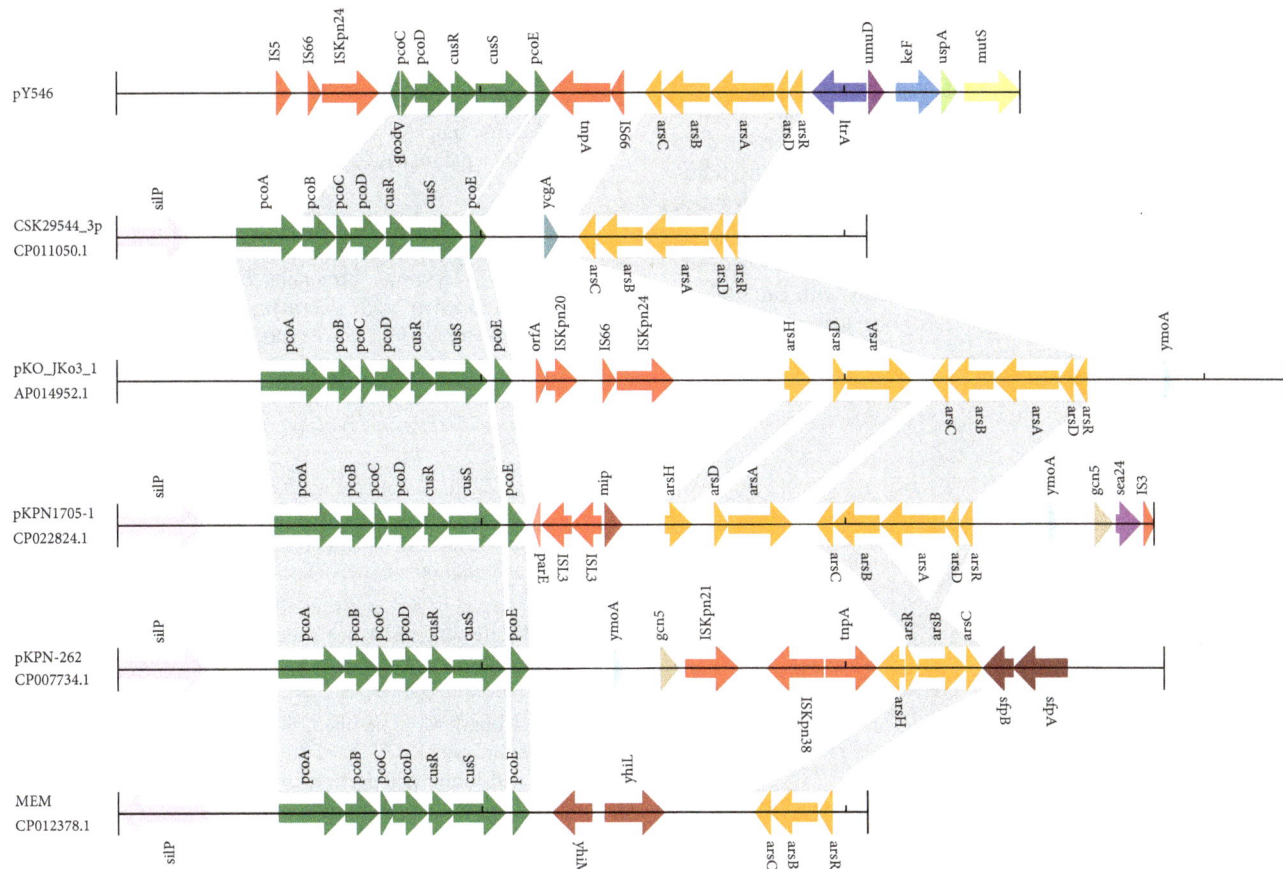

FIGURE 5: Comparative analysis of the copper and arsenic resistance gene clusters of pY546 and other sequences from different bacteria. The homologous gene clusters among plasmids pY546, CSK29544_3P (CP011050.1), pKO_JKo3_1 (AP014952.1), pKPN1705-1 (CP022824.1), pKPN-262 (CP007734.1), and MEM (CP012378.1) with the copper resistance gene clusters in bottle green, and the arsenic resistance gene cluster in orange. Annotated coding sequences are displayed as arrows. Coding sequences are colored based on their assigned gene functions.

of arsenic resistance determinants on *Enterobacteriaceae* plasmids, especially the IncH-type plasmids, has been described before [9]. Three prototypes of *ars* operons, including the three-, four-, and five-gene arsenic resistance determinants, namely, *arsABC*, *arsABCD*, and *arsABCDR*, respectively, have been well documented, although novel resistance mechanisms have also been described [39]. The *arsABCDR* operon is related to resistance to arsenic-derived compounds, including arsine, arsenic, arsenite, and arsenate. Heavy metal resistance genes or gene clusters have been widely identified in different genera of both gram-positive and gram-negative bacteria and are encoded on both chromosomes and plasmids. In this work, on the plasmid pY546, we found four clusters of genes conferring resistance to heavy metals, such as arsenic (*arsABCDR*), tetrathionate (*ttrABCDRS*), and tellurite (*terCDEF*) as well as an incomplete copper resistance operon (*pcoBCDE/cusRS*). We must expect that bacteria have adapted to heavy metals with an increasing frequency.

isolates; of these, *bla*$_Z$ has not yet been reported in *E. cloacae*. Furthermore, whole genome analysis of the *bla*$_{MIR}$-carrying *E. cloacae* strain Y546 demonstrated that the strain harbored a large plasmid carrying a variety of gene clusters and genes, such as heavy metal resistance gene clusters (e.g., the *pco*, *ars*, *ter*, and *ttr* operons), conferring resistance to antimicrobials. Comparative genomics analysis showed that the sequences sharing the highest similarity to pY546 were plasmids from *K. pneumoniae* strains (44-49% similarity) and the chromosome of *E. coli* (40-41% similarity) and that the sequence fragments with the highest similarity to heavy metal resistance gene clusters on pY546 were from other plasmids and other chromosome sequences. The colocalization of antibiotic resistance genes and heavy metal resistance genes in the genomes of clinical pathogens, which may facilitate the persistence, coselection, and dissemination of these genes between different bacterial species or genera, is alarming and needs further surveillance.

5. Conclusion

In this work, through high-throughput sequencing, we identified twelve β-lactamase genotypes in 212 clinical *E. cloacae*

Additional Points

Highlights. Twelve β-lactamase genotypes, including *bla*$_{SHV}$,

bla_{TEM}, bla_{DHA}, bla_{CTX-M}, bla_Z, bla_{VEB}, bla_{KLUC}, bla_{MIR}, bla_{SFO}, $bla_{AZECL-29}$, bla_{OXA}, and bla_{ACT}, were identified in 212 *E. cloacae* genomes. bla_Z was found in *E. cloacae* for the first time. bla_{MIR} was identified in six *E. cloacae* strains that were not clonally related. The gene clusters related to resistance to heavy metals (such as copper, arsenic, and tellurite) were identified to be encoded on the pY546 plasmid. The sequences with the highest similarity to pY546 were on plasmids from *K. pneumoniae* strains and on *E. coli* chromosomes; in addition, the sequence fragments with the highest similarity to the heavy metal resistance gene clusters on pY546 were from other sources.

Conflicts of Interest

The authors declare that there are no conflicts of interest in this work.

Authors' Contributions

Chongyang Wu and Chaoqin Lin contributed equally to this work.

Acknowledgments

The work was funded by grants from the National Natural Science Foundation of China (31500109, 81501808, and 80215049) and the Science and Technology Foundation of Wenzhou City (Y20170205).

References

[1] M. L. Mezzatesta, F. Gona, and S. Stefani, "Enterobacter cloacae complex: clinical impact and emerging antibiotic resistance," *Future Microbiology*, vol. 7, no. 7, pp. 887–902, 2012.

[2] N. Petrosillo, M. Vranić-Ladavac, C. Feudi et al., "Spread of Enterobacter cloacae carrying bla NDM-1, bla CTX-M-15, bla SHV-12 and plasmid-mediated quinolone resistance genes in a surgical intensive care unit in Croatia," *Journal of Global Antimicrobial Resistance*, vol. 4, pp. 44–48, 2016.

[3] S. Wang, S.-Z. Xiao, F.-F. Gu et al., "Antimicrobial susceptibility and molecular epidemiology of clinical *Enterobacter cloacae* bloodstream isolates in Shanghai, China," *Plos One*, vol. 12, no. 12, article e0189713, 2017.

[4] K. Zhou, W. Yu, X. Cao et al., "Characterization of the population structure, drug resistance mechanisms and plasmids of the community-associated Enterobacter cloacae complex in China," *The Journal of Antimicrobial Chemotherapy*, vol. 73, no. 1, pp. 66–76, 2018.

[5] D. Balasubramanian, H. Kumari, and K. Mathee, "*Pseudomonas aeruginosa* AmpR: an acute–chronic switch regulator," *Pathogens and Disease*, vol. 73, no. 2, pp. 1–14, 2014.

[6] A. D. Wales and R. H. Davies, "Co-selection of resistance to antibiotics, biocides and heavy metals, and its relevance to foodborne pathogens," *Antibiotics*, vol. 4, no. 4, pp. 567–604, 2015.

[5] D. Balasubramanian, H. Kumari, and K. Mathee, "*Pseudomonas aeruginosa* AmpR: an acute–chronic switch regulator," *Pathogens and Disease*, vol. 73, no. 2, pp. 1–14, 2014.

[6] A. D. Wales and R. H. Davies, "Co-selection of resistance to antibiotics, biocides and heavy metals, and its relevance to foodborne pathogens," *Antibiotics*, vol. 4, no. 4, pp. 567–604, 2015.

[7] D. Deus, C. Krischek, Y. Pfeifer et al., "Comparative analysis of the susceptibility to biocides and heavy metals of extended-spectrum β-lactamase-producing *Escherichia coli* isolates of human and avian origin, Germany," *Diagnostic Microbiology and Infectious Disease*, vol. 88, no. 1, pp. 88–92, 2017.

[8] C. J. Slipski, G. G. Zhanel, and D. C. Bay, "Biocide selective TolC-independent efflux pumps in Enterobacteriaceae," *The Journal of Membrane Biology*, vol. 251, no. 1, pp. 15–33, 2018.

[9] Y. Zhai, Z. He, Y. Kang et al., "Complete nucleotide sequence of pH11, an IncHI2 plasmid conferring multi-antibiotic resistance and multi-heavy metal resistance genes in a clinical *Klebsiella pneumoniae* isolate," *Plasmid*, vol. 86, pp. 26–31, 2016.

[10] M. A. Argudín, B. Lauzat, B. Kraushaar et al., "Heavy metal and disinfectant resistance genes among livestock-associated methicillin-resistant *Staphylococcus aureus* isolates," *Veterinary Microbiology*, vol. 191, pp. 88–95, 2016.

[11] S. Z. Abbas, M. Rafatullah, K. Hossain, N. Ismail, H. A. Tajarudin, and H. P. S. Abdul Khalil, "A review on mechanism and future perspectives of cadmium-resistant bacteria," *International journal of Environmental Science and Technology*, vol. 15, no. 1, pp. 243–262, 2018.

[12] A. Waseem, J. Arshad, F. Iqbal, A. Sajjad, Z. Mehmood, and G. Murtaza, "Pollution status of Pakistan: a retrospective review on heavy metal contamination of water, soil, and vegetables," *BioMed Research International*, vol. 2014, Article ID 813206, 29 pages, 2014.

[13] W. Deng, Y. Quan, S. Yang et al., "Antibiotic resistance in *Salmonella* from retail foods of animal origin and its association with disinfectant and heavy metal resistance," *Microbial Drug Resistance*, vol. 24, no. 6, pp. 782–791, 2018.

[14] T. Xu, J. Ying, X. Yao et al., "Identification and characterization of two novel bla(KLUC) resistance genes through large-scale resistance plasmids sequencing," *PLoS One*, vol. 7, no. 10, article e47197, 2012.

[15] J. P. R. Furlan and E. G. Stehling, "Detection of β-lactamase encoding genes in feces, soil and water from a Brazilian pig farm," *Environmental Monitoring and Assessment*, vol. 190, no. 2, p. 76, 2018.

[16] L. Haubert, I. S. Kroning, M. A. Iglesias, and W. P. da Silva, "First report of the Staphylococcus aureus isolate from subclinical bovine mastitis in the South of Brazil harboring resistance gene dfrG and transposon family Tn916-1545," *Microbial Pathogenesis*, vol. 113, pp. 242–247, 2017.

[17] S. D. Lahiri, M. R. Johnstone, P. L. Ross, R. E. McLaughlin, N. B. Olivier, and R. A. Alm, "Avibactam and class C β-lactamases: mechanism of inhibition, conservation of the binding pocket, and implications for resistance," *Antimicrobial Agents and Chemotherapy*, vol. 58, no. 10, pp. 5704–5713, 2014.

[18] L. X. Zhu, Z. W. Zhang, D. Liang et al., "Multiplex asymmetric PCR-based oligonucleotide microarray for detection of drug resistance genes containing single mutations in Enterobacteriaceae," *Antimicrobial Agents and Chemotherapy*, vol. 51, no. 10, pp. 3707–3713, 2007.

[19] G. H. Van Domselaar, P. Stothard, S. Shrivastava et al., "BASys: a web server for automated bacterial genome annotation," *Nucleic Acids Research*, vol. 33, Supplement 2, pp. W455–W459, 2005.

[20] K. Katoh and D. M. Standley, "MAFFT multiple sequence alignment software version 7: improvements in performance and usability," *Molecular Biology and Evolution*, vol. 30, no. 4, pp. 772–780, 2013.

[21] P. Siguier, J. Perochon, L. Lestrade, J. Mahillon, and M. Chandler, "ISfinder: the reference centre for bacterial insertion sequences," *Nucleic Acids Research*, vol. 34, no. 90001, pp. D32–D36, 2006.

[22] M. Remm, C. E. V. Storm, and E. L. L. Sonnhammer, "Automatic clustering of orthologs and in-paralogs from pairwise species comparisons," *Journal of Molecular Biology*, vol. 314, no. 5, pp. 1041–1052, 2001.

[23] P. J. A. Cock, T. Antao, J. T. Chang et al., "Biopython: freely available Python tools for computational molecular biology and bioinformatics," *Bioinformatics*, vol. 25, no. 11, pp. 1422-1423, 2009.

[24] S. Roussel, B. Felix, N. Vingadassalon et al., "Staphylococcus aureus strains associated with food poisoning outbreaks in France: comparison of different molecular typing methods, including MLVA," *Frontiers in Microbiology*, vol. 6, article 882, 2015.

[25] T. Tsubouchi and Y. Kaneko, "Draft genome sequence of the arsenic-resistant bacterium *Brevundimonas denitrificans* TAR-002^T," *Genome Announcements*, vol. 5, no. 47, 2017.

[26] K. Bush and G. A. Jacoby, "Updated functional classification of beta-lactamases," *Antimicrobial Agents and Chemotherapy*, vol. 54, no. 3, pp. 969–976, 2010.

[27] C. Liu, S. Qin, H. Xu et al., "New Delhi metallo-β-lactamase 1(NDM-1), the dominant carbapenemase detected in carbapenem-resistant *Enterobacter cloacae* from Henan Province, China," *PLoS One*, vol. 10, no. 8, article e0135044, 2015.

[28] K. Hayakawa, T. Miyoshi-Akiyama, T. Kirikae et al., "Molecular and epidemiological characterization of IMP-type metallo-β-lactamase-producing *Enterobacter cloacae* in a large tertiary care hospital in Japan," *Antimicrobial Agents and Chemotherapy*, vol. 58, no. 6, pp. 3441–3450, 2014.

[29] E. Juhasz, L. Janvari, A. Toth, I. Damjanova, A. Nobilis, and K. Kristof, "Emergence of VIM-4- and SHV-12-producing Enterobacter cloacae in a neonatal intensive care unit," *International Journal of Medical Microbiology*, vol. 302, no. 6, pp. 257–260, 2012.

[30] T. Naas, G. Cuzon, A. L. Robinson et al., "Neonatal infections with multidrug-resistant ESBL-producing E. cloacae and K. pneumoniae in neonatal units of two different hospitals in Antananarivo, Madagascar," *BMC Infectious Diseases*, vol. 16, no. 1, p. 275, 2016.

[31] X. Jiang, Y. Ni, Y. Jiang et al., "Outbreak of infection caused by Enterobacter cloacae producing the novel VEB-3 beta-lactamase in China," *Journal of Clinical Microbiology*, vol. 43, no. 2, pp. 826–831, 2005.

[32] H. Deng, H.-B. Si, S.-Y. Zeng et al., "Prevalence of extended-spectrum cephalosporin-resistant Escherichia coli in a farrowing farm: ST1121 clone harboring IncHI2 plasmid contributes to the dissemination of blaCMY-2," *Frontiers in Microbiology*, vol. 6, 2015.

[33] F. I. Mohd Khari, R. Karunakaran, R. Rosli, and S. Tee Tay, "Genotypic and phenotypic detection of AmpC β-lactamases in *Enterobacter* spp. isolated from a teaching hospital in Malaysia," *PLoS One*, vol. 11, no. 3, article e0150643, 2016.

[34] G. A. Papanicolaou, A. A. Medeiros, and G. A. Jacoby, "Novel plasmid-mediated beta-lactamase (MIR-1) conferring resistance to oxyimino- and alpha-methoxy beta-lactams in clinical isolates of Klebsiella pneumoniae," *Antimicrobial Agents and Chemotherapy*, vol. 34, no. 11, pp. 2200–2209, 1990.

[35] Y. Maeyama, Y. Taniguchi, W. Hayashi et al., "Prevalence of ESBL/AmpC genes and specific clones among the third-generation cephalosporin-resistant Enterobacteriaceae from canine and feline clinical specimens in Japan," *Veterinary Microbiology*, vol. 216, pp. 183–189, 2018.

[36] Y. Xie, L. Tian, G. Li et al., "Emergence of the third-generation cephalosporin-resistant hypervirulent *Klebsiella pneumoniae* due to the acquisition of a self-transferable bla_{DHA-1}-carrying plasmid by an ST23 strain," *Virulence*, vol. 9, no. 1, pp. 838–844, 2018.

[37] T. J. Lawton, G. E. Kenney, J. D. Hurley, and A. C. Rosenzweig, "The CopC family: structural and bioinformatic insights into a diverse group of periplasmic copper binding proteins," *Biochemistry*, vol. 55, no. 15, pp. 2278–2290, 2016.

[38] M. W. Gilmour, N. R. Thomson, M. Sanders, J. Parkhill, and D. E. Taylor, "The complete nucleotide sequence of the resistance plasmid R478: defining the backbone components of incompatibility group H conjugative plasmids through comparative genomics," *Plasmid*, vol. 52, no. 3, pp. 182–202, 2004.

[39] N. S. Chauhan, S. Nain, and R. Sharma, "Identification of arsenic resistance genes from marine sediment metagenome," *Indian Journal of Microbiology*, vol. 57, no. 3, pp. 299–306, 2017.

Genome-Wide Identification and Characterization of Fox Genes in the Honeybee, *Apis cerana*, and Comparative Analysis with Other Bee Fox Genes

Hongyi Nie,[1] Haiyang Geng,[1] Yan Lin,[1] Shupeng Xu,[1] Zhiguo Li,[1] Yazhou Zhao,[1,2] and Songkun Su ⓘ[1]

[1]*College of Bee Science, Fujian Agriculture and Forestry University, Fuzhou 350002, China*
[2]*Institute of Apiculture, Chinese Academy of Agricultural Sciences, Beijing 100093, China*

Correspondence should be addressed to Songkun Su; susongkun@zju.edu.cn

Academic Editor: Marco Gerdol

The forkhead box (Fox) gene family, one of the most important families of transcription factors, participates in various biological processes. However, Fox genes in Hymenoptera are still poorly known. In this study, 14 Fox genes were identified in the genome of *Apis cerana*. In addition, 16 (*Apis mellifera*), 13 (*Apis dorsata*), 16 (*Apis florea*), 17 (*Bombus terrestris*), 16 (*Bombus impatiens*), and 18 (*Megachile rotundata*) Fox genes were identified in their genomes, respectively. Phylogenetic analyses suggest that FoxA is absent in the genome of *A. dorsata* genome. Similarly, FoxG is missing in the genomes *A. cerana* and *A. dorsata*. Temporal expression profiles obtained by quantitative real-time PCR revealed that Fox genes have distinct expression patterns in *A. cerana*, especially for three genes ACSNU03719T0 (AcFoxN4), ACSNU05765T0 (AcFoxB), and ACSNU07465T0 (AcFoxL2), which displayed high expression at the egg stage. Tissue expression patterns showed that *FoxJ1* is significantly higher in the antennae of *A. cerana* and *A. mellifera* compared to other tissues. These results may facilitate a better understanding of the potential physiological functions of the Fox gene family in *A. cerana* and provide valuable information for a comprehensive functional analysis of the Fox gene family in Hymenopterans.

1. Introduction

The forkhead box (Fox) belongs to a large and diverse group of transcription factor families. It is characterized by a highly conserved forkhead (FKH) DNA-binding domain, which consists of approximately 100 residues with three α-helices, three β-sheets, and two "wing" regions that flank the third β-sheet [1–4]. The Fox gene family was initially identified in the embryo of *Drosophila* [5]. Over the past two decades, numerous Fox genes have been identified from a wide variety of taxa and have been classified into 23 subfamilies from FoxA to FoxS [6–8]. Increasing evidences show that Fox genes mediate a multitude of physiological functions. For instance, the FoxO family is related to longevity, metabolism, development, tumor suppression, immunity, and mediation of insulin [1, 9–12]. FoxJ1 contributes to motile cilia formation and colorectal cancer progression [13, 14]. FoxJ2 is specifically expressed in meiotic spermatocytes in adult mouse testes and controls meiosis during spermatogenesis in male mice [15]. In *Drosophila melanogaster*, loss of FoxL1 affects the salivary gland position and morphology during embryonic development [16]. FoxA functions as a pioneer factor to facilitate androgen receptor transactivation and prostate cancer growth [17]. Several Fox members (FoxJ1, FoxM1, FoxO1, and FoxO3) modulate neurogenesis in adults [18]. FoxN2/3 is crucial for the formation of the larval skeleton in sea urchin [19]. In humans, a series of studies have reported that Fox genes are associated with developmental disorders and diseases, including cancer, Parkinson's disease, autism spectrum disorder, ocular abnormalities, defects in immune regulation and function, and deficiencies in language acquisition [20].

The exploration of the functions of the Fox genes has been extensively carried out in mammals, especially in humans.

After the publication of several insect genomes, Fox genes have been identified in species such as *A. mellifera* (16), *D. melanogaster* (20), *Aedes aegypti* (18), *Anopheles gambiae* (20), *Bombyx mori* (17), *Heliconius melpomene* (18), and *Danaus plexippus* (19), and their functions have been revealed at the same time [19, 21–24]. However, Fox genes in Hymenopterans remain poorly understood. Honeybees are social insects, which are important model organisms for neurobiological, developmental, and sociobiological studies [25–27]. However, Fox genes of honeybees have not been intensively investigated. Western *(A. mellifera)* and eastern *(A. cerana)* honeybees are the two most important species in the genus *Apis*. In 2008, FoxP was identified in the brain of *A. mellifera*, and its expression in the worker brain was increased after eclosion, suggesting a role for FoxP in the developing and maturation of worker brains [28]. Phylogenetic analyses identified 14 Fox genes in the genome of *A. mellifera* [29]. Similar analyses on *A. cerana* have not been conducted so far.

A. cerana differs in several biological traits compared with *A. mellifera*. Eastern bees exhibit efficient adaptations for collecting sporadic nectar in mountains or forests and extreme weather conditions. They also show varroa mite resistance, cooperative group-level defense, long-haul flights, and effective hygienic behaviors [30–33]. The genomic sequence of *A. cerana* was published in 2015 and has provided a wealth of information for understanding the molecular basis of social behavior and eusocial evolution [34].

In this study, we described all Fox genes of seven Hymenoptera species: *A. cerana, A. mellifera, A. dorsata, A. florea, B. terrestris, B. impatiens,* and *M. rotundata*. The phylogenetic analysis of these genes was performed with reference to the Fox genes of three other insect species in which these genes were already characterized, *B. mori, D. melanogaster,* and *Danaus plexippus* [21]. Moreover, we provide a more detailed analysis for *A. cerana*, presenting the structural, spatial, and temporal expression profiles of its Fox genes. These data will pave the way for further molecular studies on the biological traits of *A. cerana*. Moreover, given the model status of the honeybee is for biological research, the knowledge gained in our study will be valuable for further explorations of other social insects.

2. Materials and Methods

2.1. Fox Gene Identification in A. cerana and Other Bees. To identify the complete list of the *A. cerana* Fox genes, known Fox proteins sequences of *D. melanogaster, B. mori,* and *D. plexippus* were used as query sequences to search for the homologous sequences against the *A. cerana* database using a basic local alignment search tool (BLAST, v2.3.0+), with an *E* value threshold of 10^{-10}. Known Fox amino acid sequences of three species were downloaded from Flybase (http://flybase.org/), SilkDB (http://silkworm. genomics.org.cn/), and MonarchBase (http://monarchbase. umassmed.edu/). Redundant sequences were removed. To eliminate false-positive proteins, the domains of the candidate sequences were predicted using the Pfam online server (http://pfam.xfam.org/) to check whether they harbor a

FKH domain. Only sequences with such clear FKH domains were retained. The same procedure was applied to search Fox family members in the protein databases of *A. mellifera, A. dorsata, A. florea, B. terrestris, B. impatiens,* and *M. rotundata*. Predicted protein sequences of *A. mellifera* were downloaded from BeeBase (http://hymenopteragenome.org/ beebase/). The other translated protein sequences belonging to *A. cerana, A. dorsata, A. florea, B. terrestris, B. impatiens,* and *M. rotundata* were downloaded from the National Center for Biotechnology Information (NCBI) FTP site (ftp://ftp.ncbi.nih.gov/genomes/).

The molecular weight of AcFox was predicted using ExPASY (http://www.expasy.org/) [35], and the structures were predicted with GSDS 2.0 (http://gsds.cbi.pku. edu.cn/) [36].

2.2. Phylogenetic Analysis. The complete protein sequences of Fox from the 10 species *(D. melanogaster, B. mori, D. plexippus, A. cerana, A. mellifera, A. dorsata, A. florea, B. terrestris, B. impatiens,* and *M. rotundata)* were initially aligned using Clustal W (V1.83) [37]. Protein sequences and accession numbers are provided in Supplemental File 1. The phylogenetic tree for Fox was constructed using the neighbor-joining (NJ) method in the program MEGA7.0. The Poisson model, pairwise deletion of gaps, and uniform rates were set in the NJ tree reconstruction. The accuracy of the tree topology was tested by bootstrap analysis with 1000 resampling replicates.

2.3. Multiple Sequence Alignment. A multiple sequence alignment was constructed with the Clustal W multiple alignment program in BioEdit 7.05. Then, alignment was artificially edited and visualized using GeneDoc software. Finally, the secondary structures of the FKH domain were predicted with SWISS-MODEL (https://www.swissmodel. expasy.org/) [38].

2.4. Bee Rearing and Sample Collection. Bees of *A. cerana* and *A. mellifera* were bred at the College of Bee Science, Fujian Agriculture and Forestry University, during the autumn of 2016. For *A. cerana*, eggs at different developmental stages were collected directly from combs and then mixed as egg specimens. For collection of larvae and pupae, we used a method previously described [39], which allows to determine the precise time of oviposition. Briefly, queens of *A. cerana* were caged at the comb using an excluder cage (7.2 × 5.1 × 2.2 cm) for 6 h, which made them lay eggs in the confinement region. The queen was then removed to another confinement cage in the hive while the eggs developed within the laying area, which remained enclosed. In this way, several areas for larvae and pupae collection were established within the hive, for which the precise timing of egg laying and development was known. For temporal expression analysis, brood combs of *A. cerana* were maintained in their original colonies, and samples were collected at 24 h intervals from hatching to imago. Newly emerged workers (NEW), nurses, and foragers were collected as previous methods [40, 41]. NEW were obtained from capped brood frames freed from any other bees that were placed in an incubator for 3-4 h. This period was sufficient for the emergency of the new bees. Nurses were workers entering the cells containing larvae. Foragers were workers returning to the colony with pollen loads on their hind legs. After specimen collection, tissue samples of the three bee categories (antennae, wings, midgut, head, thorax, abdomen, front leg, middle leg, and hind leg) were dissected on ice. The number of bees collected within each category/tissue combination was sufficient to ensure the analyses on Fox expression.

For *A. mellifera*, adult workers were caught randomly at the hive entrance and dissected to obtain samples of the nine types of tissue described above. The antennae and wings of about 50 worker bees were dissected, and the other seven types of tissues (midgut, head, thorax, abdomen, front leg, middle leg, and hind leg) from 10 worker bees were separated, respectively. All samples were snap frozen in liquid nitrogen and stored at −80°C prior to RNA isolation.

2.5. Expression Patterns Based on qRT-PCR. The total RNA of various samples was isolated using TransZol UP (TransGen Biotech) according to manufacturer's instructions. RNA was detected in agarose gel, and the concentration and purity was measured using the Nanodrop 2000 spectrophotometer (Thermo Fisher Scientific, Waltham, MA, USA). All quality results of extracted RNA are listed in Figure S1. RNA has three bands in electrophoresis, and the $OD_{260/280}$ ratio was in the range of 1.9~2.1. Then, the RNA was purified with absolute alcohol and treated with DNaseI. Two micrograms of the total RNA was used to synthesize cDNA with PrimeScript RT reagent kits (RR037A, Takara). qRT-PCR was performed using an ABI7500 real-time PCR system (Applied Biosystems) as previously described [42]. The relative expression of each gene was calculated with $2^{-\triangle\triangle Ct}$ [43]. The *A. cerana actin-related protein 1 (ACSNU01044T0/ HM640276)* was used as an internal control for normalization of sample loading due to its previous application as referece genes in *Apis cerana* [44]. The primers are listed in Table S1. Each sample was triplicated, and all data were presented as the mean ± standard deviation.

Statistical analyses were performed using GraphPad Prism 5.0. Student's *t*-test was used to evaluate statistical significance (***$p < 0.001$).

3. Results

3.1. Identification of Fox Genes. A BLAST (v2.3.0+) search against the genome of *A. cerana, A. mellifera, A. dorsata, A. florea, B. terrestris, B. impatiens,* and *M. rotundata* was performed using known Fox protein sequences of *B. mori* (17 Fox proteins), *D. melanogaster* (20 Fox proteins), and *D. plexippus* (19 Fox proteins) to identify all Fox homologs. Fourteen Fox genes were identified in the *A. cerana* genome (Table 1). The molecular weight of each *A. cerana* Fox protein was predicted by an online tool termed Compute pI/Mw in the ExPASy website (http://web.expasy.org/ compute_pi/). Fox genes were named according to the nomenclature proposed by Song et al. and Kaestner et al. [21, 45]. For *A. mellifera, A. dorsata, A. florea, B. terrestris, B. impatiens,* and *M. rotundata,* 16, 13, 16, 17, 16, and 18

Fox genes were, respectively, identified in their genomes (Table S2).

3.2. Phylogenetic Analysis of the Fox Genes. To further elucidate the phylogenetic relationship among insect Fox proteins, the Fox genes of all 7 bee species were compared with the known insect Fox proteins of *B. mori, D. melanogaster,* and *D. plexippus* employing a phylogenetic approach. A total of 166 insect Fox sequences were used to construct the phylogenetic tree (Figure 1). All amino acid sequences are listed in Supplemental File 1. The topologies of the subclades were similar to those reported in a previous study on Fox proteins in other insect species including *B. mori* and *D. melanogaster* [21, 23]. The phylogenetic analysis showed that the AcFox genes could be divided into 13 subfamilies (Figure 1). Notably, the two members of the FoxL2 subfamilies, which are found in 7 bee species, were not assigned to the same cluster, suggesting great differences in the sequences of different members within the FoxL2 subfamilies. Interestingly, we found no FoxJ2 subfamily in the 7 bee species investigated, whereas it was present in *B. mori* and *D. plexippus*; FoxG was absent in *A. cerana* and *A. dorsata*, but was present in all other insects investigated. The FoxA subfamily was specifically absent in *A. dorsata*, but it was present in all the other species investigated (Figure 1 and Table S2).

3.3. Multiple Sequence Alignment and Structure Analysis of the A. cerana Fox Proteins. All Fox proteins contain a highly conserved FHK domain with approximately 100 residues (DNA-binding domain). Most of this domain consists of three α-helices, three β-sheets, and two wing regions flanking the β-sheets [3, 4]. Multiple sequence alignment was performed with the FKH domains to assess the sequence conservation of AcFoxs, (Figure S3). Six of the proteins (ACSNU00036T0, ACSNU05765T0, ACSNU05909T0, ACSNU06581T0, ACSNU02354T0, and ACSNU01008T0) possessed four α-helices and three β-sheets. Four proteins (ACSNU03344T0, ACSNU04239T0, ACSNU07465T0, and ACSNU08427T0) consisted of three α-helices and three β-sheets. ACSNU05860T0 and ACSNU02483T0 harbored four α-helices and one β-sheet. ACSNU00997T0 comprised four α-helices and two β-sheets. ACSNU03719 had two α-helices and one β-sheet. Although these Fox proteins differed in the number of α-helices and β-sheets, they all exhibited the canonical FKH domain.

The structures of the 14 AcFox genes showed a high degree of complexity with exon numbers ranging from two to nineteen (Figure 2). All the AcFox proteins contained only one FKH domain, except for the AcFoxA, AcFoxK, and AcFoxP proteins, which contained an extra N-terminal FKH region, FHA domain, and FOXP coiled-coil domain, respectively (Figure 3). These genes might have additional functions relative to other AcFox proteins.

3.4. Spatial and Temporal Expression Profiles of the A. cerana Genes. Expression profiles of *A. cerana* Fox genes in the nine tissue types collected for the three bee categories defined (NEW, nurses, and foragers) were obtained using quantitative real-time PCR (qRT-PCR) (Figure 4).

TABLE 1: Summary of Fox genes in the *Apis cerana*.

Subject ID	Fox subfamily	Scaffold and interval	Exon number	Length (bp)	Predicted molecular weight (Da)
ACSNU05909T0	AcFoxA	Scaffold_0060(+): 536,171–537,866	3	2439	89,108.77
ACSNU05765T0	AcFoxB	Scaffold_0056(+): 967,977–969,846	4	1098	40,956.31
ACSNU02354T0	AcFoxC	Scaffold_0013(+): 2,876,905–2,932,333	6	1335	44,159.55
ACSNU00036T0	AcFoxD	Scaffold_0001(+): 1,036,746–1,045,196	6	1395	51,329.14
ACSNU01008T0	AcFoxF	Scaffold_0005(+): 464,006–467,605	8	1173	42,614.24
ACSNU04239T0	AcFoxJ1	Scaffold_0034(+): 1,044,883–1,059,544	13	873	32,344.77
ACSNU08427T0	AcFoxK	Scaffold_0135(+): 374,918–375,848	2	1803	65,700.29
ACSNU00997T0	AcFoxL1	Scaffold_0005(+): 387,351–393,959	18	798	30,899.18
ACSNU06581T0	AcFoxL2	Scaffold_0075(+): 438,361–441,354	2	1692	61,603.24
ACSNU07465T0	AcFoxL2	Scaffold_0100(−): 552,954–555,827	5	1602	59,830.75
ACSNU03344T0	AcFoxN3	Scaffold_0024(−): 313,506–317,738	8	1026	38,832.03
ACSNU03719T0	AcFoxN4	Scaffold_0028(+): 112,830–119,805	14	1152	43,596.57
ACSNU02483T0	AcFoxO	Scaffold_0015(−): 81,499–86,777	4	1005	37,405.76
ACSNU05860T0	AcFoxP	Scaffold_0059(−): 145,652–152,625	19	1512	56,200.78

Gene expression patterns differed among tissues. Interestingly, *ACSNU04239T0 (FoxJ1)* was significantly more expressed in the antennae of all three categories of *A. cerana*. All genes, except for *ACSNU07465T0*, were more expressed in the thorax of nurses than in that of NEW and foragers. Of the 14 genes, 10 exhibited higher expression in the hind legs of nurses than in those of NEW and foragers. In nurses, the expression level of two genes (*ACSNU00036T0* and *ACSNU08427T0*) was twice as higher in the hind legs as in the middle and the front legs; thus, they might have a role in the formation of the bee's corbiculae. Finally, three genes (*ACSNU07465T0*, *ACSNU02483T0*, and *ACSNU05860T0*) were more expressed in the middle legs of foragers compared with their front and hind legs, which may suggest a role in pollen unloading. *ASCNU05909T0* was highly expressed in the midgut and thorax of nurses, but poorly expressed in other tissues of NEW, nurses, and foragers. In addition, *ACSNU05860T0* was more expressed in the wings of nurses and less expressed in other tissues of the three categories. In honeybees, workers exhibit age-related division of labor. Young bees, such as nursing workers, are engaged in brood care, whereas the oldest bees (foragers) are responsible for collecting nectar and pollen outside the hive [46, 47]. In the present study, the high expression of most FOX genes in the tissues of nurses implied that they might have evolved distinct functions for the behaviors of nurses.

The temporal expression of all Fox genes was analyzed along with 20 development stages, from egg hatching to adult stage, including NEW, nurses, and foragers (Figure 5). Three genes (*ACSNU03719T0*, *ACSNU05765T0*, and *ACSNU07465T0*) were highly expressed at egg stage, suggesting that they might play a pivotal role in embryonic development. The expression of two other genes (*ACSNU01008T0* and *ACSNU02483T0*) was very weak at egg stage but increased dramatically on day 1 of larval stage and then gradually decreased from day 2 to day 6. This suggests that they might be crucial for larval growth of *A. cerana*.

3.5. High Levels of FoxJ1 Expression in the Antennae of A. mellifera Antennae. FoxJ1 was highly expressed in the antennae of all three categories of *A. cerana* (NEW, nurses, and foragers). By contrast, it was only slightly or not expressed in all other tissues. To determine whether *FoxJ1* was also highly expressed in the antennae of *A. mellifera*, the expression profile of *FoxJ1* (GB50277/LOC100576194), the homologous gene of AcFoxJ1 of *A. cerana*, was examined by qRT-PCR in the same nine different tissue types as in *A. cerana* (Figure 6). The results indicate that *FoxJ1* is also highly expressed in the antennae of adult *A. mellifera*, which is consistent with the profiles of AcFoxJ1 found in *A. cerana*.

4. Discussion

In the present study, our goal was to characterize the Fox genes of *A. cerana* and to perform a similar analysis in other 6 bee species. In this way, we aimed at achieving the first systematic comparison of Fox genes in Hymenoptera. We found 14 Fox genes in the genome of *A. cerana* and 16, 13, 16, 17, 16, and 18 Fox genes in the genomes of *A. mellifera*, *A. dorsata*, *A. florea*, *Bombus terrestris*, *B. impatiens*, and *M. rotundata*, respectively. Temporal expression profiles obtained by quantitative real-time PCR revealed that Fox genes have distinct expression patterns in *A. cerana*, especially for three genes (*ACSNU03719T0*, *ACSNU05765T0*, and *ACSNU07465T0*), which displayed high expression at the egg stage. Tissue expression patterns showed that *FoxJ1* is highly expressed in the antennae of *A. cerana* and *A. mellifera* compared to other tissues.

It was reported that Fox genes of *M. musculus* and *Homo sapiens* could be classified into 23 subfamilies designed as FoxA to FoxS [21]. Nevertheless, the subfamilies FoxE, FoxH, FoxI, FoxQ, FoxR, and FoxS were completely absent from all seven species of Hymenoptera, two species of Lepidoptera (*B. mori* and *D. plexippus*), and one species of Diptera (*D. melanogaster*) (Tables S2). Thus, the functions mediated

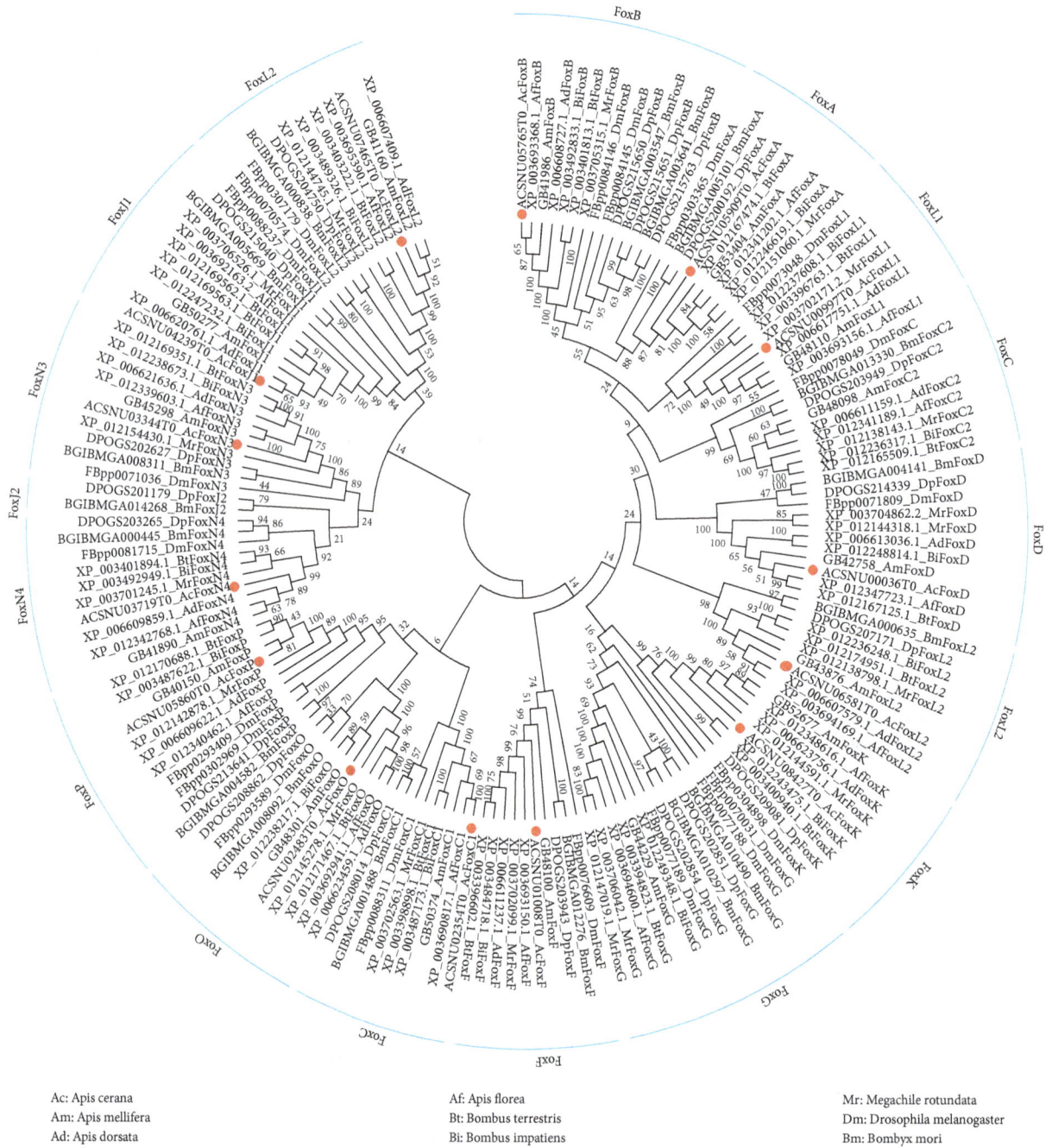

Ac: Apis cerana
Am: Apis mellifera
Ad: Apis dorsata

Af: Apis florea
Bt: Bombus terrestris
Bi: Bombus impatiens

Mr: Megachile rotundata
Dm: Drosophila melanogaster
Bm: Bombyx mori
Dp: Danaus plexippus

FIGURE 1: Phylogenetic tree of Fox including the 7 bee species considered in this work (*A. cerana, A. mellifera, A. dorsata, A. florea, B. terrestris, B. impatiens,* and *M. rotundata*) and the 3 other insect species used as reference (*B. mori, D. melanogaster,* and *D. plexippus*). Colors represent different species. Red dots indicate *A. cerana* Fox genes. Protein sequences and accession numbers are provided in Supplemental File 1.

by these genes were either inessential or they were compensated by other proteins in these species. FoxG disappeared from the *A. cerana* and *A. dorsata* genomes, while FoxA was only absent in the genome of *A. dorsata*. Therefore, FoxA and FoxG might be potential molecular markers for distinguishing *A. dorsata* and/or *A. cerana* from other bee species.

Previous studies showed that FoxJ1 mediates multiple physiological processes, especially ciliogenesis, embryonic development, spontaneous autoimmunity inhibition, and

malignancy [13, 14, 48–52]. FoxJ1 was mostly reported in mammals, and its functions in insects are virtually unknown. In this study, *FoxJ1* was more expressed in the antennae of adult workers of *A. cerana* and *A. mellifera*, and the temporal expression pattern showed that *AcFoxJ1 (ACSNU04239T0)* was highly expressed from day 4 after cell sealing to NEW stage, which was consistent with the development of antennae. Meanwhile, we found that it was also expressed at the nurse and forager stages, and a higher level of expression

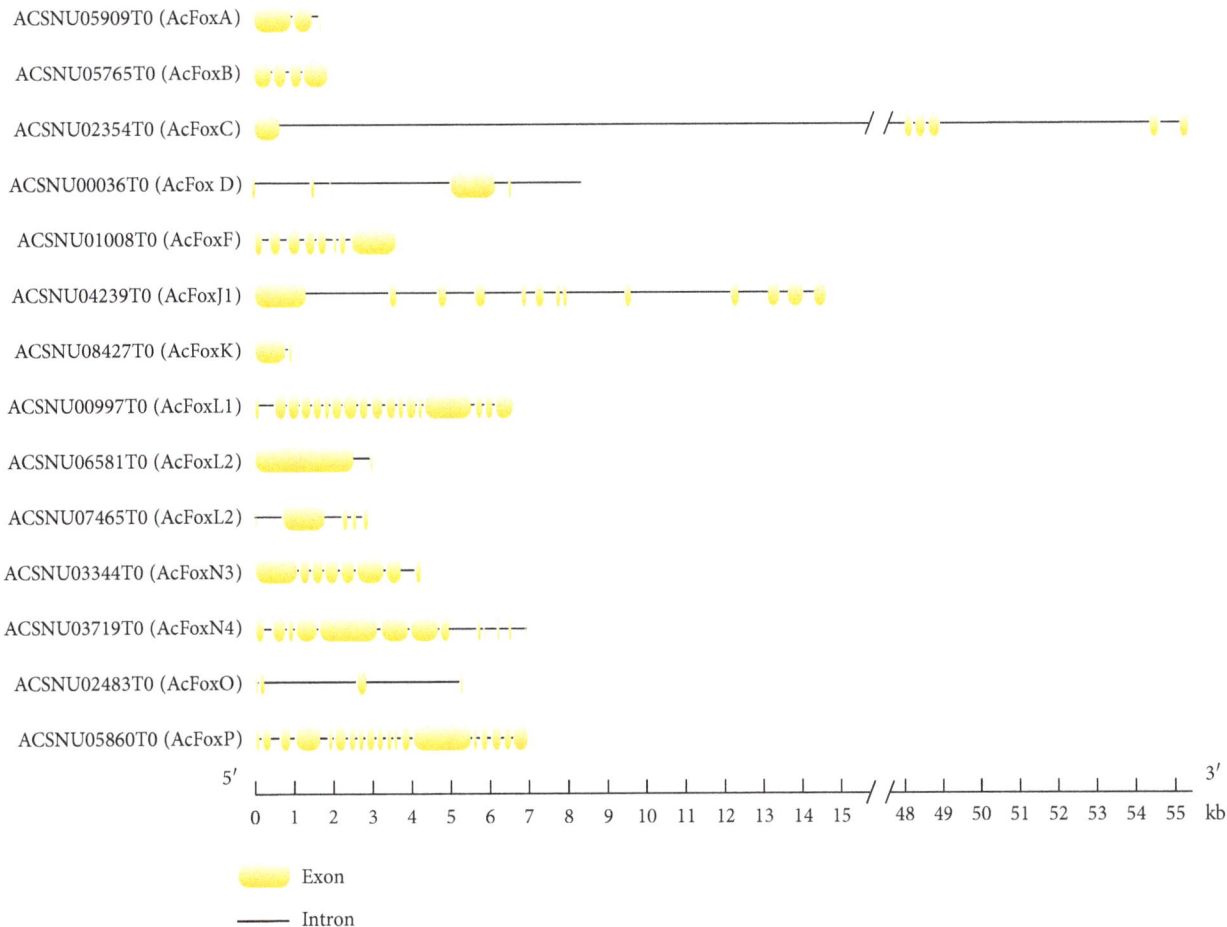

FIGURE 2: Structures of Fox genes in *A. cerana*. The putative structures of the AcFox genes consist of different numbers of exons and introns.

FIGURE 3: Domains of the Fox proteins in *A. cerana*. Conserved domains of the AcFox protein: dark cyan, forkhead N-terminal region; blue, forkhead domain; purple, FHA domain; green, FOXP coiled-coil domain.

was detected in foragers of *A. cerana*. In the honeybee, as in other insects, the antennae are the principal olfactory organs and are covered by sensilla that host olfactory receptors detecting chemical signals in the environment. In this study, *FoxJ1* was highly expressed in the antennae of *A. cerana* of NEW, nurses, and foragers, and high expression was observed from day 4 after cell sealing, followed by a continuous expression in the nurses and foragers (Figures 5 and 6). Tissue-specific genes may be relevant for the specific physiological functions of the corresponding tissues. Therefore, FoxJ1 might be involved either in the development of antennae or in the detection of specific volatiles at a certain stage. FoxJ1 was discovered in all insect genomes investigated so far, including Hymenoptera, Lepidoptera, and Diptera (Table S2). A previous study [53] showed that *AmFoxJ1 (LOC100576194)*, the homologous gene of *AcFoxJ1*, has a high expression with RPKM (reads per kilobase per million mapped reads) values of 149.2, 129.6,

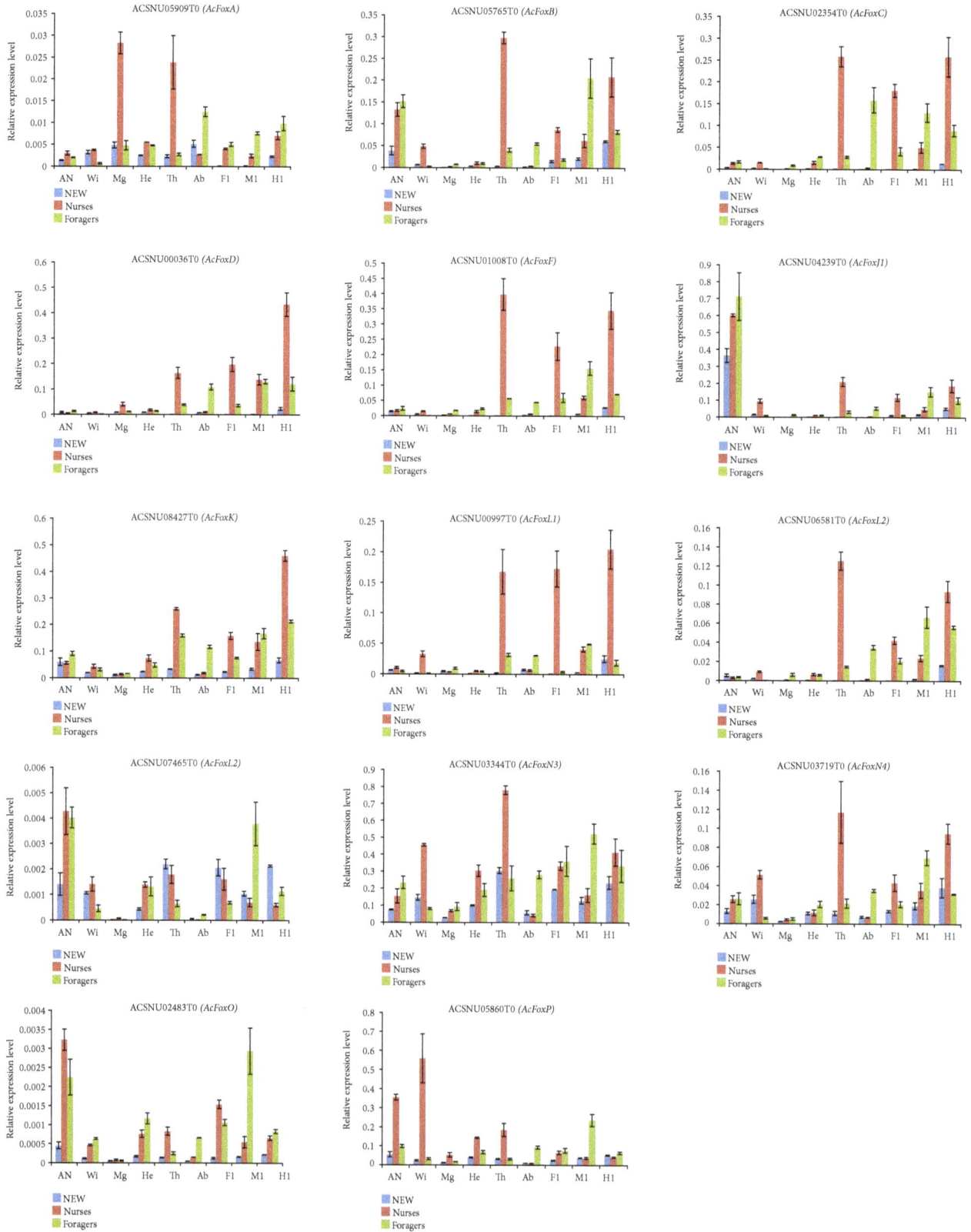

FIGURE 4: The tissue expression profiles of *A. cerana* Fox family genes in the development of newly emerged workers (NEW), nurses, and foragers. The cDNA templates were derived from the antennae (An), wings (Wi), midgut (Mg), head (He), thorax (Th), abdomen (Ab), front legs (Fl), middle legs (Ml), and hind legs (Hl).

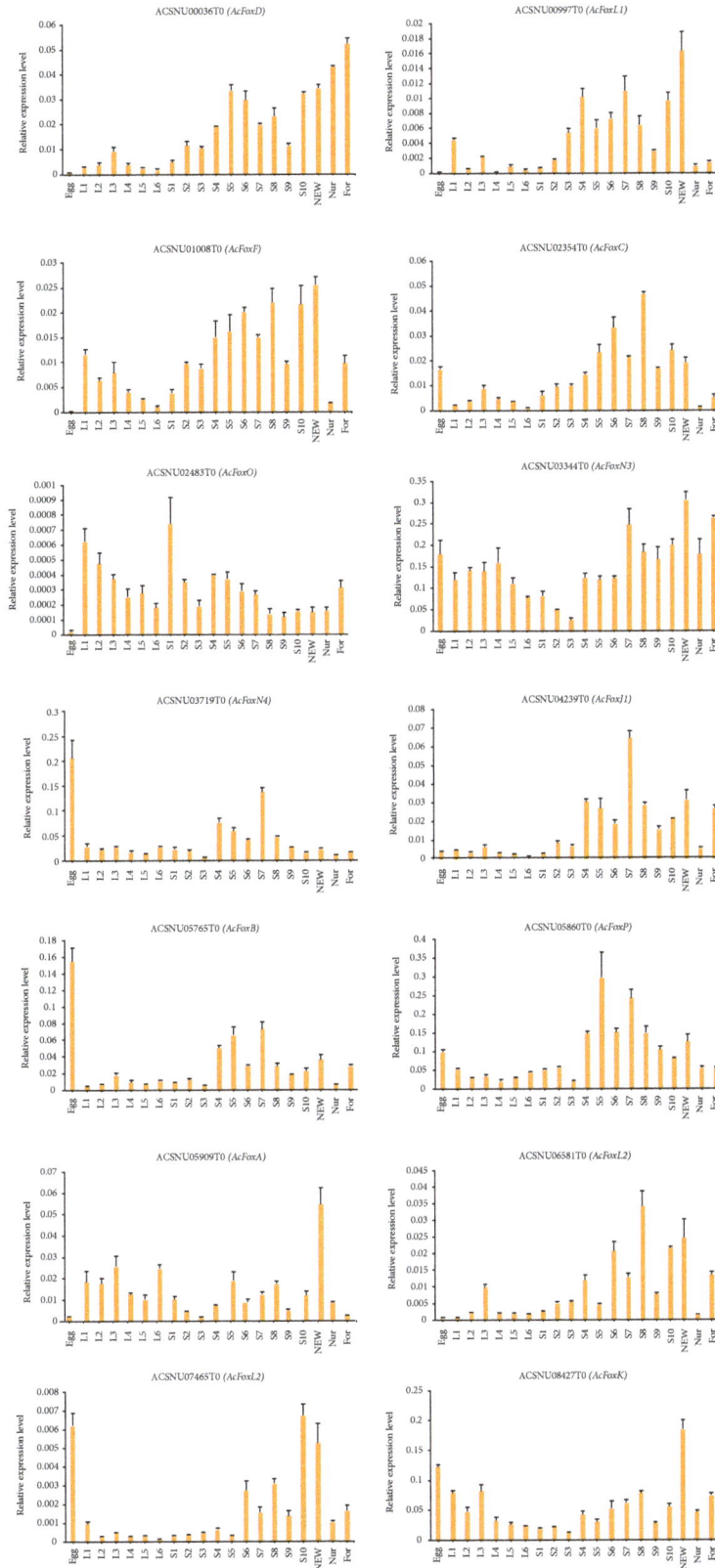

FIGURE 5: Analysis of the temporal expression pattern of 14 Fox genes in *A. cerana*. Egg: mix eggs from day 1 to 3; L1–L6: day of larval stage, from day 1 to day 6; S1–S10: day of pupal stage after cell sealing, from day 1 to day 10; NEW: newly emerged workers; Nur: nurses; For: foragers.

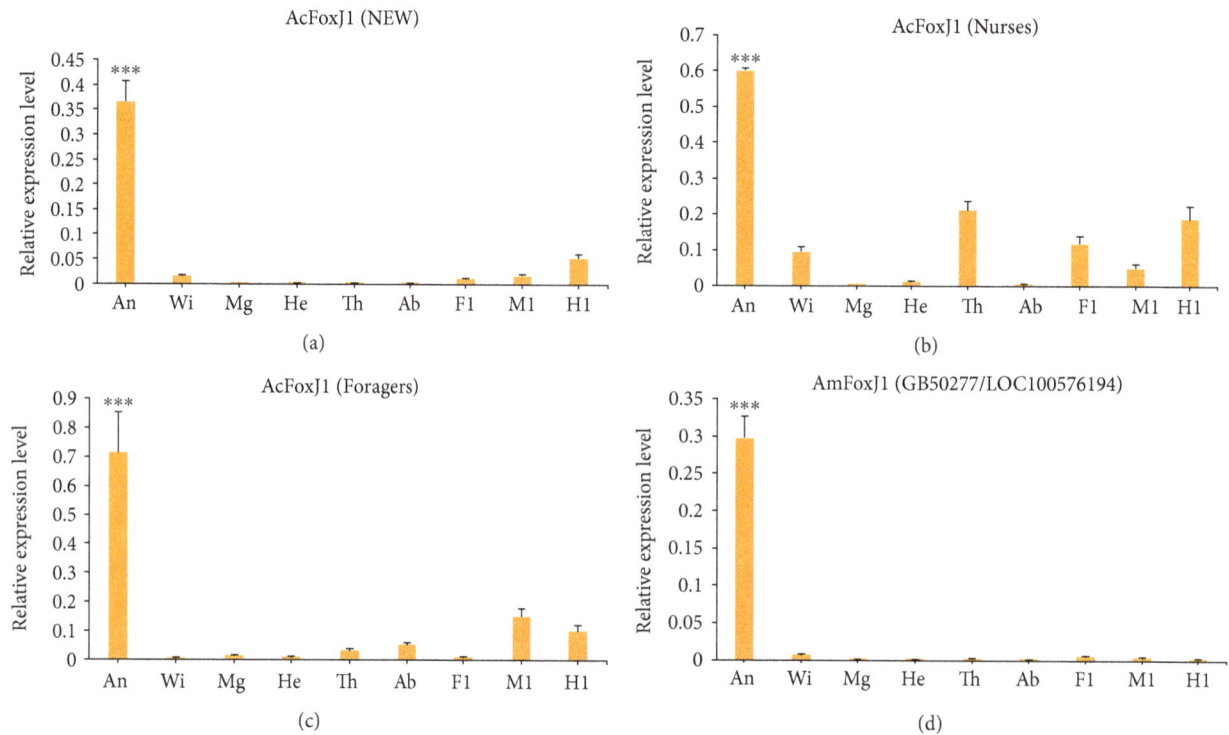

FIGURE 6: Expression patterns of *FoxJ1* in *A. cerana* and *A. mellifera*. (a, b, and c) The tissue expression profiles of *A. cerana* FoxJ1 from newly emerged workers (NEW), nurses, and foragers, respectively. (d) The tissue expression profiles of FoxJ1 in adult *A. mellifera*. ***$p < 0.001$.

and 111.4 at the antennae of NEW, nurses, and foragers, respectively. Because FoxJ1 was abundantly expressed in the antennae of *A. cerana* and *A. mellifera*, it could exhibit a similar trend in the antennae of other insects. If this were the case, FoxJ1 would be a promising target gene for the study of insect olfactory systems. Additional work is necessary to investigate this topic.

In 2010, 14 Fox genes were identified in the *A. mellifera* genome [29]. All of them, except for FoxQ2, were also identified in *A. cerana* in the present study. Our results showed that FoxQ is absent in seven species of Hymenoptera, one species of Diptera, and two species of Lepidoptera. Yet, it is present in *M. musculus* and *H. sapiens* (Table S2). Only one member of both FoxL2 and FoxC subfamilies was discovered in previous research on *A. mellifera*, but we identified two members under each subfamily both in *A. mellifera* and *A. cerana*. Hence, we discovered two novel Fox genes belonging to these two subfamilies. More importantly, we identified the FoxJ1 gene in *A. mellifera*, which was present in all species investigated. The difference with previous study [29] may be due to the updating of the *A. mellifera* genome annotation from releases 102 to 103.

5. Conclusion

In this study, we identified several candidate Fox genes for further functional studies in the honeybee as well as in other insect species. For instance, FoxJ1 with its high level of expression in the antennae constitutes an attractive target for future research addressing its role in insect olfaction. Experiments combining RNAi-based knockdown of this

gene are feasible in the honeybee. The coupling of such an approach with behavioral experiments studying olfactory learning and discrimination could reveal the role of this gene in this behavioral context. Similar studies could be conceived for other Fox genes not only in *A. mellifera* and *A. cerana* but also in other insect species, thus enabling an integrative understanding of Fox genes in several aspects of insect biology.

Abbreviations

Fox: Forkhead box
FKH: Forkhead
qRT-PCR: Quantitative real-time PCR
NEW: Newly emerged workers.

Conflicts of Interest

The authors declare that there is no conflict of interests regarding the publication of this paper.

Authors' Contributions

Songkun Su and Hongyi Nie conceived and designed the experiments; Haiyang Geng performed the experiments; Hongyi Nie, Haiyang Geng, and Shupeng Xu analyzed the data; Hongyi Nie and Yan Lin wrote the paper; Zhiguo Li contributed the reagents; Yazhou Zhao contributed the materials. Hongyi Nie and Haiyang Geng contributed equally to this work.

Acknowledgments

This work was supported by the scientific research program funded by the Fujian Provincial Department of Education for Provincial University (no. JK2017014), by the educational and scientific research program for young and middle-aged instructors of Fujian Province (no. JAT160161), and by the earmarked fund for Modern Agro-industry Technology Research System (no. CARS-45-KXJ3). The authors thank Professor Shaokang Huang from the College of Bee Science, Fujian Agriculture and Forestry University, for helpful suggestions and improvement of the manuscript. Also, the authors thank Professor Dr. Martin Giurfa from the Research Center on Animal Cognition, University Paul Sabatier, for his valuable suggestion and language editing which have greatly improved the manuscript.

Supplementary Materials

Figure S1: RNA electrophoresis photo of different developmental stages and different tissues in *A. cerana*. (A) RNA electrophoresis photo of different developmental stages in *A. cerana*. Egg: mix eggs from day 1 to day 3; L1–L6: day of larval from day 1 to day 6; S1–S10: day of individuals after sealing from day 1 to day 10; NEW: newly emerged workers; Nur: nurses; For: foragers. (B) RNA electrophoresis photo of different tissues on the development of newly emerged workers (NEW), nurses, and foragers. The different tissues are as follows: antennae (An), wing (Wi), midgut (Mg), head (He), thorax (Th), abdomen (Ab), front leg (Fl), middle leg (Ml), and hind leg (Hl). Figure S2: qRT-PCR melting curve of 14 primers using different templates of cDNA. (A) Melting curve of 14 primers using templates of 9 tissues of NEW. These genes are ACSNU00036T0, ACSNU00997T0, ACSNU01008T0, ACSNU02354T0, ACSNU02483T0, ACSNU03344T0, ACSNU03719T0, ACSNU04239T0, ACSNU05765T0, ACSNU05860T0, ACSNU05909T0, ACSNU06581T0, ACSNU07465T0, and ACSNU08427T0, respectively. Each primer was detected using nine different tissues, including antennae, wing, midgut, head, thorax, abdomen, front leg, middle leg, and hind leg. (B) Melting curve of 14 primers using templates of 9 tissues of nurses. Each primer was detected using nine different tissues. (C) Melting curve of 14 primers using templates of 9 tissues of foragers. Each primer was detected using nine different tissues. (D) Melting curve of 14 primers using templates of 20 different developmental stages. Figure S3: protein sequence alignment of *A. cerana* forkhead domains and their secondary structure. (A) The alignment of the amino acid sequences of *A. cerana* forkhead domains. (B) The secondary structure of *A. cerana* forkhead domains. The secondary structural elements are indicated with black cylinders and arrows representing α-helices and β-strands, respectively. Table S1: list of primer pairs used in qRT-PCR. Table S2: distribution of the members of the Fox gene subfamilies in *A. cerana* and other species. Supplemental File 1: accession numbers and protein sequences of Fox members in the phylogenetic analysis. *(Supplementary Materials)*

References

[1] M. Monsalve and Y. Olmos, "The complex biology of FOXO," *Current Drug Targets*, vol. 12, no. 9, pp. 1322–1350, 2011.

[2] M. M. Brent, R. Anand, and R. Marmorstein, "Structural basis for DNA recognition by FoxO1 and its regulation by post-translational modification," *Structure*, vol. 16, no. 9, pp. 1407–1416, 2008.

[3] K. L. Tsai, C. Y. Huang, C. H. Chang, Y. J. Sun, W. J. Chuang, and C. D. Hsiao, "Crystal structure of the human FOXK1a-DNA complex and its implications on the diverse binding specificity of winged helix/forkhead proteins," *Journal of Biological Chemistry*, vol. 281, no. 25, pp. 17400–17409, 2006.

[4] K. S. Gajiwala and S. K. Burley, "Winged helix proteins," *Current Opinion in Structural Biology*, vol. 10, no. 1, pp. 110–116, 2000.

[5] D. Weigel, G. Jürgens, F. Küttner, E. Seifert, and H. Jäckle, "The homeotic gene *fork* head encodes a nuclear protein and is expressed in the terminal regions of the drosophila embryo," *Cell*, vol. 57, no. 4, pp. 645–658, 1989.

[6] B. A. Benayoun, S. Caburet, and R. A. Veitia, "Forkhead transcription factors: key players in health and disease," *Trends in Genetics*, vol. 27, no. 6, pp. 224–232, 2011.

[7] K. R. Wotton and S. M. Shimeld, "Comparative genomics of vertebrate fox cluster loci," *BMC Genomics*, vol. 7, no. 1, pp. 271–271, 2006.

[8] Q. Tu, C. T. Brown, E. H. Davidson, and P. Oliveri, "Sea urchin forkhead gene family: phylogeny and embryonic expression," *Developmental Biology*, vol. 300, no. 1, pp. 49–62, 2006.

[9] S. L. Peng, "FoxO in the immune system," *Oncogene*, vol. 27, no. 16, pp. 2337–2344, 2008.

[10] D. R. Calnan and A. Brunet, "The FoxO code," *Oncogene*, vol. 27, no. 16, pp. 2276–2288, 2008.

[11] K. C. Arden, "FOXO animal models reveal a variety of diverse roles for FOXO transcription factors," *Oncogene*, vol. 27, no. 16, pp. 2345–2350, 2008.

[12] A. Barthel, D. Schmoll, and T. G. Unterman, "FoxO proteins in insulin action and metabolism," *Trends in Endocrinology & Metabolism*, vol. 16, no. 4, pp. 183–189, 2005.

[13] M. Stauber, M. Weidemann, O. Dittrich-Breiholz et al., "Identification of FoxJ1 effectors during ciliogenesis in the foetal respiratory epithelium and embryonic left-right organiser of the mouse," *Developmental Biology*, vol. 423, no. 2, pp. 170–188, 2017.

[14] K. Liu, J. Fan, and J. Wu, "Forkhead box protein J1 (FOXJ1) is overexpressed in colorectal cancer and promotes nuclear translocation of β-catenin in SW620 cells," *Medical Science Monitor*, vol. 23, pp. 856–866, 2017.

[15] H. Miao, C. X. Miao, N. Li, and J. Han, "FOXJ2 controls meiosis during spermatogenesis in male mice," *Molecular Reproduction and Development*, vol. 83, no. 8, pp. 684–691, 2016.

[16] C. D. Hanlon and D. J. Andrew, "Drosophila FoxL1 non-autonomously coordinates organ placement during embryonic development," *Developmental Biology*, vol. 419, no. 2, pp. 273–284, 2016.

[17] Y. Zhao, D. J. Tindall, and H. Huang, "Modulation of androgen receptor by FoxA1 and FoxO1 factors in prostate cancer," *International Journal of Biological Sciences*, vol. 10, no. 6, pp. 614–619, 2014.

[18] E. C. Genin, N. Caron, R. Vandenbosch, L. Nguyen, and B. Malgrange, "Concise review: forkhead pathway in the

control of adult neurogenesis," *Stem Cells*, vol. 32, no. 6, pp. 1398–1407, 2014.

[19] H. K. Rho and D. R. Mcclay, "The control of FoxN2/3 expression in sea urchin embryos and its function in the skeletogenic gene regulatory network," *Development*, vol. 138, no. 5, pp. 937–945, 2011.

[20] M. L. Golson and K. H. Kaestner, "Fox transcription factors: from development to disease," *Development*, vol. 143, no. 24, pp. 4558–4570, 2016.

[21] J. B. Song, Z. Q. Li, X. L. Tong et al., "Genome-wide identification and characterization of fox genes in the silkworm, *Bombyx mori*," *Functional & Integrative Genomics*, vol. 15, no. 5, pp. 511–522, 2015.

[22] I. A. Hansen, D. H. Sieglaff, J. B. Munro et al., "Forkhead transcription factors regulate mosquito reproduction," *Insect Biochemistry and Molecular Biology*, vol. 37, no. 9, pp. 985–997, 2007.

[23] H. H. Lee and M. Frasch, "Survey of forkhead domain encoding genes in the *Drosophila* genome: classification and embryonic expression patterns," *Developmental Dynamics*, vol. 229, no. 2, pp. 357–366, 2004.

[24] F. Mazet, J.-K. Yu, D. A. Liberles, L. Z. Holland, and S. M. Shimeld, "Phylogenetic relationships of the fox (forkhead) gene family in the bilateria," *Gene*, vol. 316, pp. 79–89, 2003.

[25] G. C. Galizia, D. Eisenhardt, and M. Giurfa, *Honeybee Neurobiology and Behavior: a Tribute to Randolf Menzel*, Springer, New York, 2012.

[26] D. Begna, B. Han, M. Feng, Y. Fang, and J. Li, "Differential expressions of nuclear proteomes between honeybee (*Apis mellifera* L.) queen and worker larvae: a deep insight into caste pathway decisions," *Journal of Proteome Research*, vol. 11, no. 2, pp. 1317–1329, 2012.

[27] A. Zayed and G. E. Robinson, "Understanding the relationship between brain gene expression and social behavior: lessons from the honey bee," *Annual Review of Genetics*, vol. 46, no. 1, pp. 591–615, 2012.

[28] T. Kiya, Y. Itoh, and T. Kubo, "Expression analysis of the *FoxP* homologue in the brain of the honeybee, *Apis mellifera*," *Insect Molecular Biology*, vol. 17, no. 1, pp. 53–60, 2008.

[29] S. M. Shimeld, B. Degnan, and G. N. Luke, "Evolutionary genomics of the fox genes: origin of gene families and the ancestry of gene clusters," *Genomics*, vol. 95, no. 5, pp. 256–260, 2010.

[30] H. Liu, Z.-L. Wang, L.-Q. Tian et al., "Transcriptome differences in the hypopharyngeal gland between western honeybees (*Apis mellifera*) and eastern honeybees (*Apis cerana*)," *BMC Genomics*, vol. 15, no. 1, p. 744, 2014.

[31] P. Xu, M. Shi, and X.-x. Chen, "Antimicrobial peptide evolution in the Asiatic honey bee Apis cerana," *PLoS One*, vol. 4, no. 1, article e4239, 2009.

[32] Y.-S. Peng, Y. Fang, S. Xu, and L. Ge, "The resistance mechanism of the Asian honey bee, *Apis cerana* Fabr., to an ectoparasitic mite, *Varroa jacobsoni* Oudemans," *Journal of Invertebrate Pathology*, vol. 49, no. 1, pp. 54–60, 1987.

[33] M. Ono, I. Okada, and M. Sasaki, "Heat production by balling in the Japanese honeybee, *Apis cerana japonica* as a defensive behavior against the hornet, *Vespa simillima xanthoptera* (Hymenoptera: Vespidae)," *Experientia*, vol. 43, no. 9, pp. 1031–1034, 1987.

[34] D. Park, J. W. Jung, B.-S. Choi et al., "Uncovering the novel characteristics of Asian honey bee, *Apis cerana*, by whole genome sequencing," *BMC Genomics*, vol. 16, no. 1, 2015.

[35] P. Artimo, M. Jonnalagedda, K. Arnold et al., "ExPASy: SIB bioinformatics resource portal," *Nucleic Acids Research*, vol. 40, no. W1, pp. W597–W603, 2012.

[36] B. Hu, J. Jin, A.-Y. Guo, H. Zhang, J. Luo, and G. Gao, "GSDS 2.0: an upgraded gene feature visualization server," *Bioinformatics*, vol. 31, no. 8, pp. 1296-1297, 2015.

[37] D. G. Higgins, J. D. Thompson, and T. J. Gibson, "[22] Using CLUSTAL for multiple sequence alignments," *Methods in Enzymology*, vol. 266, pp. 383–402, 1996.

[38] M. Biasini, S. Bienert, A. Waterhouse et al., "SWISS-MODEL: modelling protein tertiary and quaternary structure using evolutionary information," *Nucleic Acids Research*, vol. 42, no. W1, pp. W252–W258, 2014.

[39] R. Fleig and K. Sander, "Embryogenesis of the honeybee *Apis mellifera* L. (Hymenoptera : Apidae): an sem study," *International Journal of Insect Morphology and Embryology*, vol. 15, no. 5-6, pp. 449–462, 1986.

[40] H. J. McQuillan, A. B. Barron, and A. R. Mercer, "Age- and behaviour-related changes in the expression of biogenic amine receptor genes in the antennae of honey bees (*Apis mellifera*)," *Journal of Comparative Physiology A*, vol. 198, no. 10, pp. 753–761, 2012.

[41] F. Liu, W. Li, Z. Li, S. Zhang, S. Chen, and S. Su, "High-abundance mRNAs in *Apis mellifera*: comparison between nurses and foragers," *Journal of Insect Physiology*, vol. 57, no. 2, pp. 274–279, 2011.

[42] H. Nie, C. Liu, T. Cheng et al., "Transcriptome analysis of integument differentially expressed genes in the pigment mutant (*quail*) during molting of silkworm, *Bombyx mori*," *PLos One*, vol. 9, no. 4, article e94185, 2014.

[43] K. J. Livak and T. D. Schmittgen, "Analysis of relative gene expression data using real-time quantitative PCR and the $2^{-\Delta\Delta C_T}$ method," *Methods*, vol. 25, no. 4, pp. 402–408, 2001.

[44] L. Liu, Z. Gong, X. Guo, and B. Xu, "Cloning, structural characterization and expression analysis of a novel lipid storage droplet protein-1 (LSD-1) gene in Chinese honeybee (*Apis cerana cerana*)," *Molecular Biology Reports*, vol. 39, no. 3, pp. 2665–2675, 2012.

[45] K. H. Kaestner, W. Knöchel, and D. E. Martínez, "Unified nomenclature for the winged helix/forkhead transcription factors," *Genes & Development*, vol. 14, no. 2, pp. 142–146, 2000.

[46] D. J. Schulz and G. E. Robinson, "Octopamine influences division of labor in honey bee colonies," *Journal of Comparative Physiology A: Sensory, Neural, and Behavioral Physiology*, vol. 187, no. 1, pp. 53–61, 2001.

[47] Y. Le Conte, A. Mohammedi, and G. E. Robinson, "Primer effects of a brood pheromone on honeybee behavioural development," *Proceedings of the Royal Society of London B: Biological Sciences*, vol. 268, no. 1463, pp. 163–168, 2001.

[48] P. K. Jackson and L. D. Attardi, "p73 and FoxJ1: programming multiciliated epithelia," *Trends in Cell Biology*, vol. 26, no. 4, pp. 239-240, 2016.

[49] M. K. Y. Siu, E. S. Y. Wong, D. S. H. Kong et al., "Stem cell transcription factor NANOG controls cell migration and invasion via dysregulation of E-cadherin and FoxJ1 and contributes to adverse clinical outcome in ovarian cancers," *Oncogene*, vol. 32, no. 30, pp. 3500–3509, 2013.

[50] C. Cruz, V. Ribes, E. Kutejova et al., "Foxj1 regulates floor plate cilia architecture and modifies the response of cells to sonic hedgehog signalling," *Development*, vol. 137, no. 24, pp. 4271–4282, 2010.

[51] B. Demircan, L. M. Dyer, M. Gerace, E. K. Lobenhofer, K. D. Robertson, and K. D. Brown, "Comparative epigenomics of human and mouse mammary tumors," *Genes, Chromosomes and Cancer*, vol. 48, no. 1, pp. 83–97, 2009.

[52] L. Lin, S. L. Brody, and S. L. Peng, "Restraint of B cell activation by Foxj1-mediated antagonism of NF-κB and IL-6," *The Journal of Immunology*, vol. 175, no. 2, pp. 951–958, 2005.

[53] H. Nie, S. Xu, C. Xie et al., "Comparative transcriptome analysis of *Apis mellifera* antennae of workers performing different tasks," *Molecular Genetics and Genomics*, vol. 293, no. 1, pp. 237–248, 2018.

5

Circular RNAs: Biogenesis, Function, and a Role as Possible Cancer Biomarkers

Luka Bolha,[1] **Metka Ravnik-Glavač,**[1,2] **and Damjan Glavač**[1]

[1]*Department of Molecular Genetics, Institute of Pathology, Faculty of Medicine, University of Ljubljana, Ljubljana, Slovenia*
[2]*Institute of Biochemistry, Faculty of Medicine, University of Ljubljana, Ljubljana, Slovenia*

Correspondence should be addressed to Damjan Glavač; damjan.glavac@mf.uni-lj.si

Academic Editor: Davide Barbagallo

Circular RNAs (circRNAs) are a class of noncoding RNAs (ncRNAs) that form covalently closed continuous loop structures, lacking the terminal 5′ and 3′ ends. CircRNAs are generated in the process of back-splicing and can originate from different genomic regions. Their unique circular structure makes circRNAs more stable than linear RNAs. In addition, they also display insensitivity to ribonuclease activity. Generally, circRNAs function as microRNA (miRNA) sponges and have a regulatory role in transcription and translation. They may be also translated in a cap-independent manner *in vivo*, to generate specific proteins. In the last decade, next-generation sequencing techniques, especially RNA-seq, have revealed great abundance and also dysregulation of many circRNAs in various diseases, suggesting their involvement in disease development and progression. Regarding their high stability and relatively specific differential expression patterns in tissues and extracellular environment (e.g., body fluids), they are regarded as promising novel biomarkers in cancer. Therefore, we focus this review on describing circRNA biogenesis, function, and involvement in human cancer development and address the potential of circRNAs to be effectively used as novel cancer diagnostic and prognostic biomarkers.

1. Introduction

Noncoding RNAs (ncRNAs) represent a large, complex, and heterogeneous group of RNA molecules that can be classified into two major classes, according to the size of their transcripts: the small ncRNAs (<200 bp) and the long ncRNAs (lncRNAs) (>200 bp) [1, 2]. The small ncRNAs include microRNAs (miRNAs), small nuclear RNAs (snRNAs), PIWI-interacting RNAs (piRNAs), small interfering RNAs (siRNAs), small nucleolar RNAs (snoRNAs), and others [1], whereas lncRNAs comprise long intergenic ncRNAs (lincRNAs), intronic ncRNAs, macroRNAs, sense ncRNA, antisense RNAs, and others [2]. In addition, circular RNAs (circRNAs) have been recently identified as a relatively large class of ncRNAs, which are widespread and abundant in a variety of eukaryotic organisms and involved in multiple biological processes [3, 4]. CircRNAs may vary in length significantly, most of them being longer than 200 nt. However, some exonic and intronic circRNAs were shown to be shorter than 200 or even 100 nt [5].

CircRNAs form covalently closed continuous loop structures, without terminal 5′ caps and 3′ polyadenylated tails. They are generated by alternative splicing of pre-mRNA transcripts, where an upstream splice acceptor is joined to a downstream splice donor, in the process of back-splicing [6–8]. In the last decade, RNA-seq and other next-generation sequencing techniques have enabled a significant breakthrough in circRNA discovery, leading to the identification and characterization of a large number of circRNAs in humans and other eukaryotes [9, 10]. Several research groups have demonstrated a conservation of circRNA expression across mammals. In addition, circRNAs appear to be stably expressed in a cell/tissue-dependent and developmental stage-specific manner [3, 9, 11, 12]. Emerging evidence reveals the importance of circRNA involvement in regulating gene expression at transcriptional and posttranscriptional levels, and, furthermore, dysregulation in circRNA expression correlates with irregularities in developmental processes and various disease states, including cancer [13–16]. Regarding the observed correlation between altered circRNA expression profiles and a cancer patient's clinical characteristics and circRNA's structural features that enable their abundance and stability in various biological samples, we focus this review on describing circRNA involvement in cancer development. In addition, we address the potential of circRNAs to be effectively used as novel biomarkers in cancer.

2. Biogenesis of CircRNAs

The majority of circRNAs originate from exons of protein-coding genes, frequently consisting of 1–5 exons [9]. However, they may be also formed from intronic, noncoding, antisense, $3'$ UTR, $5'$ UTR, or intergenic genomic regions [9, 17]. CircRNAs are generated by a spliceosome-mediated pre-mRNA back-splicing which connects a downstream splice donor site ($5'$ splice site) to an upstream acceptor splice site ($3'$ splice site) [18]. Similar to canonical (linear) splicing, back-splicing appears to be extensively regulated by canonical cis-acting splicing regulatory elements and trans-acting splicing factors. However, the regulation process of back-splicing in controlling circRNA production fundamentally differs from that of linear splicing, where the same combinations of splicing regulatory elements and factors have distinct or even opposite activity [18]. In addition, a single gene locus can produce various circRNAs through alternative back-splice site selection, when compared to canonical splicing of linear RNAs [19]. Generally, circRNAs can be generated by canonical and noncanonical splicing [20]. Regarding their biogenesis from different genomic regions, circRNAs can be categorized into four types, as determined by RNA-seq: exonic circRNAs (ecircRNAs) [9, 11, 12], circular intronic RNAs (ciRNAs) [17], retained-intron or exon-intron circRNAs (EIciRNAs) [11, 21], and intergenic circRNAs [9]. Schematic representation of ecircRNA, ciRNA, EIciRNA, and intergenic circRNA biogenesis is shown in Figure 1.

EcircRNAs are the most abundant circRNA type, accounting for over 80% of identified circRNAs, and are predominantly located in the cytoplasm [3, 9, 12], though the exact process of nuclear export remains to be elucidated. It has been suggested that ecircRNAs may escape from the nucleus during mitosis [22]. Although the exact mechanism of circRNAs biogenesis remains unclear, three models of circRNA formation have been proposed, including lariat-driven circularization (exon skipping) [12] (Figure 1(b)), intron pairing-driven circularization [12] (Figure 1(c)), and resplicing-driven circularization [23] (Figure 1(f)). It has been demonstrated that exon circularization depends on several genomic features, essential for promoting circularization. In general, exons comprising circRNAs are longer than average exons, which is especially notable for single-exon circRNAs, being approximately 3-fold longer, when compared to other expressed exons [3, 12]. In addition, several transcriptome analyses indicated a significant correlation between the presence of flanking intronic regions, containing the reverse complementary sequences (e.g., Alu elements) that may promote intron pairing, and exon circularization [12, 24, 25]. Normally, flanking introns containing inverted tandem repeats, involved in back-splicing and circRNA

production, tend to be longer than introns generally, but some can be shorter than average [11]. It has also been demonstrated that relatively short (30–40 nt) inverted repeats are sufficient for intron base pairing and subsequent circRNA formation [26]. However, not all intronic tandem repeats can support exon circularization. In some cases, increased stability of intron base pairing prevented circRNA formation [26]. During the biogenesis process of circRNAs, introns may not be spliced out completely but are retained between the encircled exons in the newly generated circRNA. This phenomenon results in the formation of EIciRNAs [21]. The important role of RNA-binding proteins (RBPs) has been demonstrated in the regulation of circRNA production, which can act as trans-acting activators or inhibitors of the circRNA formation mechanism. Quaking (QKI) and muscleblind (MBL/MBNL1) proteins can bind to specific sequence motifs of flanking introns on linear pre-mRNA sequences, thus linking the two flanking introns together, promoting cycling and subsequent circRNA generation [27, 28] (Figure 1(d)). The process is similar to an intron pairing-driven circularization model, only that here, RBPs, after binding to specific putative binding sites, dimerize which leads to pre-mRNA looping. Conversely, the RNA-editing enzyme adenosine deaminase acting on RNA (ADAR) antagonizes circRNA production by direct binding and weakening RNA duplexes (e.g., inverted Alu repeats), through the action of adenosine-to-inosine (A-to-I) editing [25] (Figure 1(e)). High ADAR expression destabilizes intron base pairing interactions, impairing pre-mRNA looping and decreasing the likelihood of pre-mRNA circularization and circRNA formation, for a subset of circRNAs [25, 29].

CiRNA formation differs from that of ecircRNA and EIciRNA (Figure 1(g)). Stable ciRNAs can be formed, when intron lariats escape the usual intron debranching and subsequent degradation, following the canonical spliceosome-mediated pre-mRNA splicing [17]. CiRNA biogenesis depends mainly on a 7 nt GU-rich element near the $5'$ splice site and an 11 nt C-rich element near the branch point site. During back-splicing, the two elements bind into a lariat-like intermediate, containing the excised exons and introns, and are cut out by the spliceosome [17, 30]. Then, generated stable lariats undergo $3'$ tail degradation, which results in the formation of the final ciRNA molecule [17]. Generally, ciRNAs may be sensitive to RNA debranching enzymes and can be distinguished from ecircRNAs by the presence of a $2'$–$5'$ junction, a residue of the lariat structure, which is evidently absent in ecircRNAs [17]. CiRNAs, along with EIciRNAs, are predominantly located in the nucleus and are believed to be involved in regulating expression of local genes in cis [17, 21]. In addition, sequence analyses have shown a weak but significant enrichment of conserved nucleotides between few ciRNAs and intergenic circRNAs [9]. However, there is currently very little information on the overall characteristics and biogenesis processes of intergenic circRNAs.

Despite a good deal of information on circRNA biogenesis, relatively little is known about the metabolic processing of these molecules within cells. Since circRNAs are abundant and highly stable and show resistance to

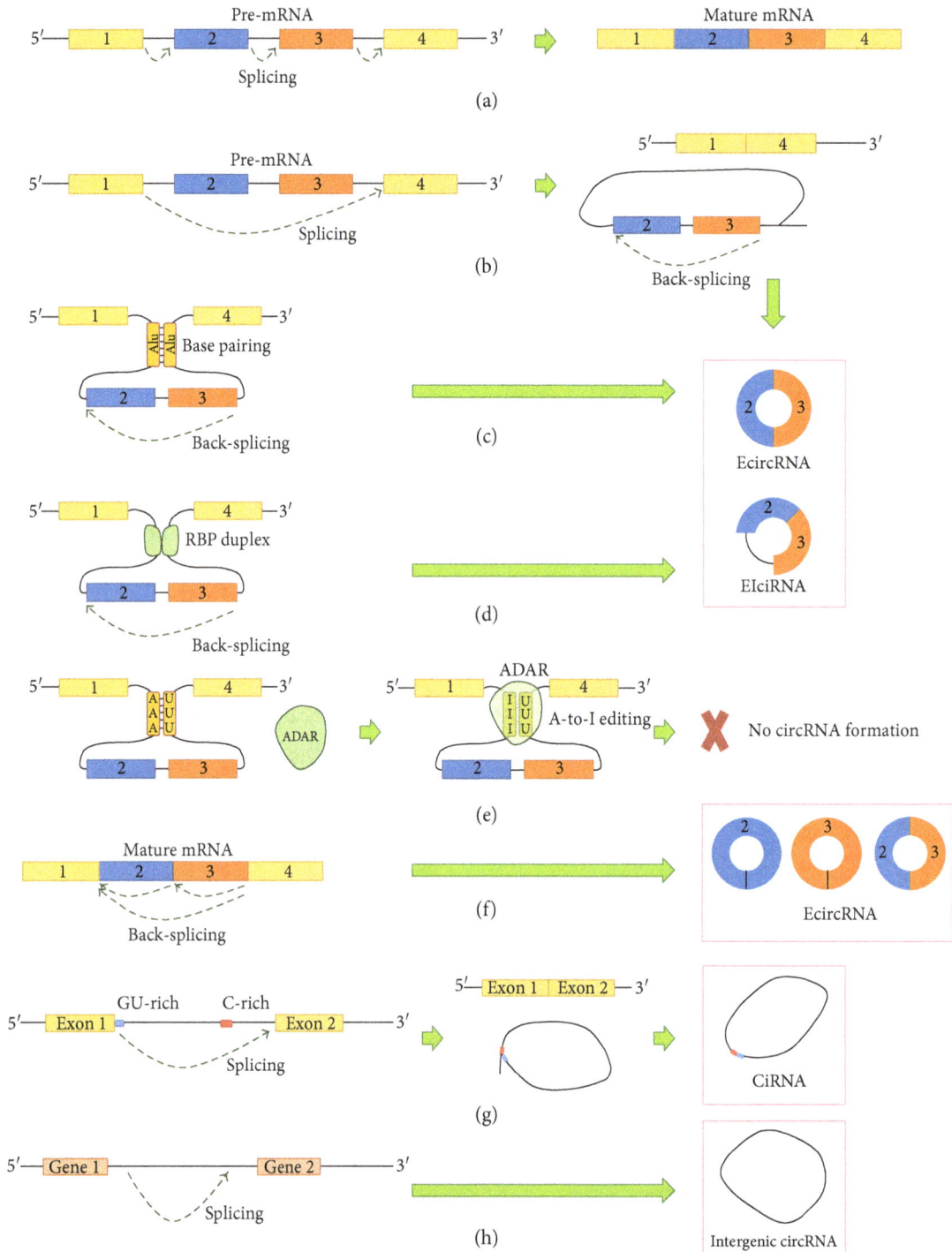

FIGURE 1: Schematic representation of circRNA biogenesis. (a) Canonical pre-mRNA splicing, yielding a mature mRNA molecule. (b) Lariat-driven circularization (exon skipping). Following canonical splicing, exons in exon-containing lariats undergo back-splicing and circularization, which results in the formation of ecircRNA or EIciRNA molecules. (c) Intron pairing-driven circularization, utilizing coupling of flanking introns by direct base pairing between *cis*-acting regulatory elements that contain reverse complementary sequences (e.g., Alu repeats). Intron pairing is followed by back-splicing and exon circularization. (d) CircRNA biogenesis, mediated by *trans*-acting factors, such as RNA-binding proteins (RBPs) (e.g., QKI, MBL/MBNL1) that bind to specific sequence motifs of flanking introns on linear pre-mRNA, dimerize, and facilitate back-splicing and exon circularization. (e) Regulation of circRNA biogenesis by the RNA-editing enzyme ADAR. ADAR destabilizes intron base pairing interactions through the action of adenosine-to-inosine (A-to-I) editing, which impairs pre-mRNA looping and diminishes exon circularization. (f) Resplicing-driven circularization. EcircRNAs may be formed from mature mRNA exons that undergo back-splicing and circularization. (g) Formation of ciRNAs from intron lariats that escape the usual intron debranching and degradation, following the canonical pre-mRNA splicing. (h) Formation of intergenic circRNAs. This figure is adapted from Wang et al. [151].

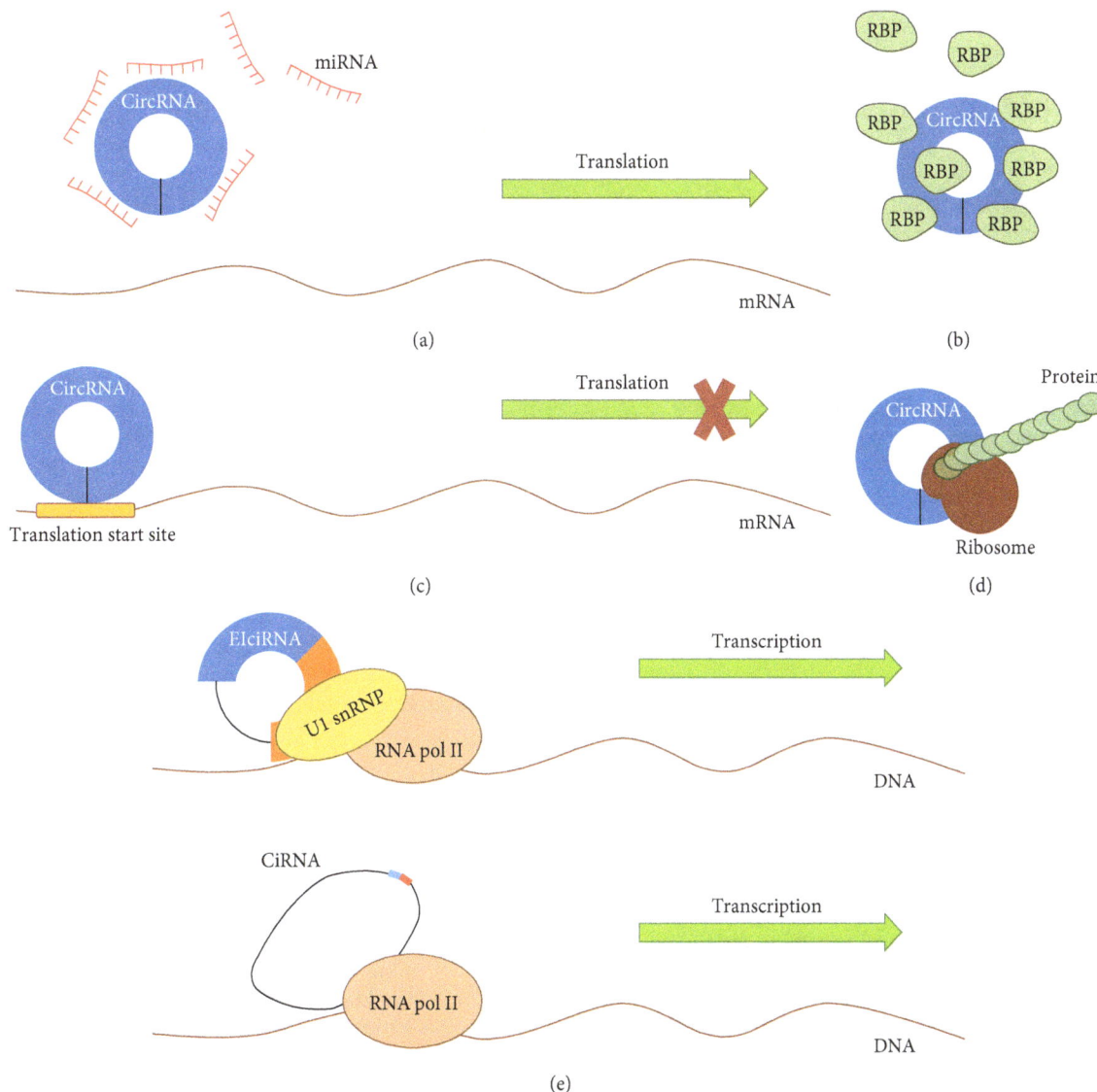

FIGURE 2: Schematic representation of circRNA functions. (a) CircRNAs may act as miRNA sponges by competing for miRNA binding sites, diminishing the effect of miRNA-mediated regulatory activities. (b) CircRNAs may act as protein sponges, by binding RNA-binding proteins (RBPs). (c) Some circRNAs may regulate protein expression by sequestering mRNA translation start sites. (d) CircRNAs may be translated to form functional proteins. (e) CircRNAs (e.g., EIciRNAs and ciRNAs) may interact with transcription complexes and enhance the expression of their parental genes.

exonucleases (e.g., RNase R) [9, 31], they may accumulate in the cell with a possible toxic effect. A study employing three cell lines has shown that the aggregated excessive circRNAs could be effectively eliminated from cells via released vesicles such as exosomes and microvesicles [32]. An additional study has also demonstrated similar results, where excessive circRNAs were enriched over their linear isoforms within extracellular vesicles, when compared to the producing cells [33].

3. Function of CircRNAs

CircRNAs have important functions in regulating gene expression and may act as miRNA sponges, RBP sponges, and regulators of transcription and translation [21, 34, 35].

Also, several circRNAs have shown the ability to be translated into peptides [18, 36–39]. Schematic representation of these circRNA functions is shown in Figure 2.

miRNAs are approximately 22 nt long ncRNAs, involved in posttranscriptional regulation of gene expression, which act by direct binding to specific target sites within UTRs of mRNAs [40, 41]. The result of such miRNA-mediated mRNA targeting is either blockage of the translation process or complete degradation of the bound mRNA molecule [41]. CircRNAs can act like miRNA sponges by competing for miRNA binding sites (Figure 2(a)), thus diminishing the effect of miRNA-mediated regulatory activities (e.g., posttranscriptional repression) [9, 34]. Indeed, it has been demonstrated that overexpression of miRNA sponge-acting circRNAs increases the expression of miRNA targets, whereas knockdown of these circRNAs had the opposite effect [34]. The human cerebellar degeneration-related protein 1 antisense (CDR1as) circRNA, also known as a circular RNA sponge for miR-7 (ciRS-7), is the most well characterized circRNA with miRNA sponge function. It contains more than 70 selectively conserved miRNA target sites and associates with Argonaute (AGO) proteins in a miR-7-dependent manner [34]. Overexpression of ciRS-7 leads to a significant decrease in miR-7 activity and results in increasing the miR-7 target gene expression level [34]. CiRS-7 is highly expressed in mammalian brain during neuronal development [29]. Studies in mice have revealed an overlapping coexpression of ciRS-7 and miR-7 in brain tissues, which indicates that this circRNA may be crucial for normal neuronal development [34]. In addition to miR-7 binding sites, ciRS-7 contains an additional binding site for miR-671. The combination of ciRS-7 and miR-671 triggers the linearization and AGO2-mediated cleavage of ciRS-7, which enables the release of the absorbed miR-7 molecules [42]. Mouse testis-specific sex-determining region Y (*Sry*) linear isoform is expressed in the developing genital ridge and has a fundamental role as a transcription factor in sex determination [43]. However, in adult testes, the circular form of *Sry* is expressed, with its circularization being dictated by promotor usage and dependent on intron pairing-driven circularization [43, 44]. *Sry* circRNA serves as the miR-138 sponge and contains 16 miR-138 binding sites [34]. Furthermore, cir-ZNF609 [45], mm9_circ_012559 [46], and circRNAs from the human C_2H_2 zinc finger gene family [47] have also been assumed to function as miRNA sponges. Overall, when compared to linear miRNA sponges, circRNAs have proven themselves as more stable and therefore more effective [48, 49]. Also, their expression is not affected upon miRNA binding [34]. On the other hand, several studies implied that most circRNAs do not act like miRNA sponges, since the majority does not have more miRNA binding sites than colinear mRNAs [47, 50].

It has been demonstrated that circRNAs can bind to several RBPs (Figure 2(b)), including AGO [9, 34], RNA polymerase II [17], QKI [27], EIF4A3 [51], and MBL [28]. In addition, some ecircRNAs can store, sort, or localize RBPs and presumably regulate the RBP function by acting as competing elements, in a similar way as they affect miRNA activity [9, 35].

CircRNAs form a large group of transcriptional and posttranscriptional regulators, and their direct involvement in regulating gene expression has been demonstrated in several studies [9, 17, 21]. CircRNAs (e.g., EIciRNAs and ciR-NAs) abundant in the nucleus displayed little target miRNA binding sites, and knockdown of these circRNAs commonly resulted in reduced expression of their parental genes [17, 21]. EIciRNAs, such as circEIF3J and circPAIP2, interact with U1 small nuclear ribonucleoprotein (snRNP) and RNA polymerase II and upregulate their parental genes in *cis* [21] (Figure 2(e)). Thus, these findings strongly imply that circRNAs could function as scaffolds for RBPs regulating transcription, as it had been determined for several lncRNAs [52, 53]. During splicing, circRNAs and their corresponding linear isoforms may compete with each other for biogenesis

[28]. However, the generated circular RNA forms may promote both circRNA and mRNA expression [21]. In addition, some circRNAs may also regulate protein expression by sequestering mRNA translation start sites [22] (Figure 2(c)).

It was shown that peptides could be translated from ecircRNAs both *in vitro* and *in vivo*, when RNA molecules contained the internal ribosomal entry site (IRES) elements or prokaryotic ribosome binding sites [36, 54, 55] (Figure 2(d)). A recent study has strongly supported the fact that a subset of *Drosophila* endogenous circRNAs has the ability to be translated *in vivo* in a cap-independent manner. Many of these translating ribosome-associated circRNAs (ribo-circRNAs) shared the start codon with the hosting gene, had evolutionary conserved stop codons, and encoded at least one identifiable ribo-circRNA-specific protein domain, implying these proteins are functional [37]. Another endogenous protein-coding circRNA, circ-ZNF609, has been identified in murine and human myoblasts. Circ-ZNF609 is involved in regulating myoblast proliferation and is generated from the second exon of its host gene. Its open reading frame (ORF) contains a start codon, common with the linear transcript, and an in-frame stop codon, created upon circularization. Circ-ZNF609 is translated into a protein in a splicing-dependent and cap-independent manner. However, molecular activity of circ-ZNF609-derived proteins still needs to be determined [38]. It was also shown that the N^6-methyladenosine (m^6A) RNA modification promotes endogenous circRNA translation in human cells. The study revealed that m^6A motifs are enriched in many circRNAs and enable protein translation in a cap-independent manner, involving the m^6A reader YTHDF3 and translation initiation factors eIF4G2 and eIF3A [39]. In addition, association of many circRNAs with polysomes has been determined [37–39]. These emerging insights into circRNA protein-coding ability and their further characterization may eventually reveal a vast assortment of thus far uncharacterized proteins and throw light on the processes they are involved in. Thus, it is getting progressively clearer that beside their initially proposed regulatory role as ncRNAs, circRNAs may also represent a novel type of protein-coding RNA.

4. CircRNA Databases and Detection Tools

4.1. CircRNA Databases. There are several existing circRNA databases that summarize and integrate the data obtained from large-scale circRNA identification studies, utilizing next-generation sequencing technology, which were performed by different research groups. These databases enable a transparent and comprehensive view on the spatiotemporal presence and function of circRNAs in various biological processes in different organisms, tissues, and cell types. The 9 most acknowledged circRNA databases, containing information on human and animal circRNAs, are presented in Table 1, along with their main features.

The circ2Traits database is a compiled collection of data on circRNAs, categorized according to their potential association with specific human diseases. CircRNAs are grouped based on the number of disease-associated SNPs,

TABLE 1: Databases containing circRNA data.

Name	URL	Description	References
circ2Traits	http://gyanxet-beta.com/circdb/	A database containing information on disease-associated circRNAs and their complete putative interaction networks with miRNAs, mRNAs, and lncRNAs in specific diseases.	[56]
circBase	http://www.circbase.org/	A collection of merged and unified datasets of circRNAs, with evidence supporting their expression	[57]
CircInteractome	https://circinteractome.nia.nih.gov/	A web tool designed for predicting and mapping RBP and miRNA binding sites on reported circRNAs	[51]
CircNet	http://circnet.mbc.nctu.edu.tw/	A database providing information on known and novel circRNAs, circRNA-miRNA-gene regulatory networks, and tissue-specific circRNA expression profiles	[58]
CIRCpedia	http://www.picb.ac.cn/rnomics/circpedia/	A database holding information on identified and annotated back-splicing and alternative splicing in circRNAs from human, mouse, fly, and worm samples	[19]
circRNADb	http://reprod.njmu.edu.cn/circrnadb	A comprehensive database for human circRNAs with protein-coding annotations	[59]
starBase v2.0	http://starbase.sysu.edu.cn/	A database for decoding predicted interaction networks between lncRNAs, miRNAs, circRNAs, mRNAs, and RBPs from large-scale CLIP-seq data	[60]
deepBase v2.0	http://deepbase.sysu.edu.cn/	A platform for annotating, discovering, and characterizing small ncRNAs, lncRNAs, and circRNAs from next-generation sequencing data	[61]
TSCD	http://gb.whu.edu.cn/TSCD/	An integrated database designed for depositing features of human and mouse tissue-specific circRNAs	[62]

AGO interaction sites, and their potential interaction with disease-associated miRNAs, as determined by genome-wide association studies. circ2Traits also stores complete putative miRNA-circRNA-mRNA-lncRNA interaction networks for each of the described diseases [56].

The circBase database provides merged and unified datasets of circRNAs and the evidence supporting their expression. It is a database repository, containing circRNA data from various samples of several species, including human, mouse, fruit fly, and nematode. circBase enables exploring of public circRNA datasets and also provides custom python scripts, needed to discover novel circRNA from user's own (RiboMinus) RNA-seq data. All circRNA transcripts deposited in circBase have been annotated, their putative splice forms predicted and, where applicable, alignments of reads spanning head-to-tail junctions provided [57].

CircInteractome is a web tool designed for mapping RBP and miRNA binding sites on human circRNAs. CircInteractome also enables the identification of potential circRNAs that act as RBP sponges, designing junction-spanning primers for specific circRNA detection, designing siRNAs for circRNA silencing, and identifying potential IRES [51]. However, CircInteractome displays limited ability to predict RBP and miRNA interactions when circRNAs form secondary or tertiary structures. Thus, experimental validation is often needed to reliably verify RBP and miRNA functional sites [51].

The CircNet database provides information on novel circRNAs, integrated miRNA target networks, expression profile of circRNA isoforms, genomic annotation of circRNA isoforms, and sequence features of circRNA isoforms. CircNet also provides tissue-specific circRNA expression profiles, circRNA-miRNA-gene regulatory networks, and a thorough expression analysis of previously reported and novel circRNAs. Furthermore, CircNet generates an integrated regulatory network that illustrates the regulation between circRNAs, miRNAs, and genes [58].

CIRCpedia is an integrative database, which employs the CIRCexplorer2 characterization pipeline for identifying and annotating back-splicing and alternative splicing in circRNAs across different cell lines. Identified back-splicing and alternative splicing in circRNAs, together with novel exons, are formatted and classified for being easily searched, browsed, and downloaded. Currently, CIRCpedia contains information on circRNAs from human, mouse, fly, and worm samples [19].

The circRNADb database is a comprehensive database for human circRNAs with protein-coding annotations. circRNADb provides detailed information on circRNA genomic properties, exon splicing, genome sequences, annotated protein-coding potential, IRES, ORF, and corresponding references [59].

The starBase v2.0 database has been developed to systematically identify RNA-RNA and protein-RNA interaction networks from large-scale CLIP-seq datasets, generated by independent studies. starBase v2.0 distinctive features facilitate annotation, graphic visualization, analysis, and discovery of miRNA-mRNA, miRNA-circRNA, miRNA-pseudogene, miRNA-lncRNA, and protein-RNA interaction networks and RBP binding sites [60].

deepBase v2.0 is a platform for annotating and discovering small ncRNAs, lncRNAs, and circRNAs from next-generation sequencing data. deepBase v2.0 provides a set of tools to decode evolution, spatiotemporal expression patterns, and functions of diverse ncRNAs across 19 species from 5 clades, including human, mouse, fruit fly, and nematode. The platform also provides an integrative, interactive, and versatile web graphical interface to display multidimensional data and facilitates transcriptomic research and the discovery of novel ncRNAs [61].

TSCD (Tissue-Specific CircRNA Database) is an integrated database designed for depositing features of human and mouse tissue-specific circRNAs. TSCD provides a global view on tissue-specific circRNAs and holds information on their genomic location and conservation [62].

4.2. CircRNA Detection Tools. The detection and integration of newly discovered circRNAs into circRNA databases are predominantly dependent on complex bioinformatics analyses of large-scale RNA-seq data. In order to identify and annotate novel circRNAs, RNA-seq data undergoes thorough and rigorous analysis utilizing various state-of-the-art circRNA detection tools and software packages. A comprehensive overview and evaluation of 11 different circRNA detection computational pipelines have been summarized in a recent review by Zeng et al. [63]. Among the described circRNA detection tools, which were also compared with regard to their precision and sensitivity, were circRNA_finder [64], CIRCexplorer [24], DCC [65], find_circ [9], UROBORUS [66], PTESFinder [67], KNIFE [68], CIRI [69], MapSplice [70], segemehl [71], and NCLscan [72]. In general, these circRNA detection tools can be divided into two categories, based on the different strategies used for circRNA identification, according to the dependency on genome annotation [63]. In the pseudo-reference-based [73] or candidate-based approach [22], all possible combinations of candidate circRNAs are constructed. Each candidate comprises two well-annotated exons in which the exon order is topologically inconsistent with the reference genome. The candidate is regarded as a circRNA if at least one RNA-seq read has been identified, which maps to its noncolinear junction site. The strategy relies on candidate circRNAs that are constructed from preexisting gene models and does not detect circRNAs from unannotated transcripts [22, 73]. Conversely, the fragmented-based [73] or segmented read approach [22] does not rely on genome annotation. In this approach, RNA-seq reads are mapped to genomic locations de novo. Reads that cannot be mapped directly are split into two or more segments, and each segment is mapped separately to the reference genome. CircRNA back-splicing junctions are identified, when segmented reads are mapped in a noncolinear manner [22, 73]. As described by Zeng et al. [63], the tools that use the pseudo-reference-based approach to detect circRNAs include PTESFinder, KNIFE, and NCLscan, whereas circRNA_finder, CIRCexplorer, DCC, find_circ, UROBORUS, CIRI, MapSplice, and segemehl utilize the fragmented-based approach. In addition, several integrated tools aimed to identify circRNAs with a protein-coding potential have been developed, including CircPro [74], IRESite [75], CPAT [76], Pfam 31.0 [77], PhyloCSF [78], and ORF Finder from the NCBI database. Also, the TopHat-Fusion algorithm has been designed to detect circRNAs that are derived from gene fusion events [79].

Despite several useful state-of-the-art circRNA detection tools exist, their regular upgrades and the introduction of novel, improved methods, with even higher precision and sensitivity, are a prerequisite to overcome current and future challenges in circRNA identification and characterization studies.

5. CircRNAs in Cancer

Current knowledge about the involvement of circRNAs in cancer development and progression is limited, and the role of circRNAs as miRNA sponges has been proposed as the most frequent mechanism of circRNA activity in tumor cells [80]. Generally, miRNAs are included in various cell processes, including cellular differentiation, development, proliferation, and apoptosis, where they play an important role as regulators of gene expression [81]. These miRNA-mediated processes are frequently deregulated in cancer and can contribute to cancer initiation and progression [81]. Since many circRNAs regulate miRNA action through sponge-like binding (several are presented in Table 2), dysregulation in circRNA expression may affect their interaction with tumor-associated miRNAs, indicating an important role of circRNAs in regulating cancer.

There is emerging evidence that miR-7 can directly downregulate oncogenes in a variety of cancers [82, 83]. miR-7 has been shown to be involved in suppressing melanoma [84], breast cancer [85], glioma [86], gastric cancer [87], liver cancer [88], non-small-cell lung cancer (NSCLC) [89], colorectal cancer [90], and other cancer types. Since circRNA ciRS-7 acts as a miR-7 sponge, quenching the activity of miR-7 may increase the expression levels of miR-7 target oncogenes, resulting in a decreased tumor suppression [83]. Thereby, the ciRS-7/miR-7 axis is likely involved in cancer-related pathways and cancer development and progression [83]. However, despite its potential oncogenic properties, only few studies have revealed ciRS-7 involvement in cancer development through miR-7 binding. CiRS-7-mediated oncogenic activity, acting partly through targeting miR-7, was recently demonstrated in cancer tissues of hepatocellular carcinoma (HCC) patients. When compared to healthy adjacent tissues, ciRS-7 expression levels were significantly upregulated and ciRS-7 expression inversely correlated with miR-7 [91]. In addition, knockdown of ciRS-7 in HCC cell lines suppressed cell invasion and proliferation through miR-7 targeting [91]. Similarly, elevated expression of ciRS-7 was determined in colorectal cancer tissues, when compared to those of the adjacent normal mucosa, which was positively associated with tumor size, T stage, lymph node metastasis, and poor overall survival [92]. Downregulation of ciRS-7 increased miR-7 expression and significantly suppressed colorectal cancer cell proliferation and invasion. As

TABLE 2: CircRNAs associated with cancer.

CircRNA	Gene symbol	Cancer type	Expression	Fold change[a]	Function	References
CiRS-7/CDR1as*	CDR1	HCC	Up	NA	Sponge: miR-7	[91]
		HCC*	Down[b]	NA	Sponge: miR-7	[137]
		Colorectal	Up	NA	Sponge: miR-7	[92]
		Colorectal*	Up	2.4	Sponge: miR-7	[132]
		GBM	Down	3.5	Target of miR-671-5p Association with cell proliferation and migration	[94]
Circ-Foxo3	FOXO3	Breast	Down	NA	Sponge: miR-22, miR-136, miR-138, miR-149, miR-433, miR-762, miR-3614-5p, miR-3622b-5p Association with apoptosis-related proteins Foxo3, MDM2, p53, and Puma and cell cycle proteins CDK2 and p21	[95–97]
Circ-ABCB10	ABCB10	Breast	Up	5.0–10.0	Sponge: miR-1271	[98]
Hsa_circ_001569	ABCC1	Colorectal	Up	NA	Sponge: miR-145	[99]
		HCC	Up	NA	Promoting tumor growth	[100]
CircHIPK3	HIPK3	HCC	Up	NA	Sponge: miR-124	[101]
		Bladder	Down	4.6	Sponge: miR-558	[103]
Cir-ITCH	ITCH	ESCC	Down	NA	Sponge: miR-7, miR-17, miR-214 Inhibition of the Wnt/β-catenin pathway	[104]
		Colorectal	Down	NA	Sponge: miR-7, miR-20a Inhibition of the Wnt/β-catenin pathway	[105]
		Lung	Down	NA	Sponge: miR-7, miR-214 Inhibition of the Wnt/β-catenin pathway	[106]
Hsa_circ_0067934	PRKCI	ESCC	Up	8.8	Promoting cell proliferation and migration	[107]
Circ-ZEB1.5 Circ-ZEB1.19 Circ-ZEB1.17 Circ-ZEB1.33	ZEB1	Lung	Down	NA	Sponge: miR-200a-3p	[58]
CircMYLK	MYLK	Bladder	Up	NA	Sponge: miR-29a-3p	[109]
CircRNA_100290	SLC30A7	OSCC	Up	6.9	Sponge: miR-29 family	[110]
Hsa_circ_0016347	KCNH1	Osteosarcoma	Up	NA	Sponge: miR-214	[111]
Hsa_circ_0001564	CANX	Osteosarcoma	Up	NA	Sponge: miR-29c-3p	[114]
CircRNA_100269	LPHN2	Gastric	Down	NA	Sponge: miR-630	[116]
Hsa_circ_0020397	DOCK1	Colorectal	Up	NA	Sponge: miR-138	[117]
Hsa_circ_0000069	STIL	Colorectal	Up	≥1.0	Promoting cell proliferation, invasion, and migration	[118]
Circ-TTBK2	TTBK2	Glioma	Up	NA	Sponge: miR-217	[121]
cZNF292	ZNF292	Glioma	Up	NA	Promoting cell proliferation and tube formation	[122]
f-circRNA	PML/ RARα[c]	APL	Up	NA	Promoting cell proliferation, transformation, and tumorigenesis	[123]
CircTCF25*	TCF25	Bladder	Up	21.4	Sponge: miR-103a-3p, miR-107	[129]
Hsa_circ_001988*	FBXW7	Colorectal	Down	NA	ND	[130]
Hsa_circRNA_103809*	ZFR	Colorectal	Down	3.6	Sponge: miR-511-5p, miR-130b-5p, miR-642a-5p, miR-532-3p, miR-329-5p	[131]
Hsa_circRNA_104700*	PTK2	Colorectal	Down	4.2	Sponge: miR-141-5p, miR-500a-5p, miR-509-3p, miR-619-3p, miR-578	[131]
CircRNA_100876*	RNF121	NSCLC	Up	1.2	Sponge: miR-136	[133]

TABLE 2: Continued.

CircRNA	Gene symbol	Cancer type	Expression	Fold change[a]	Function	References
Hsa_circ_0001649*	SHPRH	HCC*	Down	NA	Sponge: miR-1283, miR-4310, miR-182-3p, miR-888-3p, miR-4502, miR-6811, miR-6511b-5p, miR-1972 Promoting metastasis	[127]
		Gastric*	Down	NA	ND	[141]
Hsa_circ_0005075*	EIF4G3	HCC	Up	NA	Sponge: miR-23b-5p, miR-93-3p, miR-581, miR-23a-5p Promoting cell adhesion	[128]
CircZKSCAN1*	ZKSCAN1	HCC	Down	NA	Inhibition of cellular growth, migration, and invasion	[138]
Hsa_circ_0005986	PRDM2	HCC	Down	2.9	Sponge: miR-129-5p	[139]
Hsa_circ_0004018*	SMYD4	HCC	Down	NA	Sponge: miR-30e-5p, miR-647, miR-92a-5p, miR-660-3p, miR-626	[140]
Hsa_circ_002059*	KIAA0907	Gastric	Down	NA	ND	[16]
Hsa_circ_0000096*	HIAT1	Gastric	Down	NA	Sponge: miR-224, miR-200a Inhibition of cell growth and migration	[142]
Hsa_circ_0001895*	PRRC2B	Gastric	Down	NA	ND	[143]
Hsa_circ_0006633*	FGGY	Gastric	Down	NA	ND	[144]
Hsa_circ_0000190*	CNIH4	Gastric	Down	NA	ND	[145]
Hsa_circ_0003159*	CACNA2D1	Gastric	Down	NA	ND	[146]
CircPVT1*	PVT1	Gastric	Up	NA	Sponge: miR-125 family Promoting cell proliferation	[148]
Hsa_circ_100855*	C11orf80	LSCC	Up	10.5	ND	[150]
Hsa_circ_104912*	DENND1A	LSCC	Down	4.7	ND	[150]

HCC: hepatocellular carcinoma; GBM: glioblastoma multiforme; ESCC: esophageal squamous cell carcinoma; OSCC: oral squamous cell carcinoma; APL: acute promyelocytic leukemia; NSCLC: non-small-cell lung cancer; LSCC: laryngeal squamous cell cancer. Up: upregulated; down: downregulated. NA: not available (data is presented in a graphical format in the original report). ND: not determined. *Potential cancer biomarker. [a]Fold change values, relative to normal controls. [b]Expression levels were not statistically significant. [c]One or more f-circRNAs were generated from PML/RARα fusion gene, a product of the most recurrent cancer-associated aberrant chromosomal translocation in APL. In addition, other chromosomal translocations may also generate f-circRNAs.

demonstrated, ciRS-7 blocked miR-7 activity and positively regulated the expression of EGFR and IGF-1R oncogenes, indicating that the ciRS-7/miR-7 axis was associated with colorectal cancer progression. However, ciRS-7 may also regulate colorectal cancer progression through other mechanisms than as a miR-7 sponge [92]. Conversely, miR-7 overexpression has also been associated with upregulated oncogenes in several tumor cell lines and advanced colorectal cancer tissues, when compared to healthy controls [93]. This suggests that miR-671-mediated degradation of ciRS-7 may diminish the ciRS-7-mediated miR-7 inhibition and enhance miR-7 levels in tumor cells [83]. As a consequence, such miR-671 action may contribute to the increase in downstream target oncogenes (e.g., EGFR and XIAP) and promote vascularization, metastasis, and amplification of tumor cells [83]. It was also demonstrated that overexpression of miR-671-5p in glioblastoma multiforme (GBM) biopsies and cell lines increased the migration and proliferation rates of GBM cells [94]. Furthermore, overexpression of miR-671-5p negatively correlated with the expression of ciRS-7, CDR1, and VSNL1, which implied that the miR-671-5p/CDR1as/CDR1/VSNL1 axis was functionally altered in GBM [94].

The tumor suppressor gene FOXO3 encodes two ncRNAs, the pseudogene Foxo3P and circ-Foxo3, both of which may act as miRNA sponges. Foxo3P and circ-Foxo3 were highly expressed in noncancerous cells and could function as miRNA sponges for several cancer-associated miRNAs, including miR-22, miR-136, miR-138, miR-149, miR-433, miR-762, miR-3614-5p, and miR-3622b-5p. However, among the two, circ-Foxo3 appeared to possess a stronger sponging effect on these miRNAs [95]. In human breast cancer cell lines, Foxo3P and circ-Foxo3 promoted the translation of Foxo3 mRNA by binding regulatory miRNAs and increased Foxo3-mediated apoptosis. Also, mouse xenograft models for breast cancer showed arrested tumor growth in the presence of circ-Foxo3 due to apoptosis, induced through combining circ-Foxo3, Foxo3P, and Foxo3 activity, when compared to controls [95]. Ectopic expression of circ-Foxo3 can result in the formation of the circ-Foxo3-p21-CDK2 ternary complex, arising from binding circ-Foxo3 to cell cycle proteins CDK2 and p21. The circ-Foxo3-p21-CDK2 ternary complex can suppress cell cycle progression and inhibit tumor growth [96]. In addition, expression of circ-Foxo3 was significantly increased during breast cancer cell apoptosis, where circ-Foxo3 effectively bound to p53 and MDM2 proteins. As demonstrated, elevated expression of circ-Foxo3 increased Foxo3 protein levels but repressed p53 activity by promoting MDM2-induced p53 ubiquitination and subsequent degradation. Overexpression of circ-Foxo3 decreased the interaction between Foxo3 and MDM2, increased Foxo3 activity, and promoted cell apoptosis, through upregulating Puma expression [97]. Beside circ-Foxo3, circ-ABCB10 is another circRNA associated with breast cancer. Circ-ABCB10 was significantly upregulated in breast cancer tissues, and its function as a sponge for miR-1271 has been determined. Furthermore, in vitro circ-ABCB10 knockdown suppressed proliferation and increased apoptosis of breast cancer cells [98].

Hsa_circ_001569 was significantly overexpressed in colorectal cancer tissues and was positively correlated with the degree of clinical features (TNM stage) [99]. Hsa_circ_001569 may act as a miR-145 sponge and represses the transcriptional activities of miR-145, enabling upregulation of miR-145 target genes E2F5, BAG4, and FMNL2. Thus, it acts as a positive regulator in cell proliferation and invasion of colorectal cancer [99]. Elevated expression levels of hsa_circ_001569 were also determined in HCC tissues, when compared to adjacent normal tissues. As in colorectal cancer, expression of hsa_circ_001569 correlated with tumor differentiation and TNM stages in HCC. The inhibitory effect on HCC cell proliferation and tumor growth through hsa_circ_001569 silencing was also demonstrated [100].

Among many dysregulated circRNAs in several cancer types, a significant upregulation of circHIPK3 in HCC has been demonstrated, when compared with its expression in matched normal tissues [101]. CircHIPK3 could bind to 9 miRNAs with its 18 potential binding sites, including the tumor-suppressive miR-124, thus inhibiting its activity. Furthermore, circHIPK3 silencing significantly inhibited human cancer cell proliferation [101]. By inhibiting miR-124 activity, circHIPK3 might influence the proliferation of tumor cells in prostate cancer through several oncogenes, including iASPP [102]. CircHIPK3 was also significantly downregulated in bladder cancer tissues and cell lines, where it negatively correlated with cancer grade, invasion, and lymph node metastasis [103]. Mechanistic studies revealed that circHIPK3 abundantly sponged miR-558 and suppressed heparanase (HPSE) expression, which is involved in regulating tumor invasion and metastasis. In addition, overexpression of circHIPK3 effectively inhibited migration, invasion, and angiogenesis of bladder cancer cells in vitro and suppressed bladder cancer growth and metastasis in vivo, mainly through targeting the miR-558/heparanase axis [103].

CircRNA cir-ITCH may act as a miRNA sponge for cancer-associated miR-7, miR-17, and miR-214 in esophageal squamous cell carcinoma (ESCC) [104] and miR-7 and miR-20a in colorectal cancer [105] and as a sponge for miR-7 and miR-214 in lung cancer [106]. Cir-ITCH expression was downregulated in ESCC, colorectal, and lung cancer tissues, when compared to adjacent peritumoral tissues. In all three cancer types, cir-ITCH activity could increase the level of ITCH protein, a regulator of several tumor-associated proteins, which is involved in the inhibition of the Wnt/β-catenin signaling pathway. Therefore, cir-ITCH likely plays an inhibitory role in ESCC and colorectal and lung cancer, through promoting ITCH-mediated ubiquitination and subsequent proteasome-mediated degradation of phosphorylated Dvl2 scaffold protein, which impairs the canonical Wnt/β-catenin signaling [104–106].

Involvement of another circRNA in promoting ESCC has been recently demonstrated. Hsa_circ_0067934 was significantly overexpressed in ESCC tissues, when compared to adjacent healthy tissues, and its expression positively correlated with tumor differentiation, T stage, and TNM stage [107]. In vitro studies revealed that hsa_circ_0067934 promoted ESCC cell proliferation, and its presence in the cytoplasm suggested that hsa_circ_0067934 was involved in posttranscriptional regulation of the ESCC cell cycle [107]. However, the exact molecular function of hsa_circ_0067934 still needs to be determined. Nevertheless, it would be interesting to assess the possibility of correlation between hsa_circ_0067934 and cir-ITCH in the development of ESCC, due to their apparent contrary modes of action.

Identification of four circRNAs associated with lung cancer has been performed, based on computational predictions utilizing transcriptome sequencing datasets, by using the CircNet database. As demonstrated, circRNAs circ-ZEB1.5, circ-ZEB1.19, circ-ZEB1.17, and circ-ZEB1.33 were upregulated in normal lung tissues, when compared to lung cancer samples, and are presumably implicated in lung cancer suppression by binding to miR-200a-3p [58], which has been reported to target ZEB1 and to promote cancer initiation [108]. Similarly, bioinformatics approaches utilizing correlated coexpression networks of bladder cancer revealed a probable interaction between lncRNA H19 and circRNA circMYLK, demonstrating their ability to competitively bind to miR-29a-3p. Such miR-29a-3p targeting might increase the expression of DNMT3B, VEGFA, and ITGB1 oncogenes, which suggested a possible involvement of H19 and circMYLK in the development, growth, and metastasis of bladder cancer [109].

CircRNA_100290 was upregulated and coexpressed with CDK6, a member of the cyclin-dependent kinase family, in oral squamous cell carcinoma (OSCC) tissues, when matched with noncancerous tissue samples [110]. CircRNA_100290 could directly bind to miR-29 family members, including miR-29a, miR-29b, and miR-29c. Since CDK6 has been determined as the direct target of miR-29b, circRNA_100290 evidently regulates CDK6 expression through sponging miR-29. Furthermore, knockdown of circRNA_100290 decreased the expression of CDK6, induced G1/S arrest, inhibited proliferation of OSCC cell lines in vitro, and decreased the growth of tumors in vivo. Thus, circRNA_100290 likely functions as a regulator of cell cycle and cell proliferation [110].

Involvement of two circRNAs in regulating osteosarcoma has been demonstrated recently. However, no correlation between the two in promoting osteosarcoma has been determined yet. Hsa_circ_0016347 was significantly upregulated in osteosarcoma tissues and cell lines, when compared to adjacent nontumor tissues and normal osteoblasts, and has been shown to sponge miR-214 [111], which is a known tumor promoter in osteosarcoma [112, 113]. By inhibiting miR-214 activity, hsa_circ_0016347 increased the expression level of caspase-1, a direct target of miR-214, thus enabling the formation of favorable tumor microenvironment and promoting proliferation, invasion, and metastasis of osteosarcoma cells [111]. Also, overexpression of hsa_circ_0016347 increased either the size or the number of pulmonary metastasis tumors [111]. In addition to hsa_circ_0016347, a significant upregulation of hsa_circ_0001564 has been determined in osteosarcoma tissues and cell lines, which acted as a miR-29c-3p sponge [114]. Through inhibiting miR-29c-3p activity, hsa_circ_0001564 promoted tumorigenesis of osteosarcoma by regulating cell cycle and proliferation of osteosarcoma cells [114].

As demonstrated before, circRNA_100269 has been included in a group of circRNAs constituting the four-circRNA-based classifier, which was used to predict the early recurrence of stage III gastric cancer after radical surgery [115]. Further analysis has revealed a significantly downregulated level of circRNA_100269 in gastric cancer tissues, than in the corresponding adjacent healthy tissues, which correlated with histological subtypes and the node invasion number [116]. The study suggested that circRNA_100269 inhibits gastric cancer cell proliferation via inhibiting miR-630 activity, whose expression was negatively correlated with that of circRNA_100269 [116]. However, no confirmation of such circRNA_100269 action has been performed in vivo.

A negative correlation between expression profiles of miR-138 and hsa_circ_0020397 was determined in colorectal cancer cells, where hsa_circ_0020397 was significantly upregulated [117]. As determined, hsa_circ_0020397 acted as a miR-138 sponge and promoted the expression of miR-138 targets TERT and PD-L1, which promoted viability and invasion of colorectal cancer cells and inhibited their apoptosis [117]. In addition, circRNA hsa_circ_0000069 was also associated with colorectal cancer and was significantly upregulated in colorectal cancer tissues and cell lines, when compared to healthy controls [118]. Elevated expression of hsa_circ_0000069 correlated with the tumor TNM stage and could promote colorectal cancer cell proliferation, invasion, and migration in vitro [118]. However, a more detailed mechanism of hsa_circ_0000069 function still needs to be determined.

miR-217 is a tumor-suppressive miRNA, associated with various cancer types, including epithelial ovarian cancer [119] and gastric cancer [120]. In glioma, miR-217 negatively correlated with the pathological grades of tumors and exerted tumor-suppressive activity in glioma cells [121]. CircRNA circ-TTBK2 was significantly upregulated in glioma tissues and cell lines and acted as a miR-217 sponge. By sequestering miR-217 activity, circ-TTBK2 enabled higher expression of oncogenic proteins HNF1β and Derlin-1, which promoted cell proliferation, migration, and invasion, while inhibiting apoptosis of glioma cells [121]. As demonstrated, miR-217 expression was negatively regulated by circ-TTBK2 expression in an AGO2-dependent manner and there was a reciprocal repression feedback loop between circ-TTBK2 and miR-217 [121]. In addition, circ-TTBK2 knockdown in combination with miR-217 overexpression led to tumor regression in vivo [121].

In addition to circRNAs that predominantly act like miRNA sponges, few circRNAs appear not to function in such a manner (Table 2). It has been demonstrated that downregulation of cZNF292 suppresses human glioma tube formation via the Wnt/β-catenin signaling pathway. Thus, cZNF292 downregulation also resulted in inhibition of glioma cell proliferation and cell cycle progression [122]. However, the exact mechanism of cZNF292 activity still needs to be determined. Furthermore, chromosomal translocations may give rise to oncogenic fusion proteins that are often involved in the onset and progression of various cancers. Such cancer-associated chromosomal translocations may also result in the formation of fusion circRNAs (f-circRNAs), which are produced from transcribed exons of genes affected by these oncogenic translocations [123]. Among several distinctive chromosomal translocations in leukemia, PML/RARα is the most frequent translocation in acute promyelocytic leukemia (APL) [124], which can generate one or more f-circRNAs from this fusion gene [123]. In addition, MLL/AF9 aberrant translocation also generated several f-circRNAs in APL [123]. As demonstrated, f-circRNAs in combination with other oncogenic stimuli, including oncogenic fusion proteins, played an important role in promoting APL cell proliferation, transformation, and tumorigenesis progression in vivo [123]. In addition to APL, f-circRNAs have also been identified in SK-NEP-1 sarcoma and H3122 lung cancer cell lines [123]. Thus, this study strongly implied that f-circRNAs may have a potential diagnostic and therapeutic value in cancer.

In addition to the above-listed circRNAs, several research groups have identified a vast assortment of circRNAs, by using RNA-seq and other next-generation sequencing techniques that are likely involved in mechanisms which promote various cancers. The majority of the generated data can be obtained from several circRNA databases, which are presented in Table 1. Despite candidate cancer-specific circRNAs are getting discovered on a regular basis, the data currently remains insufficient to definitely associate individual circRNAs with a specific mechanism promoting a certain cancer type. However, it has recently become clear that circRNAs may represent promising biomarkers for various cancer types.

6. CircRNAs as Cancer Biomarkers

CircRNAs are abundant and highly stable molecules, exhibiting high cell/tissue and developmental stage specificity [9, 11]. The unique circular structure makes circRNAs insensitive to ribonucleases and enables them to exist intact in various tissues and body fluids. It has been shown that circRNAs may be stably expressed and present in relatively high quantities in human blood [125], saliva [126], and exosomes [33]. These characteristics make circRNAs ideal candidates as noninvasive biomarkers for cancer diagnosis, prognosis, and treatment. In addition, some circRNAs may correlate with age, gender, TNM stage, metastasis, and tumor size as it was determined for gastric cancer [16], HCC [127, 128], and colorectal cancer [99], additionally implying their suitability as cancer biomarkers. A number of circRNAs that have been associated with human cancer are presented in Table 2. From these circRNAs, several have been tested for their diagnostic performance and may eventually become novel biomarkers for cancer diagnosis in clinical practice.

6.1. CircRNAs as Biomarkers for Bladder Cancer.

CircRNA circTCF25 was found to be highly expressed in bladder cancer tissues, when compared to healthy controls. The analysis was performed by using a total of 40 paired snap-frozen bladder carcinoma and matched paracarcinoma tissue samples. The study showed that circTCF25 promotes proliferation and metastasis of urinary bladder carcinoma by acting as a sponge for miR-103a-3p and miR-107, which resulted in upregulated CDK6 expression [129]. The data also suggested that circTCF25 may be a new promising biomarker for bladder cancer [129]. However, the diagnostic performance of this circRNA still needs to be determined.

6.2. CircRNAs as Biomarkers for Colorectal Cancer.

CircRNA hsa_circ_001988 has been identified in colorectal cancer and has been significantly downregulated in colorectal cancer tissues, when compared to those of the matched normal mucosa ($n = 31$) [130]. Evaluation of the diagnostic performance of hsa_circ_001988 has shown its sensitivity of 68.0% and specificity of 73.0%. The receiver operating characteristic curve (ROC) analysis showed an area under the ROC curve (AUC) of 0.788, indicating that hsa_circ_001988 may become a novel potential biomarker in the diagnosis of colorectal cancer [130].

CircRNAs hsa_circRNA_103809 and hsa_circRNA_104700 were also significantly downregulated in colorectal cancer tissues, where hsa_circRNA_103809 correlated with lymph node metastasis and TNM stage and hsa_circRNA_104700 with distal metastasis [131]. Analysis of both circRNAs was performed on 170 paired colorectal cancer tissues and matched adjacent noncancerous tissue samples. The evaluated diagnostic performances for hsa_circRNA_103809 (AUC 0.699) and hsa_circRNA_104700 (AUC 0.616) indicated that both circRNAs may serve as reliable biomarkers for colorectal cancer [131]. However, beside their dysregulation and putative miRNA binding site determination [131], the exact mechanisms of function for both circRNAs in colorectal cancer development have not been elucidated yet.

Despite its assumed involvement in promoting various cancer types, mainly due to its ability to sponge miR-7, the clinical significance of ciRS-7 in colorectal cancer was only recently demonstrated. CiRS-7 was significantly upregulated in tumor tissues of colorectal cancer patients and correlated with advanced tumor stage, tumor depth, and metastasis [132]. The study included a training cohort comprised of 153 primary colorectal cancer tissues and 44 matched normal mucosa tissues and an additional independent validation cohort ($n = 165$). Correlation of upregulated ciRS-7 expression levels with poor patient survival strongly suggested that ciRS-7 might serve as a novel prognostic biomarker in colorectal cancer patients [132]. *In vitro* experiments revealed that ciRS-7 inhibited miR-7 activity and activated the EGFR/RAF1/MAPK pathway, which linked ciRS-7 activity with colorectal cancer progression and aggressiveness [132]. Regarding the data obtained from the study, ciRS-7 suppression could increase the expression levels of miR-7 and reduce EGFR-RAF1 activity. Thus, therapeutic targeting of ciRS-7 might represent a potential treatment option for patients with colorectal cancer [132].

In addition, elevated expression levels of circRNA circ-KLDHC10 in serum samples of colorectal cancer patients were determined, when compared to those in healthy controls ($n = 11$ for both sample groups). Since circ-KLDHC10 was abundant in exosomes, it has the potential to serve as a novel circulating biomarker for colorectal cancer [33]. However, its oncogenic activity and diagnostic performance in colorectal cancer still need to be determined.

6.3. CircRNAs as Biomarkers for Non-Small-Cell Lung Cancer (NSCLC).

CircRNA_100876 was significantly upregulated in NSCLC tissues, when compared to their paired adjacent nontumorous tissues ($n = 101$), and its elevated levels closely correlated with lymph node metastasis and advanced tumor staging in NSCLC [133]. As determined in a previous study, circRNA_100876 could regulate MMP-13 expression through inhibiting miR-136 activity and thus participated in chondrocyte extracellular matrix degradation [134]. Since MMP-13 is often overexpressed in lung cancer and can increase the risk of metastasis [135, 136], circRNA_100876 might be involved in tumor cell growth, progression, and metastasis in NSCLC, by regulating MMP-13 expression as a miRNA sponge [133]. The Kaplan-Meier survival analysis showed significantly shorter overall survival times of NSCLC patients with elevated circRNA_100876 expression levels, when compared to patients with low expression levels of circRNA_100876. Therefore, circRNA_100876 could be gradually used as a novel prognostic biomarker for NSCLC [133].

6.4. CircRNAs as Biomarkers for Hepatocellular Carcinoma (HCC).

Hsa_circ_0001649 was significantly downregulated in HCC tissues, when compared to paired adjacent healthy liver tissues ($n = 89$), and its expression levels correlated with tumor size and the occurrence of tumor embolus in HCC [127]. Hsa_circ_0001649 may play a role in tumorigenesis and metastasis of HCC through sponge-like activity toward several miRNAs, including miR-1283, miR-4310, miR-182-3p, miR-888-3p, miR-4502, miR-6811, miR-6511b-5p, and miR-1972 [127]. The evaluated diagnostic performance (sensitivity 81.0%; specificity 69.0%; and AUC 0.63) indicated that hsa_circ_0001649 might serve as a novel potential biomarker for HCC, with relatively high degrees of accuracy, specificity, and sensitivity [127].

Another circRNA associated with HCC is hsa_circ_0005075, which was significantly upregulated in HCC tissues, when compared to paired adjacent normal liver tissues ($n = 30$) [128]. Hsa_circ_0005075 correlated with tumor size and showed a great diagnostic potential with a sensitivity of 83.3%, specificity of 90.0%, and AUC of 0.94 [128]. In addition, the circRNA-miRNA-mRNA interaction network revealed that hsa_circ_0005075 could potentially interact with miR-23b-5p, miR-93-3p, miR-581, and miR-23a-5p. The study assumed that through its miRNA sponge-like activity, hsa_circ_0005075 may participate in regulating cell adhesion during HCC development, which is involved in cancer cell proliferation, invasion, and metastasis [128].

The relationship between ciRS-7 and clinical features of HCC was also demonstrated. In the study, ciRS-7 expression was upregulated in 39.8% ($n = 43$) and downregulated in 60.2% ($n = 65$) tissues of HCC patients, when compared to matched nontumor tissues ($n = 108$) [137]. Even though ciRS-7 expression was slightly higher in HCC tissues, the overall ciRS-7 expression levels were downregulated and not significantly different from those in healthy controls [137], which was in contrast with a previous study describing ciRS-7 involvement in HCC [91]. However, upregulated ciRS-7 expression significantly correlated with patient age, serum AFP levels, and hepatic microvascular invasion (MVI), which suggested ciRS-7 expression may be associated with deterioration and metastasis of HCC [137]. Also, ciRS-7 could promote MVI by inhibiting miR-7 and disrupting the PIK3CD/p70S6K/mTOR pathway [137]. The ROC curve analysis showed that ciRS-7 was related to MVI in HCC tissues with an AUC of 0.68, implying ciRS-7 expression level could predict MVI. Considering these results, the study indicated ciRS-7 may not be a key factor in HCC tumorigenesis, but rather a risk factor for MVI in HCC [137].

The zinc finger family gene *ZKSCAN1* can generate linear *ZKSCAN1* mRNA and circular circZKSCAN1 isoforms, both of which were associated with different regulatory roles in the development of HCC, mostly through inhibiting growth, migration, and invasion of HCC cells [138]. CircZKSCAN1 was significantly downregulated in HCC tissues, when compared to paired adjacent healthy tissues ($n = 102$), and its expression levels correlated with tumor numbers, cirrhosis, vascular invasion, MVI, and tumor grade [138]. The ROC analysis showed the AUC of circZKSCAN1 was 0.834 with a sensitivity of 82.2% and specificity of 72.4%, indicating circZKSCAN1 could be used as a biomarker to effectively differentiate cancerous tissues from adjacent noncancerous tissues in HCC [138].

In addition, two relatively recently identified tumor-suppressive circRNAs were associated with clinical characteristic of patients with HCC. Low expression levels of hsa_circ_0005986 correlated with chronic hepatitis B family history, tumor diameters, MIV, and Barcelona Clinic Liver Cancer staging system (BCLC) stage [139]. The analysis was performed on 81 paired HCC and matched nontumorous tissue samples. As determined, hsa_circ_0005986 regulated the HCC cell cycle and proliferation, by acting as a miR-129-5p sponge and through promoting *Notch1* gene expression [139]. However, despite the study suggested hsa_circ_0005986 could be used as a novel HCC biomarker, no diagnostic performance of this circRNA has been performed. Similar to hsa_circ_0005986, the decreased expression levels of hsa_circ_0004018 in HCC tissues correlated with AFP level, tumor diameters, differentiation, BCLC stage, and TNM stage, when compared to those in paired paratumorous tissues ($n = 102$) [140]. miRNA target prediction analysis revealed that hsa_circ_0004018 could sponge

miR-30e-5p, miR-647, miR-92a-5p, miR-660-3p, and miR-626, additionally implying its role in tumorigenesis of HCC [140]. The evaluated diagnostic performance (sensitivity 0.716; specificity 0.815; and AUC 0.848) along with its HCC stage-specific expression profile highlighted hsa_circ_0004018 as a suitable biomarker for HCC diagnosis, capable of distinguishing HCC tissues from healthy and chronic hepatitis tissues [140].

6.5. CircRNAs as Biomarkers for Gastric Cancer. Significantly downregulated expression profiles of hsa_circ_002059 have been determined in gastric cancer tissues, when compared to paired adjacent nontumor tissues ($n = 101$) [16]. In addition, hsa_circ_002059 levels in plasma were significantly different between 36 paired plasma samples from pre- and postoperative gastric cancer patients. Also, lower expression levels of hsa_circ_002059 were significantly correlated with a patient's distal metastasis, TNM stage, gender, and age. Evaluated diagnostic performance of hsa_circ_002059 has shown its sensitivity of 81.0%, specificity of 62.0%, and AUC of 0.73, indicating hsa_circ_002059 represents a potential stable biomarker for gastric cancer [16].

Beside its role as a potential biomarker in HCC, hsa_circ_0001649 has also been associated with diagnosis of gastric cancer. Hsa_circ_0001649 was significantly downregulated in gastric cancer tissues, when compared to their paired paracancerous histological normal tissues ($n = 76$), and its expression levels correlated with pathological differentiation [141]. Analysis of hsa_circ_0001649 serum expression levels between paired pre- and postoperative serum samples ($n = 20$) of gastric cancer patients showed that hsa_circ_0001649 was significantly upregulated in serum after surgery. Also, hsa_circ_0001649 expression levels were more significantly decreased in poor and undifferentiated tumors than in well-differentiated ones, indicating its potential negative correlation with gastric cancer pathological differentiation [141]. The estimated diagnostic value of hsa_circ_0001649 determined by the ROC analysis showed the AUC of 0.834, with a sensitivity and specificity of 0.711 and 0.816, respectively [141]. These results suggest hsa_circ_0001649 may become a novel noninvasive biomarker for early detection of primary gastric cancer.

Hsa_circ_0000096 is a tumor-suppressive circRNA that affects gastric cancer cell growth and migration through suppressing the expression levels of cell cycle-associated (cyclin D1, CDK6) and migration-associated (MMP-2, MMP-9) proteins. Furthermore, hsa_circ_0000096 may also interact with 17 types of miRNA, including miR-224 and miR-200a [142]. Hsa_circ_0000096 was significantly downregulated in gastric cancer tissues (compared to paired adjacent nontumorous tissues; $n = 101$), and cell lines and its aberrant expression correlated with invasion and TNM stage [142]. The ROC analysis showed the AUC of hsa_circ_0000096 was 0.82. Intriguingly, the AUC was increased to 0.91 when a combination of hsa_circ_0000096 and a previously described hsa_circ_002059 was used [142]. Thus, these results suggest that hsa_circ_0000096 could be used independently or in combination with hsa_circ_002059 for effective diagnosis of gastric cancer.

In addition, four circRNAs have been recently proposed as biomarkers for gastric cancer, all of which were statistically significantly downregulated in gastric cancer tissues, correlated with different clinical characteristics, and showed an excellent diagnostic potential with relatively high accuracy, specificity, and sensitivity [143–146]. Hsa_circ_0001895 expression levels were downregulated in 69.8% ($n = 67$) gastric cancer tissues, compared to paired adjacent normal tissues ($n = 96$), and significantly correlated with cell differentiation, Borrmann type, and tissue CEA expression. The evaluated diagnostic performance of hsa_circ_0001895 showed the AUC was up to 0.792 with a sensitivity and specificity of 67.8% and 85.7%, respectively. The optimal cutoff value was 9.53 [143]. Interestingly, better sensitivity and specificity were obtained with the use of hsa_circ_0001895, when compared to common gastric cancer biomarkers CEA, CA19-9, and CA72-4, which showed only 20.1–27.6% sensitivity individually or 48.2% when used in combination [147]. This indicates hsa_circ_0001895 may be effectively used for screening and predicting the prognosis of gastric cancer [143]. Downregulation of hsa_circ_0006633 was associated with cancer distal metastasis and tissue CEA levels. In the study, 96 paired gastric cancer tissues and their adjacent nontumorous tissues were used. The evaluated diagnostic performance (sensitivity 0.60; specificity 0.81; and AUC 0.741) and increased hsa_circ_0006633 levels in plasma samples suggested that this circRNA could be used as a novel noninvasive biomarker for screening gastric cancer [144]. Hsa_circ_0000190 levels in gastric cancer tissues were correlated with tumor diameter, lymphatic metastasis, distal metastasis, TNM stage, and CA19-9 levels. However, in plasma samples, hsa_circ_0000190 correlated only with CEA levels [145]. The analysis was performed by using 104 paired gastric cancer tissues and their adjacent nontumor tissues, 104 plasma samples from gastric cancer patients, and 104 plasma samples from healthy controls. The diagnostic potential of hsa_circ_0000190 was determined in tissue (sensitivity of 0.721, specificity of 0.683, and AUC of 0.75) and plasma (sensitivity of 0.414, specificity of 0.875, and AUC of 0.60) samples. When tissue and plasma hsa_circ_0000190 were combined, the AUC was increased to 0.775, with a sensitivity and specificity of 0.712 and 0.750, respectively [145]. Moreover, when compared to CEA and CA19-9, hsa_circ_0000190 had a much higher sensitivity and specificity in the screening of gastric cancer [145]. Low hsa_circ_0003159 tissue levels in gastric cancer patients correlated with gender, distal metastasis, and TNM stage. The analysis was performed on 108 paired gastric cancer tissues and adjacent nontumorous tissue samples. The determined sensitivity and specificity were 0.852 and 0.565, respectively. The AUC of hsa_circ_0003159 was 0.75, which indicated that hsa_circ_0003159 may be also used as a biomarker for the diagnosis of gastric cancer [146].

In contrast to the above-listed circRNAs, circPVT1 expression levels were significantly upregulated in gastric cancer cells and tissues, when compared to paired healthy controls ($n = 187$) [148]. As demonstrated, circPVT1 may promote gastric cancer proliferation through acting as a sponge toward the miR-125 family members [148]. The

Kaplan-Meier analysis confirmed that circPVT1 could serve as an independent prognostic biomarker for the overall survival and disease-free survival of patients with gastric cancer. Furthermore, the combined detection of expressed circPVT1 and its linear isoform from the *PVT1* oncogene enhanced the prognosis of patients with gastric cancer [148]. In addition, statistically significant upregulation of hsa_circ_0058246 was detected in tumor specimens of gastric cancer patients with poor clinical outcomes ($n = 43$). Also, patients who suffered recurrence of gastric cancer ($n = 12$) had a significant increase in hsa_circ_0058246 expression levels [149]. However, further studies are needed to convincingly demonstrate the suitability of hsa_circ_0058246 in diagnosis and prognosis of gastric cancer.

6.6. CircRNAs as Biomarkers for Laryngeal Squamous Cell Cancer (LSCC). Microarray and subsequent qRT-PCR analyses of laryngeal squamous cell cancer (LSCC) tissues have shown hsa_circ_100855 as the most upregulated and hsa_circ_104912 as the most downregulated circRNA in LSCC [150]. Hsa_circ_100855 levels were significantly higher in LSCC tissues and in patients with T3-4 stage, neck nodal metastasis, or advanced clinical stage. Conversely, hsa_circ_104912 levels were significantly lower in LSCC tissues than in corresponding adjacent nonneoplastic tissues. The study included 4 matched samples of LSCC tissues and corresponding adjacent nonneoplastic tissues for microarray analysis and 52 matched cancerous and noncancerous tissues for qRT-PCR analysis. Despite no diagnostic performance of either circRNAs has been determined, hsa_circ_100855 and hsa_circ_104912 may both serve as potential biomarkers and therapeutic targets for LSCC [150].

7. Conclusions and Perspectives

CircRNAs appear to be stably expressed in a cell/tissue-dependent and developmental stage-specific manner and have been shown to be dysregulated in different cancers. They are generally more stable than miRNAs and lncRNAs, several of which are currently recognized as relatively well-established biomarkers in cancer diagnosis. However, before circRNAs could be routinely used as effective biomarkers for early cancer diagnosis and prognosis, several important issues should be addressed, including the determination of their diagnostic performances for specific cancer types. As demonstrated above, several circRNAs have shown satisfactory diagnostic performances in distinguishing tumor from healthy tissues and between specific cancer types. However, suitable circRNAs as independent cancer biomarkers have not been identified yet. CircRNAs could be used in combination with RNA-based and other conventional cancer biomarkers, such as CEA, CA125, CA153, PSA, and AFP, for more specific cancer diagnosis and accurate cancer prognosis. In addition, their stability and abundance in exosomes suggest that circRNAs may represent a new class of exosome-based noninvasive cancer biomarkers. CircRNAs may also have a potential in targeted cancer treatment, where they could be utilized as sponges to bind to aberrantly expressed regulatory RNAs and proteins (e.g., RBPs), thus diminishing their oncogenic activity. However, to achieve this, further insights into circRNA's molecular mechanisms and functions in circRNA-mediated diseases are a prerequisite, before such treatments may become applicable. Also, identification of dysregulated circRNAs in other body fluids, such as urine and cerebrospinal fluid, may be beneficial for noninvasive cancer diagnosis. To conclude, circRNAs represent promising novel biomarkers for various cancer types and have a great potential to be effectively used in clinical practice in the near future.

Conflicts of Interest

The authors declare that they have no competing interests.

References

[1] J. Sana, P. Faltejskova, M. Svoboda, and O. Slaby, "Novel classes of non-coding RNAs and cancer," *Journal of Translational Medicine*, vol. 10, no. 1, p. 103, 2012.

[2] G. St Laurent, C. Wahlestedt, and P. Kapranov, "The landscape of long non-coding RNA classification," *Trends in Genetics*, vol. 31, no. 5, pp. 239–251, 2015.

[3] J. Salzman, C. Gawad, P. L. Wang, N. Lacayo, and P. O. Brown, "Circular RNAs are the predominant transcript isoform from hundreds of human genes in diverse cell types," *PLoS One*, vol. 7, no. 2, article e30733, 2012.

[4] S. P. Barrett and J. Salzman, "Circular RNAs: analysis, expression and potential functions," *Development*, vol. 143, no. 11, pp. 1838–1847, 2016.

[5] E. Lasda and R. Parker, "Circular RNAs: diversity of form and function," *RNA*, vol. 20, no. 12, pp. 1829–1842, 2014.

[6] S. P. Barrett, P. L. Wang, and J. Salzman, "Circular RNA biogenesis can proceed through an exon-containing lariat precursor," *eLife*, vol. 4, article e07540, 2015.

[7] C. Schindewolf, S. Braun, and H. Domdey, "*In vitro* generation of a circular exon from a linear pre-mRNA transcript," *Nucleic Acids Research*, vol. 24, no. 7, pp. 1260–1266, 1996.

[8] S. Starke, I. Jost, O. Rossbach et al., "Exon circularization requires canonical splice signals," *Cell Reports*, vol. 10, no. 1, pp. 103–111, 2015.

[9] S. Memczak, M. Jens, A. Elefsinioti et al., "Circular RNAs are a large class of animal RNAs with regulatory potency," *Nature*, vol. 495, no. 7441, pp. 333–338, 2013.

[10] P. L. Wang, Y. Bao, M. C. Yee et al., "Circular RNA is expressed across the eukaryotic tree of life," *PLoS One*, vol. 9, no. 3, article e90859, 2014.

[11] J. Salzman, R. E. Chen, M. N. Olsen, P. L. Wang, and P. O. Brown, "Cell-type specific features of circular RNA expression," *PLoS Genetics*, vol. 9, no. 9, article e1003777, 2013.

[12] W. R. Jeck, J. A. Sorrentino, K. Wang et al., "Circular RNAs are abundant, conserved, and associated with ALU repeats," *RNA*, vol. 19, no. 2, pp. 141–157, 2013.

[13] C. E. Burd, W. R. Jeck, Y. Liu, H. K. Sanoff, Z. Wang, and N. E. Sharpless, "Expression of linear and novel circular forms of an *INK4/ARF*-associated non-coding RNA correlates with atherosclerosis risk," *PLoS Genetics*, vol. 6, no. 12, article e1001233, 2010.

[14] A. Bachmayr-Heyda, A. T. Reiner, K. Auer et al., "Correlation of circular RNA abundance with proliferation – exemplified with colorectal and ovarian cancer, idiopathic lung fibrosis,

[15] W. J. Lukiw, "Circular RNA (circRNA) in Alzheimer's disease (AD)," *Frontiers in Genetics*, vol. 4, p. 307, 2013.

[16] P. Li, S. Chen, H. Chen et al., "Using circular RNA as a novel type of biomarker in the screening of gastric cancer," *Clinica Chimica Acta*, vol. 444, pp. 132–136, 2015.

[17] Y. Zhang, X. O. Zhang, T. Chen et al., "Circular intronic long noncoding RNAs," *Molecular Cell*, vol. 51, no. 6, pp. 792–806, 2013.

[18] Y. Wang and Z. Wang, "Efficient backsplicing produces translatable circular mRNAs," *RNA*, vol. 21, no. 2, pp. 172–179, 2015.

[19] X. O. Zhang, R. Dong, Y. Zhang et al., "Diverse alternative back-splicing and alternative splicing landscape of circular RNAs," *Genome Research*, vol. 26, no. 9, pp. 1277–1287, 2016.

[20] Y. Dong, D. He, Z. Peng et al., "Circular RNAs in cancer: an emerging key player," *Journal of Hematology & Oncology*, vol. 10, no. 1, p. 2, 2017.

[21] Z. Li, C. Huang, C. Bao et al., "Exon-intron circular RNAs regulate transcription in the nucleus," *Nature Structural & Molecular Biology*, vol. 22, no. 3, pp. 256–264, 2015.

[22] W. R. Jeck and N. E. Sharpless, "Detecting and characterizing circular RNAs," *Nature Biotechnology*, vol. 32, no. 5, pp. 453–461, 2014.

[23] T. Kameyama, H. Suzuki, and A. Mayeda, "Re-splicing of mature mRNA in cancer cells promotes activation of distant weak alternative splice sites," *Nucleic Acids Research*, vol. 40, no. 16, pp. 7896–7906, 2012.

[24] X. O. Zhang, H. B. Wang, Y. Zhang, X. Lu, L. L. Chen, and L. Yang, "Complementary sequence-mediated exon circularization," *Cell*, vol. 159, no. 1, pp. 134–147, 2014.

[25] A. Ivanov, S. Memczak, E. Wyler et al., "Analysis of intron sequences reveals hallmarks of circular RNA biogenesis in animals," *Cell Reports*, vol. 10, no. 2, pp. 170–177, 2015.

[26] D. Liang and J. E. Wilusz, "Short intronic repeat sequences facilitate circular RNA production," *Genes & Development*, vol. 28, no. 20, pp. 2233–2247, 2014.

[27] S. J. Conn, K. A. Pillman, J. Toubia et al., "The RNA binding protein quaking regulates formation of circRNAs," *Cell*, vol. 160, no. 6, pp. 1125–1134, 2015.

[28] R. Ashwal-Fluss, M. Meyer, N. R. Pamudurti et al., "circRNA biogenesis competes with pre-mRNA splicing," *Molecular Cell*, vol. 56, no. 1, pp. 55–66, 2014.

[29] A. Rybak-Wolf, C. Stottmeister, P. Glazar et al., "Circular RNAs in the mammalian brain are highly abundant, conserved, and dynamically expressed," *Molecular Cell*, vol. 58, no. 5, pp. 870–885, 2015.

[30] G. J. Talhouarne and J. G. Gall, "Lariat intronic RNAs in the cytoplasm of *Xenopus tropicalis* oocytes," *RNA*, vol. 20, no. 9, pp. 1476–1487, 2014.

[31] H. Suzuki, Y. Zuo, J. Wang, M. Q. Zhang, A. Malhotra, and A. Mayeda, "Characterization of RNase R-digested cellular RNA source that consists of lariat and circular RNAs from pre-mRNA splicing," *Nucleic Acids Research*, vol. 34, no. 8, article e63, 2006.

[32] E. Lasda and R. Parker, "Circular RNAs co-precipitate with extracellular vesicles: a possible mechanism for circRNA clearance," *PLoS One*, vol. 11, no. 2, article e0148407, 2016.

and normal human tissues," *Scientific Reports*, vol. 5, no. 1, p. 8057, 2015.

[33] Y. Li, Q. Zheng, C. Bao et al., "Circular RNA is enriched and stable in exosomes: a promising biomarker for cancer diagnosis," *Cell Research*, vol. 25, no. 8, pp. 981–984, 2015.

[34] T. B. Hansen, T. I. Jensen, B. H. Clausen et al., "Natural RNA circles function as efficient microRNA sponges," *Nature*, vol. 495, no. 7441, pp. 384–388, 2013.

[35] M. W. Hentze and T. Preiss, "Circular RNAs: splicing's enigma variations," *The EMBO Journal*, vol. 32, no. 7, pp. 923–925, 2013.

[36] J. T. Granados-Riveron and G. Aquino-Jarquin, "The complexity of the translation ability of circRNAs," *Biochimica et Biophysica Acta (BBA) - Gene Regulatory Mechanisms*, vol. 1859, no. 10, pp. 1245–1251, 2016.

[37] N. R. Pamudurti, O. Bartok, M. Jens et al., "Translation of circRNAs," *Molecular Cell*, vol. 66, no. 1, pp. 9–21.e7, 2017.

[38] I. Legnini, G. Di Timoteo, F. Rossi et al., "Circ-ZNF609 is a circular RNA that can be translated and functions in myogenesis," *Molecular Cell*, vol. 66, no. 1, pp. 22–37.e9, 2017.

[39] Y. Yang, X. Fan, M. Mao et al., "Extensive translation of circular RNAs driven by N^6-methyladenosine," *Cell Research*, vol. 27, no. 5, pp. 626–641, 2017.

[40] D. P. Bartel, "MicroRNAs: target recognition and regulatory functions," *Cell*, vol. 136, no. 2, pp. 215–233, 2009.

[41] D. P. Bartel, "MicroRNAs: genomics, biogenesis, mechanism, and function," *Cell*, vol. 116, no. 2, pp. 281–297, 2004.

[42] T. B. Hansen, E. D. Wiklund, J. B. Bramsen et al., "miRNA-dependent gene silencing involving Ago2-mediated cleavage of a circular antisense RNA," *The EMBO Journal*, vol. 30, no. 21, pp. 4414–4422, 2011.

[43] B. Capel, A. Swain, S. Nicolis et al., "Circular transcripts of the testis-determining gene *Sry* in adult mouse testis," *Cell*, vol. 73, no. 5, pp. 1019–1030, 1993.

[44] R. A. Dubin, M. A. Kazmi, and H. Ostrer, "Inverted repeats are necessary for circularization of the mouse testis *Sry* transcript," *Gene*, vol. 167, no. 1-2, pp. 245–248, 1995.

[45] L. Peng, G. Chen, Z. Zhu et al., "Circular RNA ZNF609 functions as a competitive endogenous RNA to regulate AKT3 expression by sponging miR-150-5p in Hirschsprung's disease," *Oncotarget*, vol. 8, no. 1, pp. 808–818, 2017.

[46] K. Wang, B. Long, F. Liu et al., "A circular RNA protects the heart from pathological hypertrophy and heart failure by targeting miR-223," *European Heart Journal*, vol. 37, no. 33, pp. 2602–2611, 2016.

[47] J. U. Guo, V. Agarwal, H. Guo, and D. P. Bartel, "Expanded identification and characterization of mammalian circular RNAs," *Genome Biology*, vol. 15, no. 7, p. 409, 2014.

[48] M. S. Ebert, J. R. Neilson, and P. A. Sharp, "MicroRNA sponges: competitive inhibitors of small RNAs in mammalian cells," *Nature Methods*, vol. 4, no. 9, pp. 721–726, 2007.

[49] F. C. Tay, J. K. Lim, H. Zhu, L. C. Hin, and S. Wang, "Using artificial microRNA sponges to achieve microRNA loss-of-function in cancer cells," *Advanced Drug Delivery Reviews*, vol. 81, pp. 117–127, 2015.

[50] X. You, I. Vlatkovic, A. Babic et al., "Neural circular RNAs are derived from synaptic genes and regulated by development and plasticity," *Nature Neuroscience*, vol. 18, no. 4, pp. 603–610, 2015.

[51] D. B. Dudekula, A. C. Panda, I. Grammatikakis, S. De, K. Abdelmohsen, and M. Gorospe, "CircInteractome: a web tool for exploring circular RNAs and their interacting

proteins and microRNAs," *RNA Biology*, vol. 13, no. 1, pp. 34–42, 2016.

[52] J. L. Rinn and H. Y. Chang, "Genome regulation by long noncoding RNAs," *Annual Review of Biochemistry*, vol. 81, no. 1, pp. 145–166, 2012.

[53] T. R. Mercer, M. E. Dinger, and J. S. Mattick, "Long noncoding RNAs: insights into functions," *Nature Reviews Genetics*, vol. 10, no. 3, pp. 155–159, 2009.

[54] R. Perriman and M. Ares Jr., "Circular mRNA can direct translation of extremely long repeating-sequence proteins in vivo," *RNA*, vol. 4, no. 9, pp. 1047–1054, 1998.

[55] C. Y. Chen and P. Sarnow, "Initiation of protein synthesis by the eukaryotic translational apparatus on circular RNAs," *Science*, vol. 268, no. 5209, pp. 415–417, 1995.

[56] S. Ghosal, S. Das, R. Sen, P. Basak, and J. Chakrabarti, "Circ2Traits: a comprehensive database for circular RNA potentially associated with disease and traits," *Frontiers in Genetics*, vol. 4, p. 283, 2013.

[57] P. Glazar, P. Papavasileiou, and N. Rajewsky, "circBase: a database for circular RNAs," *RNA*, vol. 20, no. 11, pp. 1666–1670, 2014.

[58] Y. C. Liu, J. R. Li, C. H. Sun et al., "CircNet: a database of circular RNAs derived from transcriptome sequencing data," *Nucleic Acids Research*, vol. 44, no. D1, pp. D209–D215, 2016.

[59] X. Chen, P. Han, T. Zhou, X. Guo, X. Song, and Y. Li, "circRNADb: a comprehensive database for human circular RNAs with protein-coding annotations," *Scientific Reports*, vol. 6, no. 1, article 34985, 2016.

[60] J. H. Li, S. Liu, H. Zhou, L. H. Qu, and J. H. Yang, "starBase v2.0: decoding miRNA-ceRNA, miRNA-ncRNA and protein-RNA interaction networks from large-scale CLIP-Seq data," *Nucleic Acids Research*, vol. 42, no. D1, pp. D92–D97, 2014.

[61] L. L. Zheng, J. H. Li, J. Wu et al., "deepBase v2.0: identification, expression, evolution and function of small RNAs, LncRNAs and circular RNAs from deep-sequencing data," *Nucleic Acids Research*, vol. 44, no. D1, pp. D196–D202, 2016.

[62] S. Xia, J. Feng, L. Lei et al., "Comprehensive characterization of tissue-specific circular RNAs in the human and mouse genomes," *Briefings in Bioinformatics*, article bbw081, 2016.

[63] X. Zeng, W. Lin, M. Guo, and Q. Zou, "A comprehensive overview and evaluation of circular RNA detection tools," *PLoS Computational Biology*, vol. 13, no. 6, article e1005420, 2017.

[64] J. O. Westholm, P. Miura, S. Olson et al., "Genome-wide analysis of *Drosophila* circular RNAs reveals their structural and sequence properties and age-dependent neural accumulation," *Cell Reports*, vol. 9, no. 5, pp. 1966–1980, 2014.

[65] J. Cheng, F. Metge, and C. Dieterich, "Specific identification and quantification of circular RNAs from sequencing data," *Bioinformatics*, vol. 32, no. 7, pp. 1094–1096, 2016.

[66] X. Song, N. Zhang, P. Han et al., "Circular RNA profile in gliomas revealed by identification tool UROBORUS," *Nucleic Acids Research*, vol. 44, no. 9, article e87, 2016.

[67] O. G. Izuogu, A. A. Alhasan, H. M. Alafghani, M. Santibanez-Koref, D. J. Elliott, and M. S. Jackson, "PTESFinder: a computational method to identify post-transcriptional exon shuffling (PTES) events," *BMC Bioinformatics*, vol. 17, no. 1, p. 31, 2016.

[68] L. Szabo, R. Morey, N. J. Palpant et al., "Statistically based splicing detection reveals neural enrichment and tissue-specific induction of circular RNA during human fetal development," *Genome Biology*, vol. 16, no. 1, p. 126, 2015.

[69] Y. Gao, J. Wang, and F. Zhao, "CIRI: an efficient and unbiased algorithm for *de novo* circular RNA identification," *Genome Biology*, vol. 16, no. 1, p. 4, 2015.

[70] K. Wang, D. Singh, Z. Zeng et al., "MapSplice: accurate mapping of RNA-seq reads for splice junction discovery," *Nucleic Acids Research*, vol. 38, no. 18, article e178, 2010.

[71] S. Hoffmann, C. Otto, G. Doose et al., "A multi-split mapping algorithm for circular RNA, splicing, trans-splicing and fusion detection," *Genome Biology*, vol. 15, no. 2, article R34, 2014.

[72] T. J. Chuang, C. S. Wu, C. Y. Chen, L. Y. Hung, T. W. Chiang, and M. Y. Yang, "NCLscan: accurate identification of non-co-linear transcripts (fusion, *trans*-splicing and circular RNA) with a good balance between sensitivity and precision," *Nucleic Acids Research*, vol. 44, no. 3, article e29, 2016.

[73] I. Chen, C. Y. Chen, and T. J. Chuang, "Biogenesis, identification, and function of exonic circular RNAs," *Wiley Interdisciplinary Reviews RNA*, vol. 6, no. 5, pp. 563–579, 2015.

[74] X. Meng, Q. Chen, P. Zhang, and M. Chen, "CircPro: an integrated tool for the identification of circRNAs with protein-coding potential," *Bioinformatics*, vol. 33, no. 20, pp. 3314–3316, 2017.

[75] M. Mokrejs, T. Masek, V. Vopalensky, P. Hlubucek, P. Delbos, and M. Pospísek, "IRESite—a tool for the examination of viral and cellular internal ribosome entry sites," *Nucleic Acids Research*, vol. 38, Supplement 1, pp. D131–D136, 2010.

[76] L. Wang, H. J. Park, S. Dasari, S. Wang, J. P. Kocher, and W. Li, "CPAT: coding-potential assessment tool using an alignment-free logistic regression model," *Nucleic Acids Research*, vol. 41, no. 6, article e74, 2013.

[77] R. D. Finn, P. Coggill, R. Y. Eberhardt et al., "The Pfam protein families database: towards a more sustainable future," *Nucleic Acids Research*, vol. 44, no. D1, pp. D279–D285, 2016.

[78] M. F. Lin, I. Jungreis, and M. Kellis, "PhyloCSF: a comparative genomics method to distinguish protein coding and non-coding regions," *Bioinformatics*, vol. 27, no. 13, pp. i275–i282, 2011.

[79] D. Kim and S. L. Salzberg, "TopHat-Fusion: an algorithm for discovery of novel fusion transcripts," *Genome Biology*, vol. 12, no. 8, article R72, 2011.

[80] Y. Wang, Y. Mo, Z. Gong et al., "Circular RNAs in human cancer," *Molecular Cancer*, vol. 16, no. 1, p. 25, 2017.

[81] R. J. Taft, K. C. Pang, T. R. Mercer, M. Dinger, and J. S. Mattick, "Non-coding RNAs: regulators of disease," *The Journal of Pathology*, vol. 220, no. 2, pp. 126–139, 2010.

[82] J. Zhao, Y. Tao, Y. Zhou et al., "MicroRNA-7: a promising new target in cancer therapy," *Cancer Cell International*, vol. 15, no. 1, p. 103, 2015.

[83] T. B. Hansen, J. Kjems, and C. K. Damgaard, "Circular RNA and miR-7 in cancer," *Cancer Research*, vol. 73, no. 18, pp. 5609–5612, 2013.

[84] K. M. Giles, R. A. M. Brown, M. R. Epis, F. C. Kalinowski, and P. J. Leedman, "miRNA-7-5p inhibits melanoma cell migration and invasion," *Biochemical and Biophysical Research Communications*, vol. 430, no. 2, pp. 706–710, 2013.

[85] X. Kong, G. Li, Y. Yuan et al., "MicroRNA-7 inhibits epithelial-to-mesenchymal transition and metastasis of breast cancer cells via targeting FAK expression," *PLoS One*, vol. 7, no. 8, article e41523, 2012.

[86] W. Wang, L. X. Dai, S. Zhang et al., "Regulation of epidermal growth factor receptor signaling by plasmid-based microRNA-7 inhibits human malignant gliomas growth and metastasis in vivo," *Neoplasma*, vol. 60, no. 03, pp. 274–283, 2013.

[87] X. Zhao, W. Dou, L. He et al., "MicroRNA-7 functions as an anti-metastatic microRNA in gastric cancer by targeting insulin-like growth factor-1 receptor," *Oncogene*, vol. 32, no. 11, pp. 1363–1372, 2013.

[88] Y. Fang, J. Xue, Q. Shen, J. Chen, and L. Tian, "MicroRNA-7 inhibits tumor growth and metastasis by targeting the phosphoinositide 3-kinase/Akt pathway in hepatocellular carcinoma," *Hepatology*, vol. 55, no. 6, pp. 1852–1862, 2012.

[89] S. Xiong, Y. Zheng, P. Jiang, R. Liu, X. Liu, and Y. Chu, "MicroRNA-7 inhibits the growth of human non-small cell lung cancer A549 cells through targeting BCL-2," *International Journal of Biological Sciences*, vol. 7, no. 6, pp. 805–814, 2011.

[90] N. Zhang, X. Li, C. W. Wu et al., "MicroRNA-7 is a novel inhibitor of YY1 contributing to colorectal tumorigenesis," *Oncogene*, vol. 32, no. 42, pp. 5078–5088, 2013.

[91] L. Yu, X. Gong, L. Sun, Q. Zhou, B. Lu, and L. Zhu, "The circular RNA Cdr1as act as an oncogene in hepatocellular carcinoma through targeting miR-7 expression," *PLoS One*, vol. 11, no. 7, article e0158347, 2016.

[92] W. Tang, M. Ji, G. He et al., "Silencing CDR1as inhibits colorectal cancer progression through regulating microRNA-7," *OncoTargets and Therapy*, vol. 10, pp. 2045–2056, 2017.

[93] Y. Nakagawa, Y. Akao, K. Taniguchi et al., "Relationship between expression of onco-related miRNAs and the endoscopic appearance of colorectal tumors," *International Journal of Molecular Sciences*, vol. 16, no. 1, pp. 1526–1543, 2015.

[94] D. Barbagallo, A. Condorelli, M. Ragusa et al., "Dysregulated miR-671-5p / CDR1-AS / CDR1 / VSNL1 axis is involved in glioblastoma multiforme," *Oncotarget*, vol. 7, no. 4, pp. 4746–4759, 2016.

[95] W. Yang, W. W. Du, X. Li, A. J. Yee, and B. B. Yang, "Foxo3 activity promoted by non-coding effects of circular RNA and Foxo3 pseudogene in the inhibition of tumor growth and angiogenesis," *Oncogene*, vol. 35, no. 30, pp. 3919–3931, 2016.

[96] W. W. Du, W. Yang, E. Liu, Z. Yang, P. Dhaliwal, and B. B. Yang, "Foxo3 circular RNA retards cell cycle progression via forming ternary complexes with p21 and CDK2," *Nucleic Acids Research*, vol. 44, no. 6, pp. 2846–2858, 2016.

[97] W. W. Du, L. Fang, W. Yang et al., "Induction of tumor apoptosis through a circular RNA enhancing Foxo3 activity," *Cell Death and Differentiation*, vol. 24, no. 2, pp. 357–370, 2017.

[98] H. F. Liang, "Circular RNA circ-ABCB10 promotes breast cancer proliferation and progression through sponging miR-1271," *American Journal of Cancer Research*, vol. 7, no. 7, pp. 1566–1576, 2017.

[99] H. Xie, X. Ren, S. Xin et al., "Emerging roles of circRNA_001569 targeting miR-145 in the proliferation and invasion of colorectal cancer," *Oncotarget*, vol. 7, no. 18, pp. 26680–26691, 2016.

[100] H. Jin, M. Fang, Z. Man, Y. Wang, and H. Liu, "Circular RNA 001569 acts as an oncogene and correlates with aggressive characteristics in hepatocellular carcinoma," *International Journal of Clinical and Experimental Pathology*, vol. 10, no. 3, pp. 2997–3005, 2017.

[101] Q. Zheng, C. Bao, W. Guo et al., "Circular RNA profiling reveals an abundant circHIPK3 that regulates cell growth by sponging multiple miRNAs," *Nature Communications*, vol. 7, 2016.

[102] J. Chen, H. Xiao, Z. Huang et al., "MicroRNA124 regulate cell growth of prostate cancer cells by targeting iASPP," *International Journal of Clinical and Experimental Pathology*, vol. 7, no. 5, pp. 2283–2290, 2014.

[103] Y. Li, F. Zheng, X. Xiao et al., "CircHIPK3 sponges miR-558 to suppress heparanase expression in bladder cancer cells," *EMBO Reports*, vol. 18, no. 9, pp. 1646–1659, 2017.

[104] F. Li, L. Zhang, W. Li et al., "Circular RNA ITCH has inhibitory effect on ESCC by suppressing the Wnt/β-catenin pathway," *Oncotarget*, vol. 6, no. 8, pp. 6001–6013, 2015.

[105] G. Huang, H. Zhu, Y. Shi, W. Wu, H. Cai, and X. Chen, "*cir-ITCH* plays an inhibitory role in colorectal cancer by regulating the Wnt/β-catenin pathway," *PLoS One*, vol. 10, no. 6, article e0131225, 2015.

[106] L. Wan, L. Zhang, K. Fan, Z. X. Cheng, Q. C. Sun, and J. J. Wang, "Circular RNA-ITCH suppresses lung cancer proliferation via inhibiting the Wnt/β-catenin pathway," *BioMed Research International*, vol. 2016, Article ID 1579490, 11 pages, 2016.

[107] W. Xia, M. Qiu, R. Chen et al., "Circular RNA has_circ_0067934 is upregulated in esophageal squamous cell carcinoma and promoted proliferation," *Scientific Reports*, vol. 6, no. 1, article 35576, 2016.

[108] S. D. Hsu, Y. T. Tseng, S. Shrestha et al., "miRTarBase update 2014: an information resource for experimentally validated miRNA-target interactions," *Nucleic Acids Research*, vol. 42, no. D1, pp. D78–D85, 2014.

[109] M. Huang, Z. Zhong, M. Lv, J. Shu, Q. Tian, and J. Chen, "Comprehensive analysis of differentially expressed profiles of lncRNAs and circRNAs with associated co-expression and ceRNA networks in bladder carcinoma," *Oncotarget*, vol. 7, no. 30, pp. 47186–47200, 2016.

[110] L. Chen, S. Zhang, J. Wu et al., "circRNA_100290 plays a role in oral cancer by functioning as a sponge of the miR-29 family," *Oncogene*, vol. 36, no. 32, pp. 4551–4561, 2017.

[111] H. Jin, X. Jin, H. Zhang, and W. Wang, "Circular RNA hsa-circ-0016347 promotes proliferation, invasion and metastasis of osteosarcoma cells," *Oncotarget*, vol. 8, no. 15, pp. 25571–25581, 2017.

[112] Z. Xu and T. Wang, "miR-214 promotes the proliferation and invasion of osteosarcoma cells through direct suppression of LZTS1," *Biochemical and Biophysical Research Communications*, vol. 449, no. 2, pp. 190–195, 2014.

[113] W. Allen-Rhoades, L. Kurenbekova, L. Satterfield et al., "Cross-species identification of a plasma microRNA signature for detection, therapeutic monitoring, and prognosis in osteosarcoma," *Cancer Medicine*, vol. 4, no. 7, pp. 977–988, 2015.

[114] J. F. Li and Y. Z. Song, "Circular RNA hsa_circ_0001564 facilitates tumorigenesis of osteosarcoma via sponging miR-29c-3p," *Tumour Biology*, vol. 39, no. 8, article 1010428317709989, 2017.

[115] Y. Zhang, J. Li, J. Yu et al., "Circular RNAs signature predicts the early recurrence of stage III gastric cancer after radical surgery," *Oncotarget*, vol. 8, no. 14, pp. 22936–22943, 2017.

[116] Y. Zhang, H. Liu, W. Li et al., "CircRNA_100269 is downregulated in gastric cancer and suppresses tumor cell growth by targeting miR-630," *Aging*, vol. 9, no. 6, pp. 1585–1594, 2017.

[117] X. L. Zhang, L. L. Xu, and F. Wang, "Hsa_circ_0020397 regulates colorectal cancer cell viability, apoptosis and invasion by promoting the expression of the miR-138 targets TERT and PD-L1," *Cell Biology International*, vol. 41, no. 9, pp. 1056–1064, 2017.

[118] J. Guo, J. Li, C. Zhu et al., "Comprehensive profile of differentially expressed circular RNAs reveals that hsa_circ_0000069 is upregulated and promotes cell proliferation, migration, and invasion in colorectal cancer," *OncoTargets and Therapy*, vol. 9, pp. 7451–7458, 2016.

[119] J. Li, D. Li, and W. Zhang, "Tumor suppressor role of miR-217 in human epithelial ovarian cancer by targeting IGF1R," *Oncology Reports*, vol. 35, no. 3, pp. 1671–1679, 2016.

[120] H. Wang, X. Dong, X. Gu, R. Qin, H. Jia, and J. Gao, "The microRNA-217 functions as a potential tumor suppressor in gastric cancer by targeting GPC5," *PLoS One*, vol. 10, no. 6, article e0125474, 2015.

[121] J. Zheng, X. Liu, Y. Xue et al., "TTBK2 circular RNA promotes glioma malignancy by regulating miR-217/HNF1β/Derlin-1 pathway," *Journal of Hematology & Oncology*, vol. 10, no. 1, p. 52, 2017.

[122] P. Yang, Z. Qiu, Y. Jiang et al., "Silencing of cZNF292 circular RNA suppresses human glioma tube formation via the Wnt/β-catenin signaling pathway," *Oncotarget*, vol. 7, no. 39, pp. 63449–63455, 2016.

[123] J. Guarnerio, M. Bezzi, J. C. Jeong et al., "Oncogenic role of fusion-circRNAs derived from cancer-associated chromosomal translocations," *Cell*, vol. 165, no. 2, pp. 289–302, 2016.

[124] G. A. Dos Santos, L. Kats, and P. P. Pandolfi, "Synergy against PML-RARa: targeting transcription, proteolysis, differentiation, and self-renewal in acute promyelocytic leukemia," *The Journal of Experimental Medicine*, vol. 210, no. 13, pp. 2793–2802, 2013.

[125] S. Memczak, P. Papavasileiou, O. Peters, and N. Rajewsky, "Identification and characterization of circular RNAs as a new class of putative biomarkers in human blood," *PLoS One*, vol. 10, no. 10, article e0141214, 2015.

[126] J. H. Bahn, Q. Zhang, F. Li et al., "The landscape of microRNA, Piwi-interacting RNA, and circular RNA in human saliva," *Clinical Chemistry*, vol. 61, no. 1, pp. 221–230, 2015.

[127] M. Qin, G. Liu, X. Huo et al., "Hsa_circ_0001649: a circular RNA and potential novel biomarker for hepatocellular carcinoma," *Cancer Biomarkers*, vol. 16, no. 1, pp. 161–169, 2016.

[128] X. Shang, G. Li, H. Liu et al., "Comprehensive circular RNA profiling reveals that hsa_circ_0005075, a new circular RNA biomarker, is involved in hepatocellular crcinoma development," *Medicine*, vol. 95, no. 22, article e3811, 2016.

[129] Z. Zhong, M. Lv, and J. Chen, "Screening differential circular RNA expression profiles reveals the regulatory role of circTCF25-miR-103a-3p/miR-107-CDK6 pathway in bladder carcinoma," *Scientific Reports*, vol. 6, no. 1, article 30919, 2016.

[130] X. Wang, Y. Zhang, L. Huang et al., "Decreased expression of hsa_circ_001988 in colorectal cancer and its clinical

significances," *International Journal of Clinical and Experimental Pathology*, vol. 8, no. 12, pp. 16020–16025, 2015.

[131] P. Zhang, Z. Zuo, W. Shang et al., "Identification of differentially expressed circular RNAs in human colorectal cancer," *Tumour Biology*, vol. 39, no. 3, article 1010428317694546, 2017.

[132] W. Weng, Q. Wei, S. Toden et al., "Circular RNA ciRS-7—a promising prognostic biomarker and a potential therapeutic target in colorectal cancer," *Clinical Cancer Research*, vol. 23, no. 14, pp. 3918–3928, 2017.

[133] J. T. Yao, S. H. Zhao, Q. P. Liu et al., "Over-expression of circRNA_100876 in non-small cell lung cancer and its prognostic value," *Pathology, Research and Practice*, vol. 213, no. 5, pp. 453–456, 2017.

[134] Q. Liu, X. Zhang, X. Hu et al., "Circular RNA related to the chondrocyte ECM regulates MMP13 expression by functioning as a miR-136 'sponge' in human cartilage degradation," *Scientific Reports*, vol. 6, 2016.

[135] X. Yu, F. Wei, J. Yu et al., "Matrix metalloproteinase 13: a potential intermediate between low expression of microRNA-125b and increasing metastatic potential of non-small cell lung cancer," *Cancer Genetics*, vol. 208, no. 3, pp. 76–84, 2015.

[136] C. P. Hsu, G. H. Shen, and J. L. Ko, "Matrix metalloproteinase-13 expression is associated with bone marrow microinvolvement and prognosis in non-small cell lung cancer," *Lung Cancer*, vol. 52, no. 3, pp. 349–357, 2006.

[137] L. Xu, M. Zhang, X. Zheng, P. Yi, C. Lan, and M. Xu, "The circular RNA ciRS-7 (Cdr1as) acts as a risk factor of hepatic microvascular invasion in hepatocellular carcinoma," *Journal of Cancer Research and Clinical Oncology*, vol. 143, no. 1, pp. 17–27, 2017.

[138] Z. Yao, J. Luo, K. Hu et al., "*ZKSCAN1* gene and its related circular RNA (circZKSCAN1) both inhibit hepatocellular carcinoma cell growth, migration, and invasion but through different signaling pathways," *Molecular Oncology*, vol. 11, no. 4, pp. 422–437, 2017.

[139] L. Fu, Q. Chen, T. Yao et al., "Hsa_circ_0005986 inhibits carcinogenesis by acting as a miR-129-5p sponge and is used as a novel biomarker for hepatocellular carcinoma," *Oncotarget*, vol. 8, no. 27, pp. 43878–43888, 2017.

[140] L. Fu, T. Yao, Q. Chen, X. Mo, Y. Hu, and J. Guo, "Screening differential circular RNA expression profiles reveals hsa_circ_0004018 is associated with hepatocellular carcinoma," *Oncotarget*, vol. 8, no. 35, pp. 58405–58416, 2017.

[141] W. H. Li, Y. C. Song, H. Zhang et al., "Decreased expression of hsa_circ_00001649 in gastric cancer and its clinical significance," *Disease Markers*, vol. 2017, Article ID 4587698, 6 pages, 2017.

[142] P. Li, H. Chen, S. Chen et al., "Circular RNA 0000096 affects cell growth and migration in gastric cancer," *British Journal of Cancer*, vol. 116, no. 5, pp. 626–633, 2017.

[143] Y. Shao, L. Chen, R. Lu et al., "Decreased expression of hsa_circ_0001895 in human gastric cancer and its clinical significances," *Tumour Biology*, vol. 39, no. 4, article 1010428317699125, 2017.

[144] R. Lu, Y. Shao, G. Ye, B. Xiao, and J. Guo, "Low expression of hsa_circ_0006633 in human gastric cancer and its clinical significances," *Tumour Biology*, vol. 39, no. 6, article 1010428317704175, 2017.

[145] S. Chen, T. Li, Q. Zhao, B. Xiao, and J. Guo, "Using circular RNA hsa_circ_0000190 as a new biomarker in the diagnosis of gastric cancer," *Clinica Chimica Acta*, vol. 466, pp. 167–171, 2017.

[146] M. Tian, R. Chen, T. Li, and B. Xiao, "Reduced expression of circRNA hsa_circ_0003159 in gastric cancer and its clinical significance," *Journal of Clinical Laboratory Analysis*, e22281, 2017.

[147] Y. Liang, W. Wang, C. Fang et al., "Clinical significance and diagnostic value of serum CEA, CA19-9 and CA72-4 in patients with gastric cancer," *Oncotarget*, vol. 7, no. 31, pp. 49565–49573, 2016.

[148] J. Chen, Y. Li, Q. Zheng et al., "Circular RNA profile identifies circPVT1 as a proliferative factor and prognostic marker in gastric cancer," *Cancer Letters*, vol. 388, pp. 208–219, 2017.

[149] Y. Fang, M. Ma, J. Wang, X. Liu, and Y. Wang, "Circular RNAs play an important role in late-stage gastric cancer: circular RNA expression profiles and bioinformatics analyses," *Tumour Biology*, vol. 39, no. 6, article 1010428317705850, 2017.

[150] L. Xuan, L. Qu, H. Zhou et al., "Circular RNA: a novel biomarker for progressive laryngeal cancer," *American Journal of Translational Research*, vol. 8, no. 2, pp. 932–939, 2016.

[151] F. Wang, A. J. Nazarali, and S. Ji, "Circular RNAs as potential biomarkers for cancer diagnosis and therapy," *American Journal of Cancer Research*, vol. 6, no. 6, pp. 1167–1176, 2016.

6

SCAR Marker for Gender Identification in Date Palm (*Phoenix dactylifera* L.) at the Seedling Stage

Fahad Al-Qurainy,[1] Abdulhafed A. Al-Ameri ⓘ,[2] Salim Khan ⓘ,[1] Mohammad Nadeem ⓘ,[1] Abdel-Rhman Z. Gaafar ⓘ,[1] and Mohamed Tarroum ⓘ[1]

[1]*Department of Botany and Microbiology, College of Science, King Saud University, Riyadh 11451, Saudi Arabia*
[2]*Department of Biology, Faculty of Education and Science, Rada'a Al-Baydha University, Al-Baydha, Yemen*

Correspondence should be addressed to Salim Khan; salimkhan17@yahoo.co.in

Academic Editor: Graziano Pesole

Date palm (*Phoenix dactylifera* L.) is cultivated in arid and semiarid regions worldwide. Given the dioecious nature of this plant, gender identification is very important at the seedling stage. Molecular markers are very effective tools that help in gender identification at this stage. A sequence characterized amplified region (SCAR) marker linked to sex-specific regions in the genome of date palm was developed. Of the 300 tested randomly amplified polymorphic DNA (RAPD) primers, only one primer (OPC-06) produced reproducible band (294 bp) in male plants. The PCR product of this primer was cloned and sequenced. The specific primers were synthesized for amplification of a 186 bp fragment in male date palm plants. These primers were validated in male and female date palm plants, wherein the designed SCAR marker was reported only in male plants and no amplification was observed in female plants. The developed SCAR marker was used with seedlings of date palm and proved very effective in identification of gender.

1. Introduction

Date palm (*Phoenix dactylifera* L.) belongs to the family Arecaceae (2n = 36) and has a socioeconomic significance. The plant is cultivated for food, fiber, and shelter in different arid and semiarid regions worldwide. It is a monocot and dioecious tree (separate male and female) and serves as an important commercial crop in Middle Eastern countries. The plant is native to the Canary Islands located in the Atlantic Ocean near the coast of Northeast Africa. Dates are a good source of energy, vitamins, and group of elements such as phosphorus, potassium, iron, manganese, selenium, zinc, and calcium [1, 2].

Research related to date palm is greatly restricted, owing to the lack of measures to identify its gender at the seedling stage. Date palm cultivation is more cost-effective through the cultivation of female plants than male plants. An increase in the number of female date palm plants per hectare may result in an increase in date production, thereby making the plantation more profitable. This has prompted the farmers to solely propagate date palm cultivars via offshoots that results in the reduction in genetic variations. The high genetic diversity is very important in plants for their survival in their natural habitat [3]. However, the seedlings produced may be either male or female, and no reproducible technique is currently available for gender determination in germinated seeds of date palm. Several efforts have been directed recently to establish a method for the early detection of seedling gender before their plantation in fields. However, no methodology has so far been developed for gender identification at the seedling stage [2].

Molecular markers based on the direct analysis of genomic DNA are used for the study of phylogenetic relationship, genetic diversity, genetic fidelity, and genotoxicity of date palm cultivars [4–7]. These markers may be useful in the study of sex determination in dioecious plants. Despite increasing research efforts on a number of different plant species, very limited information is available on the molecular basis of gender determination, and it may be very difficult to estimate the numbers of genes involved. However, in some plant species, sex-determining genes have been discovered including *Carica papaya* and *Asparagus officinalis* [8, 9]. In the last two decades, efforts have been made to understand the basis of gender determination in date palm and develop methods for gender identification at an early stage using isozymes [10], peroxidases [11] and DNA-based molecular markers using random amplified polymorphic DNA (RAPD) [12], and polymerase chain reaction-based restriction fragment length polymorphism (PCR-RFLP) [13]. The first genetic map of Khalas cultivar of date palm has been published [14] which could help in understanding the sex chromosome development. The sex chromosomes evolved from a common autosomal origin before the diversification of the extant dioecious *Phoenix* species [15]. DNA-based marker linked to sex determination locus in *Salix viminalis* was used for estimation of sex ratios in progeny [16]. In comparison with other molecular markers, RAPD markers offer advantages owing to their ease of generation and suitability for genetic polymorphism study in different plant species that lack detailed genomic sequence information [17]. However, there are some limitations with the use of RAPD markers as PCR amplification is very sensitive and depends on many factors. For more reproducible results, RAPDs may be converted into stable and reliable markers through the cloning of amplified bands, sequencing, and designing of more specific primers. Annealing of these specific primers under stringent annealing temperatures in PCR may result in the production of a single band that corresponds to genetically defined loci, sequence-characterized amplified regions (SCAR) [18]. This approach has been employed to develop several gender-linked molecular markers in dioecious plants, including *Silene latifolia* [19], *Pistacia vera* [20], *Cannabis sativa* [21], *Humulus lupulus* [22], *Actinidia chinensis* [23], *Atriplex garrettii* [24], *Carica papaya* [25], *Salix viminalis* [26], *Rumex acetosa* [27], *Mercurialis annua* [28], and *Eucommia ulmoides* [29].

Here, we developed a SCAR marker specific to male date palm plant. The male specific bands were generated through the comparative study of male and female plants using RAPD primers. The designed primers specific to male plants (SCAR primers) were used for the identification of gender at the seedling stage in date palm.

2. Materials and Methods

2.1. Plant Material Collection. Leaf samples of date palm (21 different males and females) were collected from Al-Rajhi Farm (Al-Qassim) and Agricultural Research Station (Dirab) in Saudi Arabia and stored at −80°C (Table 1).

2.2. Genomic DNA Extraction. Genomic DNA was isolated from the leaves using the modified CTAB method [30]. The leaves of date palm (200 mg) were ground into fine powder with a mortar using liquid nitrogen. The frozen powdered tissues were transferred into a 2 mL microcentrifuge tube and treated with 800 μL of preheated extraction buffer and 10 μL of RNase A (10 mg/mL) (Qiagen). To the above

TABLE 1: Date palm leaves collected from female cultivars and male plants.

S.N.	Cultivar (female)	Cultivar code	Male	Male code
1	Barhi	Ba	Male-1	M1
2	Seqae	Se	Male-2	M2
3	Sukkari	Su	Male-3	M3
4	Sabaka	Sa	Male-4	M4
5	Wannana	Wa	Male-5	M5
6	Khalas	Kh	Male-6	M6
7	Ruthana	Ru	Male-7	M7
8	Deglet Noor	Dn	Male-8	M8
9	Magdool	Mg	Male-9	M9
10	Agwa	Ag	Male-10	M10
11	Um Khashab	Uk	Male-11	M11
12	Hilaly	Hi	Male-12	M12
13	Shaishee	Sh	Male-13	M13
14	Naboot Seif	Ns	Male-14	M14
15	Ruzeiz	Rz	Male-15	M15
16	Wesaily	We	Male-16	M16
17	Sullaj	Sl	Male-17	M17
18	Thawee	Th	Male-18	M18
19	Hatmi	Ha	Male-19	M19
20	Rabeaa	Ra	Male-20	M20
21	Munif	Mu	Male-21	M21

mixture, 100 μL of 3% PVP and β-mercaptoethanol was added and followed by its incubation at 65°C for 20 min with shaking every after 5 min. The mixture was then cooled at room temperature and treated with an equal volume of chloroform and isoamyl alcohol (24:1), followed by its frequent mixing for 20 min. The mixture was subjected to centrifugation at 10,000 rpm for 10 min at room temperature. The clear upper aqueous suspension was transferred into a new microfuge tube and treated with an equal volume of ice-cold isopropanol at −20°C for 1 h. For the separation of nucleic acid, tubes were centrifuged at 10,000 rpm for 10 min. The supernatant was discarded, and the pellet was washed twice with cold 70% ethanol. The DNA pellet was dried at 37°C and dissolved in 200 μL of TE buffer (Qiagen).

2.3. RAPD Analysis. We performed PCR reaction with genomic DNA of male and female date palm plants to screen 300 decamer primers of arbitrary sequences (Operon Technologies, United States). For each primer, genomic DNA from two pools (21 female cultivars in one pool and 21 male plants in second pool) were used for PCR. PCR amplifications were performed in 20 μL reaction volumes containing 5x HOT FIREPol® Blend Master Mix ready to load (4 μL), primer (15 ng/μL), template DNA (25 ng/μL), and water. DNA amplification was performed on Applied Biosystems Veriti 96-well Thermal Cycler with the following program: first denaturation at 94°C for 5 min, followed by 40 cycles of denaturation for 1 min at 94°C, annealing at 36°C for 1 min, extension at 72°C for 1 min, and a final extension step at

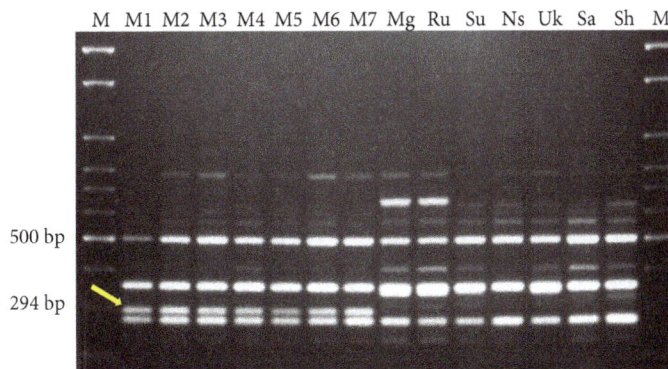

FIGURE 1: Amplification profile generated with RAPD primer (OPC-06) using genomic DNA of male and female date palm plants. Lane M, 100 bp ladder.

72°C for 5 min. Amplification products were analyzed by gel electrophoresis on 1.3% agarose gel in 1x TBE buffer (*Tris/borate/ethylenediaminetetraacetic acid*). Gels were stained with ethidium bromide and visualized under UV light. Each amplification reaction was performed using a single primer and repeated thrice to verify the reproducibility of the results.

2.4. Identification of Male Specific Band from RAPD Profile. A total of 300 decamer oligonucleotides (RAPD primers) were used with the bulk DNA of male and female plants for selection of male specific band. The male specific primer obtained in screening with bulk DNA samples was used further with individual DNA sample of male and female plants for reproducibility testing and selection of male specific band.

2.5. Cloning and Sequencing of Male Specific Band Generated in RAPD Profile. The candidate RAPD fragment specific to male plant was carefully excised from 1.2% agarose gel using a sterile gel slicer and purified with kit Wizard® SV Gel and PCR clean-up system (Promega). The male specific band was cloned and sequenced using the commercial service offered by *Macrogen* Inc. (Korea). The sequence was subjected to BLAST at NCBI database to determine its similarity with the available sequences (https://www.ncbi.nlm.nih.gov/).

2.6. Designing of SCAR Primers. The ends of the cloned RAPD fragment were used to design specific primers (SCAR primers) for the amplification of the expected size from the genomic DNA of male date palm plants. However, these specific primers were designed from the male-specific sequence using primer 3 tool (http://frodo.wi.mit.edu/) as well as Primer Select software (DNASTAR).

2.7. Validation of Developed SCAR Marker. Amplification of the genomic DNA from male and female plants was performed with SCAR primers in a 25 μL reaction volume using the master mixture. Designed primers were used for the amplification of SCAR marker using the genomic DNA of known cultivars (Table 1). The single reaction mixture contained Illustra PuReTaq Ready-To-Go PCR Beads (4 μL), 20 ng of each primer (forward and reverse), 25 ng of template DNA, and water. The PCR program followed as first denaturation at 94°C for 4 min, followed by 40 cycles

of denaturation at 94°C for 1 min, annealing at 55°C for 30 s, extension at 72°C for 1 min, and final extension at 72°C for 5 min. Amplification products were separated on 1.5% agarose gel.

2.8. Screening of Gender in Seedlings of Date Palm. The designed SCAR marker was employed for the identification of date palm gender in two-month-old seedlings. The seedlings of Khalas cultivar were screened for gender identification using designed SCAR marker.

3. Results and Discussion

3.1. RAPD Analysis. In the preliminary study, 300 arbitrary decamer RAPD primers were screened using bulk DNA from 21 female cultivars (one pool) and 21 male plants (second pool) (Table 1). Of these primers, only OPC-06 (5′-GAACGGACTC-3′) produced a band of approximately 294 bp specific to male plants. After confirmation of male specific band in bulk samples, further PCR reaction was performed using DNA of individual female and male plant as result shown in Figure 1. The presence of 294 bp band was clearly observed in all male plants but absent in female plants. This male-specific fragment amplified by the primer OPC-06 was subsequently excised, purified, cloned, and sequenced. Homology search was performed using BLAST algorithm of 294 bp sequence (Figure 2), and no similarity was found with any of the known sequences from NCBI GenBank database.

3.2. Development and Validation of SCAR Marker. A pair of SCAR primer was designed from male-specific sequence obtained in RAPD profile, to amplify a 186 bp fragment (Figure 2). PCR reaction was performed with genomic DNA of individual male and female plants with designed SCAR primer pair "ALAMERI" (ALAMERIF 5′-CGTGGG ATGAGGTAGTTTGG-3′ and ALAMERIR 5′-CTCGCG ATGCAAACCAACCAA-3′). A single, distinct bright band of size 186 bp was observed in all male plants whereas was absent in all female plants (Figures 3(a)–3(c)). Thus, RAPD marker was successfully converted into SCAR marker. The reproducibility of the developed SCAR marker was verified on 55 samples with known gender other than the previously

```
            10        20        30        40        50        60        70        80
    ····|····|····|····|····|····|····|····|····|····|····|····|····|····|····|····|
1   GAACGGACTCTGAAATGTAGGGGATTTGGCTGAGATGAGGACGCGTGAAACAGGGGATGTTTGGATGCCGTGAAACTCAC
            ALAMERI_F ⌐ CGTGGGATGAGGTAGTTTGG
            90        100       110       120       130       140       150       160
    ····|····|····|····|····|····|····|····|····|····|····|····|····|····|····|····|
81  AAGAATGGAGCTGATCCGTGGGATGAGGTAGTTTGGGATGTTGGGAAGAGGTGATTTCGAAACAGGGGAAGTGGATGAA
            170       180       190       200       210       220       230       240
    ····|····|····|····|····|····|····|····|····|····|····|····|····|····|····|····|
161 ATAGATCAACGTGGATGTGATTCCGAGGGAGGCTGGGTTGGTTCGGAAGAGGGAAGACACTGGATTTGGTTAGGATGAAA
            250       260       270       280       290
    ····|····|····|····|····|····|····|····|····|··|····|····
241 TCTAGGAAGATGCTGGGAAGATGGTTGGTTTGCATCGCGAGAGGAGTCCGTTC
                AACCAACCAAACGTAGCGCTC ⌐ ALAMERI_R
```

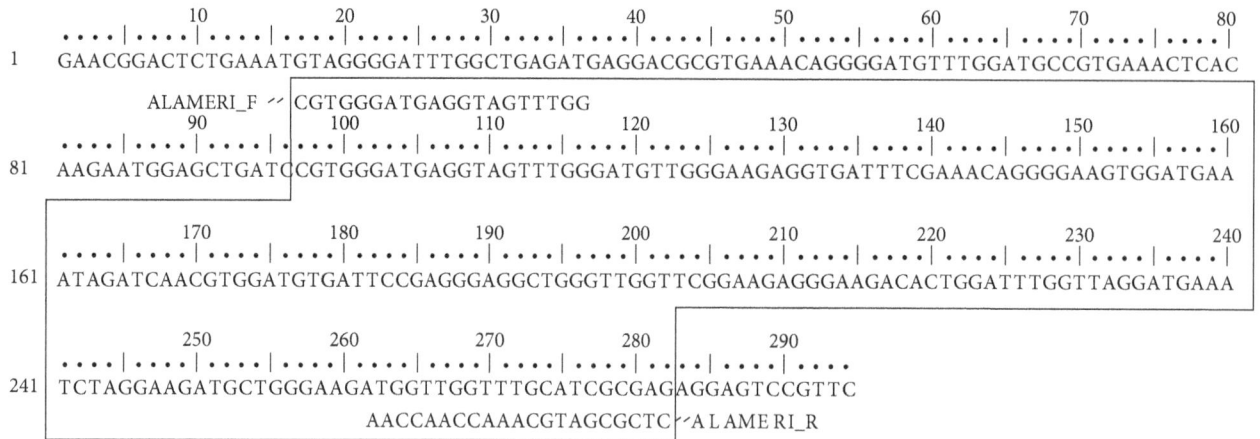

FIGURE 2: Complete DNA sequence of the cloned RAPD fragment specific to male plant of *Phoenix dactylifera*. ALAMERIF and ALAMERIR are specific forward and reverse SCAR primers.

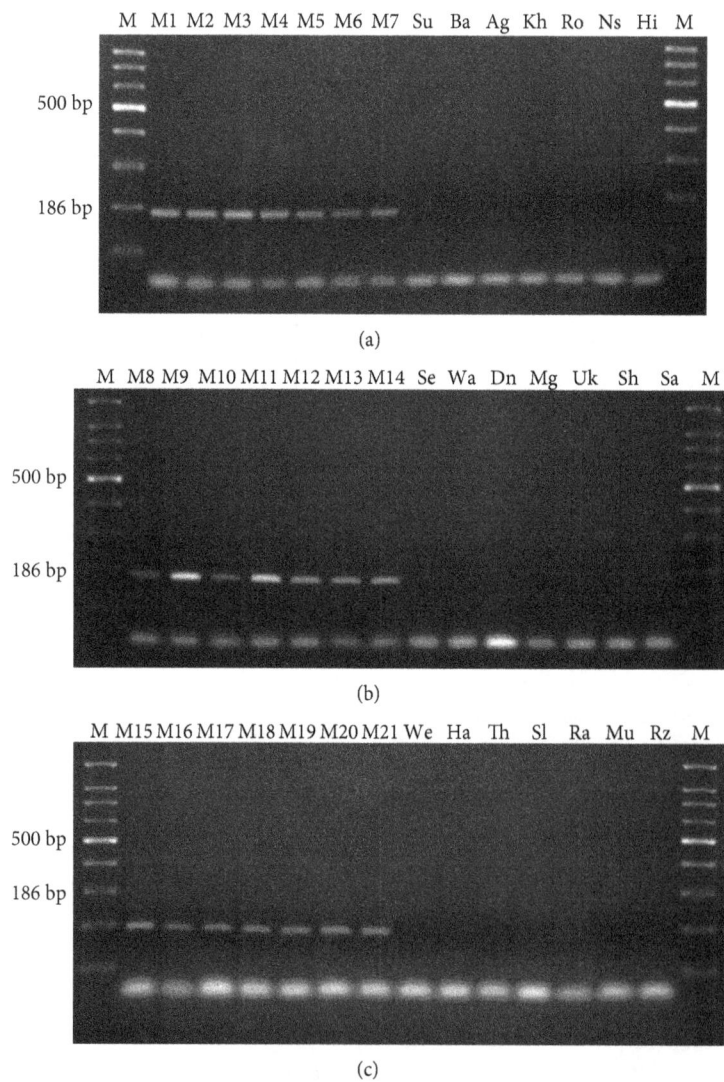

(a)

(b)

(c)

FIGURE 3: SCAR marker analysis using SCAR primers showed the amplification of a 186 bp fragment in all male plants and absent in female plants. Lane M, 100 bp ladder. (a) M1–M7 (male), Su-Hi (female). (b) M8–M14 (male), Se-Sa (female). (c) M15–M21 (male), We-Rz (female).

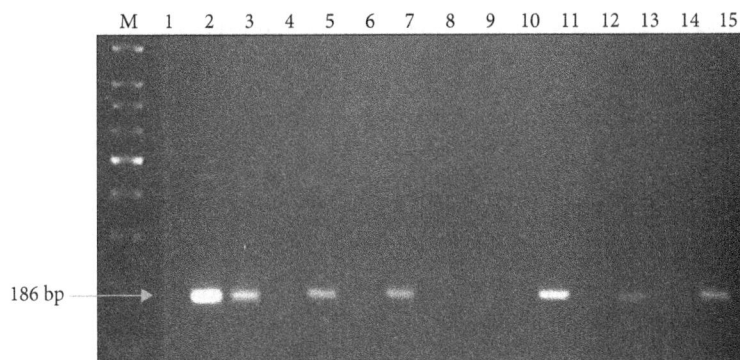

FIGURE 4: Screening of male and female plants in date palm seedlings with developed SCAR marker. Presence of band indicates male seedlings and absence of band indicates female seedlings.

tested male and female plants. The marker clearly differentiated all male from female plants based on the presence or absence of the 186 bp band. The results were reproducible owing to the longer length (20-21 base) and high Tm (60.5–61.3°C) of primers. Thus, the developed SCAR marker showed reproducible results, as specific band was obtained only in male plants.

The SCAR marker has been used for gender identification in many dioecious plant species in which male and female plants look similar at vegetative stage. Male-specific SCAR markers were developed in dioecious plants such as *Humulus scandens*, *Rumex nivalis*, and *Phoenix dactylifera* using molecular marker profiling of male and female plants [31–34]. The seeds of *Pistacia chinensis* produce biofuel. To enhance the number of female plants of *P. chinensis* for more fuel production, SCAR marker specific to female plants was developed for identification at seedling stage [35]. SCAR markers have been used to discriminate between male and female plants of *Hippophae rhamnoides* [36].

At seedling stage, all date palm plants appear similar in morphology and it is very difficult to identify as male or female among them. The presence or absence of the SCAR marker developed herein could allow differentiation between male and female plants at the seedling stage [37]. We screened two-month-old Khalas seedlings with our developed SCAR marker. The SCAR marker was present in all male plants but absent in female plants (Figure 4). The lanes 2, 3, 5, 7, 11, 13, and 15 corresponded to male seedlings, whereas lanes 1, 4, 6, 8, 9, 10, 12, and 14 had all female seedlings (Figure 4).

In conclusion, the developed SCAR marker could be used for gender identification at the seedling stage of date palms to save time, as the plant takes 5–7 years to reach its reproductive stage. Thus, plant breeders may adopt this marker as a potential tool for gender identification of date palm seedlings before their plantations in fields.

Conflicts of Interest

The authors declare that they have no conflicts of interest.

Acknowledgments

The authors extend their appreciation to the Deanship of Scientific Research at King Saud University for funding the work through the research group Project no. RGP-014.

References

[1] A. M. Shinwari, "Iron content of date fruits," *Journal of the College of Science-King-Saud-University*, vol. 18, no. 1, pp. 5–13, 1987.

[2] M. J. A. Hafiz, A. F. Shalabi, and I. D. Al. Akhal, "Chemical composition of 15 varieties of dates grown in Saudi Arabia," in *Proceedings of the 4th Conference on the Biological Aspects of Saudi Arabia*, pp. 181–194, Saudi Arabia, 1980.

[3] M. Elleuch, S. Besbes, O. Roiseux et al., "Date flesh: chemical composition and characteristics of the dietary fibre," *Food Chemistry*, vol. 111, no. 3, pp. 676–682, 2008.

[4] F. Al-Qurainy, S. Khan, F. M. Al-Hemaid, M. A. Ali, M. Tarroum, and M. Ashraf, "Assessing molecular signature for some potential date (*Phoenix dactylifera* L.) cultivars from Saudi Arabia, based on chloroplast dna sequences *rpoB* and *psbA-trnH*," *International Journal of Molecular Sciences*, vol. 12, no. 10, pp. 6871–6880, 2011.

[5] F. Al-Qurainy, S. Khan, M. Nadeem, and M. Tarroum, "SCoT marker for the assessment of genetic diversity in Saudi Arabian date palm cultivars," *Pakistan Journal of Botany*, vol. 47, no. 2, pp. 637–643, 2015.

[6] F. Al-Qurainy, S. Khan, M. Nadeem et al., "Assessing genetic fidelity in regenerated plantlets of date palm cultivars after cryopreservation," *Fresenius Environmental Bulletin*, vol. 26, no. 2a, pp. 1727–1735, 2017.

[7] F. Al-Qurainy, S. Khan, M. Tarroum, M. Nadeem, S. Alansi, and A. Alshameri, "Biochemical and genetical responses of *Phoenix dactylifera* L. to cadmium stress," *BioMed Research International*, vol. 2017, Article ID 9504057, 9 pages, 2017.

[8] N. Urasaki, K. Tarora, A. Shudo et al., "Digital transcriptome analysis of putative sex-determination genes in papaya (*Carica papaya*)," *PLoS One*, vol. 7, no. 7, article e40904, 2012.

[9] K. Murase, S. Shigenobu, S. Fujii et al., "MYB transcription factor gene involved in sex determination in *Asparagus officinalis*," *Genes to Cells*, vol. 22, no. 1, pp. 115–123, 2017.

[10] A. M. Torres and B. Tisserat, "Leaf isozymes as genetic markers in date palms," *American Journal of Botany*, vol. 67, no. 2, pp. 162–167, 1980.

[11] K. Majourhat, K. Bendiab, L. Medraoui, and M. Baaziz, "Diversity of leaf peroxidases in date palm (*Phoenix dactylifera* L.) as revealed in an example of marginal (seedling derived) palm groves," *Scientia Horticulturae*, vol. 95, no. 1-2, pp. 31–38, 2002.

[12] R. A. Younis, O. M. Ismail, and S. S. Soliman, "Identification of gender-specific DNA markers for date palm (*Phoenix dactylifera* L.) using RAPD and ISSR techniques," *Research Journal of Agriculture and Biological Sciences*, vol. 4, pp. 278–284, 2008.

[13] M. E. Al-Mahmoud, K. E. Al-Dous, K. E. Al-Azwani, and J. A. Malek, "DNA-based assays to distinguish date palm (Arecaceae) gender," *American Journal of Botany*, vol. 99, no. 1, pp. e7–e10, 2012.

[14] L. S. Mathew, M. Spannagl, A. al-Malki et al., "A first genetic map of date palm (*Phoenix dactylifera*) reveals long-range genome structure conservation in the palms," *BMC Genomics*, vol. 15, no. 1, p. 285, 2014.

[15] E. Cherif, S. Zehdi-Azouzi, A. Crabos et al., "Evolution of sex chromosomes prior to speciation in the dioecious *Phoenix* species," *Journal of Evolutionary Biology*, vol. 29, no. 8, pp. 1513–1522, 2016.

[16] C. Alstrom-Rapaport, M. Lascoux, Y. C. Wang, G. Roberts, and G. A. Tuskan, "Identification of a RAPD marker linked to gender determination in the basket willow (*Salix viminalis* L.)," *Journal of Heredity*, vol. 89, no. 1, pp. 44–49, 1998.

[17] R. W. Michelmore, I. Paran, and R. V. Kesseli, "Identification of markers linked to disease-resistance genes by bulked segregant analysis: a rapid method to detect markers in specific genomic regions by using segregating populations," *Proceedings of the National Academy of Sciences of the United States of America*, vol. 88, no. 21, pp. 9828–9832, 1991.

[18] Y. H. Zhang, V. S. Stilio, F. Rehman, A. Avery, D. Mulcahy, and R. Kesseli, "Y-chromosome specific markers and the evolution of dioecy in the genus *Silene*," *Genome*, vol. 41, no. 2, pp. 141–147, 1998.

[19] D. L. Mulcahy, N. F. Weeden, R. Kesseli, and S. B. Carroll, "DNA probes for the Y-chromosome of *Silene latifolia*, a dioecious angiosperm, gender," *Sexual Plant Reproduction*, vol. 5, no. 1, pp. 86–88, 1992.

[20] B. Yakubov, O. Barazani, and A. Golan-Goldhirsh, "Combination of SCAR primers and touchdown-PCR for sex identification in *Pistacia vera* L," *Scientia Horticulturae*, vol. 103, no. 4, pp. 473–478, 2005.

[21] O. Törjék, N. Bucherna, E. Kiss et al., "Novel male-specific molecular markers (MADC5, MADC6) in hemp," *Euphytica*, vol. 127, no. 2, pp. 209–218, 2002.

[22] A. Polley, M. W. Ganal, and E. Seigner, "Identification of sex in hop (*Humulus lupulus*) using molecular markers," *Genome*, vol. 40, no. 3, pp. 357–361, 1997.

[23] G. P. Gill, C. F. Harvey, R. C. Gardner, and L. G. Fraser, "Development of sex-linked PCR markers for gender identification in Actinidia," *Theoretical and Applied Genetics*, vol. 97, no. 3, pp. 439–445, 1998.

[24] C. F. Ruas, D. J. Fairbanks, R. P. Evans, H. C. Stutz, W. R. Andersen, and P. M. Ruas, "Male-specific DNA in the dioecious species *Atriplex garrettii* (Chenopodiaceae)," *American Journal of Botany*, vol. 85, no. 2, pp. 162–167, 1998.

[25] N. Urasaki, M. Tokumoto, K. Tarora et al., "A male and hermaphrodite specific RAPD marker for papaya (*Carica papaya* L.)," *Theoretical and Applied Genetics*, vol. 104, no. 2-3, pp. 281–285, 2002.

[26] L. E. Gunter, G. T. Roberts, K. Lee, F. W. Larimer, and G. A. Tuskan, "The development of two flanking SCAR markers linked to a sex determination locus in *Salix viminalis* L," *Journal of Heredity*, vol. 94, no. 2, pp. 185–189, 2003.

[27] H. Korpelainen, "A genetic method to resolve gender complements investigations on sex ratios in *Rumex acetosa*," *Molecular Ecology*, vol. 11, no. 10, pp. 2151–2156, 2002.

[28] D. K. Khadka, A. Nejidat, M. Tal, and A. Golan-Goldhirsh, "A., DNA markers for gender: molecular evidence for gender dimorphism in dioecious *Mercurialis annua* L," *Molecular Breeding*, vol. 9, no. 4, pp. 251–257, 2002.

[29] W.-J. Xu, B.-W. Wang, and K.-M. Cui, "RAPD and SCAR markers linked to sex determination in *Eucommia ulmoides* Oliv," *Euphytica*, vol. 136, no. 3, pp. 233–238, 2004.

[30] S. Khan, M. Irfan Qureshi, Kamaluddin, T. Alam, and M. Z. Abdin, "Protocol for isolation of genomic DNA from dry and fresh roots of medicinal plants suitable for RAPD and restriction digestion," *African Journal of Biotechnology*, vol. 6, no. 3, pp. 175–178, 2007.

[31] W. J. Gao, F. C. Sun, W. Z. Yin, Y. K. Ji, C. L. Deng, and L. D. Lu, "Clone and development of RAPD and SCAR markers linked to sex determination in the dioecious species *Humulus scandens* L," *Fen Zi Xi Bao Sheng Wu Xue Bao*, vol. 42, no. 3-4, pp. 211–216, 2009.

[32] S. F. Li, L. J. Wang, C. L. Deng, and W. J. Gao, "Identification of male-specific AFLP and SCAR markers in the dioecious plant *Humulus scandens*," *Molecular and Cellular Probes*, vol. 34, pp. 68–70, 2017.

[33] I. Stehlik and F. R. Blattner, "Sex-specific SCAR markers in the dioecious plant *Rumex nivalis* (Polygonaceae) and implications for the evolution of sex chromosomes," *Theoretical and Applied Genetics*, vol. 108, no. 2, pp. 238–242, 2004.

[34] C. Dhawan, P. Kharb, R. Sharma, S. Uppal, and R. K. Aggarwal, "Development of male-specific SCAR marker in date palm (*Phoenix dactylifera* L.)," *Tree Genetics & Genomes*, vol. 9, no. 5, pp. 1143–1150, 2013.

[35] Q. Sun, X. Yang, and R. Li, "SCAR marker for sex identification of *Pistacia chinensis* Bunge (Anacardiaceae)," *Genetics and Molecular Research*, vol. 13, no. 1, pp. 1395–1401, 2014.

[36] G. Korekar, R. K. Sharma, R. Kumar et al., "Identification and validation of sex-linked SCAR markers in dioecious *Hippophae rhamnoides* L. (Elaeagnaceae)," *Biotechnology Letters*, vol. 34, no. 5, pp. 973–978, 2012.

[37] A. A. H. Alameri, F. H. Al-Qurainy, S. Khan, M. Nadeem, and A. R. Z. Gaafar, "Method of identifying date palm gender using SCAR primers," US Patent 9, 598, 732B2, 2017.

7

Transcriptome Analysis of the Thymus in Short-Term Calorie-Restricted Mice using RNA-seq

Zehra Omeroğlu Ulu,[1] **Salih Ulu,**[1] **Soner Dogan,**[2] **Bilge Guvenc Tuna,**[3] **and Nehir Ozdemir Ozgenturk**⬤[1]

[1]*Faculty of Art and Science, Molecular Biology and Genetics, Yildiz Technical University, Istanbul, Turkey*
[2]*Department of Medical Biology, School of Medicine, Yeditepe University, Istanbul, Turkey*
[3]*Department of Medical Biophysic, School of Medicine, Yeditepe University, Istanbul, Turkey*

Correspondence should be addressed to Nehir Ozdemir Ozgenturk; nehirozdemir@yahoo.com

Academic Editor: Elena Pasyukova

Calorie restriction (CR), which is a factor that expands lifespan and an important player in immune response, is an effective protective method against cancer development. Thymus, which plays a critical role in the development of the immune system, reacts to nutrition deficiency quickly. RNA-seq-based transcriptome sequencing was performed to thymus tissues of MMTV-TGF-α mice subjected to ad libitum (AL), chronic calorie restriction (CCR), and intermittent calorie restriction (ICR) diets in this study. Three cDNA libraries were sequenced using Illumina HiSeq™ 4000 to produce 100 base pair-end reads. On average, 105 million clean reads were mapped and in total 6091 significantly differentially expressed genes (DEGs) were identified ($p < 0.05$). These DEGs were clustered into Gene Ontology (GO) categories. The expression pattern revealed by RNA-seq was validated by quantitative real-time PCR (qPCR) analysis of four important genes, which are leptin, ghrelin, Igf1, and adinopectin. RNA-seq data has been deposited in NCBI Gene Expression Omnibus (GEO) database (GSE95371). We report the use of RNA sequencing to find DEGs that are affected by different feeding regimes in the thymus.

1. Introduction

Calorie restriction (CR) is the reduction in calorie intake without inducing malnutrition [1, 2]. The two main types of calorie restriction are chronic calorie restriction (CCR) and intermittent calorie restriction (ICR). ICR refers to the application of calorie restriction in periods "on" and "off" [3–6]. Researchers have reported that CR is a more highly effective experimental manipulation for suppressing tumor development [7, 8], suppressing autoimmunity [9, 10], and extending lifespan [11] than fed ad libitum (AL) diet, in rodents [12]. Moreover, in genetically engineered animal models, several studies have shown that ICR is more effective for prevention of cancer development compared to CCR [4, 13–16], while other studies have found that ICR is less effective for cancer prevention than CCR [17–22].

The thymus, which plays an important role in the development of the immune system, is a primary lymphoid organ and a place of T-cell differentiation and maturation [23, 24]. The thymus and other lymphoid organs react to nutrition deficiency more rapidly than most of the other organs [25, 26]. Several studies show that CR potentiates thymic function [2] and can regulate thymic adiposity [27], since CR specifically inhibits the adipogenic transcription in the aging thymus [28].

Breast cancer is the most frequent cancer in women and causes mainly the death of millions of women each year. Animal studies have shown that calorie restriction prevents mammary tumor development [29–32]. MMTV-TGF-α transgenic mice have a particular value in age-related mammary tumor (MT) development studies. These mice have been reported to develop MT in their second year of life

[33] and overexpress TGF-α, epidermal growth factor, which plays a critical role in the development of the human breast cancer [34–38].

RNA sequencing (RNA-seq) is one of the applications of the next generation sequencing (NGS) technologies, with which gene expression can be measured [39–42]. It is a sensitive, fast, and efficient method for gene discovery in organisms [43, 44]. Moreover, RNA-seq has been successfully used for annotation, transcript profiling, detecting gene fusions, single-nucleotide polymorphism (SNP) discovery, and detecting alternatively spliced RNA forms [40, 45–49].

In the present study, we used RNA-seq technology to perform a comparative transcriptome analysis of the MMTV-TGF-α female mouse thymus tissues. The mice were subjected to AL, CCR (85% of AL-fed mice), and ICR (3 weeks AL fed, 1 week 40% of AL-fed mice) diets from 10 weeks of age to 17 weeks of age or 18 weeks of age. The aim of this study is to determine the differences in the gene expression profiles of the thymus tissue due to different feeding regimes.

2. Materials and Methods

2.1. Animals and Experimental Design. In this experiment, MMTV-TGF-α (C57/BL6) female mice were used. These mice overexpress human TGF-α, a part of epidermal growth factor receptor (EGFR)/ErbB cascade which is known to play a role in the development of human breast cancers [34–38]. Mice colonies were maintained using a breeding protocol and genotyping assay at Yeditepe University Animal Facility, as previously described [13]. At 10 weeks of age, female MMTV-TGF-α mice were assigned to one of the following dietary groups: ad libitum (AL), chronic caloric restricted (CCR), and intermittent caloric restricted (ICR). The CCR group received 85% of the daily food consumption of AL mice, in other words, 15% caloric restriction were applied to them. The ICR group was fed AL for 3 weeks, and then for the following week, 60% caloric restriction compared to AL was applied for one week. Mice diets (Altromin TPF1414) were purchased from Kobay AS (Ankara, Turkey). All mice had free access to water. Body weights were measured weekly, and food intakes were determined daily, for all mice. The health statuses of the animals were checked by an expert veterinarian on a regular basis, at least once a week. Mice were euthanized after overnight fasting, at the age of 17 or 18 weeks old. Mice in the ICR group were euthanized after three weeks of AL feeding. Thymus tissues were collected in liquid nitrogen and stored at minus 80°C, until used.

2.2. RNA Extraction, Library Construction, and Sequencing. Total RNA was extracted from the thymus tissue samples of three different individual mice in each different diet group, using Trizol extraction method (Invitrogen, Carlsbad, CA, USA) according to the manufacturer's protocol, then further purified with RNeasy columns (Qiagen). The concentration of each RNA sample was measured with BioSpec-nano UV-VIS specthrophometer (Shimadzu, Kyoto, Japan). The integrity of the RNA was assessed by the Agilent 2100

Bioanalyzer system (Agilent Technologies, Santa Clara, CA). The mRNA sequencing libraries for RNA-seq were constructed with the TruSeq RNA Sample Prep Kit (Illumina, San Diego, CA), according to the manufacturer's instructions. Pair-end (2×100 bp) sequencing was performed using an Illumina HiSeq 4000 Sequencing System (Illumina) at Beijing Genomics Institute (BGI). RNA-seq data has been deposited in NCBI Gene Expression Omnibus (GEO) database under accession number GSE95371.

2.3. RNA-seq Data Processing. Raw pair-end reads were subjected to quality control, and clean reads were filtered with FASTX-Toolkit (http://hannonlab.cshl.edu/fastx_toolkit/index.html). Clean reads were mapped to GRCm38 mouse assembly in the Ensemble database using Tophat 2.0.13 [50]. Gene expression levels were quantified by Cufflinks 2.2.1 and normalized by the fragments per kilobase of transcript per million fragments mapped method (FPKM). The differentially expressed genes (DEGs) were identified with Cuffdiff, a part of Cufflinks package [51]. The DEGs were filtered employing the false discovery rate (FDR) correction, Fisher's exact test, and a fold-change method, exhibiting a corrected p value not greater than 0.05 (p value < 0.05) along with a fold change value not less than 2.0. All of the data produced by a Cuffdiff analysis was visualized and integrated with R [52].

Gene Ontology (GO) is an international, standardized, gene-function classification system. GO enrichment analysis identifies all of the GO terms that are significantly enriched in DEGs compared with the genome background and filters the DEGs that correspond to biological functions. According to this method, all DEGs have been mapped to GO terms in the database (http://www.geneontology.org/); gene numbers have been calculated for every term, using a hypergeometric distribution compared with the genome background [43]. We mapped all the DEGs obtained from these libraries (p value < 0.05) to GO database, to classify for enriched GO terms.

2.4. RNA-seq Data Validation by qPCR. After the analysis of RNA-seq data, four of the important genes for calorie restriction studies, leptin (Lep), ghrelin (Ghr), insulin-like growth factor 1 (Igf1), and adinopectin (Adipoq), were selected and their differential expression results were validated by using quantitative real-time PCR (qPCR). All qPCR reactions were run in replicates using GM SYBR Green qPCR kit and conducted on the LightCycler Nano real-time system (Roche, Switzerland). The PCR conditions were as follows: denaturation at 95°C for 2 min followed by 45 cycles of amplification (95°C for 20 s, 58°C for 30 s, and 72°C for 45 s). Primer sequences can be found in Table 1. The $2^{-\Delta CT}$ method was used to calculate relative gene expression levels in each sample, and Gapdh was used as an internal control.

3. Results

3.1. Summary of RNA Isolation, RNA-seq Libraries, and Mapping. Nine total RNA isolations were done, and three sequencing libraries were constructed from thymus tissues

TABLE 1: The information of the primer pairs used for the analysis of gene expression levels by qPCR.

Gene name	Primer	Product size (bp)
Leptin	5′GGT TGT CCA GGG TTG ATC TC 3′ 5′GTG GGA GAC AGG GTT CTA CT3′	110 bp
Ghrl	5′GCT GTC TTC AGG CAC CAT CT3′ 5′TTC TCT GCT GGG CTT TCT GG5′	113 bp
Igf1	5′CAA GTC CAG AGA GGA AGC TAT G3′ 5′CCG AGA GGT GGA GTG ATT TG3′	155 bp
AdipoQ	5′GCA CGA GGG ATG CTA CTG TT3′ 5′CAC AAG TTC CCT TGG GTG GA3′	127 bp
Gapdh	5′ACT CCA CTC ACG GCA AAT TC3′ 5′CAG TAG ACT CCA CGA CAT ACT C3′	150 bp

TABLE 2: The statistical results for the AL, CCR, and ICR diet groups' libraries.

Diet groups	Raw reads	Clean reads	Read length (bp)	Clean bases	GC (%)
AL	39,760,624	34,926,546	100	3,492,654,600	49.63
CCR	39,760,624	34,991,272	100	3,499,127,200	48.4
ICR	47,826,710	35,113,458	100	3,511,345,800	47.91

of MMTV-TGF-α mice in the AL, CCR, and ICR diet groups for RNA-seq. Three libraries were subsequently sequenced on an Illumina HiSeq 4000 Sequencing System (Illumina) generated about 127 million raw reads. After filtering adapter sequences, contamination, and low-quality sequences, a total of 105 million clean reads were finally produced (Table 2).

Clean reads were mapped to GRCm38 mouse assembly in the Ensemble database with TopHat. As a result, mapping ratios (mapped reads/all reads) of 91.9%–94.9% were attained for all the three sequencing libraries.

3.2. Differentially Expressed Genes (DEGs) and Gene Ontology (GO) Analysis.
To detect transcriptomic changes in gene expression, gene expression levels were computed by Cufflinks 2.2.1 and normalized to FPKM values. 44,426 detected genes were quantified with corrected FPKM values. According to the results, the number of isoforms, TSS (transcription start site), CDS (coding sequences), and promoters were quantified 138,066, 80,392, 48,900, and 133,278 in three sequencing libraries, respectively. Corrected FPKM values were subjected to analysis of DEGs; a total of 6091 significantly differentially expressed genes were identified in three different diet groups, by using corrected p value < 0.05 as the filter (Table 3).

The 2821, 2825, and 445 significantly differentially expressed genes were detected between the diet groups AL-CCR, CCR-ICR, and AL-ICR, respectively ($p < 0.05$). The numbers of significantly differentially expressed genes (DEGs) and numbers of isoforms, TSS, and CDS between the diet groups are shown in Table 3. According to these results, 916, 1877, and 200 genes were upregulated and 1905,

TABLE 3: Number of genes differentially expressed between diet groups.

Diet groups	AL-CCR	CCR-ICR	AL-ICR
Number of significantly DEGs	2821	2825	445
Number of isoforms	1686	1637	331
Number of TSS	2391	2314	461
Number of CDS	1739	1614	316

948, and 245 genes were downregulated in DEGs between the diet groups AL and CCR, CCR and ICR, and AL and ICR, respectively.

For better understanding of gene function, DEGs obtained from three libraries were further subjected to Gene Ontology (GO) functional enrichment analysis, which provided biological terms to identify gene products in three perspectives: cellular components, biological processes, and molecular functions. DEGs obtained between the AL and CCR, AL and ICR, and CCR and ICR diet groups ($p < 0.05$) were classified according to three main categories of GO terms via http://www.geneontology.org/, as shown in Figure 1.

The assigned functions of DEGs covered a broad range of GO categories. In the molecular function category, catalytic activity (GO:0003824) (978 unigenes, 34.6%) and binding (GO:0005488) (882 unigenes, 31.2%) in significant 2821 DEGs between AL and CCR (AL-CCR), catalytic activity (GO:0003824) (170 unigenes, 38%) and binding (GO:0005488) (183 unigenes, 41.1%) in significant 445 DEGs between AL and ICR (AL-ICR), and

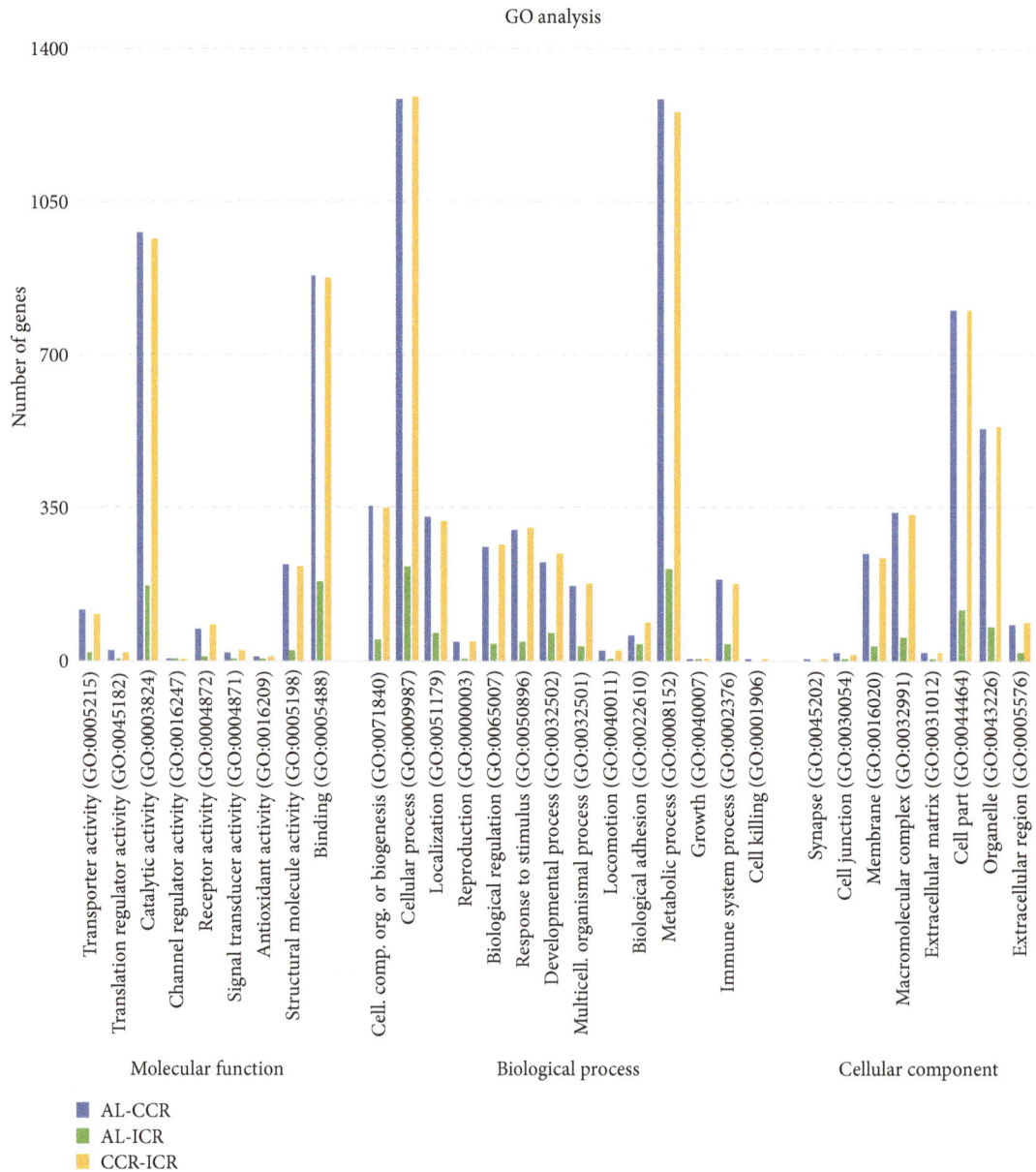

FIGURE 1: Clasifications and numbers of DEGs into GO terms (AL-CCR: DEGs between AL and CCR; AL-ICR: DEGs between AL and ICR; and CCR-ICR: DEGs between CCR and ICR).

catalytic activity (GO:0003824) (966 unigenes, 34.1%) and binding (GO:0005488) (875 unigenes, 31%) in significant 2825 DEGs between CCR and ICR (CCR-ICR) were prominently represented.

GO results showed that in the top two abundant GO terms in biological process, categories were cellular process (GO:0009987) and metabolic process (GO:0008152). In cellular process (GO:0009987), 1283 unigenes (45.4%), 216 unigenes (48.5%), and 1288 unigenes (45.5%) were involved in significant DEGs AL-CCR, AL-ICR, and CCR-ICR, respectively. Also, in metabolic process (GO:0008152), 1285 unigenes (45.5%), 212 unigenes (47.6%), and 1255 (44.4%) were involved in significant DEGs AL-CCR, AL-ICR, and CCR-ICR, respectively.

In the category of cellular component, 802, 119, and 802 unigenes were located in the cell parts, and 531, 78, and 537 unigenes were located in the organelle parts.

Besides, the expression of genes grouped as "immune system process" GO term (0002376) based on analyzed transcriptome reveals that 188 of 2821, 36 of 445, and 176 of 2825 genes were differentially expressed between AL-CCR, AL-ICR, and CCR-ICR diet groups, respectively. These DEGs were shown in Figure 2 according to the three different diet groups.

3.3. Validation of Gene Expression. To validate the results of differentially expressed genes in transcriptome sequencing, leptin (Lep), ghrelin (Ghr), insulin-like growth factor 1

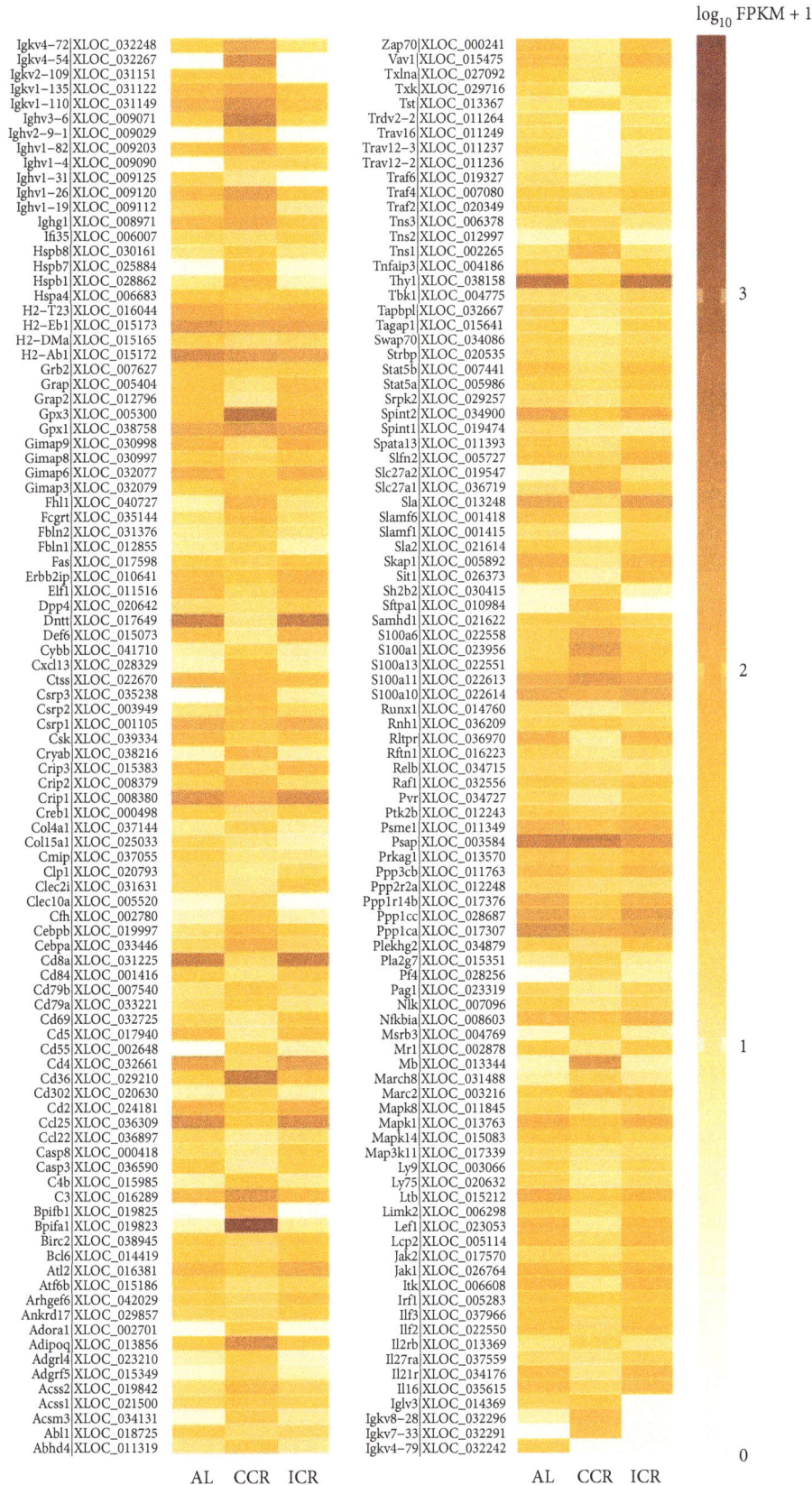

FIGURE 2: Heat map showing the expression profiles of significantly differentially expressed genes between the AL, CCR, and ICR diet groups involved in immune response processes (GO: 0002376).

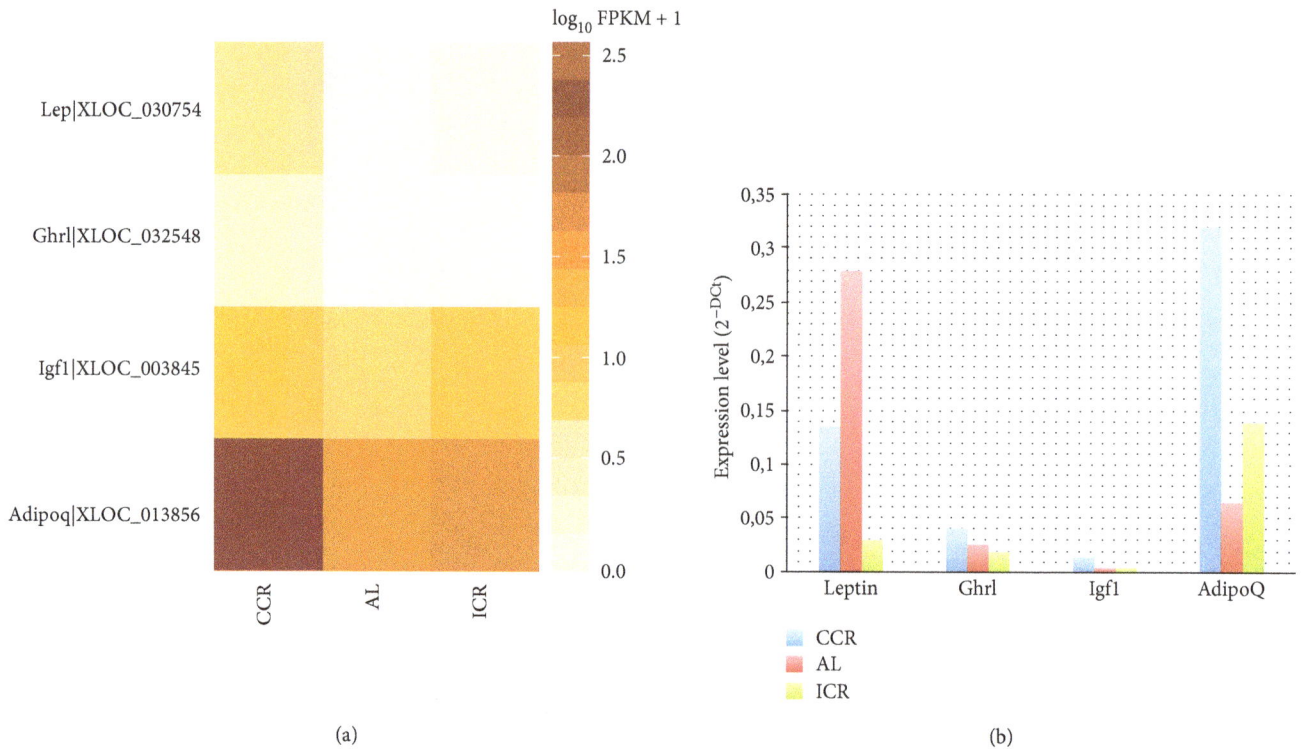

(a)

(b)

FIGURE 3: (a) Heat map showing the expression profiles of leptin, ghrelin, Igf1, and adiponectin genes in the CCR, AL, and ICR groups revealed by RNA-seq. (b) qPCR validation showing the expression levels of leptin, ghrelin, Igf1, and adiponectin genes in the CCR, AL, and ICR groups.

(Igf1), and adinopectin (Adipoq) genes were selected because they were differentially expressed in each diet group. Also, important roles of these genes in adipogenesis, nutrition metabolism, and tumor development have been reported [15, 53–56]. The results of RNA-seq (Figure 3(a)) and qPCR (Figure 3(b)) for four differentially expressed genes were shown in Figure 3. According to the results of RNA-seq, Lep, Ghr, and Adipoq were upregulated in the CCR diet group compared with the ICR and AL diet groups but Igf1 was upregulated in the CCR and ICR diet groups compared with the AL diet group (Figure 3(a)). According to the qPCR results, Ghr, Igf1, and Adipoq were upregulated in the CCR diet group compared with the ICR and AL diet groups but Lep was upregulated in the AL diet group compared with other diet groups (Figure 3(b)).

4. Discussion

Calorie restriction (CR) is a strong metabolic intervention that induces a state of chronic negative energy balance and robustly expands mean and maximal lifespan in experimental animals [57, 58]. There are many studies from animal models that suggest that calorie restriction has a significant impact on various arms of the immune system. Most of the reports suggest that CR improves many parameters of immune responses [2, 28, 56]. There are two major calorie restriction applications: chronic calorie restriction (CCR) and intermittent calorie restriction (ICR). Although CCR method is commonly applied in most studies, there are a

limited number of studies that have been reported for ICR. In many experimental animal studies, ICR and CCR have been reported to have anticancer effects. Some of the studies show that compared to the AL group, CCR is more effective, while the others showed that ICR is more effective [59].

The thymus is one of the major lymphoid organs in the immune system, and several studies have reported that calorie restriction (CR) potentiates thymic function [2, 28, 60]. RNA were isolated from thymus tissue of ad libitum, chronic calorie restriction, and intermittent calorie restriction diet MMTV-TGF-α mice from 10 weeks of age to 17 weeks of age or 18 weeks of age. RNA-seq resulted in an average of ~127 million raw reads, 100 bp reads per sample with average of ~95 million mapped reads. 6091 significantly differentially expressed genes (DEGs) were obtained between three diet groups. The results of DEGs were annotated into molecular function, cellular component, and biological process GO terms.

According to the DEG analysis, there are 2821, 2825, and 445 significantly differentially expressed genes between the diet groups AL-CCR, CCR-ICR, and AL-ICR, respectively. These results show that calorie restriction and/or the types of calorie consumption has a great effect on gene expression in the thymus. Compared to the AL group, although only 15% of calorie restriction was applied to the CCR diet group, 2821 DEGs were shown between the AL and CCR diet groups. Despite the fact that, compared to the AL diet group, only one-week long calorie restriction (60%) was applied to the ICR group, 445 significantly DEGs were determined between the AL and ICR groups.

The studies have reported that adinopectin (Adipoq), leptin (Lep), ghrelin (Ghr), and insulin-like growth factor (Igf1) hormones have an important role in the cancer development and immunity [15, 54, 56]. Adinopectin is a significant hormone to initiate insulin sensitivity; its levels rise during CR [61, 62]. Leptin decreases the level of stress hormones and rises thyroid activity and thyroid-hormone levels [63]. Because CR upregulates stress hormones and downregulates thyroid hormones, leptin levels decline in CR [63, 64]. Ghrelin also regulates immune function by reducing proinflammatory cytokines [56]. In the present study, increased Igf1 levels were detected in the CCR and ICR diet groups, compared to the AL group. Although many studies have reported reduced level of Igf1 with calorie restriction [65–67], there are other reports that show either no change or increase in Igf1 levels with calorie restriction [68, 69]. In this context, current results support our previous findings which was done using the same mouse model [70]. In our previous study, we also reported increased levels of Igf1 and IGFBP3 protein expressions in mammary fat pad tissue of mice of the CCR and ICR diet groups, compared to the AL group at 37 weeks of age [70]. In addition, Igf1 gene expression levels were similar among all diet groups. It should also be noticed that 18 weeks of age is considered to be young for a mouse model. Therefore, it is not unexpected to see higher Igf1 gene expression levels in the calorie-restricted group which needs more growth factors. With respect to our RNA-seq and qPCR results, adinopectin gene expression level increases in the CCR diet group. According to RNA-seq results, leptin gene expression level is highest in the CCR diet group, but according to the qPCR results, it has highest expression level in the AL diet group. Ghrelin gene expression was only observed in the CCR group; this result indicates CCR effects developing immune function positively.

The current studies suggest that CR improves many parameters of immune responses [2, 28, 56, 71, 72], such as responses of T-cells to mitogens, natural killer cell activity, and the ability of mononuclear cells to produce proinflammatory cytokines. According to the GO analysis results, the number of immune response of DEGs was higher in between the AL-CCR and CCR-ICR diet groups, than between the AL-ICR groups. This shows that the expression of immune system genes was regulated up or downregulated by chronic diet.

Results from present study indicate that RNA-seq is a powerful tool to analyze transcriptomes, to study gene expression profiles and to compare distinct stages of different conditions. According to our findings, differences of gene expression have occurred in AL, CCR, and ICR diet types and these results will provide new clues for caloric restriction, immune system, and cancer development studies.

Conflicts of Interest

The authors declare that they have no conflicts of interest.

Acknowledgments

This work was supported by Yildiz Technical University Bilimsel Araştırma Projeleri Koordinatorlugu (BAPK) (Grant no. 2016-01-07-DOP01). The authors thank Dr. Margot P. Cleary for generously donating the breeding colony of the MMTV-TGF-α transgenic mice. The animal experiments were supported by TUBITAK Grant 114S429 (Soner Dogan). The authors thank Ilker Coban, Munevver Burcu Cicekdal, Busra Kazan, and Mustafa Erhan Ozer for overseeing the breeding and genotyping protocols to obtain the experimental animals. The authors also thank the veterinarian and animal technicians who handle the animals at Yeditepe University Animal Facility (YUDETAM).

References

[1] A. Nozad, M. B. Safari, E. Saboory et al., "Caloric restriction and formalin-induced inflammation: an experimental study in rat model," *Iranian Red Crescent Medical Journal*, vol. 17, no. 6, article e22590, 2015.

[2] J. Nikolich-Zugich and I. Messaoudi, "Mice and flies and monkeys too: caloric restriction rejuvenates the aging immune system of non-human primates," *Experimental Gerontology*, vol. 40, no. 11, pp. 884–893, 2005.

[3] S. D. Hursting, J. A. Lavigne, D. Berrigan, S. N. Perkins, and J. C. Barrett, "Calorie restriction, aging, and cancer prevention: mechanisms of action and applicability to humans," *Annual Review of Medicine*, vol. 54, no. 1, pp. 131–152, 2003.

[4] M. P. Cleary, X. Hu, M. E. Grossmann et al., "Prevention of mammary tumorigenesis by intermittent caloric restriction: does caloric intake during refeeding modulate the response?," *Experimental Biology and Medicine*, vol. 232, no. 1, pp. 70–80, 2007.

[5] S. Dogan, X. Hu, Y. Zhang, N. J. Maihle, J. P. Grande, and M. P. Cleary, "Effects of high fat diet and/or body weight on mammary tumor leptin and apoptosis signaling pathways in MMTV-TGF-α mice," *Breast Cancer Research*, vol. 9, no. 6, article R91, 2007.

[6] W. Jiang, Z. Zhu, and H. J. Thompson, "Effects of physical activity and restricted energy intake on chemically induced mammary carcinogenesis," *Cancer Prevention Research*, vol. 2, no. 4, pp. 338–344, 2009.

[7] R. H. Weindruch and R. L. Walford, *The Retardation of Aging and Disease by Dietary Restriction*, Charles C. Thomas, Springfield, IL, USA, 1988.

[8] S. D. Hursting and F. W. Kari, "The anti-carcinogenic effects of dietary restriction: mechanisms and future directions," *Mutation Research*, vol. 443, no. 1-2, pp. 235–249, 1999.

[9] X. Luan, W. Zhao, B. Chandrasekar, and G. Fernandes, "Calorie restriction modulates lymphocyte subset phenotype and increases apoptosis in MRLlpr mice," *Immunology Letters*, vol. 47, no. 3, pp. 181–186, 1995.

[10] D. A. Troyer, B. Chandrasekar, J. L. Barnes, and G. Fernandes, "Calorie restriction decreases platelet-derived growth factor (PDGF)-A and thrombin receptor mRNA expression in autoimmune murine lupus nephritis," *Clinical & Experimental Immunology*, vol. 108, no. 1, pp. 58–62, 1997.

[11] M. Ogura, H. Ogura, S. Ikehara, M. L. Dao, and R. A. Good, "Decrease by chronic energy intake restriction of cellular proliferation in the intestinal epithelium and lymphoid organs in autoimmunity-prone mice," *Proceedings of the National Academy of Sciences of the United States of America*, vol. 86, no. 15, pp. 5918–5922, 1989.

[12] H. L. Poetschke, D. B. Klug, S. N. Perkins, T. Y. Wang, E. R. Richie, and S. D. Hursting, "Effects of calorie restriction on thymocyte growth, death and maturation," *Carcinogenesis*, vol. 21, no. 11, pp. 1959–1964, 2000.

[13] M. P. Cleary, M. K. Jacobson, F. C. Phillips, S. C. Getzin, J. P. Grande, and N. J. Maihle, "Weight-cycling decreases incidence and increases latency of mammary tumors to a greater extent than does chronic caloric restriction in mouse mammary tumor virus-transforming growth factor-α female mice," *Cancer Epidemiology, Biomarkers & Prevention*, vol. 11, pp. 836–843, 2002.

[14] O. P. Rogozina, M. J. Bonorden, J. P. Grande, and M. P. Cleary, "Serum insulin-like growth factor-I and mammary tumor development in *Ad libitum*-fed, chronic calorie-restricted, and intermittent calorie-restricted MMTV-TGF-α mice," *Cancer Prevention Research*, vol. 2, no. 8, pp. 712–719, 2009.

[15] S. Dogan, O. P. Rogozina, A. E. Lokshin, J. P. Grande, and M. P. Cleary, "Effects of chronic vs. intermittent calorie restriction on mammary tumor incidence and serum adiponectin and leptin levels in MMTV-TGF-α mice at different ages," *Oncology Letters*, vol. 1, no. 1, pp. 167–176, 2010.

[16] O. P. Rogozina, K. J. Nkhata, E. J. Nagle, J. P. Grande, and M. P. Cleary, "The protective effect of intermittent calorie restriction on mammary tumorigenesis is not compromised by consumption of a high fat diet during refeeding," *Breast Cancer Research and Treatment*, vol. 138, no. 2, pp. 395–406, 2013.

[17] D. Berrigan, S. N. Perkins, D. C. Haines, and S. D. Hursting, "Adult-onset calorie restriction and fasting delay spontaneous tumorigenesis in p53-deficient mice," *Carcinogenesis*, vol. 23, no. 5, pp. 817–822, 2002.

[18] M. J. Bonorden, O. P. Rogozina, C. M. Kluczny et al., "Intermittent calorie restriction delays prostate tumor detection and increases survival time in TRAMP mice," *Nutrition and Cancer*, vol. 61, no. 2, pp. 265–275, 2009.

[19] M. P. Cleary and M. E. Grossmann, "The manner in which calories are restricted impacts mammary tumor cancer prevention," *Journal of Carcinogenesis*, vol. 10, no. 1, p. 21, 2011.

[20] S. Lanza-Jacoby, G. Yan, G. Radice, C. LePhong, J. Baliff, and R. Hess, "Calorie restriction delays the progression of lesions to pancreatic cancer in the LSL-Kras^{G12D}; Pdx-1/Cre mouse model of pancreatic cancer," *Experimental Biology and Medicine*, vol. 238, no. 7, pp. 787–797, 2013.

[21] N. K. Mizuno, O. P. Rogozina, C. M. Seppanen, D. J. Liao, M. P. Cleary, and M. E. Grossmann, "Combination of intermittent calorie restriction and eicosapentaenoic acid for inhibition of mammary tumors," *Cancer Prevention Research*, vol. 6, no. 6, pp. 540–547, 2013.

[22] K. A. Pape-Ansorge, J. P. Grande, T. A. Christensen, N. J. Maihle, and M. P. Cleary, "Effect of moderate caloric restriction and/or weight cycling on mammary tumor incidence and latency in MMTV-Neu female mice," *Nutrition and Cancer*, vol. 44, no. 2, pp. 162–168, 2002.

[23] S. E. Wong, T. A. Papenfuss, A. Heger et al., "Transcriptomic analysis supports similar functional roles for the two thymuses of the tammar wallaby," *BMC Genomics*, vol. 12, no. 1, p. 420, 2011.

[24] D. Ribatti, E. Crivellato, and A. Vacca, "Miller's seminal studies on the role of thymus in immunity," *Clinical and Experimental Immunology*, vol. 144, no. 3, pp. 371–375, 2006.

[25] G. T. Keusch, C. S. Wilson, and S. D. Waksal, "Nutrition, host defenses and the lymphoid system," in *Advances in Host Defense Mechanisms*, J. I. Gallin and A. S. Fauci, Eds., vol. 2, pp. 275–359, Raven, New York, NY, USA, 1983.

[26] A. Gartner, W. T. Castellon, G. Gallon, and F. Simondon, "Total dietary restriction and thymus, spleen, and phenotype and function of splenocytes in growing mice," *Nutrition*, vol. 8, no. 4, pp. 258–265, 1992.

[27] Y. H. Youm, H. Yang, Y. Sun et al., "Deficient ghrelin receptor mediated signaling compromises thymic stromal cell microenvironment by accelerating thymic adiposity," *The Journal of Biological Chemistry*, vol. 284, no. 11, pp. 7068–7077, 2009.

[28] H. Yang, Y. H. Youm, and V. D. Dixit, "Inhibition of thymic adipogenesis by caloric restriction is coupled with reduction in age-related thymic involution," *The Journal of Immunology*, vol. 183, no. 5, pp. 3040–3052, 2009.

[29] M. J. Dirx, M. P. Zeegers, P. C. Dagnelie, T. Van den Bogaard, and P. A. Van den Brandt, "Energy restriction and the risk of spontaneous mammary tumors in mice: a meta-analysis," *International Journal of Cancer*, vol. 106, no. 5, pp. 766–770, 2003.

[30] M. W. Pariza, "Fat, calories, and mammary carcinogenesis: net energy effects," *The American Journal of Clinical Nutrition*, vol. 45, Supplement 1, pp. 261–263, 1987.

[31] M. J. Tucker, "The effect of long-term food restriction on tumours in rodents," *International Journal of Cancer*, vol. 23, no. 6, pp. 803–807, 1979.

[32] C. A. Gillette, Z. Zhu, K. C. Westerlind, C. L. Melby, P. Wolfe, and H. J. Thompson, "Energy availability and mammary carcinogenesis: effects of calorie restriction and exercise," *Carcinogenesis*, vol. 18, no. 6, pp. 1183–1188, 1997.

[33] S. A. Halter, P. Dempsey, Y. Matsui et al., "Distinctive patterns of hyperplasiain transgenic mice with mouse mammary tumor virus transforming growth factor (alpha). Characterization of mammary gland and skin proliferations," *The American Journal of Pathology*, vol. 140, no. 5, pp. 1131–1146, 1992.

[34] Y. Matsui, S. A. Halter, J. Holt, B. L. Hogan, and R. J. Coffey, "Development of mammary hyperplasia and neoplasia in MMTV-TGFα transgenic mice," *Cell*, vol. 61, no. 6, pp. 1147–1155, 1990.

[35] J. Lundy, A. Schuss, D. Stanick, E. S. Mccormack, S. Kramer, and J. M. Sorvillo, "Expression of neu protein, epidermal growth factor receptor and transforming growth factor alpha in breast cancer. Correlation with clinicopathologic parameters," *The American Journal of Pathology*, vol. 138, no. 6, pp. 1527–1534, 1991.

[36] P. A. Murray, P. Barrett-lee, M. Travers, Y. Luqmani, T. Powles, and R. C. Coombes, "The prognostic significance of transforming growth factors in human breast cancer," *British Journal of Cancer*, vol. 67, no. 6, pp. 1408–1412, 1993.

[37] T. Rajkumar and W. J. Gullick, "The type I growth factor receptors in human breast cancer," *Breast Cancer Research and Treatment*, vol. 29, no. 1, pp. 3–9, 1994.

[38] L. Panico, A. D'Antonio, G. Salvatore et al., "Differential immunohistochemical detection of transforming growth factor α, amphiregulin and crIPto in human normal and malignant breast tissues," *International Journal of Cancer*, vol. 65, no. 1, pp. 51–56, 1996.

[39] Y. Benjamini and D. Yekutieli, "The control of the false discovery rate in multiple testing under dependency," *The Annals of Statistics*, vol. 29, pp. 1165–1188, 2011.

[40] A. Mortazavi, B. A. Williams, K. Mccue, L. Schaeffer, and B. Wold, "Mapping and quantifying mammalian transcriptomes by RNA-Seq," *Nature Methods*, vol. 5, no. 7, pp. 621–628, 2008.

[41] Z. Wang, M. Gerstein, and M. Snyder, "RNA-Seq: a revolutionary tool for transcriptomics," *Nature Reviews Genetics*, vol. 10, no. 1, pp. 57–63, 2009.

[42] V. Costa, C. Angelini, I. De Feis, and A. Ciccodicola, "Uncovering the complexity of transcriptomes with RNA-Seq," *Journal of Biomedicine and Biotechnology*, vol. 2010, Article ID 853916, 19 pages, 2010.

[43] G. F. Liu, H. J. Cheng, W. You, E. L. Song, X. M. Liu, and F. C. Wan, "Transcriptome profiling of muscle by RNA-Seq reveals significant differences in digital gene expression profiling between Angus and Luxi cattle," *Animal Production Science*, vol. 55, no. 9, pp. 1172–1178, 2015.

[44] S. Ren, Z. Peng, J. H. Mao et al., "RNA-seq analysis of prostate cancer in the Chinese population identifies recurrent gene fusions, cancer-associated long noncoding RNAs and aberrant alternative splicings," *Cell Research*, vol. 22, no. 5, pp. 806–821, 2012.

[45] F. Ozsolak and P. M. Milos, "RNA sequencing: advances, challenges and opportunities," *Nature Reviews Genetics*, vol. 12, no. 2, pp. 87–98, 2011.

[46] Z. Y. Wang, B. P. Fang, J. Y. Chen et al., "*De novo* assembly and characterization of root transcriptome using Illumina paired-end sequencing and development of cSSR markers in sweet potato (*Ipomoea batatas*)," *BMC Genomics*, vol. 11, p. 726, 2011.

[47] S. Zenoni, A. Ferrarini, E. Giacomelli et al., "Characterization of transcriptional complexity during berry development in *Vitis vinifera* using RNA-Seq," *Plant Physiology*, vol. 152, no. 4, pp. 1787–1795, 2010.

[48] Q. Tang, X. J. Ma, C. M. Mo et al., "An efficient approach to finding *Siraitia grosvenorii* triterpene biosynthetic genes by RNA-seq and digital gene expression analysis," *BMC Genomics*, vol. 12, no. 1, p. 343, 2011.

[49] S. S. Yang, Z. J. Tu, F. Cheung et al., "Using RNA-Seq for gene identification, polymorphism detection and transcript profiling in two alfalfa genotypes with divergent cell wall composition in stems," *BMC Genomics*, vol. 12, no. 1, p. 199, 2011.

[50] S. Andrews, C. Trapnell, L. Pachter, and S. L. Salzberg, "TopHat: discovering splice junctions with RNA-Seq," *Bioinformatics*, vol. 25, no. 9, pp. 1105–1111, 2009.

[51] C. Trapnell and S. L. Salzberg, "How to map billions of short reads onto genomes," *Nature Biotechnology*, vol. 27, no. 5, pp. 455–457, 2009.

[52] R Core Team, *R: a Language and Environment for Statistical Computing, R Foundation for Statistical Computing*, Vienna, Austria, 2014.

[53] S. Dogan, A. Ray, and M. P. Cleary, "The influence of different calorie restriction protocols on serum pro-inflammatory cytokines, adipokines and IGF-I levels in female C57BL6 mice: short term and long term diet effects," *Meta Gene*, vol. 12, pp. 22–32, 2017.

[54] L. Delort, A. Rossary, M. Farges, M. Vasson, and F. Caldefie-Chézet, "Leptin, adipocytes and breast cancer: focus on inflammation and anti-tumor immunity," *Life Sciences*, vol. 140, pp. 37–48, 2015.

[55] O. P. Rogozina, M. J. L. Bonorden, C. N. Seppanen, J. P. Grande, and M. P. Cleary, "Effect of chronic and intermittent calorie restriction on serum adiponectin and leptin and mammary tumorigenesis," *Cancer Prevention Research*, vol. 4, no. 4, pp. 568–581, 2011.

[56] V. D. Dixit, "Adipose-immune interactions during obesity and caloric restriction: reciprocal mechanisms regulating immunity and health span," *Journal of Leukocyte Biology*, vol. 84, no. 4, pp. 882–892, 2008.

[57] J. L. Barger, R. L. Walford, and R. Weindruch, "The retardation of aging by caloric restriction: its significance in the transgenic era," *Experimental Gerontology*, vol. 38, no. 11-12, pp. 1343–1351, 2003.

[58] N. A. Bishop and L. Guarente, "Genetic links between diet and lifespan: shared mechanisms from yeast to humans," *Nature Reviews Genetics*, vol. 8, no. 11, pp. 835–844, 2007.

[59] Y. Chen, L. Ling, G. Su et al., "Effect of intermittent versus chronic calorie restriction on tumor incidence: a systematic review and meta-analysis of animal studies," *Scientific Reports*, vol. 6, no. 1, article 33739, 2016.

[60] M. J. White, C. M. Beaver, M. R. Goodier et al., "Calorie restriction attenuates terminal differentiation of immune cells," *Frontiers in Immunology*, vol. 7, p. 667, 2016.

[61] U. Meier and A. M. Gressner, "Endocrine regulation of energy metabolism: review of pathobiochemical and clinical chemical aspects of leptin, ghrelin, adiponectin, and resistin," *Clinical Chemistry*, vol. 50, no. 9, pp. 1511–1525, 2004.

[62] U. B. Pajvani and P. E. Scherer, "Adiponectin: systemic contributor to insulin sensitivity," *Current Diabetes Reports*, vol. 3, no. 3, pp. 207–213, 2003.

[63] N. Barzilai and G. Gupta, "Revisiting the role of fat mass in the life extension induced by caloric restriction," *The Journals of Gerontology Series A*, vol. 54, no. 3, pp. B89–B96, 1999.

[64] G. Légrádi, C. H. Emerson, R. S. Ahima, J. S. Flier, and R. M. Lechan, "Leptin prevents fasting-induced suppression of prothyrotropin-releasing hormone messenger ribonucleic acid in neurons of the hypothalamic paraventricular nucleus," *Endocrinology*, vol. 138, no. 6, pp. 2569–2576, 1997.

[65] C. R. Breese, R. L. Ingram, and W. E. Sonntag, "Influence of age and long-term dietary restriction on plasma insulin-like growth factor-1 (IGF-1), IGF-1 gene expression, and IGF-1 binding proteins," *Journal of Gerontology*, vol. 46, no. 5, pp. B180–B187, 1991.

[66] S. E. Dunn, F. W. Kari, J. French et al., "Dietary restriction reduces insulin-like growth factor I levels, which modulates apoptosis, cell proliferation, and tumor progression in p53 deficient mice," *Cancer Research*, vol. 57, no. 21, pp. 4667–4672, 1997.

[67] R. J. Klement and M. K. Fink, "Dietary and pharmacological modification of the insulin/IGF-1 system: exploiting the full repertoire against cancer," *Oncogenesis*, vol. 5, no. 2, article e193, 2016.

[68] L. Fontana, D. T. Villareal, and S. K. Das, "Effects of 2-year calorie restriction on circulating levels of IGF-1, IGF-binding proteins and cortisol in nonobese men and women: a randomized clinical trial," *Aging Cell*, vol. 15, no. 1, pp. 22–27, 2016.

[69] L. Fontana, E. P. Weiss, D. T. Villareal, S. Klein, and J. O. Holloszy, "Long-term effects of calorie or protein restriction on serum IGF-1 and IGFBP-3 concentration in humans," *Aging Cell*, vol. 7, no. 5, pp. 681–687, 2008.

[70] S. Dogan, A. C. Johannsen, J. P. Grande, and M. P. Cleary, "Effects of intermittent and chronic calorie restriction on mammalian target of rapamycin (mTOR) and IGF-I

signaling pathways in mammary fat pad tissues and mammary tumors," *Nutrition and Cancer*, vol. 63, no. 3, pp. 389–401, 2011.

[71] D. W. Trott, G. D. Hensonb, M. H. Hoa, S. A. Allisona, L. A. Lesniewskia, and A. J. Donatoa, "Age-related arterial immune cell infiltration in mice is attenuated by caloric restriction or voluntary exercise," *Experimental Gerontology*, 2016, In Press.

[72] S. Fabbiano, N. Sua'rez-Zamorano, D. Rigo et al., "Caloric restriction leads to browning of white adipose tissue through type 2 immune signaling," *Cell Metabolism*, vol. 24, pp. 434–446, 2016.

8

Transcriptome Comparison Reveals Distinct Selection Patterns in Domesticated and Wild Agave Species, the Important CAM Plants

Xing Huang [iD],[1] Bo Wang,[2] Jingen Xi,[1] Yajie Zhang,[3] Chunping He,[1] Jinlong Zheng,[1] Jianming Gao,[4] Helong Chen,[4] Shiqing Zhang,[4] Weihuai Wu,[1] Yanqiong Liang,[1] and Kexian Yi [iD][1]

[1]Environment and Plant Protection Institute, Chinese Academy of Tropical Agricultural Sciences, Haikou 571101, China
[2]National Key Laboratory of Crop Genetic Improvement, Huazhong Agricultural University, Wuhan, Hubei 430070, China
[3]Hainan Climate Center, Haikou 570203, China
[4]Institute of Tropical Bioscience and Biotechnology, Chinese Academy of Tropical Agricultural Sciences, Haikou 571101, China

Correspondence should be addressed to Xing Huang; hxalong@gmail.com and Kexian Yi; yikexian@126.com

Academic Editor: Giandomenico Corrado

Agave species are an important family of crassulacean acid metabolism (CAM) plants with remarkable tolerance to heat and drought stresses (*Agave deserti*) in arid regions and multiple agricultural applications, such as spirit (*Agave tequilana*) and fiber (*Agave sisalana*) production. The agave genomes are commonly too large to sequence, which has significantly restricted our understanding to the molecular basis of stress tolerance and economic traits in agaves. In this study, we collected three transcriptome databases for comparison to reveal the phylogenic relationships and evolution patterns of the three agave species. The results indicated the close but distinctly domesticated relations between *A. tequilana* and *A. sisalana*. Natural abiotic and biotic selections are very important factors that have contributed to distinct economic traits in agave domestication together with artificial selection. Besides, a series of candidate unigenes regulating fructan, fiber, and stress response-related traits were identified in *A. tequilana*, *A. sisalana*, and *A. deserti*, respectively. This study represents the first transcriptome comparison within domesticated and wild agaves, which would serve as a guidance for further studies on agave evolution, environmental adaptation, and improvement of economically important traits.

1. Introduction

Agave species assembled an important group of crassulacean acid metabolism (CAM) plants with remarkable tolerance to heat and drought stresses in arid regions [1]. CAM plants usually have a higher efficiency in water use than C3 and C4 plants [2]. For this reason, CAM plants brought a great chance to enhance sustainable production of food and bioenergy under the background of limited freshwater resources and global climate change [3]. As a traditional cultivated CAM plant, *Agave tequilana* has been used for the production of distilled spirit tequila for centuries [4]. Further, it also shows a great potential in bioenergy production [5]. Besides, *Agave sisalana* has also been widely cultivated as a cash crop

for fiber production in tropical regions [6]. As a native wild plant in the Sonoran Desert regions of the Southwestern United States and Northwestern Mexico, *Agave deserti* has successfully survived in a severe environment within elevation ranges that experience both hot, dry summers and occasional freezing temperatures in winter [7, 8]. This capacity of high tolerance to multiple stresses has important values to the improvement of the main food crops. These economic and stress-tolerant features have made agave a model CAM crop system for a hot and dry/droughty/xeric environment [9].

However, few reports have revealed the physiology and molecular basis of agaves, especially in *A. sisalana*. To date, *A. tequilana*-related studies mainly focused on fructan, with the obvious purpose of improving fructan production and

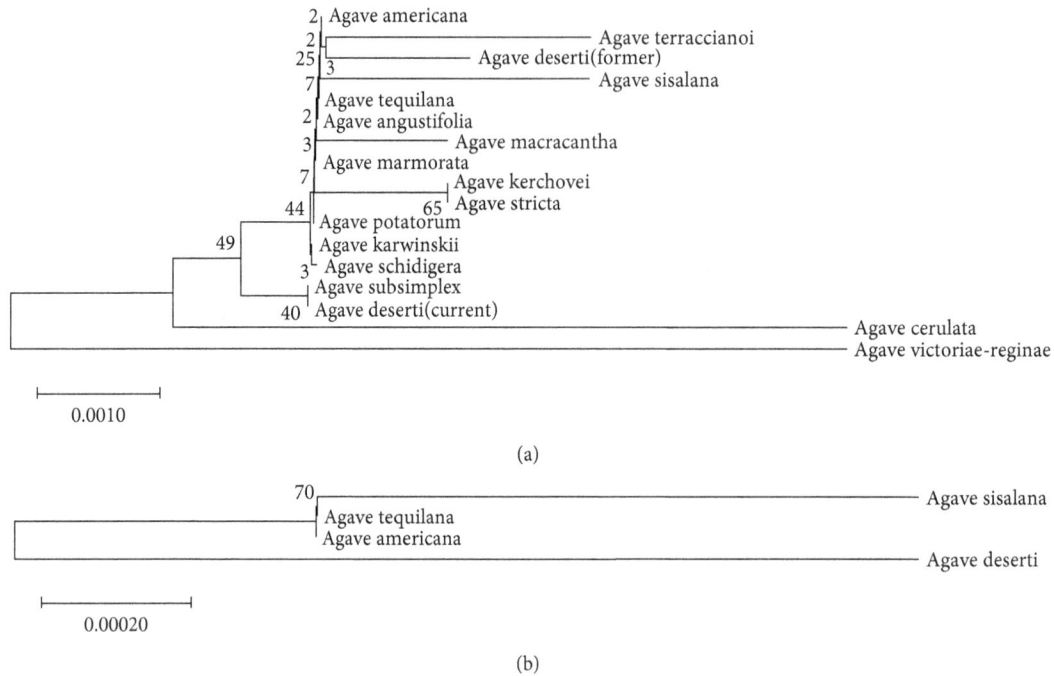

FIGURE 1: Phylogenetic analysis of chloroplast sequences in Agave species. (a) The trnL+trnL-trnF sequences (about 900 bp) were used to construct a phylogenetic tree. (b) The sequences of *A. tequilana* (GAHU01110124), *A. deserti* (GAHT01022741), and *A. americana* (KX519714) were from NCBI. *A. sisalana* (CL7065.Contig2) sequence and the other 13 sequences were from previous studies [15, 24].

TABLE 1: Summary of the transcriptomes for the 3 Agave species.

Species	*A. tequilana*	*A. deserti*	*A. sisalana*
Total sequenced high-quality data	293.5 Gbp	184.7 Gbp	11.3 Gbp
Number of Agave unigenes	204,530	128,869	131,422
Sum length of Agave unigenes	204.9 Mbp	125.0 Mbp	77.6 Mbp
N50 length	1387 bp	1323 bp	861 bp

application to generate bioenergy [5, 10, 11]. *A. deserti*-related studies were mainly conducted for its ecological and physiological adaptation to a severe environment, which is highly valuable for the improvement of the main food crops [12]. Besides, a series of saponin-related researches have been reported, while few reports were related to fiber in *A. sisalana* [13, 14]. These three agave species are closely related species but with different remarkable biological features. A previous study has revealed the phylogenetic relations according to their trnL sequences [15]. The result showed that they were closely related species in spite of different origins for *A. tequilana* (Jalisco), *A. deserti* (the Sonoran Desert), and *A. sisalana* (Chiapas) [4, 6, 7]. Agaves have very large genomes, which has significantly restricted their genome assembly and limited our understanding to their evolution pattern [16]. In other crops, accessible genomes and genome-wide association analysis have revealed many economically important traits for crop improvement [17–19]. Recently, the development of next-generation sequencing has brought a new direction for gene-related studies without the restriction of genome data [20, 21]. Furthermore, transcriptome

comparison has also been conducted for evolution analysis and searching economically important traits in some genome unavailable crops [22]. In this study, we selected three transcriptome databases for comparison to reveal the phylogeny and evolution pattern of the three agave species [23, 24]. Those genes related to species-specific traits would be also identified and evaluated for their importance in agronomy production and environmental adaptation of agaves. This study represents the first transcriptome comparison within domesticated and wild agaves, which would serve as a guidance for further studies on agave evolution, environmental adaptation, and improvement of economically important traits.

2. Materials and Methods

2.1. Phylogenetic Analysis. Phylogenetic analysis was conducted by MEGA 5.0 software with the minimum-evolution method [25]. The methods and parameters were according to the previous study [15]. The bootstrap method was employed for confidence in nodes with 1000 replicates. Partial chloroplast sequences for *A. tequilana* and *A. deserti* were

(a)

(b)

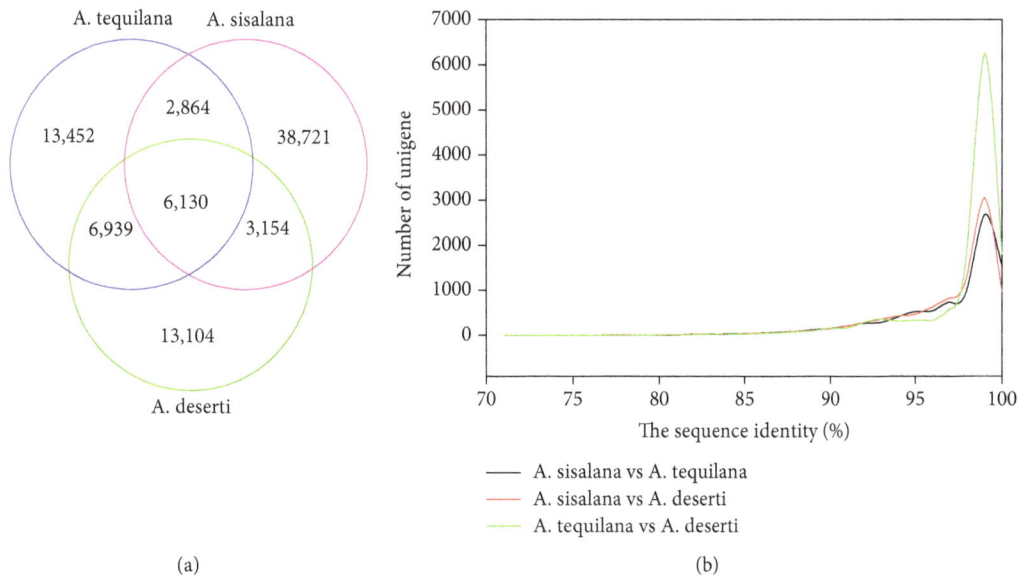

FIGURE 2: Sequence comparisons between domesticated *A. tequilana* and *A. sisalana* and wild *A. deserti* transcriptomes. (a) The sketch map showing 6130 unigene terms that were identified within the three transcriptomes. (b) The identity distribution of all orthologous unigene pairs.

downloaded from NCBI (https://www.ncbi.nlm.nih.gov/) with the accession numbers GAHU01110124 and GAHT01022741, respectively. Partial chloroplast sequences for *A. sisalana* were obtained from a previous transcriptome database [24]. The whole chloroplast sequence of *A. americana* was obtained from NCBI under the accession number KX519714. The trnL+trnL-trnF sequences (about 900 bp) for the 4 and other 14 agave species were also downloaded according to the previous study [15].

2.2. Transcriptome Data and Gene Annotation.

Three agave transcriptome databases were according to previous studies [23, 24]. Transcriptome assembly of the Illumina sequence was performed by Rnnotator for each species, respectively [26]. Only unigenes with corresponding predicted proteins in *A. deserti* transcriptome were used to search orthologous genes from *A. tequilana* and *A. sisalana* transcriptomes by the BBH method, respectively [27]. These orthologous unigenes in three transcriptomes were annotated in the public databases: NCBI nonredundant protein (Nr) and nonredundant nucleotide (Nt) databases (http://www.ncbi.nlm.nih.gov/), Swiss-Prot (http://www.ebi.ac.uk/uniprot/), and Gene Ontology (GO) (http://www.geneontology.org/), respectively.

2.3. Ka/Ks Analysis.

It represents positive selection when the ratio of nonsynonymous (Ka) to synonymous nucleotide substitutions (Ks) is significantly higher than 1, whereas the ratios significantly less than or equal to 1 are subjected to purifying or neutral selection [28]. The Ka, Ks, and Ka/Ks values were estimated by the Codeml model of the program of phylogenetic analysis by maximum likelihood (PAML) between *A. deserti* unigenes with orthologous unigenes in *A. tequilana* or *A. sisalana*, respectively [29].

2.4. In Silico Gene Expression Analysis.

The expression pattern of positive selected unigenes was subjected to in silico gene expression analysis in agave leaves. The reads per kilobase per million mapped read (RPKM) value of these unigenes in *A. deserti* and *A. tequilana* were calculated by RSEM software according to the previous study [23, 30]. The RPKM data was further normalized with two reference genes (tubulin and serine/threonine-protein phosphatase) in each agave species [31].

2.5. Selection Pressure Detection and Protein Structure Modeling.

To detect the selection pressure on positive selection unigenes, Ka/Ks ratios were calculated in sliding window (30 bp under a step size of 6 bp) by using DnaSP 5.0 [32]. Translated protein sequences of positive selection unigenes were used for structure modeling by Swiss-Model (https://swissmodel.expasy.org/) [33].

3. Results

3.1. Phylogeny of Agave Species.

Phylogenetic analysis was conducted to reveal the phylogenic relation for agave species in this study. We obtained chloroplast sequences for *A. tequilana* (GAHU01110124), *A. sisalana* (CL7065.Contig2), and *A. deserti* (GAHT01022741), by using blast against three agave transcriptomes [23, 24]. The chloroplast sequence of *A. americana* was from GenBank under the accession number KX519714 (http://www.ncbi.nlm.nih.gov/genbank/). The trnL+trnL-trnF sequences (about 900 bp) from the 4 and other 13 agave species [15] were employed for phylogenetic analysis to reveal their evolutionary pattern. The results indicated that *A. tequilana*, *A. sisalana*, *A. americana*, and *A. deserti* (former, DQ500894 + DQ500928) sequences were grouped together, while the *A. deserti* (GAHT01022741) sequence was in another group (Figure 1(a)). We further

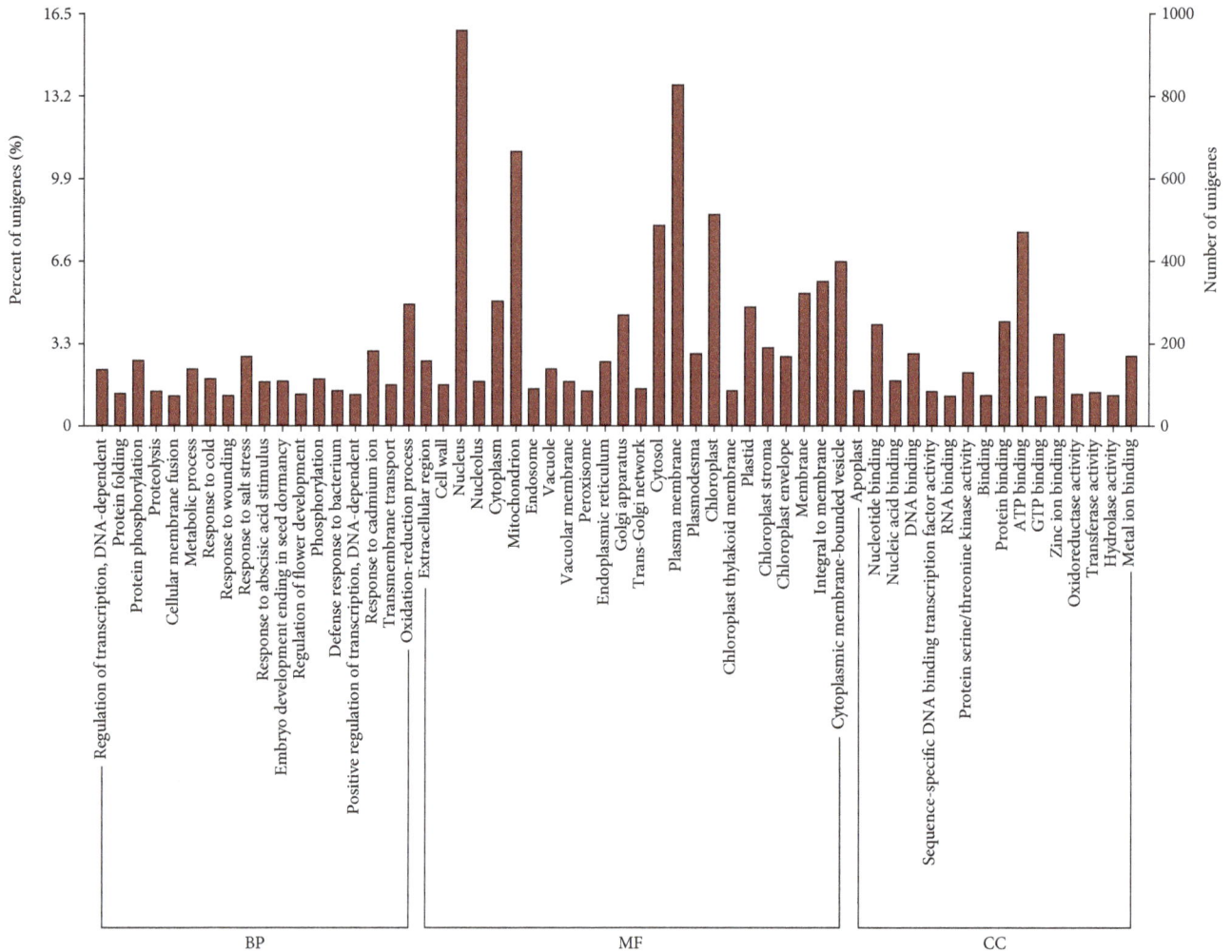

FIGURE 3: Gene ontology classifications of 6130 orthologous unigenes. The results are clustered in the three main categories as biological process (BP), cellular component (CC), and molecular function (MF).

selected the four chloroplast sequences (about 2490 bp) for phylogenetic analysis and found that *A. deserti* (GAHT01022741) was separated with the other 3 species (Figure 1(b)).

3.2. Sequence Comparison between Agave Transcriptomes.

A summary for the three databases was described in Table 1. There were 29,367, 29,327, and 50,851 sequences employed for orthologous gene searching from *A. tequilana*, *A. deserti*, and *A. sisalana* transcriptomes, respectively. As a result, we identified 13,069 unigene pairs between *A. tequilana* and *A. deserti*, 8976 pairs between *A. sisalana* and *A. tequilana*, and 9284 pairs between *A. sisalana* and *A. deserti* (Figure 2(a)). Among these orthologous unigene pairs, more than 91% unigene pairs had an identity over 91% (Figure 2(b)). Furthermore, a total of 6130 unigene terms were obtained from the three agave transcriptomes. GO functional classification indicated that these genes were assigned to 30,405 functional terms. There were 14,915 terms in biological process (49.05%), 8546 in molecular function (28.11%), and 6944 in cellular component (22.84%) (Figure 3).

3.3. Identification of Genes Selected in the Domestication of Agaves.

The Ka, Ks, and Ka/Ks values were calculated for 6130 orthologous unigene pairs in *A. tequilana* and *A. sisalana* separately with *A. deserti* (Figure 4(a)). The correlation between Ka and Ks values was also estimated in *A. tequilana* ($r = 0.515$, $P < 0.05$) and *A. sisalana* ($r = 0.206$, $P < 0.05$) unigene pairs, respectively. 393 unigenes (6.5%) in *A. tequilana* and 262 unigenes (4.5%) in *A. sisalana* showed a Ka/Ks ratio higher than 1, while the Ka/Ks ratio of more than 90% unigenes was lower than 1 (Figure 4(b)).

The significance of the Ka/Ks value for all 6130 unigene terms was analyzed, and the results indicated that 1117 unigenes were significantly selected at least in *A. tequilana* or *A. sisalana* (*P* value < 0.05) (Supplementary Table 1). These genes were further characterized into 15 classifications according to their annotations (Table 2). Among these, 54 unigenes were subjected to positive selection (Ka/Ks > 1), while the residues of 1064 unigenes were subjected to purifying selection (Ka/Ks < 1). Furthermore, 27 and 19 unigenes were positively selected in *A. tequilana* and *A. sisalana*, respectively. 8 unigenes were positively selected in both agave species (Supplementary Table 1).

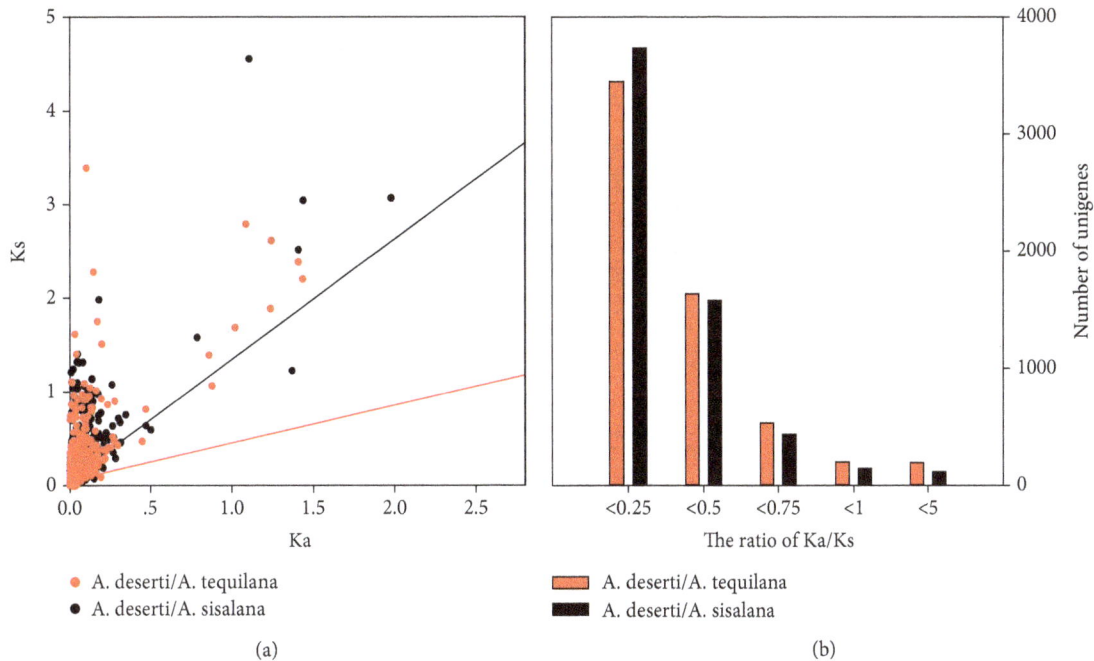

FIGURE 4: The nonsynonymous (Ka) and synonymous (Ks) nucleotide substitutions, as well as their Ka/Ks ratio. (a) The scatter diagram of Ka and Ks values of *A. tequilana* (red) and *A. sisalana* (black) compared with *A. deserti*, respectively. (b) The Ka/Ks distribution of *A. tequilana* (red) and *A. sisalana* (black) compared with *A. deserti*, respectively. The *x*-axis indicates the ratio of Ka/Ks.

TABLE 2: Characterization of significant selection unigenes.

Classification	Significant selection unigenes			
	Total	In *A. sisalana*	In *A. tequilana*	In both
Cell growth/division	78	30	37	11
Cell structure	37	12	22	3
Disease/defense	62	25	22	15
Energy	26	5	18	3
Intracellular traffic	6	2	3	1
Primary metabolism	94	27	55	12
Protein destination and storage	102	27	56	19
Protein synthesis	28	7	16	5
Secondary metabolism	56	16	33	7
Signal transduction	127	40	64	23
Transcription	82	26	40	16
Transcription factor	77	28	37	12
Transporter	56	17	31	8
Transposon	9	3	4	2
Unclassified	277	72	161	44
Total	1117	337	599	181

We further characterized the 54 unigenes subjected to positive selection and found that three unigenes were annotated as disease resistance protein (Table 3). In *A. tequilana*, four unigenes were annotated as 6G-fructosyl-transferase, d-2-hydroxyglutarate dehydrogenase, gluconokinase, and 6-phosphofructokinase, respectively, and they might be related to fructan production, while in *A. sisalana*, five unigenes were characterized as auxin-responsive protein IAA6, gibberellin receptor, PHD and RING finger protein, elongation of fatty acids protein, and zinc finger A20 and AN1 protein, respectively. These unigenes are probably related to the fiber development in *A. sisalana*. Besides, two unigenes, designated as RING-H2 finger protein and TOM1-like protein, were positively selected in both *A. tequilana* and *A. sisalana*.

3.4. In Silico Expression of Genes under Positive Selection in Agave Species. The expression patterns of genes under positive selection were analyzed according to the three transcriptome databases. In agave leaves, the 27 and 8 unigenes were grouped into two expression modes (Figure 5). However, the 19 unigenes in *A. sisalana* were not distinctly clustered into different expression modes. All these genes were differentially expressed in *A. tequilana* or *A. sisalana* when compared with *A. deserti*.

3.5. Selection Pressure and Structure Model of Putative Economic Trait-Related Genes. The sliding window analysis was used to examine the selection pressure of putative economic Trait-related genes. The results indicated the existence of different selection pressures in *A. tequilana* and *A. sisalana* genes (Figure 6). The three disease resistance genes all had a strong selection pressure in most sequence regions. Five unigenes showed a stronger selection pressure in *A. sisalana* (GAHT01109565, GAHT01002417, GAHT01054013, GAHT01038220, and GAHT01027892). Only GAHT01031288 showed a stronger selection pressure

TABLE 3: Unigenes subjected to positive selection in the domestication of Agave species.

| A. deserti accession | Accession | A. tequilana | | | Accession | A. sisalana | | | Functional annotation |
		Ka	Ks	Ka/Ks		Ka	Ks	Ka/Ks	
GAHT01106881	GAHU01106881	0.0533	0.0493	1.0803	Unigene21947	0.0084	0.0104	0.8112	6G-Fructosyltransferase
GAHT01031288	GAHU01044798	0.0133	0.0121	1.1055	Unigene12882	0.0011	0.0269	0.0428	d-2-Hydroxyglutarate dehydrogenase
GAHT01109565	GAHU01096053	0.0165	0.0084	1.9594	CL19921.Contig1	0.0444	0.2201	0.2015	Gluconokinase
GAHT01002417	GAHU01001324	0.0214	0.0106	2.0164	Unigene18337	0.0639	0.1531	0.4171	6-Phosphofructokinase
GAHT01070676	GAHU01114859	0.0876	0.0797	1.0994	Unigene374	0.0929	0.2571	0.3613	NBS-type resistance protein RGC5
GAHT01099649	GAHU01173424	0.1314	0.1049	1.2523	CL12994.Contig3	0.1753	0.2373	0.7387	Disease resistance protein RGA1
GAHT01006596	GAHU01010948	0.0167	0.0132	1.2634	CL13186.Contig4	0.023	0.0139	1.6485	RING-H2 finger protein
GAHT01048216	GAHU01097154	0.1621	0.1363	1.1894	CL18667.Contig2	0.149	0.0737	2.0217	Disease resistance protein RPS2
GAHT01055399	GAHU01077165	0.0118	0.0043	2.7653	Unigene28650	0.0188	0.0065	2.9139	TOM1-like protein
GAHT01016426	GAHU01038186	0.0143	0.1124	0.1272	CL8882.Contig1	0.0289	0.0226	1.28	Auxin-responsive protein IAA6
GAHT01005221	GAHU01003524	0.0322	0.0583	0.5528	CL2916.Contig1	0.0301	0.0107	2.8275	Gibberellin receptor
GAHT01054013	GAHU01067136	0.009	0.0092	0.9716	Unigene233	0.0081	0.0059	1.383	PHD and RING finger protein
GAHT01038220	GAHU01063926	0.0116	0	—	Unigene4300	0.0156	0.0049	3.154	Elongation of fatty acids protein
GAHT01027892	GAHU01009883	0.0022	0.0281	0.0794	Unigene28305	0.0659	0.0551	1.1956	Zinc finger A20 and AN1 protein

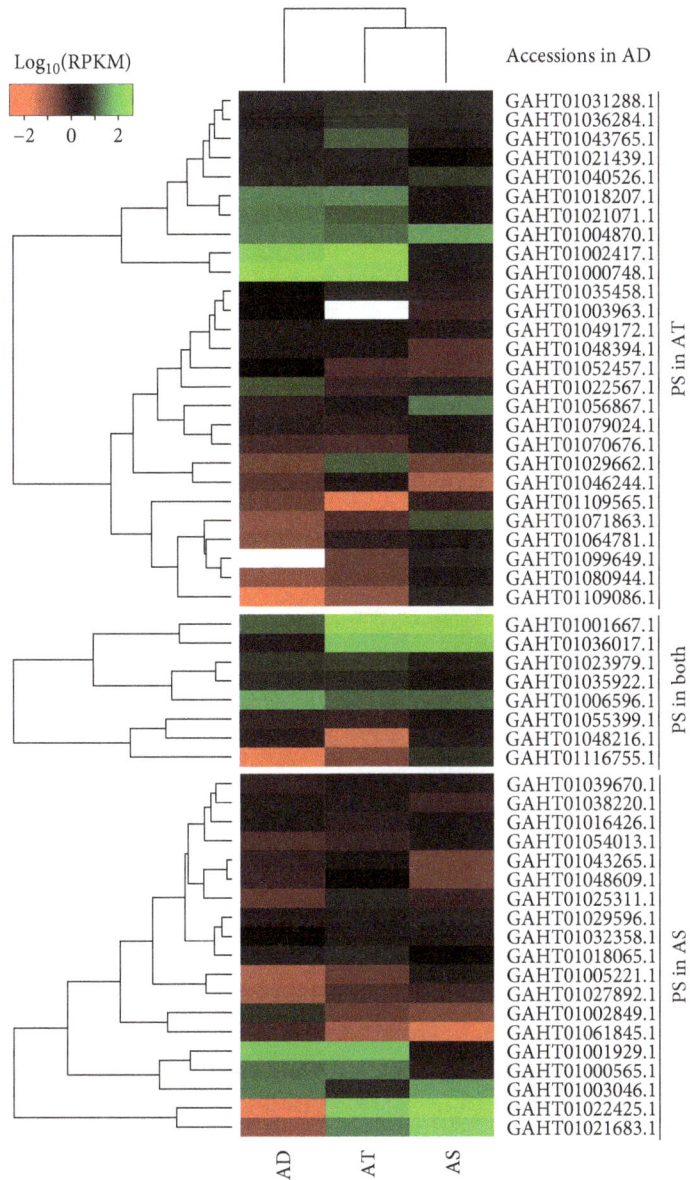

FIGURE 5: In silico expression pattern of positive selection (PS) unigenes in Agave leaves. Rows and columns in heat maps represent genes and Agave species. Species name of *A. deserti* (AD), *A. tequilana* (AT), and *A. sisalana* (AS) is shown under the heat maps. The colour bar at the top indicates the degree of expression: red—low expression; green—high expression.

in *A. tequilana*. The residue of five unigenes shared a strength-similar but region-different selection pressure.

The structure modeling for the 14 unigenes was further conducted to analyze the difference of agave proteins. According to the results, four unigenes designated as 6G-fructosyltransferase (Figure 7(a)), d-2-hydroxyglutarate dehydrogenase (Figure 7(b)), gluconokinase (Figure 7(c)), and 6-phosphofructokinase (Figure 7(d)), could match Swiss-Model sequences with identity > 30%, coverage > 75%, and appropriate description (Supplementary Table 2). Their structure models also showed significant differences in *A. tequilana* or *A. sisalana*. The unigene with a significantly distinct structure was differentially expressed when compared to their orthologous unigenes in other two species (Figure 5).

4. Discussion

A previous phylogenetic analysis suggested that *A. deserti* was closely related to *A. angustifolia* [15]. However, we compared the recently published chloroplast sequence (GAHT01022741) with the former one (DQ500894 + DQ500928) in *A. deserti* and found them having totally different sequences. It might be caused by sample collection because the two studies were separately conducted in Mexico and America. Wild agave species are usually identified by morphological traits, which is not as reliable as molecular identification. Our phylogenetic results also indicated a different evolution relation by both short and long chloroplast sequences (Figure 1). This indicated the closer evolution

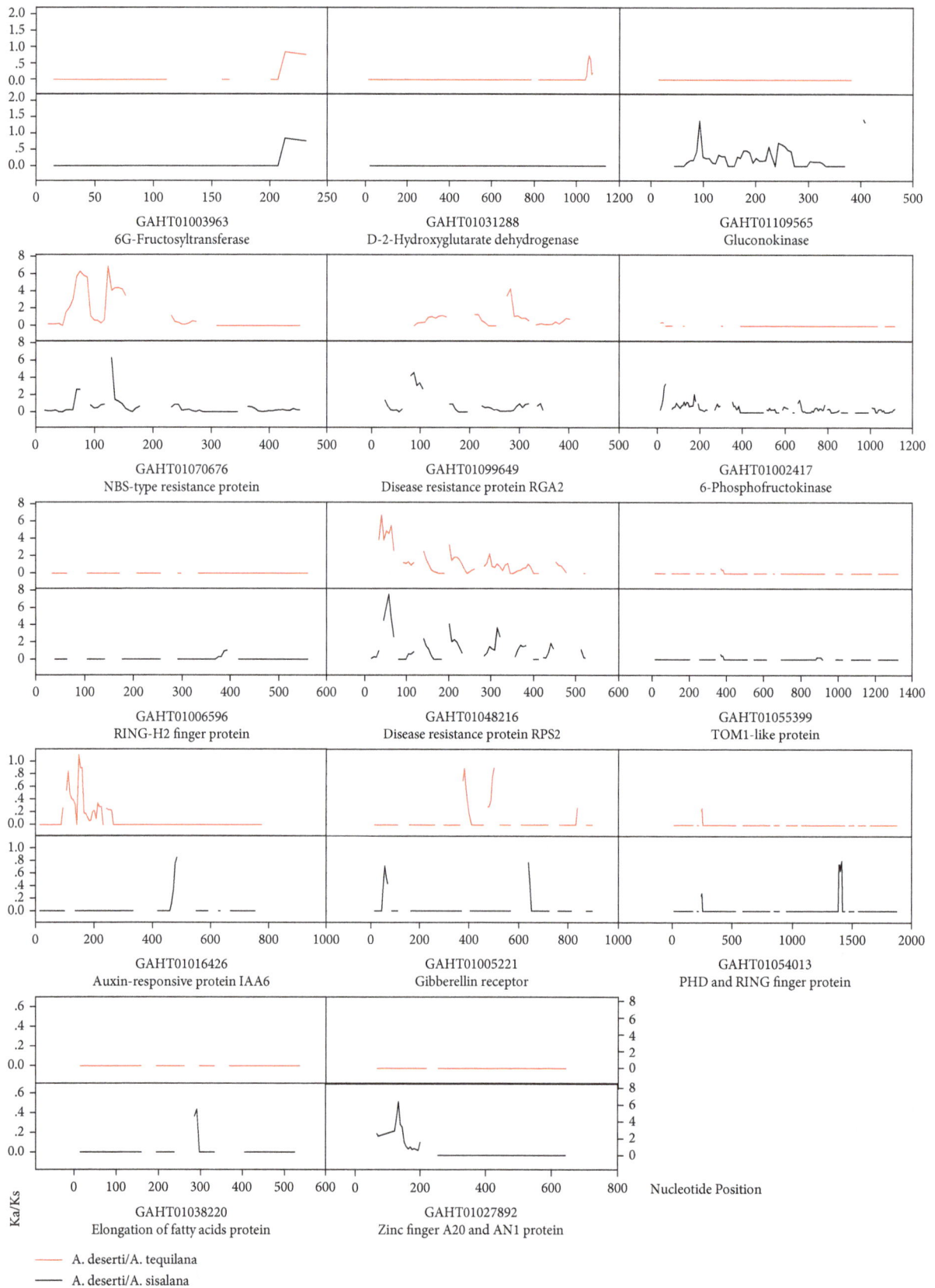

FIGURE 6: Sliding window analysis of positive selection unigenes in *A. tequilana* (red) and *A. sisalana* (black) compared with *A. deserti*, respectively. The window size is 30 bp, with a step size of 6 bp. The *x*-axis denotes the nucleotide position. The *y*-axis denotes the Ka/Ks ratio. The gaps represent Ka/Ks ratios that could not be computed.

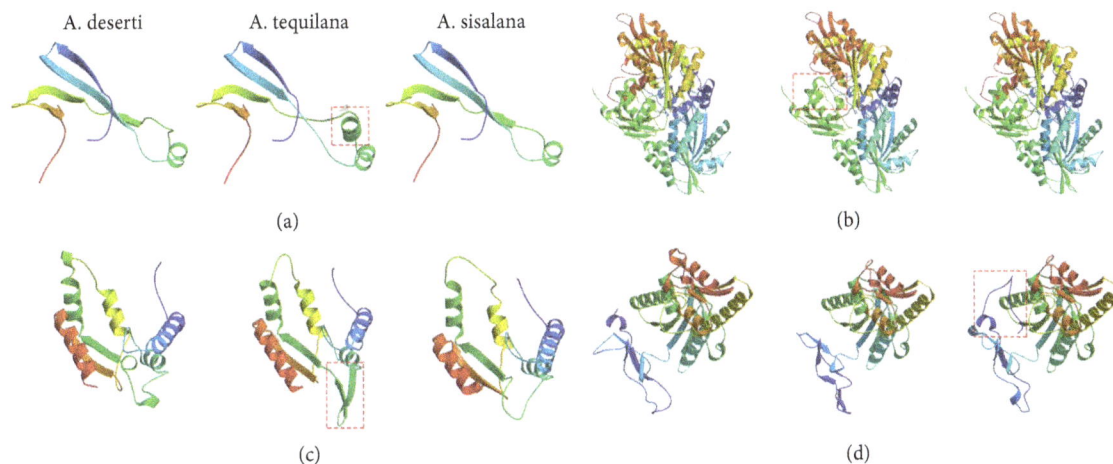

FIGURE 7: Protein structure model of 6G-fructosyltransferase (a), d-2-hydroxyglutarate dehydrogenase (b), gluconokinase (c), and 6-phosphofructokinase (d) in three Agave species by using Swiss-Model [33]. The significant structure difference within species was marked in red dotted line.

relationship between *A. tequilana* and *A. sisalana* than the evolution relationship between *A. tequilana* and *A. deserti* or between *A. sisalana* and *A. deserti*. It has been reported that *A. tequilana* and *A. sisalana* both had genetic relations with *A. angustifolia* [6, 7]. The phylogenetic result has also proved that, however, the two species actually possessed very different agronomic traits and applications. An important reason should be artificial selection even if the low-frequency serendipitous backyard hybridization would lead to distinct domestication of crops. This is a historical inheritance in Mexico and has significantly enriched the genetic diversity of crops [34].

Agave species originate from Central America with high tolerance to drought and temperature, which makes them a main and important kind of plants there [9]. Therefore, they would inevitably confront a series of abiotic and biotic stresses. Actually, we identified 62, 127, and 77 unigenes with classifications of disease/defense, signal transduction, and transcription factor, respectively. Among them, 13 significant selected unigenes were related to disease resistance and ten of them subjected to purifying selection (Supplementary Table 1). This might be accompanied with the process of agronomic Trait-derived domestication. Many disease resistance genes were also found to be lost during domestication in other crops [17, 22]. A previous study has already revealed a differentiated selection pressure on NBS-LRR genes in some agave species [35]. In this study, the sliding window analysis also showed a strong selection pressure with the three disease resistance unigenes (Figure 6). Furthermore, two of the three unigenes (GAHT01070676 and GAHT01099649) were subjected to purifying selection in *A. sisalana* (Table 3), which might be harmful for growth and development. This might also be responsible for the susceptibility to zebra disease caused by *Phytophthora nicotianae* in *A. sisalana* [36].

It has been reported that several transcription factor families play an important role in abiotic stress regulation, such as bHLH, zinc finger, MYB, AP2, NAC, WRKY, and bZIP families [37–42]. We also found 47 TFs subjected to purifying selection either in *A. tequilana* or in *A. sisalana*, from the bHLH (8), zinc finger (23), MYB (6), AP2 (5), NAC (3), WRKY (1), and bZIP (1) families (Supplementary Table 1). For agave species, drought and high temperature are the main abiotic stresses. We speculated that different habitats should be an important natural selection pressure that affected the shape and size of the three agave species [43]. The purifying selection of the 47 stress-related candidate TFs might weaken the drought tolerance of *A. tequilana* and *A. sisalana* but enhance their biomass accumulations. The complex regulation and interaction of these TFs might be the key to reveal the mechanism of the remarkable drought tolerance in *A. deserti*. Much more molecular characterizations are still needed in future studies.

It has been reported that fruit/seed-related traits were subjected to high artificial selection pressure for their economic value [18, 19]. In *A. tequilana*, the most important economic trait focuses on fructan, and several studies have conducted functional characterization for fructosyltransferase genes [10, 11, 44]. In this study, we identified a fructan-related unigene and it was subjected to positive selection in *A. tequilana* (Table 3). Besides, three carbohydrate-related unigenes were also subjected to positive selection. Their positive selection might be responsible for the improvement of fructan yield in *A. tequilana*. In *A. sisalana*, fiber is the main economic purpose for its cultivation in tropical areas. In the present study, we found 5 unigenes subjected to positive selection only in *A. sisalana* (Table 3). A previous publication has reviewed the hormonal regulation of secondary cell wall formation [45]. The zinc finger family TFs are also proved to regulate cell wall development and cellulose biosynthesis [46, 47]. Besides, the elongation of fatty acids protein plays an important role in cell elongation [48]. Therefore, we speculated that the fiber-related traits in agaves are more likely controlled by hormonal and transcriptional regulation. And there are significant differences when compared with fructan-related traits, which might be mainly controlled by metabolic regulation in *A. tequilana*. Natural fiber is commonly generated as the result of secondary cell wall thickening in the main fiber crops such as cotton, ramie, flax, and hemp [49–52]. As a constitutive structure of plant cells, there are many housekeeping and regulating genes during cell wall development, especially the secondary cell wall development [45, 53–55]. In contrast, fructans are mainly responsible for carbohydrate storage to vegetative tissues in many plant species [56]. They have also been increasingly considered protective agents against abiotic stresses [57]. However, the capacity to produce and store fructans in *A. tequilana* is much stronger than that in *A. deserti*, even if there is a much more severe environment for *A. deserti* [23, 44]. It is probably because fructan-related traits are regulated at the metabolic level. A more recent study has combined transcript, protein, and metabolite methods to reveal the molecular basis of the CAM process in agave [58]. The rapid development of high-throughput molecular methods has brought a great opportunity for the further understanding of the differences and evolution patterns among agave species.

5. Conclusion

This study represents the first transcriptome comparison within domesticated and wild agave species. The results revealed the importance of abiotic/biotic natural selection in agave evolution. Four unigenes related to fructan in *A. tequilana* and five unigenes related to fiber in *A. sisalana* were positively selected. These genes revealed the difference between *A. tequilana* and *A. sisalana* evolution, which would serve as a guidance for further studies on agave evolution, environmental adaptation, and improvement of economically important traits.

Conflicts of Interest

The authors declare no conflict of interest.

Authors' Contributions

XH and KY conceived and designed the experiments. XH analyzed the data and drafted the manuscript. BW carried out the Ka/Ks analysis. JX, JG, HC, SZ, JZ, and KY contributed to the transcriptome data. YZ revised the manuscript. CH and WW helped in the expression analysis. YL helped in the selection pressure detection. All authors read and approved the final manuscript.

Acknowledgments

We would like to thank Dr. Xiaohan Yang from the Oak Ridge National Laboratory (Oak Ridge, TN 37831-6407, USA), Dr. Thomas Nothnagel from Julius Kühn-Institut (Quedlinburg 06484, Germany), and Dr. Adel M. R. A. Abdelaziz from the Central Laboratory of Organic Agriculture, Agricultural Research Center (Giza 12619, Egypt), for the manuscript revision. This study was supported by the National Key R&D Program of China (2018YFD0201100), the earmarked fund from the China Agriculture Research System (CARS-16-E16), the National Natural Science Foundation of China (31371679, 31771849), and the Central Public-Interest Scientific Institution Basal Research Fund (2017hzs1J014, 1630042018014).

References

[1] A. M. Borland, H. Griffiths, J. Hartwell, and J. A. C. Smith, "Exploiting the potential of plants with crassulacean acid metabolism for bioenergy production on marginal lands," *Journal of Experimental Botany*, vol. 60, no. 10, pp. 2879–2896, 2009.

[2] P. S. Nobel, "Achievable productivities of certain CAM plants: basis for high values compared with C_3 and C_4 plants," *New Phytologist*, vol. 119, no. 2, pp. 183–205, 1991.

[3] X. Yang, J. C. Cushman, A. M. Borland et al., "A roadmap for research on crassulacean acid metabolism (CAM) to enhance sustainable food and bioenergy production in a hotter, drier world," *New Phytologist*, vol. 207, no. 3, pp. 491–504, 2015.

[4] M. Cedeño, "Tequila production," *Critical Reviews in Biotechnology*, vol. 15, no. 1, pp. 1–11, 1995.

[5] K. R. Corbin, C. S. Byrt, S. Bauer et al., "Prospecting for energy-rich renewable raw materials: *Agave* leaf case study," *PLoS One*, vol. 10, no. 8, article e0135382, p. 10, 2015.

[6] G. Lock, *Sisal*, Longmans, Green, London, UK, 1962.

[7] S. H. Gentry, *Agaves of Continental North America*, University Of Arizona Press, Tucson, AZ, USA, 1982.

[8] P. S. Nobel and T. L. Hartsock, "Temperature, water, and PAR influences on predicted and measured productivity of *Agave deserti* at various elevations," *Oecologia*, vol. 68, no. 2, pp. 181–185, 1986.

[9] J. R. Stewart, "*Agave* as a model CAM crop system for a warming and drying world," *Frontiers in Plant Science*, vol. 6, p. 684, 2015.

[10] C. Cortés-Romero, A. Martínez-Hernández, E. Mellado-Mojica, M. G. López, and J. Simpson, "Molecular and functional characterization of novel fructosyltransferases and invertases from *Agave tequilana*," *PLoS One*, vol. 7, no. 4, article e35878, p. 7, 2012.

[11] E. M. Suarez-Gonzalez, M. G. Lopez, J. P. Délano-Frier, and J. F. Gómez-Leyva, "Expression of the 1-SST and 1-FFT genes and consequent fructan accumulation in *Agave tequilana* and *A. inaequidens* is differentially induced by diverse (a)biotic-stress related elicitors," *Journal of Plant Physiology*, vol. 171, no. 3-4, pp. 359–372, 2014.

[12] P. S. Nobel, *Environmental Biology of Agaves and Cacti*, Cambridge University Press, Cambridge, 1988.

[13] P. Y. Chen, C. H. Chen, C. C. Kuo, T. H. Lee, Y. H. Kuo, and C. K. Lee, "Cytotoxic steroidal saponins from *Agave sisalana*," *Planta Medica*, vol. 77, no. 09, pp. 929–933, 2011.

[14] J. D. Santos, I. J. Vieira, R. Braz-Filho, and A. Branco, "Chemicals from *Agave sisalana* biomass: isolation and identification," *International Journal of Molecular Sciences*, vol. 16, no. 12, pp. 8761–8771, 2015.

[15] S. V. Good-Avila, V. Souza, B. S. Gaut, and L. E. Eguiarte, "Timing and rate of speciation in Agave (Agavaceae)," *Proceedings of the National Academy of Sciences of the United States of America*, vol. 103, no. 24, pp. 9124–9129, 2006.

[16] M. L. Robert, K. Y. Lim, L. Hanson et al., "Wild and agronomically important *Agave* species (Asparagaceae) show proportional increases in chromosome number, genome size, and genetic markers with increasing ploidy," *Botanical Journal of the Linnean Society*, vol. 158, no. 2, pp. 215–222, 2008.

[17] S. Guo, J. Zhang, H. Sun et al., "The draft genome of watermelon (*Citrullus lanatus*) and resequencing of 20 diverse accessions," *Nature Genetics*, vol. 45, no. 1, pp. 51–58, 2013.

[18] L. Lu, D. Shao, X. Qiu et al., "Natural variation and artificial selection in four genes determine grain shape in rice," *The New Phytologist*, vol. 200, no. 4, pp. 1269–1280, 2013.

[19] J. Qi, X. Liu, D. Shen et al., "A genomic variation map provides insights into the genetic basis of cucumber domestication and diversity," *Nature Genetics*, vol. 45, no. 12, pp. 1510–1515, 2013.

[20] J. Canales, R. Bautista, P. Label et al., "De novo assembly of maritime pine transcriptome: implications for forest breeding and biotechnology," *Plant Biotechnology Journal*, vol. 12, no. 3, pp. 286–299, 2014.

[21] X. Huang, J. Chen, Y. Bao et al., "Transcript profiling reveals auxin and cytokinin signaling pathways and transcription regulation during in vitro organogenesis of ramie (*Boehmeria nivea* L. gaud)," *PLoS One*, vol. 9, no. 11, article e113768, p. 9, 2014.

[22] T. Liu, S. Tang, S. Zhu, Q. Tang, and X. Zheng, "Transcriptome comparison reveals the patterns of selection in domesticated and wild ramie (*Boehmeria nivea* L. gaud)," *Plant Molecular Biology*, vol. 86, no. 1-2, pp. 85–92, 2014.

[23] S. M. Gross, J. A. Martin, J. Simpson, M. Abraham-Juarez, Z. Wang, and A. Visel, "*De novo* transcriptome assembly of drought tolerant CAM plants, *Agave deserti* and *Agave tequilana*," *BMC Genomics*, vol. 14, no. 1, p. 563, 2013.

[24] P. Wang, J. Gao, F. Yang et al., "Transcriptome of sisal leaf pretreated with Phytophthora nicotianae Breda," *Chinese J Tropical Crops*, vol. 35, pp. 576–582, 2014.

[25] K. Tamura, D. Peterson, N. Peterson, G. Stecher, M. Nei, and S. Kumar, "MEGA5: molecular evolutionary genetics analysis using maximum likelihood, evolutionary distance, and maximum parsimony methods," *Molecular Biology and Evolution*, vol. 28, no. 10, pp. 2731–2739, 2011.

[26] J. Martin, V. M. Bruno, Z. Fang et al., "Rnnotator: an automated *de novo* transcriptome assembly pipeline from stranded RNA-seq reads," *BMC Genomics*, vol. 11, no. 1, p. 663, 2010.

[27] M. Zhang and H. W. Leong, "Bidirectional best hit *r*-window gene clusters," *BMC Bioinformatics*, vol. 11, Supplement 1, p. S63, 2010.

[28] A. Doron-Faigenboim, A. Stern, I. Mayrose, E. Bacharach, and T. Pupko, "Selecton: a server for detecting evolutionary forces at a single amino-acid site," *Bioinformatics*, vol. 21, no. 9, pp. 2101–2103, 2005.

[29] Z. Yang, "PAML 4: phylogenetic analysis by maximum likelihood," *Molecular Biology and Evolution*, vol. 24, no. 8, pp. 1586–1591, 2007.

[30] B. Li and C. N. Dewey, "RSEM: accurate transcript quantification from RNA-Seq data with or without a reference genome," *BMC Bioinformatics*, vol. 12, no. 1, p. 323, 2011.

[31] M. Hu, W. Hu, Z. Xia, Z. Zhou, and W. Wang, "Validation of reference genes for relative quantitative gene expression studies in cassava (*Manihot esculenta* Crantz) by using quantitative real-time PCR," *Frontiers in Plant Science*, vol. 7, p. 680, 2016.

[32] P. Librado and J. Rozas, "DnaSP v5: a software for comprehensive analysis of DNA polymorphism data," *Bioinformatics*, vol. 25, no. 11, pp. 1451-1452, 2009.

[33] M. Biasini, S. Bienert, A. Waterhouse et al., "SWISS-MODEL: modelling protein tertiary and quaternary structure using evolutionary information," *Nucleic Acids Research*, vol. 42, no. W1, pp. W252–W258, 2014.

[34] C. E. Hughes, R. Govindarajulu, A. Robertson, D. L. Filer, S. A. Harris, and C. D. Bailey, "Serendipitous backyard hybridization and the origin of crops," *Proceedings of the National Academy of Sciences of the United States of America*, vol. 104, no. 36, pp. 14389–14394, 2007.

[35] M. C. Tamayo-Ordonez, L. C. Rodriguez-Zapata, J. A. Narvaez-Zapata et al., "Morphological features of different polyploids for adaptation and molecular characterization of CC-NBS-LRR and LEA gene families in *Agave* L.," *Journal of Plant Physiology*, vol. 195, pp. 80–94, 2016.

[36] J. Gao, Luoping, C. Guo et al., "AFLP analysis and zebra disease resistance identification of 40 sisal genotypes in China," *Molecular Biology Reports*, vol. 39, no. 5, pp. 6379–6385, 2012.

[37] G. Castilhos, F. Lazzarotto, L. Spagnolo-Fonini, M. H. Bodanese-Zanettini, and M. Margis-Pinheiro, "Possible roles of basic helix–loop–helix transcription factors in adaptation to drought," *Plant Science*, vol. 223, pp. 1–7, 2014.

[38] S. Ciftci-Yilmaz and R. Mittler, "The zinc finger network of plants," *Cellular and Molecular Life Sciences*, vol. 65, no. 7-8, pp. 1150–1160, 2008.

[39] K. J. Dietz, M. O. Vogel, and A. Viehhauser, "AP2/EREBP transcription factors are part of gene regulatory networks and integrate metabolic, hormonal and environmental signals in stress acclimation and retrograde signalling," *Protoplasma*, vol. 245, no. 1-4, pp. 3–14, 2010.

[40] C. Dubos, R. Stracke, E. Grotewold, B. Weisshaar, C. Martin, and L. Lepiniec, "MYB transcription factors in *Arabidopsis*," *Trends in Plant Science*, vol. 15, no. 10, pp. 573–581, 2010.

[41] S. Puranik, P. P. Sahu, P. S. Srivastava, and M. Prasad, "NAC proteins: regulation and role in stress tolerance," *Trends in Plant Science*, vol. 17, no. 6, pp. 369–381, 2012.

[42] K. Singh, R. C. Foley, and L. Onate-Sanchez, "Transcription factors in plant defense and stress responses," *Current Opinion in Plant Biology*, vol. 5, no. 5, pp. 430–436, 2002.

[43] N. J. Kooyers, "The evolution of drought escape and avoidance in natural herbaceous populations," *Plant Science*, vol. 234, pp. 155–162, 2015.

[44] E. A. D. Dios, A. D. G. Vargas, M. L. D. Santos, and J. Simpson, "New insights into plant glycoside hydrolase family 32 in Agave species," *Frontiers in Plant Science*, vol. 6, p. 594, 2015.

[45] V. Didi, P. Jackson, and J. Hejatko, "Hormonal regulation of secondary cell wall formation," *Journal of Experimental Botany*, vol. 66, no. 16, pp. 5015–5027, 2015.

[46] W. C. Kim, J. Y. Kim, J. H. Ko, H. Kang, J. Kim, and K. H. Han, "AtC3H14, a plant-specific tandem CCCH zinc-finger protein, binds to its target mRNAs in a sequence-specific manner and affects cell elongation in *Arabidopsis thaliana*," *The Plant Journal*, vol. 80, no. 5, pp. 772–784, 2014.

[47] D. Wang, Y. Qin, J. Fang et al., "A missense mutation in the zinc finger domain of OsCESA7 deleteriously affects cellulose biosynthesis and plant growth in rice," *PLoS One*, vol. 11, no. 4, article e0153993, p. 11, 2016.

[48] Y. M. Qin, C. Y. Hu, Y. Pang, A. J. Kastaniotis, J. K. Hiltunen,

and Y.-X. Zhu, "Saturated very-long-chain fatty acids promote cotton fiber and *Arabidopsis* cell elongation by activating ethylene biosynthesis," *The Plant Cell*, vol. 19, pp. 3692–3704, 2007.

[49] M. Behr, S. Legay, E. Žižková et al., "Studying secondary growth and bast fiber development: the hemp hypocotyl peeks behind the wall," *Frontiers in Plant Science*, vol. 7, p. 1733, 2016.

[50] M. Chantreau, B. Chabbert, S. Billiard, S. Hawkins, and G. Neutelings, "Functional analyses of cellulose synthase genes in flax (*Linum usitatissimum*) by virus-induced gene silencing," *Plant Biotechnology Journal*, vol. 13, no. 9, pp. 1312–1324, 2015.

[51] J. Chen, Z. Pei, L. Dai et al., "Transcriptome profiling using pyrosequencing shows genes associated with bast fiber development in ramie (*Boehmeria nivea* L.)," *BMC Genomics*, vol. 15, no. 1, p. 919, 2014.

[52] A. M. Schubert, C. R. Benedict, J. D. Berlin, and R. J. Kohel, "Cotton fiber development-kinetics of cell elongation and secondary wall thickening," *Crop Science*, vol. 13, no. 6, pp. 704–709, 1973.

[53] D. J. Cosgrove, "Growth of the plant cell wall," *Nature Reviews Molecular Cell Biology*, vol. 6, no. 11, pp. 850–861, 2005.

[54] S. G. Hussey, E. Mizrachi, N. M. Creux, and A. A. Myburg, "Navigating the transcriptional roadmap regulating plant secondary cell wall deposition," *Frontiers in Plant Science*, vol. 4, p. 325, 2013.

[55] M. Schuetz, R. Smith, and B. Ellis, "Xylem tissue specification, patterning, and differentiation mechanisms," *Journal of Experimental Botany*, vol. 64, no. 1, pp. 11–31, 2012.

[56] C. J. Nelson and W. G. Spollen, "Fructan," *Physiologia Plantarum*, vol. 71, no. 4, pp. 512–516, 1987.

[57] R. Valluru and W. V. D. Ende, "Plant fructans in stress environments: emerging concepts and future prospects," *Journal of Experimental Botany*, vol. 59, no. 11, pp. 2905–2916, 2008.

[58] P. E. Abraham, H. Yin, A. M. Borland et al., "Transcript, protein and metabolite temporal dynamics in the CAM plant *Agave*," *Nature Plants*, vol. 2, p. 16178, 2016.

Genome-Wide Identification, Phylogeny, and Expression Analysis of ARF Genes Involved in Vegetative Organs Development in Switchgrass

Jianli Wang [ID],[1,2,3] Zhenying Wu [ID],[2] Zhongbao Shen,[3] Zetao Bai,[2] Peng Zhong,[4] Lichao Ma,[2] Duofeng Pan,[3] Ruibo Zhang,[3] Daoming Li,[3] Hailing Zhang,[3] Chunxiang Fu,[2] Guiqing Han [ID],[1,3] and Changhong Guo [ID][1]

[1]College of Life Science and Technology of Harbin Normal University, Harbin 150080, China
[2]Key Laboratory of Biofuels, Shandong Provincial Key Laboratory of Energy Genetics, Qingdao Institute of Bioenergy and Bioprocess Technology, Chinese Academy of Sciences, Qingdao 266101, China
[3]Grass and Science Institute of Heilongjiang Academy of Agricultural Sciences, Harbin, Heilongjiang 150086, China
[4]Rural Energy Research Institute of Heilongjiang Academy of Agricultural Sciences, Harbin, Heilongjiang 150086, China

Correspondence should be addressed to Guiqing Han; ccyj15@163.com and Changhong Guo; guochanghong2016@163.com

Academic Editor: Graziano Pesole

Auxin response factors (ARFs) have been reported to play vital roles during plant growth and development. In order to reveal specific functions related to vegetative organs in grasses, an in-depth study of the ARF gene family was carried out in switchgrass (*Panicum virgatum* L.), a warm-season C4 perennial grass that is mostly used as bioenergy and animal feedstock. A total of 47 putative ARF genes (*PvARFs*) were identified in the switchgrass genome (2n = 4x = 36), 42 of which were anchored to the seven pairs of chromosomes and found to be unevenly distributed. Sixteen *PvARFs* were predicted to be potential targets of small RNAs (microRNA160 and 167). Phylogenetically speaking, PvARFs were divided into seven distinct subgroups based on the phylogeny, exon/intron arrangement, and conserved motif distribution. Moreover, 15 pairs of *PvARFs* have different temporal-spatial expression profiles in vegetative organs (2nd, 3rd, and 4th internode and leaves), which implies that different *PvARFs* have specific functions in switchgrass growth and development. In addition, at least 14 pairs of *PvARFs* respond to naphthylacetic acid (NAA) treatment, which might be helpful for us to study on auxin response in switchgrass. The comprehensive analysis, described here, will facilitate the future functional analysis of ARF genes in grasses.

1. Introduction

Auxin, an essential plant hormone, plays vital roles in various aspects of plant growth and development, such as embryogenesis, organogenesis, tropic growth, shoot elongation, root architecture, flower and fruit development, tissue and organ patterning, and vascular development [1–9]. Most of these processes are controlled by auxin response genes, which are regulated at transcriptional level by *cis*-acting DNA elements in their promoter regions, including the auxin response element (AuxRE, TGTCTC), core of auxin response region (AuxRR-core, GGTCCAT), and TGA-

element (AACGAC). Of these, AuxREs are reported to be specifically bound and regulated by a class of transcription factors, called auxin response factors (ARFs) [10, 11]. ARF proteins generally contain a DNA-binding domain (DBD) in the amino (N)-terminal region, a central region that functions as an activation domain (AD) or a repression domain (RD) [12, 13], and a carboxyl (C)-terminal dimerization domain (CTD), which is a protein-protein interaction domain that mediates ARF homo- and heterodimerization and also the heterodimerization of ARF and Aux/IAA proteins, another category of auxin response regulators [12–16].

Because of their important roles in auxin signaling pathways, which are indispensable to plant growth and development, ARF gene families have been studied in many plant species. For example, there are 23 ARF transcription factors in Arabidopsis (Arabidopsis thaliana) [17], 25 in rice (Oryza sativa) [18], 39 in poplar (Populus trichocarpa) [19], 24 in Medicago truncatula [20], and 36 in maize (Zea mays) [21]. In previous studies, ARF proteins were split into three clades (clades A, B, and C) based on phylogenic relationships, which could be traced back to the origin of land plants [22]. In particular, phylogenetic analysis of the ARF gene family in many species has been widely reported. Arabidopsis ARFs were divided into four subgroups, which is in accordance with the phylogenetic classifications of ARFs in rice [18], banana (Musa acuminata L.) [23] and Salvia miltiorrhiza [24]. Maize and poplar ARFs are classified to six subgroups [25], whereas Medicago ARFs were divided into eight subgroups [20]. In general, the wide variety of ARF phylogenetic grouping patterns are based on the diversification of its gene structure and motif locations, which may be the result of gene truncation or alternative splicing [22].

Biochemical and genetic analyses have established the crucial roles of ARF genes in plant growth and development. In Arabidopsis thaliana, AtARF2 regulates floral organ abscission, leaf senescence, and seed size and weight [26–28]. AtARF5 affects vascular development and early embryo formation [29]. AtARF8 controls the uncoupling of fruit development from pollination and fertilization, and loss-of-function mutations in these genes result in seedless fruit [30]. AtARF7 and AtARF19 redundantly regulate auxin-mediated lateral root development [31]. In rice, OsARF1 is required for vegetative and reproductive development [32]. OsARF16 is essential for iron and phosphate deficiency responses in rice [33]. In addition, some ARF genes are involved in the response to abiotic stresses, such as drought, salt, or cold stress [34, 35]. Taken together, these studies have shown that the ARF gene family function in multiple signal transduction pathways to regulate multiple aspects of plant growth and development.

Switchgrass (Panicum virgatum L.) is a warm-season C4 perennial grass used as a bioenergy and animal feedstock [36, 37]. To avoid competing with food crops for arable fields, a large proportion of switchgrass fields will be located on marginal lands where various abiotic stresses, such as salt, drought, and extreme temperatures. The genome sequence of switchgrass has been published recently [38] and provides a powerful resource to identify ARF gene family members. Considering the value of switchgrass as a bioenergy and animal feedstock, we mainly focused on vegetative organs in this study.

Here, we identified 47 switchgrass ARF genes and comprehensively characterized the physical location, conserved motif architecture, and expression profile of the PvARFs. We also subdivided these 47 PvARF genes based on phylogenetic relationships based on the well-studied ARF genes in other species. To determine which ARF genes potentially work on different developmental processes, the temporal-spatial expression pattern in vegetative organs (2nd, 3rd,

and 4th internode and leaves) and expression response to auxin treatment in seedlings were determined by real-time PCR (qRT-PCR). Our works provide preliminary information about ARF genes in switchgrass and lays the foundation for the further elucidation of the biological roles of ARF genes in grasses.

2. Materials and Methods

2.1. Plant Materials and Treatments. A widely used and highly productive lowland-type switchgrass cultivar, Alamo, was grown in the greenhouse at $28 \pm 1°C$ with 16 h lighting, followed by 8 h darkness. Switchgrass development in our greenhouse was divided into three vegetative stages (V1, V2, and V3), five elongation stages (E1, E2, E3, E4, and E5), and three reproductive stages (R1, R2, and R3). Six different tissues, including the second internode (I2), the third internode (I3), the fourth internode (I4), the second leaf (L2), the third leaf (L3), and the fourth leaf (L4), were collected at the R2 stage [39].

For auxin treatments, plantlets grown in tissue culture until 20 days after rooting were incubated for 1, 2, and 3 h in hormone-supplemented $5 \mu M$ naphthylacetic acid (NAA) medium [18]. Control plants were grown in hormone-free medium. Whole seedlings were sampled from NAA-treated and control plants at the same time points. All experiments included three biological replicates. All of the samples were stored at $-80°C$.

2.2. Sequence Retrieval and Identification. The conserved ARF domain based on the Hidden Markov Model (HMM) (Pfam06507) was obtained from the Pfam protein family database (http://pfam.sanger.ac.uk/) and used as a query to search against the switchgrass genome database in Phytozome v11 (http://www.phytozome.net/). Sequences were selected for further analysis if the E value was less than 1e-10. Several coding sequences (CDS) were corrected based on the switchgrass unique transcript sequence database [38]. Peptide length, molecular weight, and isoelectric point of each PvARF were calculated using the online ExPASy program (http://www.expasy.org/).

2.3. Phylogenetic Analysis. The putative PvARF proteins from another fifteen species were used to construct a phylogenetic tree. ARF protein sequences were obtained from the public genome database Phytozome. The BlastP program was used to identify putative ARF proteins from the genomic databases of well-sequenced species, including Arabidopsis, sweet orange (Citrus sinensis), Chinese cabbage (Brassica rapa), poplar, Medicago (Medicago truncatula), cotton (Gossypium raimondii), Grandis (Eucalyptus grandis), soybean, tomato (Solanum lycopersicum), grape (Vitis vinifera), maize, rice, foxtail millet (Setaria italica), sorghum (Sorghum vulgare), and Brachypodium distachyon. Multiple sequence alignments of the full-length ARF sequences were performed using Clustal X1.83, and the edges of the alignments were manually trimmed. An unrooted neighbor-joining (bootstrap value = 1000) tree was constructed using MEGA5 and then

TABLE 1: The information of ARF family genes in switchgrass.

Gene name[a]	Gene ID[b]	ORF length (bp)[c]	Length (aa)	MW (kDa)	pI	Number of intron[e]	Chr[f]	Chr locations[g]
			Deduced polypeptide[d]					
PvARF1	Pavir.Ca02838	2685	894	98.5	5.75	13	3a	47617320–47623645
PvARF2	Pavir.Cb00190	2136	711	78	6.02	7	3b	3249145–3252725
PvARF3	Pavir.Da00065	2739	912	100.8	5.92	13	4a	1765253–1771355
PvARF4	Pavir.Db00366	2781	926	102.1	5.81	13	4b	4362788–4368192
PvARF5	Pavir.Aa03303	2721	906	99.5	5.58	13	1a	67423915–67430303
PvARF6	Pavir.Ab00451	2715	904	99.5	5.54	13	1b	4856121–4861815
PvARF7	Pavir.Ga00205	2502	833	92.8	6.03	13	7a	3171190–3179787
PvARF8	Pavir.Gb00274	1992	663	73.7	5.33	11	7b	3192562–3,197923
PvARF9	Pavir.Da01885	1296	431	48.1	8.45	10	4a	42434950–42439045
PvARF10	Pavir.Db01975	2205	734	81	5.6	13	4b	43526790–43533795
PvARF11	Pavir.Fb01896	3282	1093	121.4	6.07	12	6b	48581098–48587963
PvARF12	Pavir.Fa00483	3255	1084	120.3	6.14	12	6a	7216236–7223377
PvARF13	Pavir.J00164	3549	1182	130.9	6.29	14	contig00149	16339–25992
PvARF14	Pavir.Db00232	3231	1076	120	6.28	11	4b	3367149–3373291
PvARF15	Pavir.J32718	3246	1081	120.5	6.08	11	contig39521	1–6360
PvARF16	Pavir.Ab00366	3174	1057	117.6	6.14	11	1b	4304985–4314018
PvARF17	Pavir.Gb00117	2829	942	104.3	5.81	12	7b	1282083–1288213
PvARF18	Pavir.Ga00157	2838	945	104.5	5.77	12	7a	2852206–2858929
PvARF19	Pavir.J03524	1554	517	56.3	6.24	2	contig04638	11091–14207
PvARF20	Pavir.J37640	1548	515	56	6.16	3	contig69503	1–2970
PvARF21	Pavir.Ia01695	2040	679	74.4	9.3	4	9a	20465406–20472448
PvARF22	Pavir.Ib03238	1374	457	50.1	6.33	2	9b	52838875–52843142
PvARF23	Pavir.Da00107	2061	686	74.9	7.04	2	4a	2050051–2053504
PvARF24	Pavir.J26437	2067	688	74.8	7.27	2	contig29414	2799–7192
PvARF25	Pavir.Aa01271	1836	611	65.4	9.52	3	1a	16431915–16436701
PvARF26	Pavir.J17862	1335	444	47.7	8.55	2	contig196091	4–1855
PvARF27	Pavir.Gb00635	1851	616	66.1	8.8	2	7b	7512630–7516588
PvARF28	Pavir.J24081	2118	705	75.7	7.35	2	contig263498	76–1753
PvARF29	Pavir.J08401	2214	737	80.5	7.15	10	contig11657	2523–8428
PvARF30	Pavir.Ca00928	1716	571	63.3	9.24	8	3a	10393086–10397991
PvARF31	Pavir.Eb02716	2133	710	78.5	6.18	9	5b	59020641–59026106
PvARF32	Pavir.J35323	1065	354	39.5	7.62	6	contig52555	409–3563
PvARF33	Pavir.J38128	1104	367	40.6	7.98	7	contig73490	486–4550
PvARF34	Pavir.J17623	732	243	27.4	8.97	5	contig193925	123–2287
PvARF35	Pavir.J17853	2214	737	79.7	7.82	9	contig19600	3358–9328
PvARF36	Pavir.Eb03157	2049	682	74.1	6.66	9	5b	65190090–65198384
PvARF37	Pavir.Cb00753	2202	733	82	6.26	11	3b	15971362–15976212
PvARF38	Pavir.Ca02218	1560	519	58	6.66	6	3a	36368836–36372762
PvARF39	Pavir.J22605	1143	380	43.1	9.3	7	contig24640	4150–7685
PvARF40	Pavir.Eb03734	2430	809	90.7	6.1	13	5b	71953597–71958902
PvARF41	Pavir.Ea03860	2433	810	90.8	6.05	13	5a	61521645–61526769
PvARF42	Pavir.Ea00026	2064	687	76.6	5.58	11	5a	658461–662814
PvARF43	Pavir.Eb00045	2064	687	76.7	5.62	11	5b	655839–660061
PvARF44	Pavir.Ab01961	1440	479	53.7	7.28	12	1b	38024228–38038833
PvARF45	Pavir.Aa01676	2136	711	79.3	5.88	12	1a	22183544–22188794
PvARF46	Pavir.Gb01617	1986	661	73.5	5.77	13	7b	21374440–21379797
PvARF47	Pavir.Ga01750	1923	640	71.1	6.35	12	7a	21881663–21886932

[a]Names referred to the identified PvARF genes in switchgrass in this work. [b]The alias of each ARF gene in iTAG 2.30 genome annotation. [c]Length of open reading frame in base pairs. [d]The number of amino acids, molecular weight (kilodaltons), and isoelectric point of deduced polypeptide calculated by DNASTAR. [e]The number of intron. [f,g]Chromosome location from Phytozome (https://phytozome.jgi.doe.gov/).

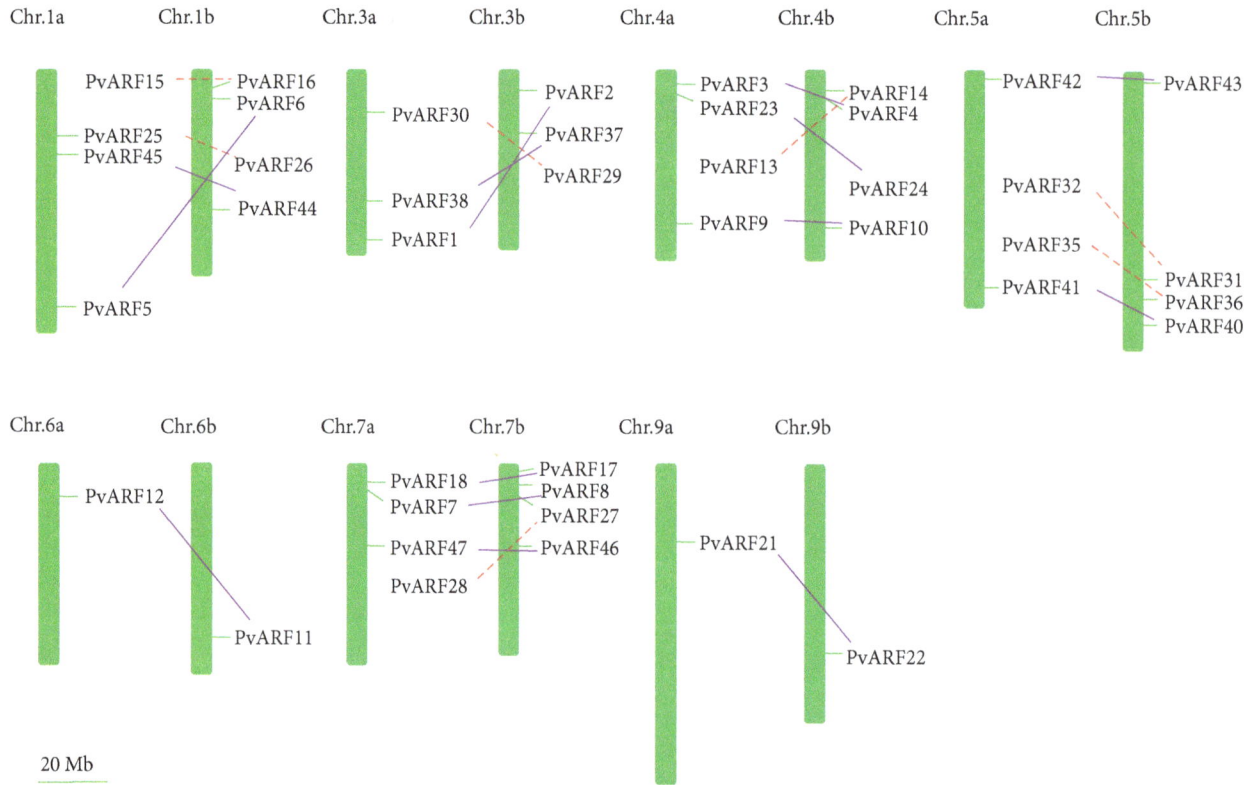

FIGURE 1: Chromosomal distribution of ARF genes in switchgrass. Distribution on the chromosomes (vertical bar) indicate the position of centromeres. The chromosome numbers (except for Chromosome 2a, 2b, 8a, and 8b) are indicated at the top of each chromosome image. The purple solid lines show the gene pairs, and the red dotted lines represent the putative gene pairs according to the sequence similarity.

manually improved by online program EvolView (http://www.evolgenius.info/evolview/).

2.4. Chromosomal Locations of PvARF Genes.
The lowland switchgrass cultivars are allotetraploid ($2n = 4x = 36$) and consist of two highly homologous subgenomes, designated as Chr.a and Chr.b [40]. Specific physical locations of each PvARF were obtained from the Phytozome database. Chromosome locations were then determined using MapChart 2.2 based on the genetic linkage map [41, 42]. Tandem gene duplicates were defined as paralogous genes located within 50 kb and separated by fewer than five nonhomologous spacer genes [43].

2.5. Gene Structure, Conserved Motif, and Cis-Acting DNA Element Analysis.
A comparison of each CDS with the corresponding genomic DNA sequence was made to determine the positions and numbers of introns and exons of each PvARF gene using the Gene Structure Display Server (http://gsds.cbi.pku.edu.cn/). Conserved motifs were analyzed using the MEME program (http://meme.nbcr.net/meme/cgi-bin/meme.cgi). Putative microRNA target sites in PvARFs were identified using the miRanda online software (http://cbio.mskcc.org/microrna_data/manual.html). Cis-acting DNA elements were analyzed using the PLACE online program (https://sogo.dna.affrc.go.jp/) [44]. Ka/Ks calculation was analyzed by PAL2NAL [45].

2.6. Gene Expression Analysis by qRT-PCR.
Probesets of PvARF genes were retrieved from public database of switchgrass (https://switchgrassgenomics.noble.org/). qRT-PCR was performed to analyze the transcript abundance of PvARFs in different switchgrass tissue samples. Plant tissue samples were ground in liquid nitrogen using a mortar and pestle. Total RNA was isolated using the TRIZOL reagent according to the manufacturer's supplied protocol (Transgen, China) and subjected to reverse transcription with Superscript PrimeScript™ RT reagent Kit (TaKaRa, China) after treatment with TURBO DNase I (TaKaRa, China). The qRT-PCR primers were designed using Primer Premier 5 (Table S1), and their specificity was verified by PCR. qRT-PCR analysis was conducted in triplicate using SYBR® Premix Ex Taq™ II (TaKaRa, Japan), with PvUBQ as a reference gene, with a Light Cycler 480 real-time PCR system (Roche, Switzerland). The qRT-PCR reactions and data analyses were performed according to previously published methods [46].

3. Results

3.1. Identification and Chromosomal Localization of Switchgrass ARFs.
To identify ARF proteins in switchgrass, the Hidden Markov Model (HMM) profile of the conserved ARF domain (Pfam06507) was used as a query to search against the publicly available switchgrass genome database

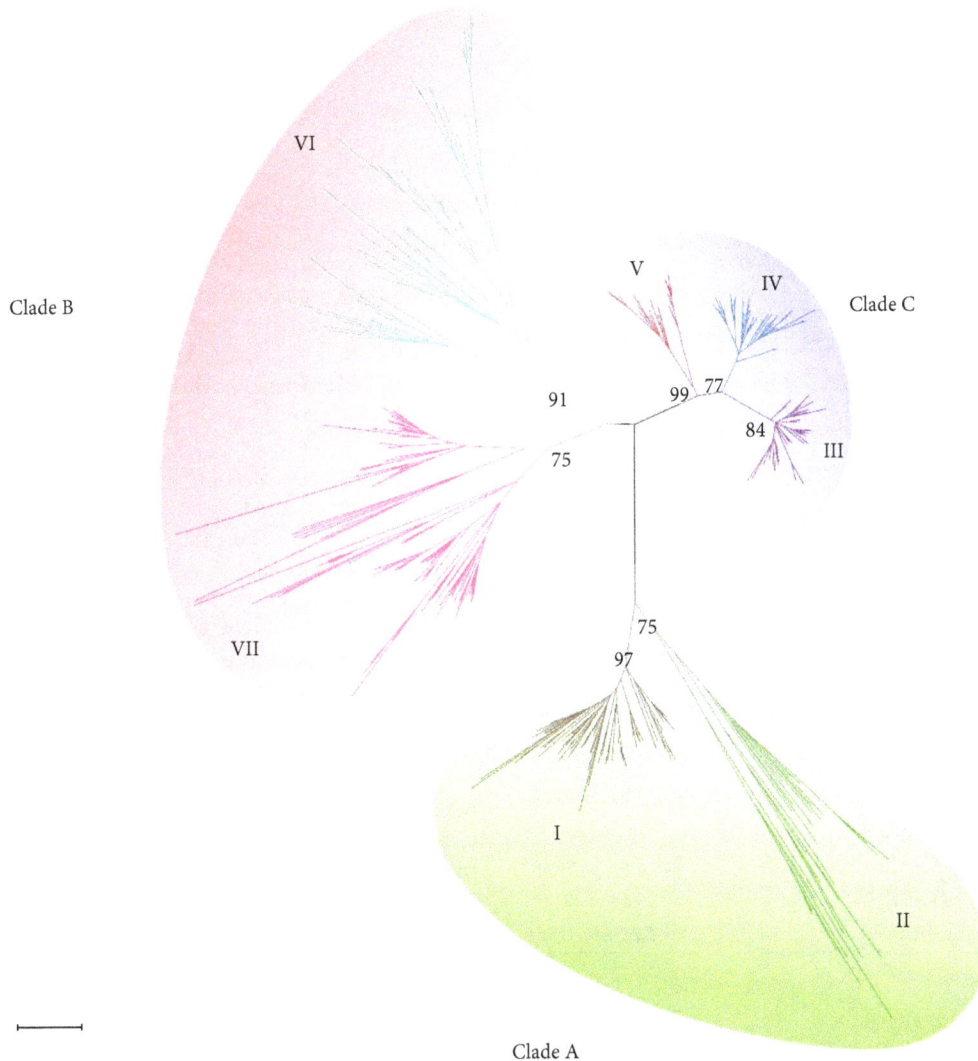

FIGURE 2: Phylogenetic analysis of the ARF proteins in switchgrass and other plant species. An unrooted neighbor-joining (bootstrap value = 1000) tree was constructed using MEGA5 on the basis of multiple alignments of conserved domain sequences of the ARF proteins from monocot species (switchgrass, foxtail millet, maize, sorghum, rice, and *Brachypodium*) and dicot species (*Arabidopsis*, sweet orange, Chinese cabbage, poplar, *Medicago*, cotton, soybean, tomato, *Grandis*, and grape). And the detailed information was listed in Table S2.

(Phytozome v11) by BlastP and tBlastN program. A total of 47 putative ARF proteins were found, and Pfam analysis confirmed that all of these proteins contain the ARF domain. The putative candidates were designated as PvARF1 to PvARF47, based on the alignments of predicted amino acid sequences. The predicted PvARF proteins ranged from 243 (PvARF34) to 1182 (PvARF13) amino acids (aa) in length and from 27.4 kDa to 130.9 kDa in molecular weight (Table 1). The isoelectric points (*pI*) ranged from 5.33 (PvARF8) to 9.52 (PvARF25) (Table 1), suggesting that different PvARFs might have roles in specific subcellular environments.

To examine the chromosomal distribution of *PvARFs*, the physical locations of the *PvARFs* on chromosomes (Chrs) were obtained through BlastN searches against the switchgrass genome database in Phytozome. Due to the allotetraploidization of switchgrass (2n = 4x = 36), the PvARF genes exist as paralogous gene pairs in the genome with only one exception, *PvARF39*, which might be lost in the

evolutionary process. Of the 47 *PvARFs*, 42 were putatively anchored onto seven of the nine switchgrass chromosomes (Figure 1), while the other five *PvARFs* (*PvARF19*, *20*, *39*, *42*, and *43*) are located on unmapped scaffolds. The chromosomal distribution and density of PvARF genes are not uniform. Chr 1, 4, 5, and 7 contain four PvARF gene pairs, respectively. Chr 3 has three pairs, Chr 6 and 9 have only one pair of *PvARFs*, and no gene is located on Chr 2 and 8. Consistent with expectations, 14 gene pairs obviously exist on the two set of chromosomes (Figure 1), while the other 7 gene pairs were putatively located on the chromosomes based on their sequence similarity. The indeed relationships among these *PvARFs* need to be explained by phylogenetic analysis.

3.2. Phylogenetic Analysis of Switchgrass ARFs. To profoundly characterize the phylogenetic relationships of ARF proteins among switchgrass and other land plants, we

TABLE 2: *Ka/Ks* calculation of ARF genes between switchgrass and rice.

Othologs	*Ka/Ks* ratio	Selection pattern
PvARF1 (2) versus *OsARF25*	1.61	Positive selection
PvARF3 (4) versus *OsARF6*	98.11	Positive selection
PvARF5 (6) versus *OsARF17*	2.19	Positive selection
PvARF8 (7) versus *OsARF12*	76.96	Positive selection
PvARF10 (9) versus *OsARF16*	99.00	Positive selection
PvARF11 (12) versus *OsARF21*	0.54	Purifying selection
PvARF14 (13) versus *OsARF19*	7.51	Positive selection
PvARF15 (16) versus *OsARF5*	9.13	Positive selection
PvARF17 (18) versus *OsARF11*	0.44	Purifying selection
PvARF19 (20) versus *OsARF13*	3.91	Positive selection
PvARF21 (22) versus *OsARF22*	14.81	Positive selection
PvARF23 (24) versus *OsARF18*	0.93	Purifying selection
PvARF25 (26) versus *OsARF8*	2.14	Positive selection
PvARF27 (28) versus *OsARF10*	0.06	Purifying selection
PvARF30 (29) versus *OsARF15*	2.81	Positive selection
PvARF31 (32) versus *OsARF2*	3.32	Positive selection
PvARF33 (34) versus *OsARF14*	3.81	Positive selection
PvARF36 (35) versus *OsARF3*	3.99	Positive selection
PvARF38 (37) versus *OsARF24*	0.29	Purifying selection
PvARF39 versus *OsARF23*	1.22	Positive selection
PvARF40 (41) versus *OsARF4*	0.61	Purifying selection
PvARF42 (43) versus *OsARF1*	1.75	Positive selection
PvARF45 (44) versus *OsARF7*	99.00	Positive selection
PvARF47 (46) versus *OsARF9*	2.38	Positive selection

selected ARFs from another 15 species, which have public genome database in Phytozome, to construct a phylogenetic tree together with PvARFs. These species include five monocots (foxtail millet, maize, sorghum, *Brachypodium*, and rice) and ten dicots (*Arabidopsis*, sweet orange, Chinese cabbage, poplar, cotton, soybean, *Medicago*, tomato, *Grandis*, and grape). Seven separate clusters of ARF proteins were defined based on the NJ tree topology and bootstrap values (higher than 50%) (Figure 2). As previously reported by Finet et al., three large groups of ARF proteins were classified as clades A, B, and C. In detail, clusters I and II in our study together make up clade C, and clusters III, IV, and V make up clade A. ARF members in these two clades are considered to be more ancient than those in clade B [22], which comprises clusters VI and VII in our study.

Considering that switchgrass is an allotetraploid plant, the gene number of PvARFs in each cluster should be approximately twice than the other monocots, especially in foxtail millet, the most closely relatives to switchgrass among the selected species (Table S2). Cluster VII, which has the largest number (11 out of 47) of PvARFs, contains six foxtail millet ARFs. Cluster I, III, IV, and VI also contain eight switchgrass and four foxtail millet ARFs, and clusters II and V have only two PvARFs, respectively (Table S2).

PvARF39 in cluster VII is most closely related to the foxtail millet ARF protein Seita.8G135700.1, which indicates that the paralog of PvARF39 has been lost or mutated gradually during the evolutionary process of switchgrass genome.

In order to comprehensively clarify the evolutionary process of the *PvARFs*, we carried out the tandem repeat duplication analysis based on the chromosomal location and phylogenetic analysis of the *PvARFs*. The results showed that no tandem repeat duplication events were found in *PvARFs*. In addition, we calculated the *Ka/Ks* analysis between *PvARFs* and *OsARFs*. Compared with rice ARF genes, 18 pairs of orthologs originated from positive selection (*Ka/Ks* ratio was larger than 1), while 6 orthologs showed purifying selection (*Ka/Ks* ratio was less than 1) (Table 2).

3.3. PvARF Gene Structures and Locations of Conserved Motif. To better understand the phylogenetic relationships of the *PvARFs*, the exon/intron arrangements were determined by aligning cDNA sequences to genomic sequences. Another phylogenetic tree was firstly constructed only using switchgrass ARF protein sequences. The PvARF genes were clearly displayed in the form of gene pairs (Figure 3(a)), which confirmed the previous speculation in the chromosomal distribution (Figure 1). All *PvARFs* have introns in the coding sequence (CDS), and the number of introns ranges from 2 to 14 (Table 1, Figure 3(b)). In particular, members belonging to clade C (clusters I and II) contain relatively fewer introns (two to four). In contrast, *PvARFs* in clade A (clusters III, IV, and V) have much more introns (11 to 14), with the exception of PvARF2, which might have lost the exons in N-terminus. The number of introns in clade B (clusters VI and VII) were ranging from 5 to 13. This variability of intron number might be correlated to the multiple functions of clade B *ARFs* in higher plants. Additionally, we further identified the putative microRNA target sites of ARF genes in switchgrass. 16 out of 47 *PvARFs* were found to contain the potential microRNA target sites. Eight *PvARFs* were predicted to be the targets of miR160 and miR167, respectively (Figure 3(b), Figure 4).

Analysis of motif locations in PvARF proteins was performed to explore structural diversity and to predict their functions. A total of 12 conserved motifs were identified using the MEME program (Figure 3(c), Figure S1). The DNA-binding domain (DBD) (motifs 1, 2, and 9 corresponding to Pfam02362) was lost in four members (*PvARF22, 25, 25*, and *38*). The ARF domain (motifs 3, 5, 8, and 11 corresponding to Pfam06507) exists in all *PvARFs*. The AUX/IAA domain (motifs 4 and 10 corresponding to cl03528) has been lost in almost all of the *PvARFs* in clusters I, II, and VI. These results confirm the phylogenetic relationship between the PvARFs in clades A/C and B and indicate that there has been functional differentiation among PvARFs in different clusters.

3.4. Expression Patterns of PvARFs in Different Organs of Switchgrass. To analyze the expression levels of *PvARFs*, we firstly acquired the probesets of the *PvARFs* in switchgrass expression atlas from the public database [38]. The results showed that *PvARFs* were expressed in root, node, internode,

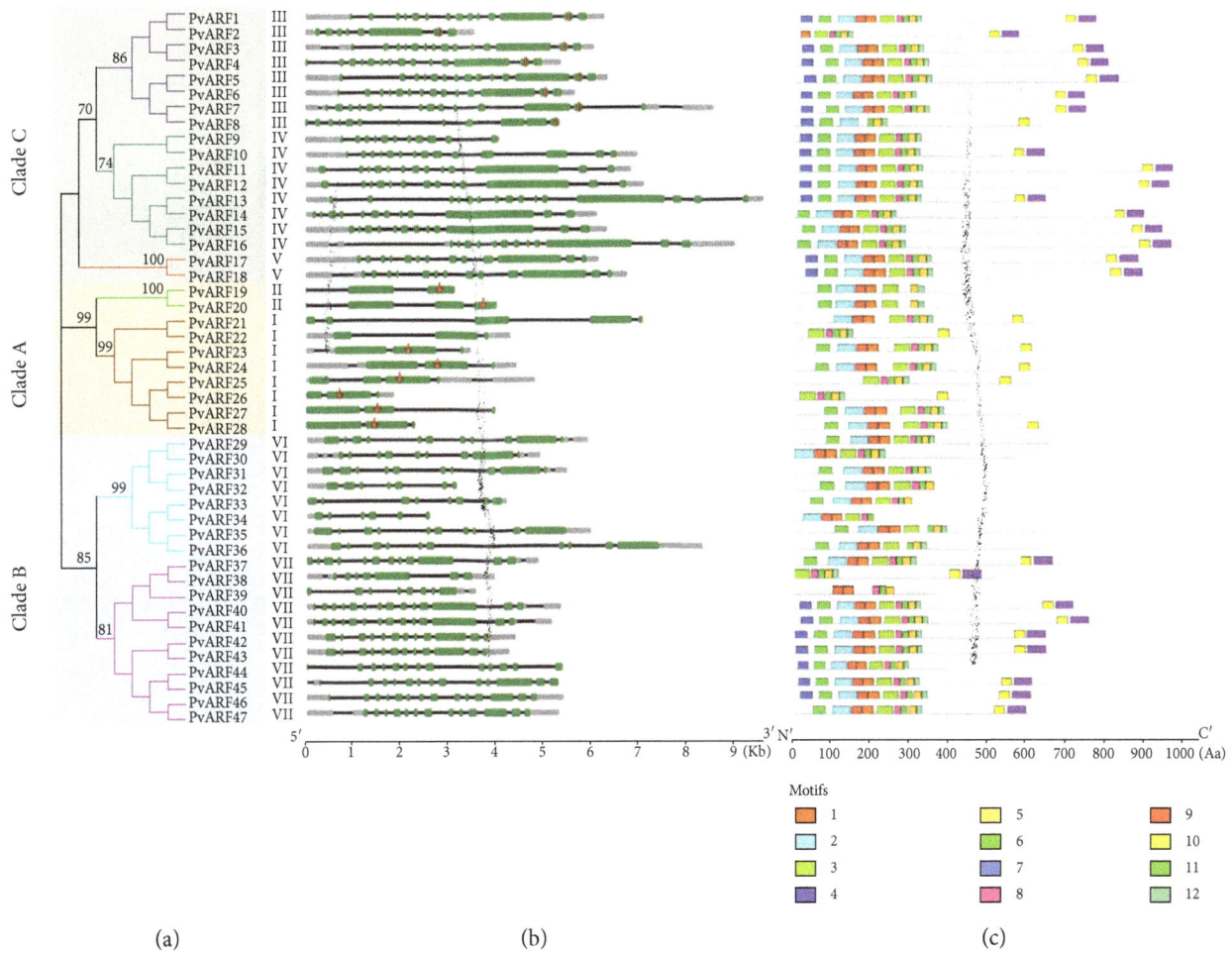

FIGURE 3: Exon/intron structure of PvARF genes. (a) Phylogenetic tree of PvARF proteins constructed using MEGA5 based on the multiple alignments of full-length amino acids. (b) Exon/intron arrangements of PvARF genes. Exons and introns are represented by green boxes (open reading frame in green, untranslated region (UTR) in gray), and black lines, respectively, and their sizes are indicated by the scale at the bottom. The red vertical bar denotes the targets of Osa-miR167a in PvARF genes; the red arrows denotes the targets of Osa-miR160a in PvARF genes. (c) Schematic representation of conserved motifs in the PvARF proteins predicted by MEME. Each motif is represented by a number in the colored box. The black lines represent nonconserved sequences. Lengths of motifs for each PvARF protein were exhibited proportionally.

FIGURE 4: Putative microRNA160 and microRNA167 targeted binding sites of the PvARF genes. (a) Putative microRNA160 targeted binding sites of the PvARF genes. (b) Putative microRNA167 targeted binding sites of the PvARF genes.

leaf, leaf sheath, flower, and seed but with different expression profile of each PvARF gene. For example, *PvARF3/4*, 11/12, and *46/47* were highly expressed in all tested organs, while *PvARF1/2*, *9/10*, *19/20*, and *33/34* were extremely lowly expressed in switchgrass (Figure 5).

Biomass yield is one of the most important criteria used to evaluate the quality of switchgrass. Vegetative organs, especially internodes and leaves, are the primary sources of biomass. Auxin is one of the most important phytohormone, which regulate the plant growth and development. To

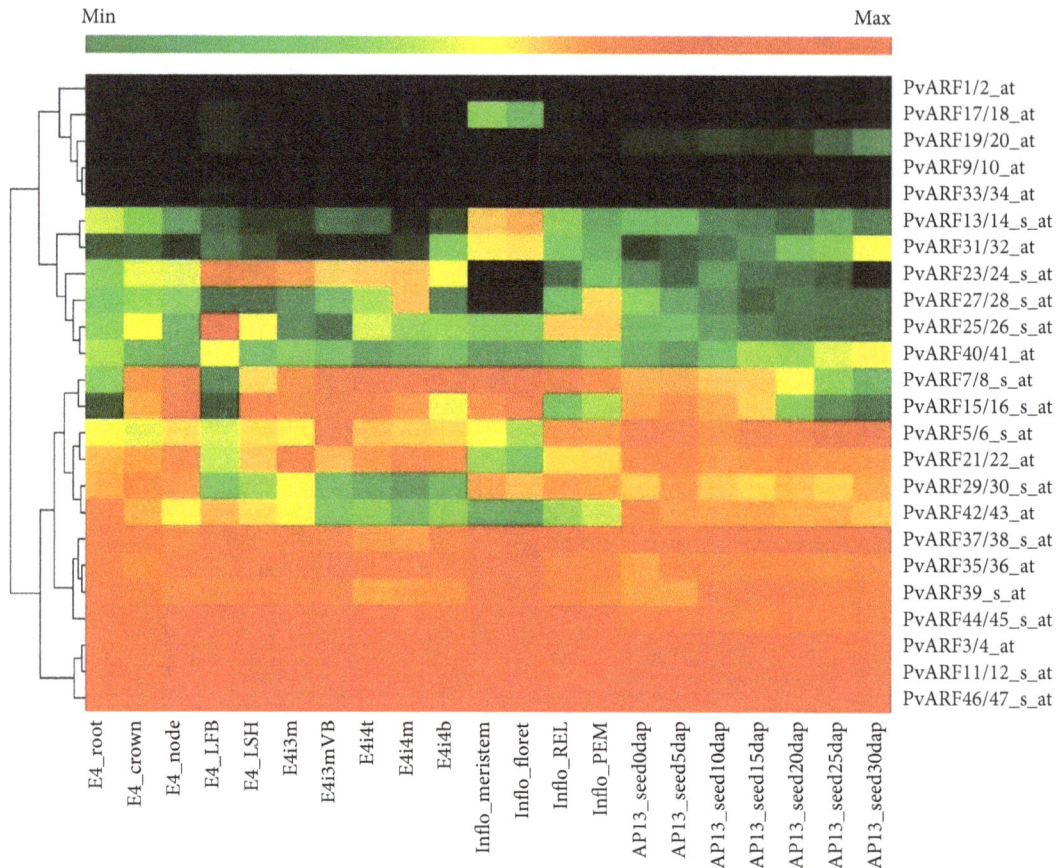

FIGURE 5: Heatmap of expression profiles of the PvARF gene pairs in different tested tissues. The data was collected from switchgrass gene atlas database. Clustering analysis was carried out using Genesis program (v1.7.6).

investigate whether and how PvARFs work on vegetative organs, we selected the second, the third, and the fourth internode and the corresponding leaves at the second reproductive (R2) stage to test the expression profile of the *PvARFs* by qRT-PCR analysis. Eight pairs of PvARF genes and *PvARF39* are not expressed or are extremely lowly expressed in the tested tissues, whereas the other 15 pairs of PvARF genes show substantial expression in internodes and leaves. In internodes, 14 pairs of *PvARFs* have higher expression levels in the upper internode (I4) than the other two internodes (I2 and I3), with one exception (*PvARF17/18*) having no obvious difference in the expression level in the three internodes (Figure 6). In leaves, ten pairs of PvARF genes (*PvARF3/4, 15/16, 21/22, 23/24, 29/30, 35/36, 37/38, 40/41, 44/45,* and *46/47)* show no significant changes in expression level in the three tested leaves. There was lower expression of *PvARF5/6, 7/8,* and *17/18* in the upper leaf (L4) than in the bottom leaves (L2 or L3), whereas *PvARF11/12* and *42/43* are more highly expressed in L4 compared to L2, and even lower expression is observed in L3 (Figure 6). These results suggest that the biosynthesis and transport of endogenous auxin in switchgrass might affect the expression profile of PvARF genes, especially in the internode.

3.5. Expression Analysis of PvARFs in Response to Auxin Treatment. In order to clarify the biofunctions of PvARF proteins, *cis*-acting DNA elements were analyzed using 2 kb

promoter sequence of *PvARFs*. The results showed that PvARFs in different clades were putatively involved in specific process. For example, PvARFs in clade A might participate in nodule formation, while clade B genes function on wounding response (Table S3). However, all of the PvARF proteins mostly tended to be involved in plant growth and development, such as phytohormone signaling, abiotic stress response, carbon metabolism, pollen development, and so on (Table S3).

As a key component of the auxin signaling pathway, ARF proteins play vital roles in auxin response. It has been reported that auxin induces or represses the expression of some ARF genes in *Arabidopsis* [31], rice [18], and maize [21]. To examine the response of PvARF genes to the exogenous auxin, one-month-old switchgrass seedlings were treated with 5 μM NAA for 0, 1, 2, and 3 hours, and the expression patterns of the *PvARFs* were determined. The qRT-PCR results revealed that auxin repressed the expression of eleven pairs of genes (*PvARF5/6, 7/8, 11/12, 15/16, 29/30, 35/36, 37/38, 40/41, 42/43, 44/45,* and *46/47*) at all three time points, whereas it induced the expression of three pairs of genes (*PvARF3/4, 23/24,* and *25/26*) at 1 hour and then reduced at the latter two time points. In contrast, *PvARF21/22* expression was not significantly affected by auxin (Figure 7). These results suggest that exogenous auxin could induce or repress the expression of most PvARF genes to regulate switchgrass growth and development.

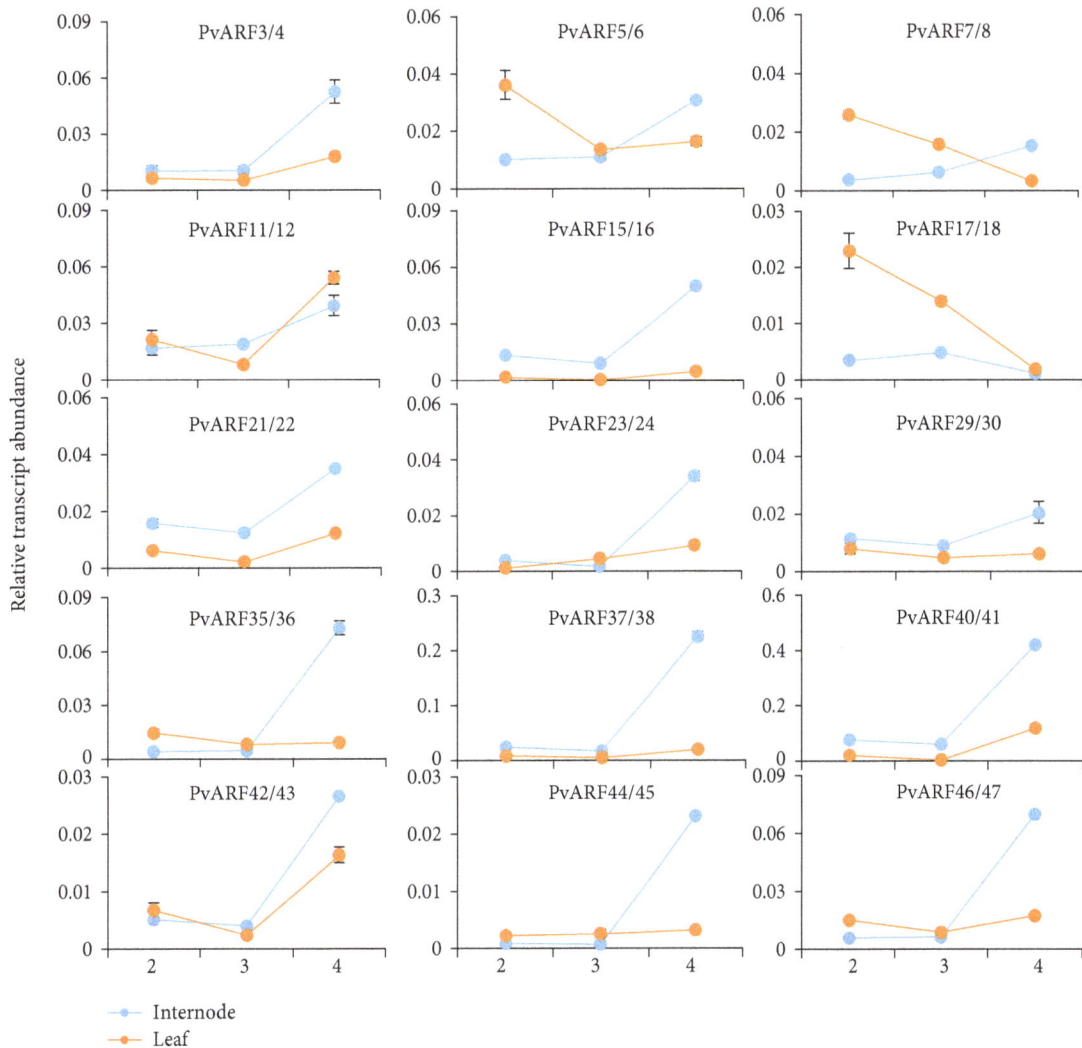

FIGURE 6: The expression of PvARF genes in the vegetative organs. The expression of PvARF genes in the second internodes (I2), the third internodes (I3), the fourth internodes (I4), the second leaves (L2), the third leaves (L3), and the fourth leaves (L4) of switchgrass. Relative transcript levels are calculated by qRT-PCR. Data are means ± SE of three separate measurements.

4. Discussion

Extensive studies have shown that *ARFs* play crucial roles in plant growth and developmental processes [10]. However, a systematic analysis of ARF gene members in switchgrass has not been done. In this study, we identified 47 PvARF genes, almost twice than Arabidopsis (23) and rice (25) [18, 31], for the reason of allopolyploidization in switchgrass evolutionary process [41]. Gene structure analysis showed that the number of exons in PvARF genes ranged from 3 to 14, while similar results were found in Arabidopsis [31], rice [18], and tomato [9], which indicates that the plant ARF gene family has highly conserved structures and potentially similar functions across dicotyledonous and monocotyledonous plant species.

Based on phylogenetic analysis, 47 switchgrass ARF genes were assigned to seven separate clusters, which was similar to the previous studies [22]. The number of ARFs of switchgrass in each cluster was about twice than those of five other monocots (maize, rice, sorghum, foxtail millet, and

Brachypodium) but not consistent with the number of ARFs in ten dicot species (*Arabidopsis*, citrus, Chinese cabbage, poplar, cotton, soybean, *Medicago*, tomato, *Grandis*, and grape), which indicates that differences in the evolution of ARF genes in monocotyledonous and dicotyledonous plants. According to the phylogenetic tree of ARFs from different species, the orthologous relationship was found to dramatically divorce. In cluster V (PvARF17/18-AtARF5-CiARF5-BrARF5-1/5-3-PoptrARF5.1/5.2-GrARF5a/5b-EgrARF5-GmARF40/47-Sl-AF5-VvARF18-ZmARF4/29-OsARF11-Seita.3G028100.1-Sobic.006G255300.1-Bradi5g25157.1), the ratio of orthologous gene number between species is 1 : 1 which suggests that the functions might be well-conserved across species. Orthologous clusters with ratios greater or smaller than 1 : 1 were also found, indicating the functional diversity of ARF genes in switchgrass.

In general, the members of a subgroup are characterized by the presence of conserved domains. According to previous studies, the ARF genes contain several conserved domains, such as motifs 1, 2, and 9 made up the DNA-binding domain,

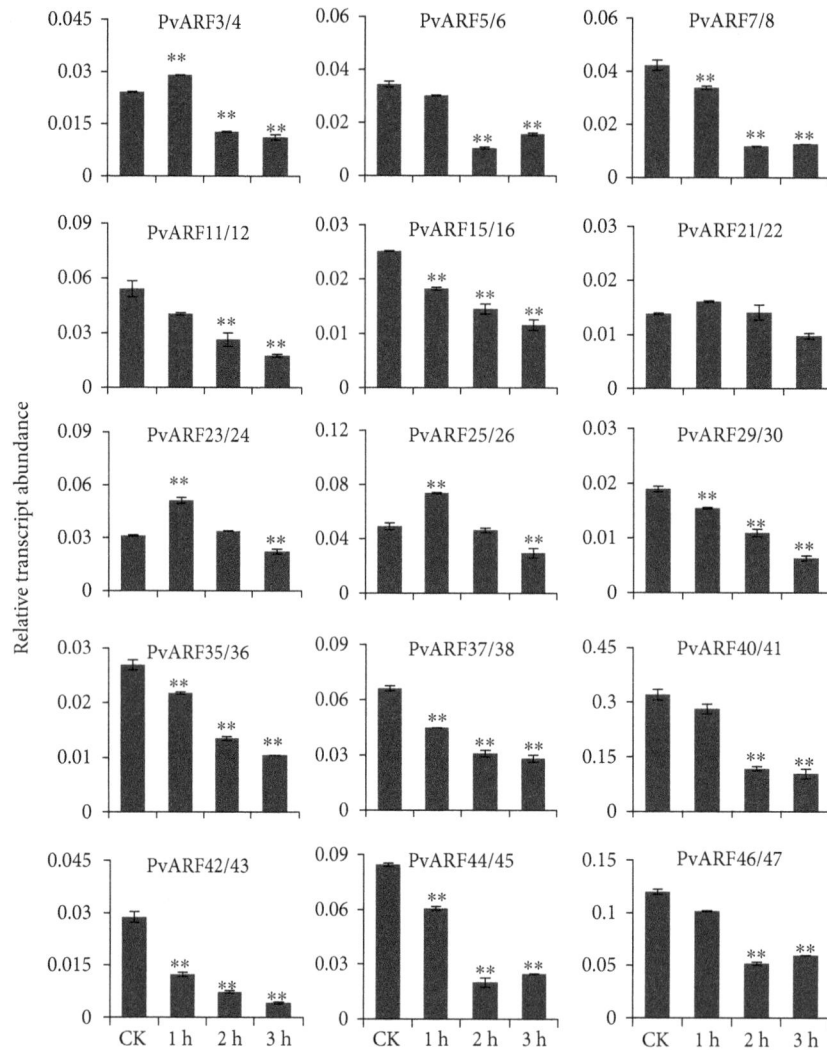

FIGURE 7: The expression of PvARF genes in response to treatment with $5\,\mu$M NAA solution for 1, 2, and 3 hours. Control plants were grown in hormone-free medium. Error bars represented variability of qRT-PCR results from three replicates. Data are means $\pm$ SE of three separate measurements. Statistically significant differences were assessed using Student's t-tests ($^{**}P < 0.01$).

motifs 3, 5, 8, and 11, which correspond to the ARF domain, and motifs 4 and 10, which are located in the C-terminus and correspond to the AUX/IAA super family domain [10]. The high level of conservation of the various motifs among different species indicates that they are involved in similar regulatory pathway. In our study, the C-terminal AUX/IAA super family domain was missing in several gene members, including *PvARF9, 19, 20, 27, 29–36, 39*, and *44*, which is consistent with the lack of this domain in *AtARF3, 13*, and *17* and *MdARF6, 8, 14, 17, 18, 20, 21*, and *28* as well as in *SlARF2, 3, 7*, and *13* [2, 3, 47]. In addition, PvARF genes which are present in the same clade and possess similar motifs might function redundantly and have similar expression patterns. For example, *PvARF5/6* and *7/8* which are members of cluster III and *PvARF37/38, 40/41, 42/43, 44/45*, and *46/47*, which are members of cluster VII exhibit similar expression patterns at the R2 stage.

Switchgrass is an important resource for bioenergy and feedstock materials, and biomass yield is the most important target in molecular breeding of switchgrass. Comprehensive

analysis of PvARF gene expression patterns helped us screen for candidate PvARF genes with potentially distinct functions in regulating vegetative organ growth and development. Taken together, 15 pairs of PvARF genes were detected to have high levels of expression in vegetative organs. Similar patterns of expression were also found in tea plants [48], apple [47], and tomato [9]. 13 of the 15 CsARF genes were expressed in root, stem, leaf, flower, and fruit [24]. Eight of the 31 MdARF genes were expressed in stem, leaf, flower, and fruit [46], and 17 SlARF genes were expressed in root, stem, leaf, flower bud, and ovary [9]. In our study, 14 pairs of *PvARFs* were more highly expressed in the I4 than in the I2 and I3, which suggests that these genes might play vital roles during the formation of young stems, and these results are consistent with the reported function of their homologous genes in *Arabidopsis* [26, 49]. However, in leaves, the expression level of most *PvARFs* did not change significantly in different developmental stages. *PvARF11/12* is most highly expressed in the fourth leaf, suggesting that these genes may play roles in leaf development like their *Arabidopsis* homologs, *ARF7* and *19* [31]. Of the 47 PvARF genes, 17 were not expressed in the tested tissues, which indicates that they might not function in these organs, or that the functions of these genes may have been lost during evolution. In general, most PvARF genes have different expression profiles in the internodes and leaves, indicating that they might be regulated by the distribution and concentration of endogenous auxin. The in-depth studies will be needed to confirm these results in future.

Because ARF proteins are transcription factors that regulate the expression of auxin response genes, we determined the response of PvARF genes to NAA treatment. The regulation of gene expression in response to auxin has been reported in *Arabidopsis* [3, 31], rice [20, 32, 34], maize [21], tomato [2], *Medicago* [20], and so on. In this study, we found that at least 14 pairs of PvARF genes were responsive to NAA treatment in seedlings but showed diverse expression patterns. Eleven pairs of PvARF genes (*PvARF5/6, 7/8, 11/12, 15/16, 29/30, 35/36, 37/38, 40/41, 42/43, 44/45,* and *46/47*) were downregulated by exogenous auxin treatment across all time points, indicating that their expression was negatively regulated by NAA, similar to their homologs in rice and maize (*OsARF5, 14,* and *21* and *ZmARF5* and *18*), of which expression levels decreased marginally in response to auxin [21, 32]. In contrast, the other three pairs of PvARF genes (*PvARF3/4, 23/24,* and *25/26*) were upregulated by auxin treatment at 1 hour point and then downregulation at later time points, indicating that NAA significantly induced the target gene in a short period of time, as their homologs in *Arabidopsis*, rice, and maize (*AtARF4, 19; OsARF1, 23;* and *ZmARF3, 8, 13, 15, 21, 27,* and *30*). In brief, expression of these ARF genes increased slightly in response to auxin [21, 31, 32], implying that these genes are potential primary auxin responsive genes. Generally, the expression level of the ARF genes was directly regulated by auxin. Considering that the endogenous auxin concentration is sufficient for plant growth and development, the extra auxin (NAA) applied exogenously might act as inhibitor of auxin response genes in our study, and the further study will be carried out in the future to clarify the mechanism of auxin response in grasses.

5. Conclusions

We identified 47 switchgrass ARF genes and established the evolutionary relationship between these genes using phylogenic, gene structure, and conserved protein motif analyses. Expression analyses revealed the potential role of PvARF genes involved in growth and development of switchgrass internodes and leaves and in response to NAA treatment in seedlings. These data provide a solid foundation for future functional characterization of ARF genes and ARF-mediated signal transduction pathway in switchgrass.

Conflicts of Interest

The authors have declared that no competing interests exist.

Authors' Contributions

Jianli Wang, Changhong Guo, and Guiqing Han conceived and designed the study. Jianli Wang and Zhenying Wu performed the laboratory experiments and the data analysis. Chunxiang Fu, Zhongbao Shen, Dequan Sun, Peng Zhong, Lichao Ma, Zetao Bai, Duofeng Pan, Ruibo Zhang, Daoming Li, and Hailing Zhang assisted in the data analysis. Jianli Wang and Zhenying Wu wrote the manuscript with assistance from Guiqing Han. All authors read and approved the final manuscript. Jianli Wang and Zhenying Wu contributed equally to this work.

Acknowledgments

This work was supported by the National Natural Science Foundation of China (Grant number 31601365), the major research project of Heilongjiang Academy of Agricultural Sciences (Germplasm Resource Renewal of Frozen Crops in Heilongjiang Province).

References

[1] J. W. Chandler, "Auxin response factors," *Plant, Cell & Environment*, vol. 39, no. 5, pp. 1014–1028, 2016.

[2] R. Kumar, A. K. Tyagi, and A. K. Sharma, "Genome-wide analysis of auxin response factor (ARF) gene family from tomato and analysis of their role in flower and fruit development," *Molecular Genetics and Genomics*, vol. 285, no. 3, pp. 245–260, 2011.

[3] J. S. Li, X. H. Dai, and Y. D. Zhao, "A role for auxin response factor 19 in auxin and ethylene signaling in Arabidopsis," *Plant Physiology*, vol. 140, no. 3, pp. 899–908, 2006.

[4] K. Ljung, "Auxin metabolism and homeostasis during plant development," *Development*, vol. 140, no. 5, pp. 943–950, 2013.

[5] K. Mockaitis and M. Estelle, "Auxin receptors and plant development: a new signaling paradigm," *Annual Review of Cell and Developmental Biology*, vol. 24, no. 1, pp. 55–80, 2008.

[6] A. Santner and M. Estelle, "Recent advances and emerging trends in plant hormone signalling," *Nature*, vol. 459, no. 7250, pp. 1071–1078, 2009.

[7] Y. H. Su, Y. B. Liu, B. Bai, and X. S. Zhang, "Establishment of embryonic shoot–root axis is involved in auxin and cytokinin response during Arabidopsis somatic embryogenesis," *Frontiers in Plant Science*, vol. 5, 2015.

[8] A. W. Woodward and B. Bartel, "Auxin: Regulation, action, and interaction," *Annals of Botany*, vol. 95, no. 5, pp. 707–735, 2005.

[9] J. Wu, F. Y. Wang, L. Cheng et al., "Identification, isolation and expression analysis of auxin response factor (ARF) genes in *Solanum lycopersicum*," *Plant Cell Reports*, vol. 30, no. 11, pp. 2059–2073, 2011.

[10] T. J. Guilfoyle and G. Hagen, "Auxin response factors," *Current Opinion in Plant Biology*, vol. 10, no. 5, pp. 453–460, 2007.

[11] E. Liscum and J. W. Reed, "Genetics of Aux/IAA and ARF action in plant growth and development," *Plant Molecular Biology*, vol. 49, no. 3/4, pp. 387–400, 2002.

[12] T. Ulmasov, G. Hagen, and T. J. Guilfoyle, "ARF1, a transcription factor that binds to auxin response elements," *Science*, vol. 276, no. 5320, pp. 1865–1868, 1997.

[13] T. Ulmasov, G. Hagen, and T. J. Guilfoyle, "Activation and repression of transcription by auxin-response factors," *Proceedings of the National Academy of Sciences*, vol. 96, no. 10, pp. 5844–5849, 1999.

[14] J. Kim, K. Harter, and A. Theologis, "Protein-protein interactions among the Aux/IAA proteins," *Proceedings of the National Academy of Sciences*, vol. 94, no. 22, pp. 11786–11791, 1997.

[15] F. Ouellet, P. J. Overvoorde, and A. Theologis, "IAA17/AXR3: biochemical insight into an auxin mutant phenotype," *The Plant Cell Online*, vol. 13, no. 4, pp. 829–842, 2001.

[16] R. Shin, A. Y. Burch, K. A. Huppert et al., "The Arabidopsis transcription factor MYB77 modulates auxin signal transduction," *The Plant Cell Online*, vol. 19, no. 8, pp. 2440–2453, 2007.

[17] G. Hagen and T. Guilfoyle, "Auxin-responsive gene expression: genes, promoters and regulatory factors," *Plant Molecular Biology*, vol. 49, no. 3/4, pp. 373–385, 2002.

[18] D. K. Wang, K. M. Pei, Y. P. Fu et al., "Genome-wide analysis of the auxin response factors (ARF) gene family in rice (*Oryza sativa*)," *Gene*, vol. 394, no. 1-2, pp. 13–24, 2007.

[19] U. C. Kalluri, S. P. Difazio, A. M. Brunner, and G. A. Tuskan, "Genome-wide analysis of Aux/IAA and ARF gene families in *Populus trichocarpa*," *BMC Plant Biology*, vol. 7, no. 1, p. 59, 2007.

[20] C. J. Shen, R. Q. Yue, T. Sun et al., "Genome-wide identification and expression analysis of auxin response factor gene family in *Medicago truncatula*," *Frontiers in Plant Science*, vol. 6, 2015.

[21] Y. J. Wang, D. X. Deng, Y. T. Shi, N. Miao, Y. L. Bian, and Z. T. Yin, "Diversification, phylogeny and evolution of auxin response factor (ARF) family: insights gained from analyzing maize ARF genes," *Molecular Biology Reports*, vol. 39, no. 3, pp. 2401–2415, 2012.

[22] C. Finet, A. Berne-Dedieu, C. P. Scutt, and F. Marletaz, "Evolution of the ARF gene family in land plants: old domains, new tricks," *Molecular Biology and Evolution*, vol. 30, no. 1, pp. 45–56, 2013.

[23] W. Hu, J. Zuo, X. W. Hou et al., "The auxin response factor gene family in banana: genome-wide identification and expression analyses during development, ripening, and abiotic stress," *Frontiers in Plant Science*, vol. 6, 2015.

[24] Z. C. Xu, A. J. Ji, J. Y. Song, and S. L. Chen, "Genome-wide analysis of auxin response factor gene family members in medicinal model plant *Salvia miltiorrhiza*," *Biology Open*, vol. 5, no. 6, pp. 848–857, 2016.

[25] C. X. Yang, M. Xu, L. Xuan, X. M. Jiang, and M. R. Huang, "Identification and expression analysis of twenty ARF genes in Populus," *Gene*, vol. 544, no. 2, pp. 134–144, 2014.

[26] C. M. Ellis, P. Nagpal, J. C. Young, G. Hagen, T. J. Guilfoyle, and J. W. Reed, "Auxin Response Factor1 and Auxin Response Factor2 regulate senescence and floral organ abscission in Arabidopsis thaliana," *Development*, vol. 132, no. 20, pp. 4563–4574, 2005.

[27] P. O. Lim, I. C. Lee, J. Kim et al., "Auxin response factor 2 (ARF2) plays a major role in regulating auxin-mediated leaf longevity," *Journal of Experimental Botany*, vol. 61, no. 5, pp. 1419–1430, 2010.

[28] M. C. Schruff, M. Spielman, S. Tiwari, S. Adams, N. Fenby, and R. J. Scott, "The AUXIN RESPONSE FACTOR 2 gene of Arabidopsis links auxin signalling, cell division, and the size of seeds and other organs," *Development*, vol. 133, no. 2, pp. 251–261, 2006.

[29] C. S. Hardtke and T. Berleth, "The Arabidopsis gene MONOPTEROS encodes a transcription factor mediating embryo axis formation and vascular development," *The EMBO Journal*, vol. 17, no. 5, pp. 1405–1411, 1998.

[30] M. Goetz, A. Vivian-Smith, S. D. Johnson, and A. M. Koltunow, "AUXIN RESPONSE FACTOR8 is a negative regulator of fruit initiation in Arabidopsis," *The Plant Cell Online*, vol. 18, no. 8, pp. 1873–1886, 2006.

[31] Y. Okushima, H. Fukaki, M. Onoda, A. Theologis, and M. Tasaka, "ARF7 and ARF19 regulate lateral root formation via direct activation of LBD/ASL genes in Arabidopsis," *The Plant Cell Online*, vol. 19, no. 1, pp. 118–130, 2007.

[32] K. A. Attia, A. F. Abdelkhalik, M. H. Ammar et al., "Antisense phenotypes reveal a functional expression of OsARF1, an auxin response factor, in transgenic rice," *Current Issues in Molecular Biology*, vol. 11, Supplement 1, pp. i29–i34, 2009.

[33] C. J. Shen, R. Q. Yue, T. Sun, L. Zhang, Y. J. Yang, and H. Z. Wang, "OsARF16, a transcription factor regulating auxin redistribution, is required for iron deficiency response in rice (*Oryza sativa* L.)," *Plant Science*, vol. 231, pp. 148–158, 2015.

[34] M. Jain and J. P. Khurana, "Transcript profiling reveals diverse roles of auxin-responsive genes during reproductive development and abiotic stress in rice," *FEBS Journal*, vol. 276, no. 11, pp. 3148–3162, 2009.

[35] S. K. Wang, Y. H. Bai, C. J. Shen et al., "Auxin-related gene families in abiotic stress response in *Sorghum bicolor*," *Functional & Integrative Genomics*, vol. 10, no. 4, pp. 533–546, 2010.

[36] B. Anderson, J. K. Ward, K. P. Vogel, M. G. Ward, H. J. Gorz, and F. A. Haskins, "Forage quality and performance of yearlings grazing switchgrass strains selected for differing digestibility," *Journal of Animal Science*, vol. 66, no. 9, pp. 2239–2244, 1988.

[37] S. B. McLaughlin and L. Adams Kszos, "Development of switchgrass (*Panicum virgatum*) as a bioenergy feedstock in the United States," *Biomass and Bioenergy*, vol. 28, no. 6, pp. 515–535, 2005.

[38] J. Y. Zhang, Y. C. Lee, I. Torres-Jerez et al., "Development of an integrated transcript sequence database and a gene expression atlas for gene discovery and analysis in switchgrass (*Panicum virgatum* L.)," *The Plant Journal*, vol. 74, no. 1, pp. 160–173, 2013.

[39] Z. Wu, Y. Cao, R. Yang et al., "Switchgrass SBP-box transcription factors PvSPL1 and 2 function redundantly to initiate side tillers and affect biomass yield of energy crop," *Biotechnology for Biofuels*, vol. 9, no. 1, p. 101, 2016.

[40] S. Yuan, B. Xu, J. Zhang et al., "Comprehensive analysis of CCCH-type zinc finger family genes facilitates functional gene discovery and reflects recent allopolyploidization event in tetraploid switchgrass," *BMC Genomics*, vol. 16, no. 1, p. 129, 2015.

[41] R. E. Voorrips, "MapChart: software for the graphical presentation of linkage maps and QTLs," *Journal of Heredity*, vol. 93, no. 1, pp. 77-78, 2002.

[42] M. Okada, C. Lanzatella, M. C. Saha, J. Bouton, R. L. Wu, and C. M. Tobias, "Complete switchgrass genetic maps reveal sub-

[42] M. Okada, C. Lanzatella, M. C. Saha, J. Bouton, R. L. Wu, and C. M. Tobias, "Complete switchgrass genetic maps reveal subgenome collinearity, preferential pairing and multilocus interactions," *Genetics*, vol. 185, no. 3, pp. 745–760, 2010.

[43] S. B. Cannon, A. Mitra, A. Baumgarten, N. D. Young, and G. May, "The roles of segmental and tandem gene duplication in the evolution of large gene families in *Arabidopsis thaliana*," *BMC Plant Biology*, vol. 4, no. 1, p. 10, 2004.

[44] K. Higo, Y. Ugawa, M. Iwamoto, and T. Korenaga, "Plant cis-acting regulatory DNA elements (PLACE) database: 1999," *Nucleic Acids Research*, vol. 27, no. 1, pp. 297–300, 1999.

[45] N. Goldman and Z. Yang, "A codon-based model of nucleotide substitution for protein-coding DNA sequences," *Molecular Biology and Evolution*, vol. 11, no. 5, pp. 725–736, 1994.

[46] Z. Y. Wu, X. Q. Xu, W. D. Xiong et al., "Genome-wide analysis of the NAC gene family in physic nut (*Jatropha curcas* L.)," *PLoS One*, vol. 10, no. 6, article e0131890, 2015.

[47] X. C. Luo, M. H. Sun, R. R. Xu, H. R. Shu, J. W. Wang, and S. Z. Zhang, "Genomewide identification and expression analysis of the ARF gene family in apple," *Journal of Genetics*, vol. 93, no. 3, pp. 785–797, 2014.

[48] Y. X. Xu, J. Mao, W. Chen et al., "Identification and expression profiling of the auxin response factors (ARFs) in the tea plant (*Camellia sinensis* (L.) O. Kuntze) under various abiotic stresses," *Plant Physiology and Biochemistry*, vol. 98, pp. 46–56, 2016.

[49] N. Fahlgren, T. A. Montgomery, M. D. Howell et al., "Regulation of AUXIN RESPONSE FACTOR3 by TAS3 ta-siRNA affects developmental timing and patterning in Arabidopsis," *Current Biology*, vol. 16, no. 9, pp. 939–944, 2006.

Common DNA Variants Accurately Rank an Individual of Extreme Height

Corinne E. Sexton,[1] Mark T. W. Ebbert,[2] Ryan H. Miller,[3] Meganne Ferrel,[1]
Jo Ann T. Tschanz,[4,5] Christopher D. Corcoran,[5,6]
Alzheimer's Disease Neuroimaging Initiative,[7] Perry G. Ridge ⓘ,[1] and John S. K. Kauwe ⓘ[1]

[1]*Department of Biology, Brigham Young University, Provo, UT 84602, USA*
[2]*Department of Neuroscience, Mayo Clinic, Jacksonville, FL 32224, USA*
[3]*Department of Oncological Sciences, University of Utah, Salt Lake City, UT 84112, USA*
[4]*Department of Psychology, Utah State University, Logan, UT, USA*
[5]*Center for Epidemiologic Studies, Utah State University, Logan, UT, USA*
[6]*Department of Mathematics and Statistics, Utah State University, Logan, UT, USA*
[7]*Alzheimer's Disease Neuroimaging Initiative, University of Southern California, Los Angeles, CA 90089, USA*

Correspondence should be addressed to John S. K. Kauwe; kauwe@byu.edu

Academic Editor: Monika Dmitrzak-Weglarz

Polygenic scores (or genetic risk scores) quantify the aggregate of small effects from many common genetic loci that have been associated with a trait through genome-wide association. Polygenic scores were first used successfully in schizophrenia and have since been applied to multiple phenotypes including multiple sclerosis, rheumatoid arthritis, and height. Because human height is an easily-measured and complex polygenic trait, polygenic height scores provide exciting insights into the predictability of aggregate common variant effect on the phenotype. Shawn Bradley is an extremely tall former professional basketball player from Brigham Young University and the National Basketball Association (NBA), measuring 2.29 meters (7'6″, 99.99999th percentile for height) tall, with no known medical conditions. Here, we present a case where a rare combination of common SNPs in one individual results in an extremely high polygenic height score that is correlated with an extreme phenotype. While polygenic scores are not clinically significant in the average case, our findings suggest that for extreme phenotypes, polygenic scores may be more successful for the prediction of individuals.

1. Introduction

Polygenic, or genetic risk, scores are aggregate measurements of the effects of multiple common genetic loci that are associated with a trait. First used in schizophrenia [1], they have been applied to many complex traits such as multiple sclerosis [2], rheumatoid arthritis [3], and cardiovascular risk [4]. However, polygenic scores are not generally expected to be clinical predictors of an individual's phenotype. For example, Machiela et al. observed that the calculated AUC for the

prediction of breast cancer from the polygenic score did not exceed 53%, which suggests that more validated variants (increased sample size) are necessary for a better prediction or that other factors besides common variants account for a large part of the disease phenotype [5]. Similarly, Evans et al. found that while adding genome-wide variant information can slightly improve prediction accuracy, it is unlikely to be used for the prediction of individual phenotypes until larger datasets can improve the number of validated associated variants [6].

Most phenotypes (e.g., height, Alzheimer's disease, Parkinson's disease, etc.) are complex and polygenic, and our understanding of the underlying biology is limited because of high data dimensionality and small sample sizes. Approximately 80% of adult height variation has been attributed to genetic factors [7–10], and common SNPs are believed to account for approximately 50% of that variation [11, 12]. The Genetic Investigation of ANthropometric Traits (GIANT) consortium recently identified 697 SNPs across 423 loci that explain 20% of adult height heritability and further demonstrated that the 2000, 3700, and 9500 most significantly associated SNPs explained 21%, 24%, and 29% of height variation [10], respectively. Using 160 of these SNPs, which explain 10% of variation in height as reported by the GIANT consortium, Chan et al. observed that weighted polygenic allele scores were as predictive as expected in the extreme height phenotypes [13]. This conclusion was also validated by Liu et al., who reported an AUC of 0.75 for a weighted allele score prediction for 180 SNPs on tall stature [14].

Shawn Bradley is an extremely tall former professional basketball player from Brigham Young University and the National Basketball Association (NBA), measuring 2.29 m (7′ 6″) tall (Figure 1) and has no known medical conditions. Mr. Bradley's height is 8.6 standard deviations (standard deviation = 6.05 cm) above the average height for US males (176.8 cm), putting him in the 99.99999th percentile [15]. While height is known to be polygenic, exceptional outliers for height and other phenotypes remain intriguing because their rarity may present exciting genetic insights. Possible explanations for their rare height may include a combination of rare genetic variants, environmental factors (e.g., diet) and an extremely rare combination of common SNPs. Here, we present evidence of a relationship between common SNPs and an extreme polygenic phenotype and demonstrate that in Mr. Bradley's specific case, the polygenic score predicts his height ranking as expected.

2. Materials and Methods

2.1. Sample Collection and Sequencing. The Cache County Study on Memory Health and Aging was initiated in 1994 [16] and consists of 5092 participants representing approximately 90% of the Cache County population aged 65 and older in 1994. Specific details about data collection, obtaining consent, and phenotyping individuals in the Cache County population were reported previously [16], and other additional information on this dataset can be found in previous reports [16, 17].

Whole genome sequences (WGS) from 809 individuals (432 males, 354 females, and 23 unknown) were obtained from the Alzheimer's Disease Neuroimaging Initiative (ADNI) database (http://adni.loni.usc.edu). ADNI is a large collaboration from several academic and private institutions, and subjects have been recruited from over 50 sites across the US and Canada. Currently, over 1500 adults (ages 55 to 90) participate, consisting of cognitively normal older individuals, people with early or late MCI, and people with

FIGURE 1: Shawn Bradley is 2.29 m (7′ 6″) tall with no known medical conditions. Mr. Bradley played basketball for Brigham Young University from 1990 to 1991. He played in the National Basketball Association from 1993–2005. Photo courtesy of BYU photography.

early stage Alzheimer's disease. For up-to-date information, see http://www.adni-info.org.

We combined WGS from ADNI with WGS for 211 individuals (82 males and 129 females) from the Cache County study. All samples were sequenced using the Illumina HiSeq technology at an average of 30x coverage. We sequenced Mr. Bradley's exome using the Ion Torrent and the Ion Ampliseq Exome Kit at an average coverage of 30x. Sequence data from all studies were mapped to the human reference genome, version GrCh37 with BWA (Burrows-Wheeler Aligner) [18]. We further genotyped Mr. Bradley using the Illumina HumanOmniExpress chip and imputed additional

SNPs using Impute2 [19] and the 1000G reference panel [20]. Subsequently, we filtered imputed SNPs with low information (info <0.4). Mr. Bradley and all individuals in the ADNI and Cache County cohorts are of Northern European ancestry [21].

SNP data from the Alzheimer's Disease Genetics Consortium (ADGC) were used to examine patterns of linkage disequilibrium. The ADGC consists of 32 studies collected over two phases that include 16,000 cases and 17,000 controls. All subjects were self-reported as being of European American ancestry. More information about this dataset can be found in the study of Naj et al. [22] and the ADGC data preparation description [23].

2.2. Analyses. The GIANT Consortium reported 22,539 genome-wide significant SNPs associated with human height. We extracted these SNPs from the ADGC data and identified unique tag SNPs within each LD block to (1) estimate the number of unique signals in the GIANT data and (2) prevent counting the same signal more than once. We identified tag SNPs using default settings in Haploview [24] for each chromosome individually ($r^2 = 0.8$). We then extracted as many of the remaining SNPs as possible from Mr. Bradley's data, the ADNI samples, and Cache County samples. We calculated an additive polygenic height score [25] for each individual and their respective ranks in the distribution of height scores. We also calculated the maximum possible score across the selected SNPs.

To estimate the number of SNPs needed to elevate Mr. Bradley's height score to the highest in the distribution, we performed a random selection of SNPs (bootstrap) at various SNP-set sizes ranging from 100 to 2000 SNPs, recalculating Mr. Bradley's height score and rank each time. We performed 1 million replicates for each SNP-set size and measured the range (minimum and maximum), first and third quartiles (25th and 75th percentiles), and the median for each SNP-set size.

We also explored the difference in height scores between the observed distribution of height scores amongst the 1020 individuals from ADNI and Cache County compared to the null distribution, assuming no evolutionary constraints. We simulated genotypes and height scores across the extracted common SNPs for 20 billion individuals. Specifically, for each SNP, we randomly chose one of three possible genotypes and calculated the simulated individual's height score.

Understanding whether Mr. Bradley's height is attributed to an increased proportion of heterozygous or homozygous genotypes associated with increased height could shed additional light on whether the SNP effects are additive or nonadditive (i.e., being homozygous has a greater effect than the sum). We tested for a difference between Mr. Bradley's genotype distribution and the average ADNI and Cache County genotype distribution using a goodness-of-fit test. Alleles with a positive effect size are associated with increased height, while alleles with a negative effect size are associated with decreased height. A significant difference that could indicate the effects on height are nonadditive, though more data from extremely tall individuals would be necessary to provide definitive evidence.

We also tested whether height scores were correlated with actual height in 407 individuals from the ADNI and Cache County datasets for each individual with both height and genetic data available. We tested for a correlation between the two using Pearson's product moment correlation coefficient, which is calculated using the R statistical package [26].

3. Results and Discussion

We tested whether a simple polygenic height score, calculated using SNPs that were statistically associated with human height in the GIANT consortium data [10], could accurately predict Mr. Bradley's height rank amongst 1020 individuals of Northern European descent. We used Haploview to identify tag SNPs for each LD block across the 22,539 GIANT SNPs to avoid counting a single signal multiple times and to estimate how many independent signals exist in the GIANT SNPs. Using the Alzheimer's Disease Genetics Consortium (ADGC) [22, 23] data with over 30,000 individuals, we identified 3428 unique signals, suggesting that most of the GIANT SNPs tag redundant effects. This is consistent with the GIANT result that most of the adult height variability explained by their SNPs is captured in the top 697 SNPs identified. After extracting genome-wide significant GIANT SNPs from Mr. Bradley's exome and SNP data and using only a single tag SNP within each linkage disequilibrium (LD) block, 2910 SNPs (2491 genotyped, 419 imputed, Supplementary Table 1) remained and were included in the analysis. These represent 2910 of the 3428 LD blocks identified across the 22,539 significant GIANT SNPs using the ADGC dataset. Each allele included in this study is estimated by the GIANT consortium to affect an individual's height by −0.14 to 0.19 millimeters.

We calculated height scores weighted by effect size (see Supplementary Table 1 for effect betas) for Mr. Bradley and 1020 individuals from the Alzheimer's Disease Neuroimaging Initiative (ADNI) and the Cache County Study on Memory Health and Aging. Because Mr. Bradley's height is 8.6 standard deviations above the average height of a male in the US, it is expected that his height score would be much higher than the average of the 1020 individuals for whom height scores were calculated. Mr. Bradley's height score (10.32), calculated using the 2910 SNPs, was ranked highest, while the next highest was 7.43 (Figure 2). The mean height score within the ADNI and Cache County data was 0.98 with a standard deviation of 2.22, making Mr. Bradley's height score 4.2 standard deviations above the mean, as expected.

In order to determine how few SNPs could be used for Mr. Bradley's height score to rank highest when compared to the ADNI and Cache County population data, we created subsets of SNPs randomly from the 2910 available SNPs and then calculated height scores for all 1020 individuals as well as Mr. Bradley. We then ranked the resulting height scores and recorded Mr. Bradley's percentile (Table 1). This procedure was replicated 1 million times for each SNP subset size. Choosing a subset of 100 SNPs randomly 1 million times, Mr. Bradley's height scores calculated from the SNP subsets range from the lowest to the highest when compared to the

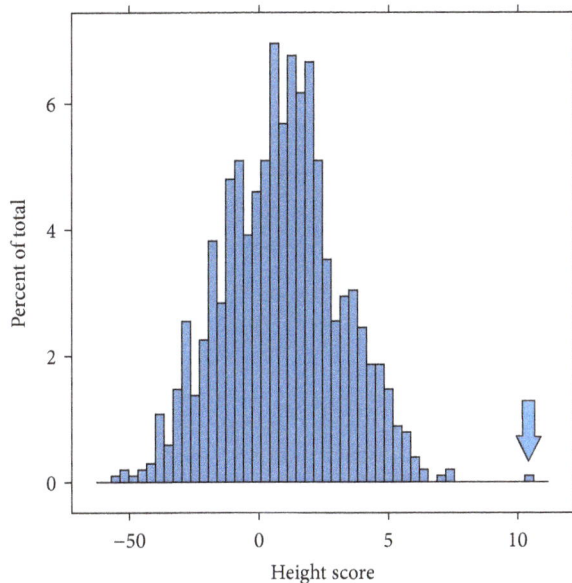

FIGURE 2: Height score distribution calculated using the 2910 SNPs. Mr. Bradley's height score (10.32, indicated by the arrow) ranked highest when compared to the 1020 individuals from ADNI and Cache County, while the next highest was 7.43. The mean height score within the ADNI and Cache County data was 0.98 with a standard deviation of 2.22, making Mr. Bradley's height score 4.2 standard deviations above the mean.

TABLE 1: Mr. Bradley's height score percentiles when compared to the population data for random subsets of SNPs.

Set size	100	250	500	750	1000	1250	1500	1750	2000
Min	0	20.4	54.4	78.8	94.3	97.1	99.2	99.6	99.6
Q1	89.6	98.2	99.8	*	*	*	*	*	*
Median	96.9	99.6	*	*	*	*	*	*	*
Q3	99.3	99.9	*	*	*	*	*	*	*
Max	*	*	*	*	*	*	*	*	*

Table 1 Shawn Bradley's height score quickly stabilizes at the highest rank as SNP-set size increases. Data are represented in percentiles. The "*" indicates that his score was the highest.

ADNI and Cache County SNP subset height scores. His median height percentile settles at 96.9. Using a subset of 250 SNPs across 1 million iterations, Mr. Bradley's median height percentile rises to 99.6 with his minimum height percentile at 20.4 and his maximum ranking highest. By using 750 SNPs, Mr. Bradley's Q1 height rank is the top of the distribution, meaning that at least 75% of the time, his height score was ranked highest in the distribution. His lowest percentile using 750 SNPs was 78.8. Randomly selecting 1500 of the 2910 SNPs, Mr. Bradley's lowest rank was in the 99.2 percentile (1017 of 1021).

We also explored the difference in height scores between the observed distribution of height scores amongst the 1020 individuals from ADNI and Cache County when compared to the null distribution, based on 20 billion simulated individuals created from ADNI and Cache County genotypes, assuming no evolutionary constraints. The mean simulated height score (−0.30) was 1.28 mm lower than the observed height score mean (0.98). The maximum simulated height score (8.37) was 1.95 mm lower than Mr. Bradley's (10.32).

We tested whether Mr. Bradley's extreme height may be caused by an increased proportion of heterozygous or homozygous genotypes using a goodness-of-fit test ($p = 1.28 \times 10^{-24}$). Mr. Bradley has an increased proportion of homozygous genotypes for alleles with a positive effect (Table 2). He has nearly identical numbers of heterozygous genotypes for positive (associated with increased height) and negative (associated with decreased height) effect sizes with 621 and 634, respectively. The additive effects on his score for the positive and negative heterozygous genotypes are approximately equal and opposite at 15.12 and −15.27, respectively, summing to −0.17. There is a large difference, however, when comparing the homozygous genotypes for alleles with a positive and negative effect. Mr. Bradley has 465 genotypes where he is homozygous for GIANT alleles with a positive effect and only 267 genotypes where he is homozygous for GIANT alleles with a negative effect. The additive effects where Mr. Bradley is homozygous for positive and negative alleles are 25.89 and −15.42, respectively. The sum of all four scores equates to his height score of 10.32. Based on these data, Mr. Bradley's height score rank is largely attributed to an excess of 198 positive-effect homozygous genotypes.

Using available height data from the ADNI and Cache County data, we tested whether the height scores calculated using the 2910 SNPs were correlated with the self-reported heights (at age 18) for the 407 individuals for which we have both height and genetic data. We failed to detect significant correlation between the two (correlation coefficient = 0.06, $p = 0.25$; Figure 3). This is consistent with the findings of the GIANT consortium. With a population of 1914 individuals, Wood et al. found a predictive $r^2 = 0.14$ for 697 SNPs (20% variation explained) [10]. It is expected that this r^2 should be stronger than the correlation coefficient in our findings because of our smaller population size of 407 individuals of the ADNI and Cache County individuals as well as the fact that the GIANT consortium identified the 697 SNPs used for prediction directly from their population of 1914 individuals.

4. Conclusions

While research has shown that height is a polygenic trait heavily influenced by common SNPs [7–12], a polygenic score that quantifies common SNP effect is generally insufficient for successful individual phenotype prediction. We demonstrate that in the case of Mr. Bradley, a rare combination of common SNPs corresponds to an extremely high polygenic score that predicts an extreme phenotype. Because Mr. Bradley is an outlier, studying his genetic makeup provides a unique context to understand the complex nature of human height. Using a simple polygenic model across approximately 2000 SNPs, we accurately predicted Mr. Bradley's height rank amongst a population of 1020 individuals.

The accurate prediction of tall individuals based on polygenic score has been found by both Chan et al. [13]

TABLE 2: Genotype counts for effect alleles in Shawn Bradley and the ADNI/Cache County populations.

	Shawn Bradley			Average across ADNI/Cache County		
	Homozygous for effect allele (additive effect on score)	Heterozygous (additive effect on score)	Homozygous for noneffect allele	Homozygous for effect allele	Heterozygous	Homozygous for noneffect allele
Positive effect	465 (25.89)	621 (15.12)	347 (NA)	428 (22.72)	552 (13.01)	479 (NA)
Negative effect	267 (−15.42)	634 (−15.27)	510 (NA)	416 (−22.28)	535 (−12.60)	497 (NA)

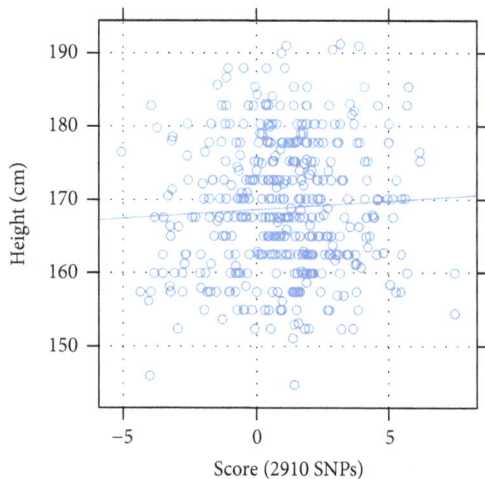

FIGURE 3: Correlation between height scores and self-reported height in the ADNI and Cache County individuals. We plotted height scores and self-reported heights (at age 18) for individuals in the ADNI and Cache County datasets and found poor correlation between the two. We also calculated the Pearson product moment correlation coefficient (correlation coefficient = 0.06, $p = 0.25$).

and Liu et al. [14], confirming that in the case of an extremely tall phenotype, such as Mr. Bradley's, polygenic scores can predict height rank. While these studies used a population of tall individuals to confirm their findings, we provide a validation of one individual polygenic height score rather than a distribution.

Mr. Bradley's height score—like his actual height—was an extreme outlier (4.2 standard deviations above the mean). This appears to be driven by an increased proportion of homozygous genotypes for SNPs associated with increased height when compared to the average ADNI and Cache County genotype values. Despite this, his height score only predicted him to be 10.32 mm taller than average. This suggests that while Mr. Bradley's extreme polygenic score could accurately rank his height amongst 1020 individuals, it does not accurately predict his actual height measurement, demonstrating that there are significant factors unaccounted for. Similarly, and as expected, this model was not able to accurately predict actual heights among the 407 ADNI and Cache County individuals for which we had both height and genetic data. These results as well as Mr. Bradley's predicted height (10.32 mm taller than average) suggest that other factors such as environmental factors [27], nonadditive individual loci [28], and both epistasis (gene by gene interactions) and gene by environment interactions [29] play a significant role in determining actual height

measurement. Recent studies of heritability in height and other complex traits suggest significant contributions of nonadditive factors [30, 31].

Height is a complex trait that may serve as an effective phenotype model for other complex traits and diseases because it is a noninvasive and easily-measured phenotype to study. By developing new models and studies to better understand all genetic contributors to an individual's height, researchers will be able to apply the methods to other complex data.

Conflicts of Interest

The authors declare that there is no conflict of interest regarding the publication of this paper.

Authors' Contributions

Corinne E. Sexton and Mark T. W. Ebbert contributed equally to this work.

Acknowledgments

The authors thank Mr. Bradley and the participants and staff of the centers that were involved in the data collection for ADNI and the Cache County study for their important contributions to this work. The data used in the preparation of this article were obtained in part from the Alzheimer's Disease Neuroimaging Initiative (ADNI) database (http://adni.loni.usc.edu). As such, the investigators within the ADNI contributed to the design and implementation of ADNI and/or provided data but did not participate in the analysis or writing of this report. A complete listing of ADNI investigators can be found at: http://adni.loni.usc.edu/wp-content/uploads/how_to_apply/ADNI_Acknowledgement_List.pdf. Whole-genome data collection and sharing for this project was funded by the Alzheimer's Disease Neuroimaging Initiative (ADNI), National Institutes of Health Grant (U01 AG024904), and DOD ADNI, Department of Defense award (W81XWH-12-2-001). ADNI is funded by the National Institute on Aging, the National Institute of Biomedical Imaging and Bioengineering, and through generous contributions from the following: Alzheimer's Association; Alzheimer's Drug Discovery Foundation; Araclon Biotech; BioClinica, Inc.; Biogen Idec Inc.; Bristol-Myers Squibb

Company; Eisai Inc.; Elan Pharmaceuticals, Inc.; Eli Lilly and Company; EuroImmun; F. Hoffmann-La Roche Ltd. and its affiliated company Genentech, Inc.; Fujirebio; GE Healthcare; IXICO Ltd.; Janssen Alzheimer Immunotherapy Research & Development, LLC.; Johnson & Johnson Pharmaceutical Research & Development LLC.; Medpace, Inc.; Merck & Co., Inc.; Meso Scale Diagnostics, LLC.; NeuroRx Research; Neurotrack Technologies; Novartis Pharmaceuticals Corporation; Pfizer Inc.; Piramal Imaging; Servier; Synarc Inc.; and Takeda Pharmaceutical Company. The Canadian Institutes of Health Research is providing funds to support ADNI clinical sites in Canada. Private sector contributions are facilitated by the Foundation for the National Institutes of Health (http://www.fnih.org). The grantee organization is the Northern Rev December 5, 2013 California Institute for Research and Education, and the study is coordinated by the Alzheimer's Disease Cooperative Study at the University of California, San Diego. ADNI data are disseminated by the Laboratory for Neuroimaging at the University of Southern California.

References

[1] S. M. Purcell, N. R. Wray, J. L. Stone et al., "Common polygenic variation contributes to risk of schizophrenia and bipolar disorder," *Nature*, vol. 460, pp. 748–752, 2009.

[2] The International Multiple Sclerosis Genetics Consortium (IMSGC), "Evidence for polygenic susceptibility to multiple sclerosis—the shape of things to come," *The American Journal of Human Genetics*, vol. 86, no. 4, pp. 621–625, 2010.

[3] E. A. Stahl, D. Wegmann, G. Trynka et al., "Bayesian inference analyses of the polygenic architecture of rheumatoid arthritis," *Nature Genetics*, vol. 44, no. 5, pp. 483–489, 2012.

[4] M. A. Simonson, A. G. Wills, M. C. Keller, and M. B. McQueen, "Recent methods for polygenic analysis of genome-wide data implicate an important effect of common variants on cardiovascular disease risk," *BMC Medical Genetics*, vol. 12, no. 1, p. 146, 2011.

[5] M. J. Machiela, C. Y. Chen, C. Chen, S. J. Chanock, D. J. Hunter, and P. Kraft, "Evaluation of polygenic risk scores for predicting breast and prostate cancer risk," *Genetic Epidemiology*, vol. 35, no. 6, pp. 506–514, 2011.

[6] D. M. Evans, P. M. Visscher, and N. R. Wray, "Harnessing the information contained within genome-wide association studies to improve individual prediction of complex disease risk," *Human Molecular Genetics*, vol. 18, no. 18, pp. 3525–3531, 2009.

[7] R. A. Fisher, "XV.—The correlation between relatives on the supposition of Mendelian inheritance," *Earth and Environmental Science Transactions of the Royal Society of Edinburgh*, vol. 52, no. 2, pp. 399–433, 1919.

[8] K. Silventoinen, S. Sammalisto, M. Perola et al., "Heritability of adult body height: a comparative study of twin cohorts in eight countries," *Twin Research and Human Genetics*, vol. 6, no. 5, pp. 399–408, 2003.

[9] P. M. Visscher, S. E. Medland, M. A. R. Ferreira et al.,

"Assumption-free estimation of heritability from genome-wide identity-by-descent sharing between full siblings," *PLoS Genetics*, vol. 2, no. 3, article e41, 2006.

[10] A. R. Wood, T. Esko, J. Yang et al., "Defining the role of common variation in the genomic and biological architecture of adult human height," *Nature Genetics*, vol. 46, no. 11, pp. 1173–1186, 2014.

[11] J. Yang, B. Benyamin, B. P. McEvoy et al., "Common SNPs explain a large proportion of the heritability for human height," *Nature Genetics*, vol. 42, no. 7, pp. 565–569, 2010.

[12] J. Yang, T. A. Manolio, L. R. Pasquale et al., "Genome-partitioning of genetic variation for complex traits using common SNPs," *Nature Genetics*, vol. 43, no. 6, pp. 519–525, 2011.

[13] Y. Chan, O. L. Holmen, A. Dauber et al., "Common variants show predicted polygenic effects on height in the tails of the distribution, except in extremely short individuals," *PLoS Genetics*, vol. 7, no. 12, article e1002439, 2011.

[14] F. Liu, A. E. J. Hendriks, A. Ralf et al., "Common DNA variants predict tall stature in Europeans," *Human Genetics*, vol. 133, no. 5, pp. 587–597, 2014.

[15] C. D. Fryar, Q. Gu, C. L. Ogden, and K. M. Flegal, "Anthropometric reference data for children and adults; United States, 2011-2014," *Vital and Health Statistics*, vol. 3, no. 392016, 2016.

[16] J. C. S. Breitner, B. W. Wyse, J. C. Anthony et al., "APOE-ε4 count predicts age when prevalence of AD increases, then declines the Cache County study," *Neurology*, vol. 53, no. 2, pp. 321–331, 1999.

[17] M. T. W. Ebbert, P. G. Ridge, A. R. Wilson et al., "Population-based analysis of Alzheimer's disease risk alleles implicates genetic interactions," *Biological Psychiatry*, vol. 75, no. 9, pp. 732–737, 2014.

[18] H. Li, "Aligning sequence reads, clone sequences and assembly contigs with BWA-MEM," 2013, http://arxiv.org/abs/1303.3997.

[19] B. N. Howie, P. Donnelly, and J. Marchini, "A flexible and accurate genotype imputation method for the next generation of genome-wide association studies," *PLoS Genetics*, vol. 5, no. 6, article e1000529, 2009.

[20] The 1000 Genomes Project Consortium, "A global reference for human genetic variation," *Nature*, vol. 526, pp. 68–74, 2015.

[21] A. R. Sharp, P. G. Ridge, M. H. Bailey et al., "Population substructure in Cache County, Utah: the Cache County study," *BMC Bioinformatics*, vol. 15, Supplement 7, pp. S8–S8, 2014.

[22] A. C. Naj, G. Jun, G. W. Beecham et al., "Common variants at MS4A4/MS4A6E, CD2AP, CD33 and EPHA1 are associated with late-onset Alzheimer's disease," *Nature Genetics*, vol. 43, no. 5, pp. 436–441, 2011.

[23] K. L. Boehme, S. Mukherjee, P. K. Crane, and J. S. Kauwe, "ADGC 1000 Genomes combined data workflow," October 2015, http://kauwelab.byu.edu/Portals/22/adgc_combined_1000G_09192014.pdf.

[24] J. C. Barrett, B. Fry, J. Maller, and M. J. Daly, "Haploview: analysis and visualization of LD and haplotype maps," *Bioinformatics*, vol. 21, no. 2, pp. 263–265, 2005.

[25] F. Dudbridge, "Power and predictive accuracy of polygenic risk scores," *PLoS Genetics*, vol. 9, no. 3, article e1003348, 2013.

[26] R Core Team, *R: A Language and Environment for Statistical Computing*, R Foundation for Statistical Computing, Vienna, Austria, 2018, http://www.R-project.org/.

[27] B. Bogin and L. Rios, "Rapid morphological change in living humans: implications for modern human origins," *Comparative Biochemistry and Physiology Part A: Molecular & Integrative Physiology*, vol. 136, no. 1, pp. 71–84, 2003.

[28] G. Su, O. F. Christensen, T. Ostersen, M. Henryon, and M. S. Lund, "Estimating additive and non-additive genetic variances and predicting genetic merits using genome-wide dense single nucleotide polymorphism markers," *PLoS One*, vol. 7, no. 9, article e45293, 2012.

[29] P. M. Visscher, J. Yang, and M. E. Goddard, "A commentary on 'Common SNPs explain a large proportion of the heritability for human height' by Yang et al. (2010)," *Twin Research and Human Genetics*, vol. 13, no. 06, pp. 517–524, 2010.

[30] T. A. Manolio, F. S. Collins, N. J. Cox et al., "Finding the missing heritability of complex diseases," *Nature*, vol. 461, no. 7265, pp. 747–753, 2009.

[31] J. Yang, J. Zeng, M. E. Goddard, N. R. Wray, and P. M. Visscher, "Concepts, estimation and interpretation of SNP-based heritability," *Nature Genetics*, vol. 49, no. 9, pp. 1304–1310, 2017.

11

Evidence of the Complexity of Gene Expression Analysis in Fish Wild Populations

Mbaye Tine

UFR des Sciences Agronomiques, de l'Aquaculture et des Technologies Alimentaires (UFR S2ATA), Universite Gaston Berger (UGB), Route de Ngallele BP 234, Saint-Louis, Senegal

Correspondence should be addressed to Mbaye Tine; mbaye.tine@ugb.edu.sn

Academic Editor: Henry Heng

The present work examines the induction of the *band 3 anion transport protein, mitogen-activated protein kinase,* and *lactate dehydrogenase,* respectively related to osmolyte transport, cell volume regulation, and energy production in the gills of two tilapia strains exposed to either freshwater or hypersaline water. Overall, genes showed similar expression patterns between strains. However, a wild population survey across a range of natural habitats and salinities did not reveal the expected patterns. Although significant, the correlations between gene expression and salinity were slightly ambiguous and did not show any link with phenotypic differences in life history traits previously reported between the same populations. The differential expression was also not associated with the population genetic structure inferred from neutral markers. The results suggest that the differential expression observed is not the result of evolutionary forces such as genetic drift or adaptation by natural selection. Instead, it can be speculated that genes responded to various abiotic and biotic stressors, including factors intrinsic to animals. This study provides clear evidence of the complexity of gene expression analysis in wild populations and shows that more attention needs to be paid when selecting candidates as potential biomarkers for monitoring adaptive responses to a specific environmental perturbation.

1. Introduction

Natural variations in environmental factors such as temperature, salinity, pH and dissolved oxygen concentration, and anthropogenic alterations of natural habitats are among the environmental stressors that have profound and diverse impacts on aquatic ecosystems [1, 2]. Climate change can lead to shifts in physicochemical factors that include salinity, temperature, dissolved oxygen levels, and water quality [3]. While salinity has been identified as a key environmental factor that may have physiological effects on aquatic species, temperature, dissolved oxygen, and water quality are also among the major environmental constraints that may have broad impact on fish biological functions such as growth and reproduction [4–6]. Changes in these physicochemical factors can be stressful for many organisms including fishes and can potentially induce adaptive stress responses [7, 8]. Organisms adapt to

environmental changes prevailing in their natural environment through fixation of changes in protein-coding DNA sequences that alter protein function and enhance individual fitness and/or via epigenetic changes that affect gene expression [9–12].

Abrupt changes in ambient conditions require fast and specific responses of preexisting molecular pathways associated with acclimation responses. Compared to other vertebrates, fishes are characterized by very plastic phenotypic responses to changing environmental conditions, which are reflected as drastic and rapid changes in gene expression in response to the external stimuli. Therefore, one might expect changes in gene expression patterns to be in the same direction as the environmental factor that has induced them. Short-term physiological acclimation is facilitated by changes at the cellular level, which include the activation of pathways and the induction of candidate genes involved in the adaptation to the environmental stress. The modulation of gene expression is, therefore, among the first stress responses available to fishes that are confronted with changes in ambient conditions. Consequently, the expression and regulation of genes related to individual fitness have been widely used to investigate the adaptation of wild populations to local environmental conditions [13].

Experimental research generally uses common garden experiments to evaluate the role of gene expression in an organisms' response to changes in environmental conditions. Such experiments are helpful in order to distinguish environmental influences from genetic influences on phenotypic variations among populations. They have been successfully used in a range of taxa, including fishes (38–40). When conducted with offspring from a single population with the same genetic background, common garden experiments can also be used to reveal which among a suite of environmental factors is responsible for interindividual variation in a given phenotypic trait of a population [14]. This is achieved by varying one environmental parameter while keeping the others constant.

The strict control of environmental parameters in common garden experiments is one of the major differences compared with studies in species' native environments, where biotic and abiotic factors cannot be controlled. Comparative and evolutionary studies of wild populations are often conducted across environmental clines, such as temperature and salinity gradients [15–18]. This implies that there is at least one environmental factor that is the most constraining for the species in the study area and that predominates over other factors. However, despite this predominance, there are many other biotic/abiotic parameters whose variations, however slight, may directly or indirectly affect the expression of certain genes. The effects of such variables may interfere with the predominant factor depending on whether the genes analysed are involved in one or several biological pathways and also on the degree of the environmental stressor to which they respond.

Studies on osmoregulation have yielded a large amount of information on the cellular and molecular modifications associated with spatiotemporal changes in salinity [19–22]. Most of these studies have focused on water- and ion-transporting proteins [23–26], and the osmoregulatory roles of these ion pumps are now well understood at both cellular and molecular levels. Acclimation to spatiotemporal changes in salinity depends on the ability of the gill epithelium to adjust NaCl secretion or absorption, depending on environmental salinity [23, 26]. However, the regulation of ion concentrations is not the only challenge with which fish are confronted. They are also confronted with changes in cellular volume induced by salinity changes. Unfortunately, the mechanisms of cellular volume regulation remain poorly understood.

The black-chinned tilapia, *Sarotherodon melanotheron*, is a euryhaline teleost widely distributed in West African aquatic ecosystems, where it is regularly exposed to a wide range of salinity values. The species is found in marine coastal waters and freshwater habitats, where salinity is constant throughout the year [27]. It is also represented in estuaries, such as the Saloum estuary in Senegal, which exhibits extreme variations in salinity, ranging from fresh to extremely hypersaline (up to 130 psu), and where seasonal variations in salinity can be considerable [28, 29]. The existence of a salinity cline across the distributional range of the species offers an excellent opportunity for investigating gene expression responses to salinity changes.

The aim of the present study was to assess the relationship between salinity and gene expression patterns in three proteins, namely, *band 3 anion transport protein* (SLC4A1), *mitogen-activated protein kinase* (MAPK), and *lactate dehydrogenase* (LDH), in relation to osmolyte transport, cell volume regulation, and energy production, respectively, in individuals of the black-chinned tilapia acclimatised to different salinities. Based on their function, these genes may play important roles in *S. melanotheron* acclimatisation to salinity variations. The SLC4A1 is a transport protein that promotes the reversible exchange of bicarbonate ion for chloride ion [30]. This anion exchanger is the major contributor of the compensation mechanism of cell swelling because it facilitates the efflux of osmolytes, including KCl and amino acids [31, 32], a mechanism that is crucial in pH and cell volume regulation. MAPKs are protein kinases that respond to several extracellular signals and participate in the regulation of various cellular activities such as gene expression, mitosis, differentiation, and cell survival/apoptosis [33, 34]. MAPKs are involved in amplification and integration of extracellular signal and are, therefore, very important in the adaptation of fish to salinity. The LDH is a key enzyme of glycolysis involved in anaerobic energy production and reversible conversion of pyruvate and NADH into lactate and NAD^+, respectively [35]. This enzyme is found abundantly in some organs including the heart, kidney, liver, muscle, and, to a lesser extent, gills, where it plays an important role in glucose homeostasis.

The three genes analysed in this study were initially identified using suppressive subtractive hybridisation (SSH) on *S. melanotheron* acclimatised to different salinities (0 and 70 psu) in controlled laboratory conditions [36]. They were selected for the current study because they are not housekeeping or reference genes, that is, genes expressed in a wide variety of tissues or cells with no or minimum variation in gene expression patterns between individuals. They were chosen based on the patterns of their expression response to ambient salinity in experimental conditions that is relevant to the ability of individuals to respond to environmental changes. Accordingly, their transcriptional responses are likely to reflect salinity differences in natural environments where this parameter is the predominant stressor.

We previously published previous works conducted on the same populations using genes encoding ion pumps and channels (Na/K-ATPase, voltage-dependent anion channel) [37, 38], genes encoding general stress proteins such as HSP70 [37] (from the same SSH libraries), and pituitary hormones (growth hormone, prolactin) [29]. There were clear relationships between the levels of mRNA expression of these genes and the environmental salinity, which can be explained

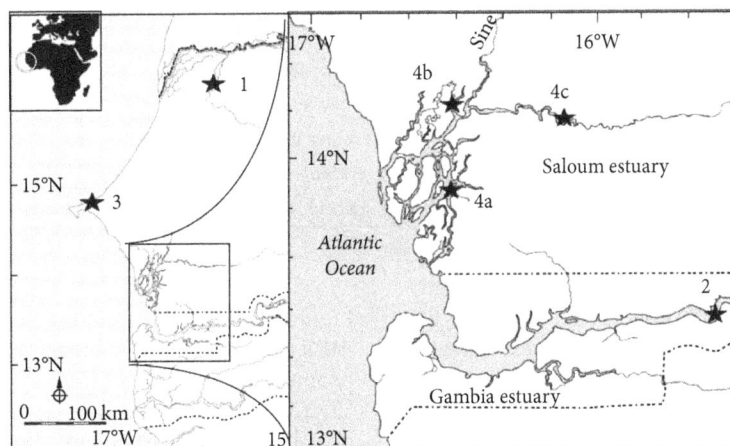

FIGURE 1: Map showing sampling locations. 1: Guiers Lake; 2: Balingho; 3: Hann Bay; 4a: Missirah; 4b: Foundiougne; 4c: Kaolack.

by the fact that these genes play major roles in the responses to osmotic stress. For this present study, we sought to test other genes (those from the same SSH libraries rather than those analysed in the first published reports) whose involvement in osmoregulatory processes has not been yet or well described. We selected genes that may be involved in osmoregulation because their involvement in physiological processes related to this function such as osmolyte transport, energy production, and cell volume and signal regulation is commonly accepted. It is likely that these physiological processes are activated during adaptive responses to biotic or abiotic stressor other than ambient salinity. Although water salinity is the most constraining abiotic factor for *S. melanotheron* in our study areas, gene expression may be influenced by biotic stressors including factors intrinsic to animals and whose effects are difficult to evaluate in wild populations.

Relative levels of the expression of these genes in the fishes' gills were first compared under experimental conditions in individuals of *S. melanotheron* whose broodstock originated from two different strains that had been collected in hypersaline water and seawater, respectively. I then investigated the hypothesis that relative gene expression in wild populations acclimatised to salinities ranging from 0 to 100 psu is correlated with salinity. The new set of genes was analysed on samples collected at the same time rather than those used in our previously published studies. The main advantage of this study compared to that of previous works is the comparison of the transcriptional responses between conditions and strains, followed by a survey of wild populations across a range of natural habitats and salinities, which constitutes an excellent approach for selecting candidates as potential biomarkers for monitoring adaptive responses to a specific environmental perturbation. The combination of experimental and wild populations also offers the advantage of allowing a better understanding of the role of gene expression in fish adaptation to extreme salinities and could also highlight the complexity of gene expression analysis in natural environments.

2. Methods

2.1. Experiments Conducted under Controlled Conditions. The original broodstock of the black-chinned tilapia (*S. melanotheron*) juveniles used in this study was collected by Dr. Abdou Mbow (IFAN) during the rainy season at Kaolack (site 4c in Figure 1) in the upper reaches of the Saloum estuary in Senegal, at a salinity of 48 psu. They were transferred to the CIRAD facilities (Montpellier) nearly 15 years ago. Fishes were transported under a joint collaborative project between IFAN and CIRAD research institutes, and no permit for transporting fish to France was required at that time. Since then, fish from the same initial broodstocks have been used in many scientific publications [36–39]. They were divided into two groups of 60 individuals each and maintained in 30 L freshwater aquaria for three days before being transferred to seawater. Each aquarium was equipped with a thermostat to maintain a 25°C constant water temperature (Table 1). The aquaria were equipped with a mechanical/biological filtration system and airstones to ensure optimal oxygen saturation. Fish were fed with commercial processed fish feed containing 40% crude proteins which was distributed six days a week. After an acclimation period of one week in seawater, the salinity at which half of the fish were held was increased to hypersaline conditions (70 psu) while in the other half, it was decreased to freshwater conditions. Salinity changes were carried out by slowly emptying the seawater and replacing it with either freshwater or hypersaline water. The commercial marine salt (Instant Ocean® salts) used to make seawater and hypersaline water contains all the components of natural seawater and was initially dissolved in a bucket with the same water as that used in the freshwater aquaria before being added to a particular aquarium. After 45 days, fish in freshwater and hypersaline water were anesthetized with 2-phenoxyethanol (1.5 mL/L of water), weighed, and killed by decapitation. Gill tissues were collected, immediately frozen in liquid nitrogen, and stored at −80°C until processing. The same procedure was followed for another black-chinned tilapia strain that was collected in seawater (Joal, Senegal). This second experiment was

TABLE 1: Sample characteristics of the black-chinned tilapia *Sarotherodon melanotheron* strains kept under different experimental conditions.

Strain	Strain origin	Experimental condition	Salinity (psu)	WT (°C)	NI
Kaolack	Upstream Saloum estuary (48 psu)	Freshwater	0	25	6
		Hypersaline water	70	25	6
Joal	Atlantic Ocean (38 psu)	Freshwater	0	25	6
		Hypersaline water	70	25	6

NI: number of individuals; WT: water temperature.

essentially conducted to validate the results from the first experiment. The genes that are differentially expressed in both strains as a result of salinity variations are most likely to be real candidates for salinity adaptation. The experimental conditions were identical to those described for the first experiment, except that the time of the year when the experiment was conducted, the size and type of the tanks, the type of food, and the developmental stage differed from those of the first experiment. In the second experiment, gill tissues were collected after an exposure period of only 10 days to freshwater and hypersaline water. This experiment was specially designed to identify genes involved in the long-term survival of fish exposed to specific salinity conditions. Such routinely expressed genes can be expected to be the target of adaptive selection in wild populations that are subjected to long-term natural acclimatisation. For this reason, samples were collected after an exposure period of 45 or 10 days to different conditions because at these times, fish can be considered already acclimatised to extreme salinities [40, 41].

2.2. Sampling Design of Natural Populations. Six natural populations (Figure 1) of *Sarotherodon melanotheron* were sampled in 2006 at the end of the dry season (May), when the salinity in the Saloum estuary was at its maximum level and has been stable for few months. Two of the sampling sites [Guiers Lake (site 1) and Hann Bay (3)] have relatively stable salinities throughout the year, while the other sites (three locations of the Saloum estuary, namely, Missirah (4a), Foundiougne (4b), and Kaolack (4c), and one location in the Gambia estuary, Balingho (2)) exhibit large spatial and seasonal variations in salinity. Fish sampling was carried out by local fishermen using cast nets. Only five fish were sampled from each cast net thrown in order to limit fish stress and prevent variability resulting from handling. Given that gene expression could conceivably be influenced by differences in the developmental or sexual stage, only size classes between 120 and 160 mm fork length with sexual stage 1 or 2 [42] corresponding to immature individuals were analysed (Table 2). All individuals were collected at the same time during the same sampling campaign and samples treated identically to avoid sampling artefacts. Gills were immediately extracted from these captured individuals and stored in RNAlater (Ambion) at 4°C for 24 h and then at −20°C until processing.

2.3. Ethical Statement. No ethical approval was required for this study. The samples used in this study were collected in accordance with good animal practice as outlined by the

French Research Institute for Exploitation of the Sea (IFRE-MER) in a training course on how to handle fish and promote their welfare under experimental conditions. The IFREMER do not approve or give a permit for studies of wild populations of fish but only provide a code of conduct to follow to minimize the suffering in experiments involving fish. Study approval by another academic ethics committee (permit number or approval ID) was not necessary as all procedures carried out with the black-chinned tilapia in this study conformed to the IFREMER recommendations. No specific permissions were required for the collection of samples in the Saloum estuary (Kaolack, Foundiougne, and Missirah locations), Guiers Lake, and Hann Bay. These ecosystems are undergoing important environmental and anthropogenic perturbations. The Senegalese authorities encourage scientists to conduct scientific projects in these areas to better understand the consequences of these environmental and anthropogenic constraints on live history traits of organisms inhabiting these areas including fishes. To encourage them to investigate these areas, the government gave free access to scientists to perform their research without any prior administrative permission. The permission for sampling in Balingho location in the Gambia estuary was verbally given by the Gambia Fisheries Department. No official written document from the Gambian Fisheries Department was required. Likewise, approval for working on the tilapia *S. melanotheron* in the study area was not required from any animal ethics committee because this species is not an endangered species. Instead, it is among the most adapted to environmental and anthropogenic changes in the Senegal and Gambia estuaries, and scientists have convinced the authorities to use it as a model for biological and ecological studies.

2.4. Environmental Data Collection. For each sampling location, salinity (psu) and temperature (°C) were measured in situ at the time and place where fish were sampled with a refractometer (ATAGO) and a thermometer, respectively (Table 2). In the Saloum and Gambia estuaries, additional salinity, temperature, and dissolved oxygen data were obtained from experimental fish sampled by the IRD research group UR 070-RAP (Dakar, Senegal). During these sampling expeditions, surface salinity was measured with an optical refractometer and bottom salinity was measured with a YSI multiparameter probe that allows simultaneous measurements of salinity, water temperature, and dissolved oxygen. The same probe was used for the measurements of surface and bottom percentage oxygen saturation. Previous studies have demonstrated that there is no significant difference between the bottom and surface water temperatures in

TABLE 2: Sample characteristics of the black-chinned tilapia *Sarotherodon melanotheron* from six wild populations acclimatised to different salinities that prevailed in their habitats.

Ecosystem	Station	Site number	GPS coordinates	Salinity (psu)	WT(°C)	NI	FL range (mm)	W range (g)
Guiers Lake	Keur Momar Sarr	1	16°15′00″N, 15°50′00″W	0	28	10	120–149	32–45
Gambia	Balingho	2	13°29′00″N, 15°37′00″W	22	29	10	122–152	42–60
Hann plage	Hann Bay	3	14°43′16″N, 17°26′13″W	37	28	10	124–161	41–80
Saloum estuary	Missirah	4a	13°40′60″N, 16°30′01″W	40	28	10	135–165	40–92
Saloum estuary	Foundiougne	4b	14°08′00″N, 16°28′00″W	60	28	10	122–145	40–52
Saloum estuary	Kaolack	4c	14°11′00″N, 16°15′00″W	100	26	10	123–147	33–51

WT: water temperature; NI: number of individuals; FL: fork length; W: weight.

these sampling locations [4, 6]. Therefore, only the temperature near the water surface was measured with a thermometer. Additional data on salinity, water temperature, and dissolved oxygen were taken from the weather station of the IRD research group UR098-FLAG (Dakar, Senegal) for Guiers Lake and Hann Bay locations (Supplementary Table 1 in Supplementary Material available online at https://doi.org/10.1155/2017/1258396).

2.5. Selection of Candidate Markers. Based on their function, these genes may also play important roles in *S. melanotheron* acclimatisation to salinity variations. The SLC4A1 is a transport protein that promotes the reversible exchange of bicarbonate ion for chloride ion [30]. This anion exchanger is the major contributor of the compensation mechanism of cell swelling because it facilitates the efflux of osmolytes, including KCl and amino acids [31, 32], a mechanism that is crucial in pH and cell volume regulation. MAPKs are protein kinases that respond to several extracellular signals and participate in the regulation of various cellular activities such as gene expression, mitosis, differentiation, and cell survival/apoptosis [33, 34]. MAPKs are involved in amplification and integration of extracellular signal and are, therefore, very important in the adaptation of fish to salinity. The LDH is a key enzyme of glycolysis involved in anaerobic energy production and reversible conversion of pyruvate and NADH into lactate and NAD^+, respectively [35]. This enzyme is found abundantly in some organs including the heart, kidney, liver, muscle, and, to a lesser extent, gills, where it plays an important role in glucose homeostasis.

There are no genes that are universally controlled by salinity. Therefore, positive controls (indicator of osmoregulation status) such as genes involved in ion uptake would be used to show a well-known salinity-regulated gene pattern. Such an indicator could be added to evaluate the stress status of fish induced by salinity variations and to validate the experimental procedures. For example, we could use the sodium-potassium ATPase (Na^+/K^+-ATPase α-subunit), a membrane protein which maintains ion gradients required for cell homeostasis, to assess the osmoregulatory status of fish in natural environments. In this study, we did not use a positive control gene because we have previously demonstrated that the mRNA expression levels of the Na^+/K^+-ATPase α-subunit in the same populations were correlated with the environmental salinity [43].

2.6. RNA Extraction, mRNA Purification, Reverse Transcription, and Real-Time PCR Analysis. Total RNA was extracted with a TRIZOL® reagent (Gibco-BRL, USA) according to the manufacturer's instructions. RNA concentrations were determined spectrophotometrically, and RNA integrity was verified by 1% TAE 1X agarose gel electrophoresis (40 mM Tris, acetate, and 1 mM EDTA). The relative expression of SLC4A1 (GenBank accession number ES881098), MAPK (ES881524), and LDH (ES881483) was analysed in both natural and experimental populations.

The eukaryotic mRNAs possess a repeat of adenosine residues at their 3′ end, which is used in the purification method of the Poly(A)Purist™ kit (Ambion). This method consists firstly in passing a total RNA solution onto a column having binding sites which are polydT oligonucleotides. These oligo (dT) are trapped in magnetic particles coupled to streptavidin and retain only the mRNAs. The aqueous phase which contains the ribosomal RNAs and the transfer RNAs is eliminated, and then, the mRNAs are eluted with ultrapure water previously autoclaved. We used this method to purify the mRNA from 2 mg of total RNA for both experimental and wild population samples.

The purified mRNAs of the gill from both experimental and wild populations are retrotranscribed into cDNA using the BD PCR-Select™ cDNA Subtraction Kit (Ambion). To synthesize the first strand of cDNA, 4 μg of mRNA was mixed with 1 μL of cDNA Synthesis Primer in a total volume of 5 μL. The whole mixture is heated at 70°C for 2 minutes to eliminate potential secondary structures. After denaturation, the mixture is immediately stabilized by cooling in ice. Then, 2 μL of 5X Fist-Strand Buffer, 1 μL of dNTP Mix (10 mM each), 1 μL of sterile water, and 1 μL of AMV Reverse Transcriptase (20 μL/μL) are added to this reaction mixture. This final mixture is incubated at 42°C for 1 h and 30 min in a PTC-100MT thermocycler (MJ Research Inc.). The second strand of cDNA is synthesized by adding the following products to the tubes containing the first-strand synthesis reaction: 48.4 μL of sterile water, 16 μL of 5X Second-Strand Buffer, 1.6 μL of dNTP Mix (10 mM), and 4 μL of 20X Second-Strand Enzyme Cocktail. The total reaction is incubated at 16°C for 2 h and 30 min, and 2 μL of the T4 is added 30 min before the end of the incubation. The synthesis of the second strand is stopped by adding 4 μL of 20X EDTA/Glycogen Mix. The double-stranded cDNA is purified by phenol/chloroform extraction, precipitated with 100% ethanol, washed with 75% ethanol, and suspended into 50 μL of nuclease-free water.

Primer3 software was used to design primers to amplify these genes.

The following forward (F) and reverse (R) primer sequences were used for qRT-PCR amplification: SLC4A1 (F: TCTGCAAAGAAGTGGCATCA; R: ATGACGCCA AGGTGACATTT), LDH (F: TGATCACCTCGTAGGCA CTG; R: AAATGTGGCTGGAGTCAACC), MAPK (F: CT GGCCCTTCAACAGAGACTG; R: CTCTTCGATGGCCTG TTTCAC), and beta-actin (F: ACAGGTCCTTACGGAT GTCG; R: CTCTTCCAGCCTTCCTTCCT).

Gene expression was quantified using quantitative real-time PCR (qRT-PCR) on a LightCycler (Roche Molecular Biomedicals) using the QuantiTect SYBR Green PCR Master Mix kit (Qiagen). Quantification of each sample was performed in a total volume of $10 \mu L$ containing $1 \mu L$ of cDNA, $0.5 \mu L$ of each primer, and 1X of SYBR Green Master Mix (Qiagen). Each qRT-PCR reaction was conducted in duplicate, with an initial denaturation step of 15 min at 95°C, followed by an amplification of the target cDNA for 40 cycles, each cycle consisting of denaturation at 95°C for 15 s, annealing between 54°C and 55°C for 15 s, and elongation at 72°C for 15 s. To determine qRT-PCR efficiency of each primer pair, standard curves were generated using five serial dilutions (1, 1/10, 1/50, 1/ 100, and 1/500) of a unique cDNA sample constituted of a pool of 6 cDNA from each salinity group or population to be analysed. The efficiencies (E) of qRT-PCR reactions were calculated from the given slope of the standard curve according to the equation $E = 10^{(-1/slope)}$. The amplification products were validated by analysing the amplicon size by means of agarose gel electrophoresis. Results are shown as changes in relative expression, normalised to the reference gene, β-actin, using the $2^{-\Delta\Delta Ct}$ method described by Kultz [34]; β-actin was previously analysed and showed no change in response to salinity acclimation.

2.7. Statistical Analyses. Gene expression data at each site were expressed as mean ± SEM. Bartlett [44] and Kolmogorov-Smirnov [45] tests were, respectively, used to test for homogeneity of variances and normality of the data. These tests showed that expression data were not normally distributed and did not have uniform variance. Therefore, only nonparametric statistical tests were applied. For each variable, a Kruskal-Wallis nonparametric analysis of variance (ANOVA) [46] was performed to reveal significant differences in means between salinity conditions and the sampling sites. The Mann–Whitney U test was applied as a post hoc test to assess differences in gene expression levels between salinity conditions and populations. Using the individual data from the sites, the correlation between salinity, temperature, and mRNA transcription levels of all genes was assessed by means of Spearman's rank correlation test [43]. These tests were performed with R Software (v.3.1.1). For all tests, a significance level of 0.05 was applied.

3. Results

3.1. Environmental Data. The salinity measured at the sampling sites during the dry season ranged from 0 to 100 psu (Table 2). Salinities at the reference sites ranged from 0 (Guiers Lake, freshwater) to 37 (Hann Bay, seawater). The salinity was markedly higher at all the sites of the Saloum estuary, up to an extreme of 100 psu at Kaolack (Table 2). The peculiarities of the study area are seasonal changes in ambient salinity and associated deterioration of water quality due to the alternation of an extended eight-month dry season (November to June) and a short four-month rainy season (July to October). In the Gambia and Saloum estuaries, salinity levels change very significantly between these two seasons [28, 29]. The salinity increased very significantly during the first months following the end of the rainy season because of intense evaporation [6]. It becomes then quite stable at the end of the dry season. During the short rainy season, by contrast, salinity in the estuaries is very unstable and can decrease considerably due to inputs of freshwater from precipitation, which drastically changes the water quality (turbidity). The poorer water quality during the rainy season represents an additional abiotic factor that could be constraining for the populations of *S. melanotheron*. However, the fish analysed in this study were collected at the end of the dry season (June) when the salinity in the estuaries had been stable for some months and the water turbidity was at the lowest level. The water temperature (Table 2) varied slightly among the sites, ranging from 26 to 29°C. The additional environmental data indicated that salinities measured at all the sites of the Saloum estuary for combined seasonal data ranged from brackish to hypersaline (up to 130 psu) in the estuary's upper reaches (Supplementary Figure 1). At Guiers Lake, freshwater conditions prevailed throughout the year whereas at Hann Bay, salinity varied slightly between 37 and 38 psu (Supplementary Table 1). Water temperatures in the Saloum and Gambia estuaries showed slight variations (Supplementary Figure 2). Average water temperatures were 24.9 and 25°C at Guiers Lake and Hann Bay, respectively (Supplementary Table 1). None of the sampling locations did reach values of percentage oxygen saturation that are limiting to the survival of fish (Supplementary Figure 3 and Table 1). Based on all these results, it can be concluded that salinity is the most constraining environmental factor for *S. melanotheron* in our study area during the dry season.

3.2. Branchial Abundance of SLC4A1, MAPK, and LDH mRNA. The efficiency of the real-time PCR analysis showed values around 1.9, which implies that the two primer pairs of each gene (including the reference gene, β-actin) amplified a single specific product. A melting curve analysis and agarose gel visualization are not shown in this study, but the products of amplification of the reference gene were previously validated by analysing the amplicon size on agarose gel electrophoresis [47]. Relative abundance in individuals from both hypersaline water and seawater strains was calculated to correct for differences in efficiency. The results revealed significant differences in SLC4A1 and MAPK expression levels between salinity conditions (Kruskal-Wallis test; $p < 0.05$) in the hypersaline water strain. Fish maintained in freshwater had significantly higher SLC4A1 mRNA levels than those acclimated to hypersaline water (Figure 2). The MAPK mRNA levels exhibited a similar pattern between

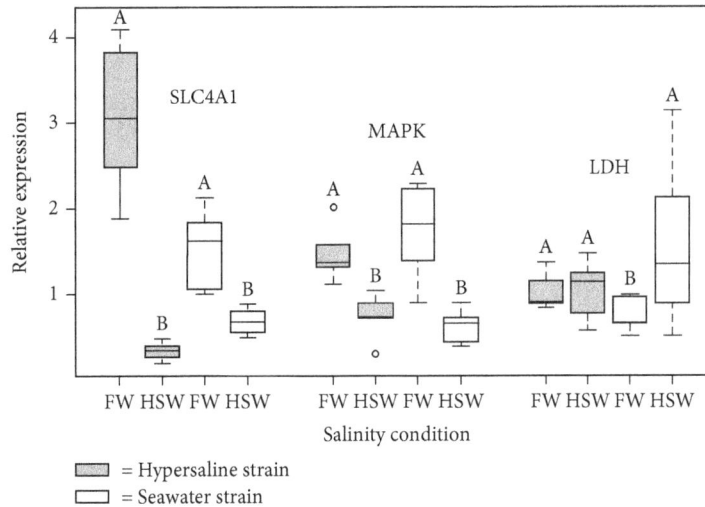

FIGURE 2: Relative expression of SLC4A1, MAPK, and LDH in fish gills. The relative expression was measured after an exposure period of 45 (hypersaline strain) or 10 (seawater strain) days to freshwater (FW) and hypersaline water (HSW). Data are illustrated as boxplots with the median represented by a horizontal line and the 25th and 75th percentiles corresponding to the bottom and top edges of the boxes. Different letters on the boxplots indicate significant difference in gene expression levels whereas the same letter indicated that there is no significant difference.

salinity conditions, being higher in freshwater and lower in hypersaline water (Figure 2). The seawater strain (Joal) showed significant differences in LDH expression levels between freshwater and hypersaline water (Kruskal-Wallis test; $p < 0.05$), but there was no significant difference between these two salinity conditions in individuals of the hypersaline (Saloum) strain (Figure 2). The comparisons between strains showed that the SLC4A1 and MAPK expression profiles are comparable between the strains (Figure 2) with elevated levels in freshwater and low levels in hypersaline water. By contrast, there are significant differences in gene expression profiles between the strains for the LDH gene with elevated levels being recorded in hypersaline water for the seawater strain.

In wild populations, the comparison of SLC4A1 expression between sampling locations revealed significant differences between populations (Kruskal-Wallis test; $p < 0.05$). The SLC4A1 mRNA levels were higher at site 3 (Figure 3(a)) than at the sampling sites of the Gambia (site 2) and Saloum (4a, 4b, and 4c) estuaries, and they were also higher than those at site 1, while the levels at this latter site were higher than those at the other sampling locations. Within the Saloum estuary, fish sampled at site 4b, whose salinity was intermediate, had lower SLC4A1 mRNA levels than fish sampled at the less saline station (4a) and the most saline sampling site (4c) (Figure 3(a)). No significant difference in SLC4A1 expression was found between sites 4a and 4c. The amounts of MAPK mRNA were significantly higher (Kruskal-Wallis test; $p < 0.05$) in freshwater, brackish water, and seawater at sites 1, 2, and 3, respectively (Figure 3(b)), in comparison to the most saline sampling sites (4a, 4b, and 4c) of the Saloum estuary. No significant difference in MAPK expression was found between sites 1, 2, and 3 (Figure 3(b)). Within the Saloum estuary, the MAPK expression levels were significantly higher at site 2 in comparison to sites 1 and 3, but there were no significant differences between these two locations.

The highest LDH mRNA levels (Kruskal-Wallis test; $p < 0.05$) were recorded at site 3 and the Saloum estuary (4a, 4b, and 4c) whereas the lowest levels were observed at the least saline locations, sites 1 and 2 (Figure 3(c)). Within the Saloum estuary, there was no significant difference in LDH expression between locations. The relative expression levels of LDH at the Saloum estuary locations were not significantly different from those at site 3 (Figure 3(c)). The amounts of LDH mRNA at site 2 were higher than those at site 1.

3.3. Correlation between Salinity, Temperature, and Gene Expression. Environmental salinity and SLC4A1 mRNA levels were negatively correlated (Spearman's rank correlation: $R = -0.44$, $p < 0.001$), and the same relationship was found for salinity and MAPK relative expression ($R = -0.57$, $p < 0.001$) (Figure 4). In contrast, LDH mRNA levels were significantly positively correlated with salinity ($R = 0.65$; $p < 0.001$) (Figure 4). There was no significant correlation between ambient temperature and mRNA levels of SLC4A1 ($R = 0.033$; $p > 0.5$), MAPK ($R = 0.029$; $p > 0.5$), and LDH ($R = -0.31$; $p > 0.5$).

4. Discussion

The results of the experiments conducted under controlled laboratory conditions revealed significant differences depending on the salinity conditions at which fish were held. In contrast, the differential expression in the natural environment is not consistent with that in the model of neutral divergence of wild populations and also does not reflect salinity-induced differences in life history traits observed between populations. These results suggest that the gene expression patterns are not a result of evolutionary forces such as genetic drift or adaptation by natural selection. They

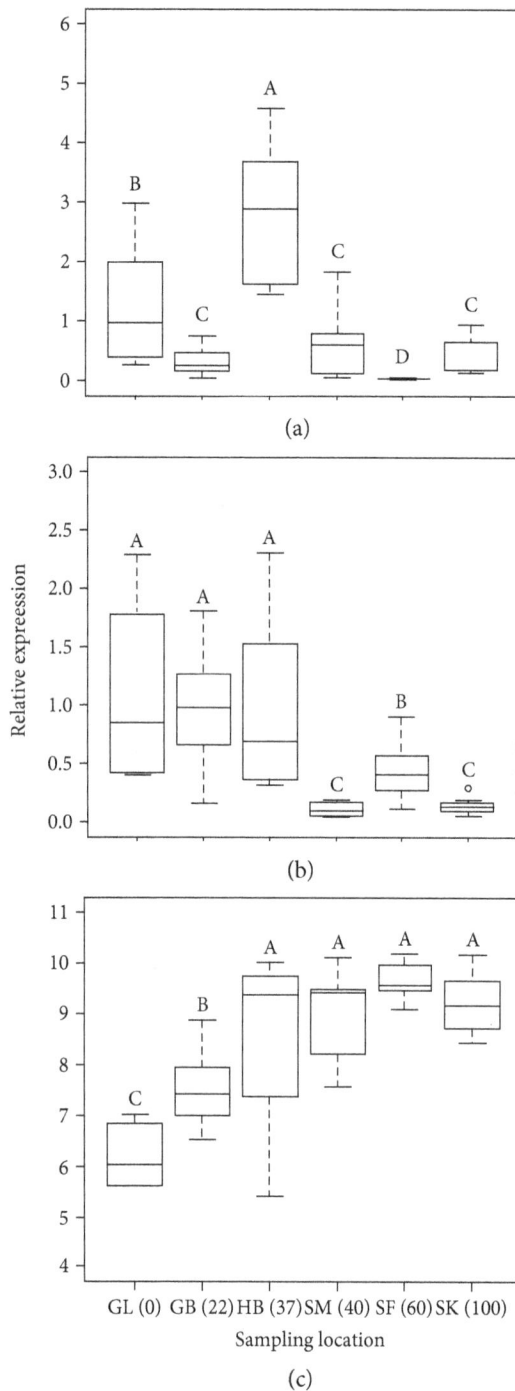

(a)

(b)

GL (0) GB (22) HB (37) SM (40) SF (60) SK (100)

Sampling location

(c)

FIGURE 3: Relative expression of SLC4A1 (a), MAPK (b), and LDH (c) in six wild populations adapted to different salinities. Data are illustrated as boxplots with the median represented by the horizontal line and the 25th and 75th percentiles corresponding to the bottom and top edges of the boxes. Different alphabetic letters above the boxplots indicate significant difference in gene expression levels whereas the same letter indicated that there is no significant difference. GL: Guiers Lake; GB: Gambia Balingho; HB: Hann Bay; SM: Saloum Missirah; SF: Saloum Foundiougne; SK: Saloum Kaolack. The values in brackets on the x-axis correspond to salinities recorded at each sampling location at the time of fish sampling.

suggest that genes responded to various environmental stressors, which may occur in combination with biotic stressors, including factors intrinsic to animals. Although gene expression variations were explored in two different strains (seawater and hypersaline water strains) established in the laboratory 15 years ago under experimental conditions identical to those described above, overall genes showed similar expression patterns. These results indicate a reproducibility but also an absence of a strain effect. By contrast, the lack of correlations between laboratory and wild conditions may reflect the adaptation of experimental fish to laboratory conditions. It is likely that individuals adapted in the laboratory for such a long time have evolved particular adaptations to the laboratory conditions plus the effect of inbreeding.

For a better understanding of how the differential expression observed here could relate to the adaptation of tilapia populations to environmental salinities, it is necessary to contrast gene expression patterns with the genetic structure of the populations inferred from neutral markers. The expectation for such comparisons is whether gene expression patterns are correlated with this genetic differentiation; this would probably indicate that differences among populations have arisen from genetic drift. The genetic divergence among populations was not analysed in this study, but previous work has shown that populations of *S. melanotheron* are strongly structured, even at microgeographical scales (i.e., within the Saloum estuary) [48]. This high-population genetic structure is essentially due to the low larval dispersal ability of this species, which results from its mouth-brooding reproductive behaviour. However, the patterns of gene expression found here do not reflect the genetic differentiation of *S. melanotheron* populations revealed by neutral markers. For example, the genetically and geographically remote populations of Saloum sites 4a, 4b, and 4c did not display any difference in SLC4A1 expression levels. Likewise, no significant difference in LDH expression levels was found between populations of sites 1 and 2, two populations that are genetically strongly differentiated. These observations indicate that the differential expression observed in this study is not the result of genetic drift. By contrast, it cannot be ruled out that this differential expression reflects the effects of natural selection, which could sort out some specific adaptive polymorphisms that may increase the relative fitness of populations in their natural environments. However, considering the extensive temporal variations of salinity conditions in the Saloum and Gambia estuaries throughout the year, the idea that divergent selection could lead to fixation of mutations seems unlikely. Indeed, the salinity in the Saloum and Gambia estuaries increases during the first months following the end of the rainy season, because of intense evaporation [6], and then becomes quite stable at the end of the dry season. Thus, during the long dry season, the estuary only exhibits minor variations in salinity. By contrast, during the short rainy season, salinity in the estuary is very unstable and can decrease considerably due to inputs of freshwater from precipitation. In such a situation, long-term adaptation as the response to environmental conditions that remain relatively constant throughout the lifetime of the species is unlikely. By contrast, unstable salinity in the rainy season

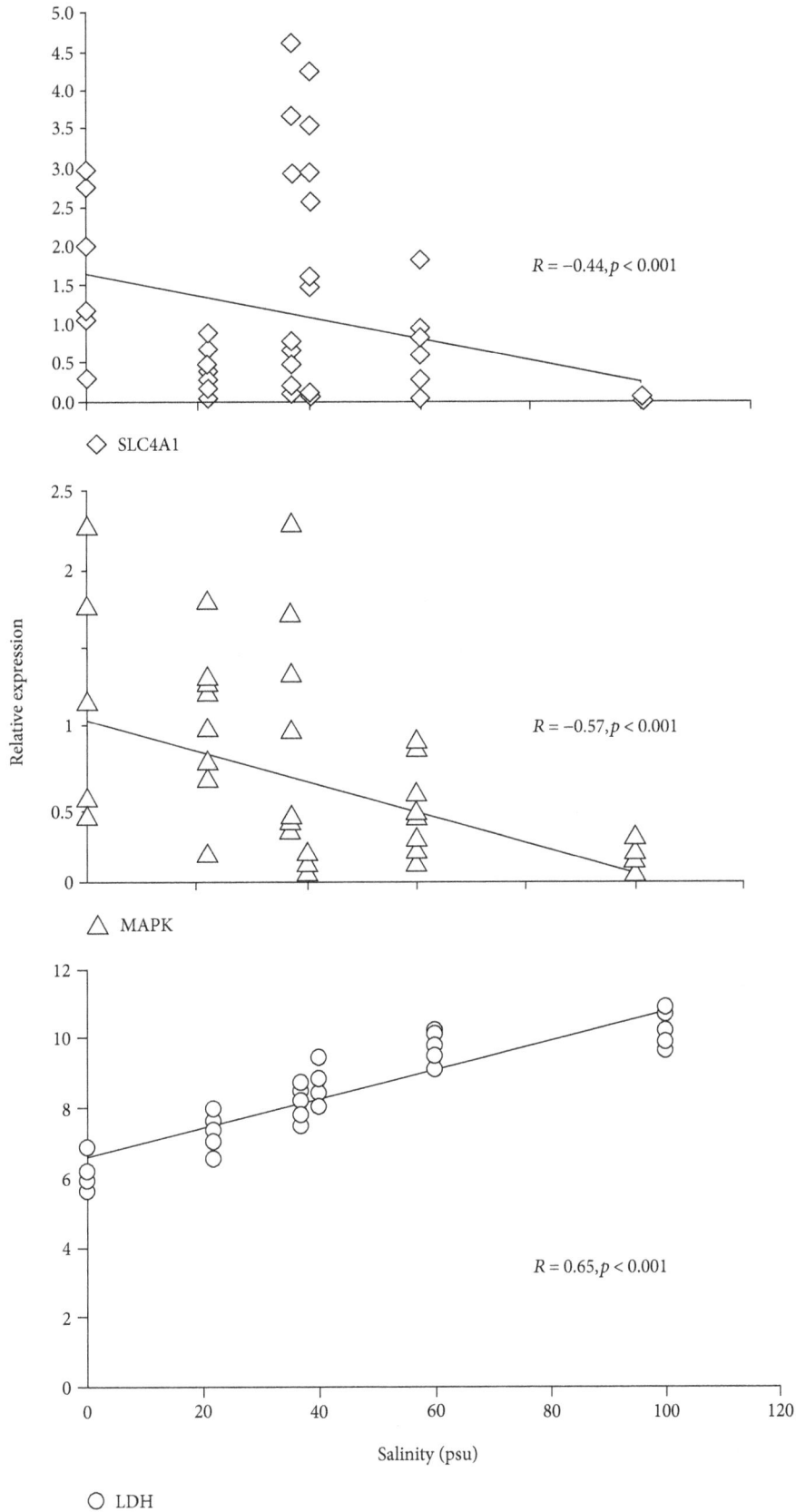

FIGURE 4: Relationship between environmental salinity and mRNA levels of SLC4A1, MAPK, and LDH. The mRNA expression levels represent the relative expression normalised to β-actin.

may represent a major stress factor that might drive selection of genotypes better suited to adjusting physiological responses to manage the consequences of stress.

Thus, the differences in gene expression between populations may result from gene plasticity in response to salinity variations. The expression of genes analysed here may be differentially regulated to respond directly to the salinity variation [49]. Therefore, it would be interesting to analyse and compare the epigenetic patterns and diversity between populations to explain differential responses in terms of gene expression to the environment. Environmentally induced phenotypic variations, known as gene expression plasticity, have been reported in several fish species [50–52]. This plasticity can be adaptive when the phenotypic variation confers higher fitness to a particular individual [53, 54]. Previous studies have shown that *S. melanotheron* populations exhibit greater fitness at intermediate salinities, where the energy cost for osmoregulation is lower compared to that at hypersaline conditions, where impaired growth and precocious reproduction were observed [28, 55]. However, differential expression patterns observed here do not reflect these differences in life history traits induced by salinity variations. Moreover, the correlations between gene expression levels and environmental salinity are somewhat ambiguous, suggesting the potential influence of factors other than ambient salinity.

The SLC4A1 gene may be induced by other factors including anthropogenic pollution, because this gene is involved in other regulatory processes, such as metal homeostasis. Likewise, the LDH expression may reflect switches in metabolic pathways used to yield the energy required for hydromineral maintenance [56–58]. Indeed, elevated LDH expression levels were seen in hypersaline water for the seawater strain, which is consistent with the elevated levels observed in all the wild populations where salinities are above 22 psu. These results point to the higher energy requirements. While the elevated LDH expression levels in the Saloum estuary (sites 4a, 4b, and 4c) may reflect energy cost for osmoregulation at high salinities, the elevated mRNA amounts observed in seawater (site 3) are somewhat difficult to explain. They may reflect energy investments in growth since *S. melanotheron* is known for having the best growth performances in seawater [55]. Similarly, elevated expression levels in seawater (site 3) for the SLC4A1 are somewhat difficult to explain. They may reflect an involvement of this gene in somatic growth, especially since the species has better growth rates in salinities around seawater. Indeed, it has been demonstrated that the MAPK/ERK pathway activation is inhibited during fasting and stimulated during refeeding [59]. The same authors have established a link between insulin-like growth factor-I (IGF-I), the activation of the MAPK/ERK, and somatic growth in fine flounder, strongly suggesting regulatory effects of MAPK pathways on fish growth. While the low MAPK expression levels when salinities are above normal seawater salinity, only observed in the Saloum populations where impaired growth was observed [28, 55], are consistent with this interpretation, the low expression in brackish water (site 2, 22 psu) indicates quite the opposite. MAPK pathways are activated by changes in biotic factors including oxidative stress, growth factors,

and cytokines [60, 61]. These internal cues are known to play important roles in transcriptional responses through feedback mechanisms, through which organisms respond to internal physiological state including nutrient and energy availability, growth rate, and pathogen infection, by activating gene expression programs. Mechanisms that directly link gene expression with internal variables differ from responses to external signals, through which gene expression is modulated directly by signal transduction pathways that translate extracellular stimuli into intracellular responses to cope with environmental variations. In natural environments where biotic factors cannot be controlled, their effects may interfere with external environmental factors and impact gene expression levels between populations. This may explain the absence of significant correlations between gene expression and ambient salinity in this study. It has been reported that many genes regulated during a stress response are not specifically induced by a given perturbation, but from a part of stress response pathways. Such unspecific stress response genes respond to various environmental stressors that may occur together and/or in combination with factors intrinsic to animals whose effects are more difficult to evaluate. Furthermore, unspecific stress response genes are also generally involved in a wide diversity of other biological functions including growth, sexual maturation, and reproduction, which may interfere with stress responses. The fact that the effects of biotic factors may interfere with external environmental factors in natural environments may explain the differences in LDH expression levels between the breeding strains and wild fish.

The main drawback of our approach is the limited number of genes analysed, but these genes are used here as proxies of the entire networks of stress response pathways. The roles of the SLC4A1 and MAPK genes have previously been investigated in the same populations in the context of adaptive plasticity of gene expression in response to salinity changes [17]. Reanalysing these genes here to address a new question has the advantage of better elucidating their adaptive roles in response to salinity variations. The experimental populations analysed here were exposed to specific treatments for different lengths of time, which may render the findings difficult to interpret with respect to population and salinity of origin and time of exposure to a particular treatment. However, both experiments indicated that two of the three genes analysed were significantly differentially expressed between the freshwater and hypersaline treatments, and in both cases, the population origin and therefore the salinities at which the fish were captured did not significantly affect the expression of the three genes. Likewise, differences between experiments in developmental stages, type of the tanks, food type, light, and time of the year do not seem to affect the fishes' physiology and, consequently, their levels of gene expression.

5. Conclusion

Genes that contribute to the specific traits for stress tolerance have been successfully studied and are still being used as biomarkers for investigating the adaptive mechanisms employed by living organisms to environmental variations. The relationship between gene expression and variation in environmental salinity in *S. melanotheron* populations along salinity gradients has been successfully explored in previous studies [17, 47, 62]. Significant correlations were found in populations between salinity and mRNA transcription levels of genes putatively involved in osmoregulation, suggesting that they play a role in salinity acclimatisation. It was also demonstrated that gene expression patterns in wild populations match salinity-related differences in life history traits previously observed in the same populations [63]. These results clearly indicated that *S. melanotheron* is an excellent model species for understanding the molecular and physiological mechanisms used by fish to cope with extreme changes in environmental salinities, particularly in the Senegal and Gambia estuaries where other factors, such as water temperature and dissolved oxygen, show considerable variations. These results also clearly demonstrate that candidate gene approaches can be applied to wild populations to identify genes and pathways involved in the acclimatisation to extreme salinities, including populations inhabiting estuarine ecosystems where the complexity of environmental conditions can pose great difficulties in the interpretation of gene expression profiles. The results of the present study show that candidate genes selected to be responsive to salinity variation under experimental conditions do not show the expected behaviour when monitored in wild samples naturally experimenting life in a similar salinity gradient. These results raise the possibility that the investigated genes are not specifically and exclusively responsive to salt stress. They can be qualified as unspecific stress response genes that responded to various environmental stressors which may occur in combination with biotic stressors including factors intrinsic to animals and whose effects are difficult to evaluate in wild populations. This study is among the first to illustrate the complexity of gene expression analyses in wild fish populations due to the inherent limitation of gene candidate identification approaches, which essentially rely on the knowledge in physiological, biochemical, and metabolic functions. These results are particularly significant because they provide an indication on how much attention needs to be paid when selecting candidate genes for studies on natural populations.

Abbreviations

SLC4A1:   Band 3 anion transport protein
MAPK: Mitogen-activated protein kinase
LDH: Lactate dehydrogenase
qRT-PCR: Quantitative real-time PCR
SEM: Standard error of mean
SSH: Suppressive subtractive hybridisation.

Additional Points

Availability of Supporting Data. All datasets supporting the results of this article are within the article and its Additional File 1. The accession numbers of any nucleic acid sequences of the three candidate markers used in this study are indicated in Section 2.6: SLC4A1 (GenBank accession number ES881098), MAPK (GenBank accession number ES881524), and LDH (GenBank accession number ES881483). *Authors' Information.* Mbaye Tine is an evolutionary biologist who performed his Doctorate Thesis in Evolutionary Biology and Ecology at the University of Montpellier II, in the department Integrative Biology of the Institute of Sciences and Evolution, under the direction of Dr. François Bonhomme (CNRS) and Dr. Jean-Dominique Durand (IRD). His research focused on genetic and physiological mechanisms underlying the adaptation of fishes to environmental variations including changes in ambient salinity. He performed his first postdoctoral research in Genomics and Evolution in the laboratory of Dr. Richard Reinhardt at the Max Plank Institute for Molecular Genetics in Berlin. Afterward, he performed a second postdoc in Genomics and Next Generation Sequencing at the Genome Centre at MIP Cologne. More specifically, he worked on the sequencing project of the European sea bass (*Dicentrarchus labrax* L.) genome, which was published in Nature Communication. He is currently doing his research at the Molecular Zoology Laboratory at Johannesburg University where he works on a project that aims at investigating thermal adaptation and the role of phenotypic plasticity in coastal invertebrates and/or fishes by studying populations of species whose ranges span across multiple South African marine biogeographic provinces.

Conflicts of Interest

Eventual conflicts of interest (including personal communications or additional permissions and related manuscripts), sources of financial support, corporate involvement, and patent holdings are disclosed.

Authors' Contributions

Mbaye Tine conceived and planed the study. He has collected the wild population samples in the natural environment with the help of Dr. Jean-Dominique Durand and Khady Diop. He has conducted the common garden experiments with the help of Dr. Jean Francois Baroiller and his team. He performed gene expression and statistical analyses, planned and coordinated the manuscript preparation, and wrote the manuscript.

Acknowledgments

The author thanks Jean-Dominique Durand and Khady Diop for their help during the sample collection in the natural environment. He is also grateful to Jean Francois Baroiller and his team for offering facilities and helping to conduct the common garden experiments. This project has been financially supported by the IRD research unit 070-RAP and the Department of Support and Training (DST). Financial supports (postdoctoral fellowship) were also obtained from the Max Planck Institute for Molecular Genetics (Berlin, Germany) and the Department of Zoology, University of Johannesburg (South Africa) for the manuscript preparation.

References

[1] T. E. Grantham, A. M. Merenlender, and V. H. Resh, "Climatic influences and anthropogenic stressors: an integrated framework for streamflow management in Mediterranean-climate California, U. S. A.," *Freshwater Biology*, vol. 55, pp. 188–204, 2010.

[2] J. L. Meyer, M. J. Sale, P. J. Mulholland, and N. L. Poff, "Impacts of climate change on aquatic ecosystem functioning and health," *Journal of the American Water Resources Association*, vol. 35, pp. 1373–1386, 1999.

[3] V. S. Kennedy, R. R. Twilley, J. A. Kleypas, J. H. Cowan Jr., and S. R. Hare, *Coastal and Marine Ecosystems & Global Climate Change: Potential Effects on U.S. Resources*, 64 pages, Pew Center On Global Climate Change, Arlington, VA 22201, USA, 2002.

[4] J.-J. Albaret, M. Simier, F. S. Darboe, J.-M. Ecoutin, J. Rafray, and L. T. de Morais, "Fish diversity and distribution in the Gambia estuary, West Africa, in relation to environmental variables," *Aquatic Living Resources*, vol. 17, pp. 35–46, 2004.

[5] E. Ojaveer and M. Kalejs, "The impact of climate change on the adaptation of marine fish in the Baltic Sea," *ICES Journal of Marine Science*, vol. 62, pp. 1492–1500, 2005.

[6] M. Simier, L. Blanc, C. Alioume, P. S. Diouf, and J. J. Albaret, "Spatial and temporal structure of fish assemblages in an "inverse estuary", the Sine Saloum system (Senegal)," *Estuarine Coastal and Shelf Science*, vol. 59, pp. 69–86, 2004.

[7] S. Brinda and S. Bragadeeswaran, "Influence of physico-chemical properties on the abundance of a few economically important juvenile fin-fishes of Vellar estuary," *Journal of Environmental Biology*, vol. 26, pp. 109–112, 2005.

[8] A. J. Jaureguizar, R. Menni, R. Guerrero, and C. Lasta, "Environmental factors structuring fish communities of the Rio de la Plata estuary," *Fisheries Research*, vol. 66, pp. 195–211, 2004.

[9] R. Blekhman, A. Oshlack, A. E. Chabot, G. K. Smyth, and Y. Gilad, "Gene regulation in primates evolves under tissue-specific selection pressures," *PLoS Genetics*, vol. 4, no. 11, article e1000271, 2008.

[10] C.-R. Lee and T. Mitchell-Olds, "Environmental adaptation contributes to gene polymorphism across the *Arabidopsis thaliana* genome," *Molecular Biology and Evolution*, vol. 29, no. 12, pp. 3721–3728, 2012.

[11] P. Morán, F. Marco-Rius, M. Megías, L. Covelo-Soto, and A. Pérez-Figueroa, "Environmental induced methylation changes associated with seawater adaptation in brown trout," *Aquaculture*, vol. 392-395, pp. 77–83, 2013.

[12] A. Varriale, "DNA methylation, epigenetics, and evolution in vertebrates: facts and challenges," *International Journal of Evolutionary Biology*, vol. 2014, Article ID 475981, 7 pages, 2014.

[13] P. F. Larsen, P. M. Schulte, and E. E. Nielsen, "Gene expression analysis for the identification of selection and local adaptation in fishes," *Journal of Fish Biology*, vol. 78, pp. 1–22, 2011.

[14] D. O. Conover and H. Baumann, "The role of experiments in understanding fishery-induced evolution," *Evolutionary Applications*, vol. 2, pp. 276–290, 2009.

[15] L. E. Eierman and M. P. Hare, "Transcriptomic analysis of candidate osmoregulatory genes in the eastern oyster *Crassostrea virginica*," *BMC Genomics*, vol. 15, p. 503, 2014.

[16] M. Telonis-Scott, A. A. Hoffmann, and C. M. Sgrò, "The molecular genetics of clinal variation: a case study of *ebony* and thoracic trident pigmentation in *Drosophila melanogaster* from eastern Australia," *Molecular Ecology*, vol. 20, pp. 2100–2110, 2011.

[17] M. Tine, B. Guinand, and J. D. Durand, "Variation in gene expression along a salinity gradient in wild populations of the euryhaline black-chinned tilapia *Sarotherodon melanotheron*," *Journal of Fish Biology*, vol. 80, pp. 785–801, 2012.

[18] A. Whitehead and D. L. Crawford, "Neutral and adaptive variation in gene expression," *Proceedings of the National Academy of Sciences*, vol. 103, no. 14, pp. 5425–5430, 2006.

[19] D. H. Evans, P. M. Piermarini, and K. P. Choe, "The multifunctional fish gill: dominant site of gas exchange, osmoregulation, acid-base regulation, and excretion of nitrogenous waste," *Physiological Reviews*, vol. 85, pp. 97–177, 2005.

[20] J. Hiroi and S. D. McCormick, "Variation in salinity tolerance, gill Na$^+$/K$^+$-ATPase, Na$^+$/K$^+$/2Cl$^-$ cotransporter and mitochondria-rich cell distribution in three salmonids *Salvelinus namaycush*, *Salvelinus fontinalis* and *Salmo salar*," *Journal of Experimental Biology*, vol. 210, pp. 1015–1024, 2007.

[21] T. Sakamoto, K. Uchida, and S. Yokota, "Regulation of the ion-transporting mitochondrion-rich cell during adaptation of teleost fishes to different salinities," *Zoological Science*, vol. 18, pp. 1163–1174, 2001.

[22] S. Varsamos, J. P. Diaz, G. Charmantier, G. Flik, C. Blasco, and R. Connes, "Branchial chloride cells in sea bass (*Dicentrarchus labrax*) adapted to fresh water, seawater, and doubly concentrated seawater," *Journal Experimental Zoology*, vol. 293, pp. 12–26, 2002.

[23] D. H. Evans, P. M. Piermarin, and W. T. W. Potts, "Ionic transport in the fish gill epithelium," *Journal of Experimental Zoology*, vol. 283, pp. 641–652, 1999.

[24] D. Kultz and G. N. Somero, "Ion transport in gills of the euryhaline fish *Gillichthys mirabilis* is facilitated by a phosphocreatine circuit," *American Journal of Physiology Regulatory, Integrative and Comparative Physiology*, vol. 268, no. 4, Part 2, pp. R1003–R1012, 1995.

[25] W. S. Marshall, "Na$^+$, Cl$^-$, Ca^{2+} and Zn^{2+} transport by fish gills: retrospective review and prospective synthesis," *Journal Experimental Zoology*, vol. 293, pp. 264–283, 2002.

[26] W. S. Marshall and S. E. Bryson, "Transport mechanisms of seawater teleost chloride cells: an inclusive model of a multifunctional cell," *Comparative Biochemistry and Physiology Part A: Molecular & Integrative Physiology*, vol. 119, pp. 97–106, 1998.

[27] T. M. Falk, G. G. Teugels, E. K. Abban, W. Villwock, and L. Renwrantz, "Phylogeographic patterns in populations of the black-chinned tilapia complex (Teleostei, Cichlidae) from coastal areas in West Africa: support fort the refuge zone theory," *Molecular Phylogenetics and Evolution*, vol. 27, pp. 81–92, 2003.

[28] J. Panfili, A. Mbow, J.-D. Durand et al., "Influence of salinity on the life-history traits of the West African black-chinned tilapia (*Sarotherodon melanotheron*): comparison between the Gambia and Saloum estuaries," *Aquatic Living Resources*, vol. 17, pp. 65–74, 2004.

[29] K. Diouf, J. Panfili, M. Labonne, C. Aliaume, J. Tomas, and T. Do Chi, "Effects of salinity on strontium: calcium ratios in the otoliths of the West African black-chinned tilapia *Sarotherodon melanotheron* in a hypersaline estuary," *Environmental Biology of Fishes*, vol. 77, pp. 9–20, 2006.

[30] S. L. Alper, "Molecular physiology of SLC4 anion exchangers," *Experimental Physiology*, vol. 91, no. 1, pp. 153–161, 2005.

[31] R. Motais, B. Fiévet, F. Borgese, and F. Garcia-Romeu, "Association of the band 3 protein with a volume-activated, anion and amino acid channel: a molecular approach," *The Journal of Experimental Biology*, vol. 200, Part 2, pp. 361–367, 1997.

[32] S. K. Pierce and J. W. Warren, "The taurine efflux portal used to regulate cell volume in response to hypoosmotic stress seems to be similar in many cell types: lessons to be learned from molluscan red blood cells," *American Zoologist*, vol. 41, pp. 710–720, 2001.

[33] J. G. Bode, P. Gatsios, S. Ludwig et al., "The mitogen-activated protein (MAP) kinase p38 and its upstream activator MAP kinase kinase 6 are involved in the activation of signal transducer and activator of transcription by hyperosmolarity," *The Journal of Biological Chimistry*, vol. 274, pp. 30222–30227, 1999.

[34] D. Kultz, "Cellular osmoregulation: beyond ion transport and cell volume," *Zoology*, vol. 104, pp. 198–208, 2001.

[35] M. Zakhartsev, T. Johansen, H. O. Pörtner, and R. Blust, "Effects of temperature acclimation on lactate dehydrogenase of cod (*Gadus morhua*): genetic, kinetic and thermodynamic aspects," *The Journal of Experimental Biology*, vol. 207, pp. 95–112, 2004.

[36] M. Tine, J. de Lorgeril, H. D'Cotta et al., "Transcriptional responses of the black-chinned tilapia *Sarotherodon melanotheron* to salinity extremes," *Marine Genomics*, vol. 1, pp. 37–46, 2008.

[37] C. Lorin-Nebel, J. C. Avarre, N. Faivre, S. Wallon, and G. Charmantier, "Osmoregulatory strategies in natural populations of the black-chinned tilapia *Sarotherodon melanotheron* exposed to extreme salinities in West African estuaries," *Journal of Comparative Physiology B*, vol. 182, pp. 771–780, 2012.

[38] M. Guèye, J. Kantoussan, and M. Tine, "The impact of environmental degradation on reproduction of the black-chinned tilapia *Sarotherodon melanotheron* from various coastal marine, estuarine and freshwater habitats," *Comptes Rendus Biologies*, vol. 336, no. 7, pp. 342–353, 2013.

[39] N. Ouattara, C. Bodinier, G. Nègre-Sadargues et al., "Changes in gill ionocyte morphology and function following transfer from fresh to hypersaline waters in the tilapia *Sarotherodon melanotheron*," *Aquaculture*, vol. 290, pp. 155–164, 2009.

[40] S. P. Kelly and N. Y. S. Woo, "The response of sea bream following abrupt hyposmotic exposure," *Journal of Fish Biology*, vol. 55, pp. 732–750, 1999.

[41] G. R. Scott, P. M. Schulte, and C. M. Wood, "Plasticity of osmoregulatory function in the killifish intestine: drinking rates, salt and water transport, and gene expression after freshwater transfer," *The Journal of Experimental Biology*, vol. 209, pp. 4040–4050, 2006.

[42] M. Legendre and J.-M. Écoutin, "Suitability of brackish water tilapia species from the Ivory Coast for lagoon aquaculture. I – reproduction," *Aquatic Living Resources*, vol. 2, pp. 71–79, 1989.

[43] C. Spearman, "The proof and measurement of association between two things," *American Journal of Psychology*, vol. 15, pp. 72–101, 1904.

[44] M. S. Bartlett, "Properties of sufficiency and statistical tests," *Proceedings of the Royal Statistical Society, Series A: Mathematical, Physical and Engineering Sciences*, vol. 160, pp. 268–282, 1937.

[45] N. Smirnov, "Table for estimating the goodness of fit of empirical distributions," *Annals of Mathematical Statistics*, vol. 19, pp. 279–281, 1948.

[46] W. A. Wallis, "Use of ranks in one-criterion analysis of variance," *Journal of American Statistic Association*, vol. 47, pp. 583–621, 1952.

[47] M. Tine, D. J. McKenzie, F. Bonhomme, and J.-D. Durand, "Salinity-related variation in gene expression in wild populations of the black-chinned tilapia from various West African coastal marine, estuarine and freshwater habitats," *Estuarine, Coastal and Shelf Science*, vol. 91, pp. 102–109, 2011.

[48] L. Pouyaud, E. Desmarais, A. Chenuil, J. F. Agnèse, and F. Bonhomme, "Kin cohesiveness and possible inbreeding in the mouth brooding tilapia *Sarotherodon melanotheron* (Pisces Cichlidae)," *Molecular Ecology*, vol. 8, pp. 803–812, 1999.

[49] T. Fuller, "The integrative biology of phenotypic plasticity," *Biology and Philosophy*, vol. 18, pp. 381–389, 2003.

[50] A. Whitehead, "Comparative genomics in ecological physiology: toward a more nuanced understanding of acclimation and adaptation," *The Journal of Experimental Biology*, vol. 215, pp. 884–891, 2012.

[51] A. Whitehead, F. Galvez, S. Zhang, L. M. Williams, and M. F. Olecsiak, "Functional genomics of physiological plasticity and local adaptation in killifish," *Journal of Heredity*, vol. 102, no. 5, pp. 499–511, 2011.

[52] A. Whitehead, J. L. Roach, S. Zhang, and F. Galvez, "Salinity- and population-dependent genome regulatory response during osmotic acclimation in the killifish (*Fundulus heteroclitus*) gill," *The Journal of Experimental Biology*, vol. 215, pp. 1293–1305, 2012.

[53] C. D. Schlichting and H. Smith, "Phenotypic plasticity: linking molecular mechanisms with evolutionary outcomes," *Evolutionary Ecology*, vol. 16, pp. 189–211, 2002.

[54] M. J. West-Eberhard, "Developmental plasticity and the origin of species differences," *Proceedings of the National Academy of Sciences*, vol. 102, Supplement 1, pp. 6543–6549, 2005.

[55] M. Tine, J. de Lorgeril, J. Panfili, K. Diop, F. Bonhomme, and J.-D. Durand, "Growth hormone and prolactin-1 gene transcription in natural populations of the black-chinned tilapia *Sarotherodon melanotheron* acclimatized to different salinities," *Comparative Biochemistry and Physiology, Part B*, vol. 147, pp. 541–549, 2007.

[56] C. B. Cowey, "Intermediary metabolism in fish with reference to output of end products of nitrogen and phosphorus," *Water Science & Technology*, vol. 31, no. 10, pp. 21–28, 1995.

[57] D. R. Jury, S. Kaveti, Z. H. Duan, B. Willard, M. Kinter, and R. Londraville, "Effects of calorie restriction on the zebrafish liver proteome," *Comparative Biochemistry and Physiology Part D: Genomics and Proteomics*, vol. 3, pp. 275–282, 2008.

[58] C. I. Kolditz, G. Paboeuf, M. Borthaire et al., "Changes induced by dietary energy intake and divergent selection for muscle fat content in rainbow trout (*Oncorhynchus mykiss*), assessed by transcriptome and proteome analysis of the liver," *BMC Genomics*, vol. 9, p. 506, 2009.

[59] I. A. Mayer, A. Verma, I. M. Grumbach et al., "The p38 MAPK pathway mediates the growth inhibitory effects of interferon-α in BCR-ABL-expressing cells," *The Journal of Biological Chemistry*, vol. 276, no. 30, pp. 28570–28577, 2001.

[60] K. J. Cowan and K. B. Storey, "Mitogen-activated protein kinases: new signaling pathways functioning in cellular

responses to environmental stress," *Journal of Experimental Biology*, vol. 206, pp. 1107–1115, 2003.

[61] D. Kultz and M. Burg, "Evolution of osmotic stress signaling via MAP kinase cascades," *Journal of Experimental Biology*, vol. 201, pp. 3015–3021, 1998.

[62] M. Tine and H. Kuhl, "Changes in calmodulin gene expression in the gills of the black-chinned tilapia *Sarotherodon melanotheron* from drainage basins with different salinities," *International Research Journal of Biotechnology*, vol. 2, no. 1, pp. 009–015, 2011.

[63] M. Tine, F. Bonhomme, D. McKenzie, and J.-D. Durand, "Differential expression of the heat shock protein Hsp70 in natural populations of the tilapia, *Sarotherodon melanotheron*, acclimatised to a range of environmental salinities," *BMC Ecology*, vol. 10, no. 1, p. 11, 2010.

12

Whole Genome Sequencing of Greater Amberjack (*Seriola dumerili*) for SNP Identification on Aligned Scaffolds and Genome Structural Variation Analysis Using Parallel Resequencing

Kazuo Araki [iD],[1,2] Jun-ya Aokic,[1] Junya Kawase,[1,2] Kazuhisa Hamada,[3] Akiyuki Ozaki [iD],[1]
Hiroshi Fujimoto,[1] Ikki Yamamoto,[1] and Hironori Usuki[1]

[1]*Research Center for Aquatic Breeding, National Research Institute of Aquaculture, Fisheries Research Agency, 224 Hiruda, Tamaki-cho, Watarai, Mie 519-0423, Japan*
[2]*Marine Biological Science, Faculty of Bio-resources, Mie University Graduate School, 1577 Kurimamachiya-cho, Tsu City, Mie 514-8507, Japan*
[3]*Marine Farm Laboratory Limited Company, 309 Takahiro Tachibaura Otsuki-cho, Hata-gun, Kochi 788-0352, Japan*

Correspondence should be addressed to Kazuo Araki; arakin@affrc.go.jp and Akiyuki Ozaki; aozaki@affrc.go.jp

Academic Editor: Hieronim Jakubowski

Greater amberjack (*Seriola dumerili*) is distributed in tropical and temperate waters worldwide and is an important aquaculture fish. We carried out de novo sequencing of the greater amberjack genome to construct a reference genome sequence to identify single nucleotide polymorphisms (SNPs) for breeding amberjack by marker-assisted or gene-assisted selection as well as to identify functional genes for biological traits. We obtained 200 times coverage and constructed a high-quality genome assembly using next generation sequencing technology. The assembled sequences were aligned onto a yellowtail (*Seriola quinqueradiata*) radiation hybrid (RH) physical map by sequence homology. A total of 215 of the longest amberjack sequences, with a total length of 622.8 Mbp (92% of the total length of the genome scaffolds), were lined up on the yellowtail RH map. We resequenced the whole genomes of 20 greater amberjacks and mapped the resulting sequences onto the reference genome sequence. About 186,000 nonredundant SNPs were successfully ordered on the reference genome. Further, we found differences in the genome structural variations between two greater amberjack populations using BreakDancer. We also analyzed the greater amberjack transcriptome and mapped the annotated sequences onto the reference genome sequence.

1. Introduction

With the improvements in next generation sequencing technologies in the past few years, many genome projects of aquaculture fishes have been reported, including Atlantic salmon (*Salmo salar*) [1], Atlantic cod (*Gadus morhua*) [2], rainbow trout (*Oncorhynchus mykiss*) [3], Japanese flounder (*Paralichthys olivaceus*) [4], half-smooth tongue sole (*Cynoglossus semilaevis*) [5], platyfish (*Xiphophorus maculatus*) [6], common carp (*Cyprinus carpio*) [7], and channel catfish (*Ictalurus punctatus*) [8], and molecular markers of the shared genomic loci among individuals have been obtained

for genotype-phenotype linkage analysis. Chromosome-level assemblies or assembled genome sequences integrated with genetic maps are powerful tools that enable analyses of fish genetic breeding by marker-assisted or gene-assisted selection as well as help identify functional genes for biological traits. However, there are a limited number of fish species for which chromosome-level genome assemblies are available.

Single nucleotide polymorphisms (SNPs) in whole genomes are one of the most important genomic resources for studying population diversity, conservation genetics, and functional gene identification for biological traits [9–13]. To obtain molecular markers of the shared genomic loci among individuals, many technologies have been developed to probe whole-genome polymorphisms. These techniques have allowed the synthesis of DNA probes that can be used on SNP microarrays [14], making it possible to explore genome-wide SNPs in a high-throughput manner. However, the cost of array design and application obstructs their wider use in nonmodel species, especially for economically important organisms [15]. More importantly, microarray approaches cannot discover novel SNP loci for species without reference sequences [16]. The development of state-of-the-art next generation sequencing platforms has enabled scientists to scan small variants in genomes on an unprecedented scale [17, 18]. Multiplex library strategies have been used widely to further reduce the cost per sample [19]. However, cost is still one of the biggest challenges for whole-genome resequencing in nonmodel organisms [20]. We are developing whole-genomic analysis of *Seriola* species to study how much genetic variation remains in natural fishes and to investigate the mechanism of whole genome duplication, as well as to obtain molecular markers of the shared genomic loci among individuals for genotype-phenotype linkage analysis and to identify functional genes for biological traits. We have reported radiation hybrid (RH) physical and linkage maps of yellowtail (*Seriola quinqueradiata*) [21, 22] and compared the synteny among four model fishes, because yellowtail is one of the most important fishery resources in Japan. Greater amberjack (*Seriola dumerili*) is evolutionarily closely related to yellowtail and is more widely distributed in tropical and temperate waters than yellowtail. We are now interested in how much commonality exists between the yellowtail and amberjack genomes.

One of the purposes of this study was to construct a platform for quantitative trait locus (QTL), marker-assisted selection [23, 24], and gene-assisted selection [25, 26] programs for greater amberjack breeding based on SNPs. Therefore, we carried out de novo sequencing of the greater amberjack genome (hereafter referred to as the reference genome) and detected SNPs genome-wide using next sequencing technology and high-quality genome assembly. We then resequenced 20 greater amberjack genomes and identified SNPs on the reference genome sequence and analyzed genome-wide structural variations in two greater amberjack populations. In addition, we analyzed the RNA sequences from 12 amberjack tissues (muscle, brain, eye, heart, liver, intestine, kidney, spleen, gonad, gill, fin, and bladder) and mapped the resulting sequences onto the reference genome sequence to assemble them.

2. Materials and Methods

2.1. Ethics Statement.
In Japan, field permits are not required for greater amberjack. The Institute Animal Care and Use Committee of the National Research Institute of Aquaculture (IACUC-NRIA 27004) approved the fish handling, husbandry, and sampling methods used in this study. We sampled sperm from one anesthetized male amberjack that had been cultured in the aquarium of the Komame Station of the National Research Institute of Aquaculture (Kochi, Japan). We gathered blood from eight greater amberjacks cultured in Komame Station and 12 individual greater amberjacks in the marine crop of Owasebussan Co. Ltd. (Mie, Japan). For RNA sequencing, we sampled tissues from one anesthetized greater amberjack fished in the Pacific Ocean near Mie Prefecture, Japan.

2.2. Whole-Genome Sequencing.
We extracted high molecular weight genomic DNA from the sperm of one male greater amberjack and checked the DNA quality by spectrophotometer and 2% agarose gel electrophoresis before library construction. DNA fractions of 170–300 bp (for libraries with 250 bp insert size), 450–550 bp (for libraries with 500 bp insert size), and 700–900 bp (for libraries with 800 bp insert size) were excised and eluted from the gel slices overnight at 4°C in 300 μl of elution buffer (5 : 1 [vol/vol] LoTE buffer [3 mM Tris–HCl (pH 7.5), 0.2 mM EDTA] to 7.5 M ammonium acetate) and purified using a Spin-X filter tube (Fisher Scientific, Waltham, MA, USA) and ethanol precipitation. Genome libraries were prepared using a modified paired-end tag protocol supplied by Illumina (http://prodata.swmed.edu/LepDB/Protocol/illumina_Paired-End_Sample_Preparation_Guide.pdf#search=%27pairedend+tag+protocol+Illumina%27).

Mate pair (aka jumping) libraries were constructed using 4 mg of genomic DNA with the Illumina Nextera Mate Pair library construction protocol and reagent (FC-132-1001). We amplified the 2 kbp and 5 kbp inserts with 10 cycles of PCR and the 10 kbp and 20 kbp inserts with 15 cycles of PCR. The two fractions were pooled for mate pair (100 bp) sequencing using an Illumina HiSeq 2500 system.

For sequencing on the PacBio RSII long read platform (Pacific Biosciences, Melon Park, CA, USA), the genomic DNA was sheared to 10 to 20 kb using an ultrasonicator (Covaris Inc., Woburn, MA, USA) and converted to the proprietary SMRTbell™ library format using an RS DNA Template Preparation Kit (Pacific Biosciences). SMRTbell templates were subjected to standard single-molecule real-time (SMRT) sequencing using an engineered phi29 DNA polymerase on a PacBio RS II system according to the manufacturer's protocol. We sequenced 8 Gbp of the 10 kbp library by PacBio RS II (10 kb library) using a SMRT cell.

We removed the 3′-end adaptor sequences and mate pair annular junctional adaptor sequences from the Illumina sequencing data with Cutadapt (http://cutadapt.readthedocs.io/en/stable/guide.html) and removed low-quality reads and reads shorter than 20 bp with a flexible read trimming tool, Trimmomatic (version 0.32) [27]. We assembled the paired-end short read sequences using SOAPdenovo2 assembly software [28], mapped the resulting contigs to the mate pair read sequences, and closed gaps between contigs with Platanus [29]. Then, we converted the mate pair sequences to reverse complemental sequences using the Reverse Complement tool in the FASTX-Toolkit (http://hannonlab.cshl.edu/fastx_toolkit/). We assembled the resulting contigs again using SOAPdenovo2 and mapped the PacBio RSII read data to long scaffolds using PBJelly [30] to remove sequence gaps.

2.3. Alignment of the Amberjack Genome Assembly. We mapped the amberjack genome sequence onto the yellowtail RH physical map [22]. Then, we used the sequences of the mapped markers of yellowtail for BLAST searches against the scaffolds in the amberjack genome assembly using BWA software [31]. When a query sequence had multiple hits, we selected the top hit for analysis only if its *E* value was less than half that of the second hit; all other hits were removed from further analysis. The amberjack genome assembly sequences were lined up on the yellowtail RH physical map based on the results of the BLAST searches.

2.4. Resequencing of 20 Greater Amberjack Genomes and SNP Detection. We resequenced high molecular weight genomic DNA extracted from the blood of 20 greater amberjacks. The DNA quality was checked by spectrophotometer and 2% agarose gel electrophoresis before library construction. DNA fractions of 200–300 bp (for libraries with 250 bp insert size) were excised and eluted from the gel slices and purified using a Spin-X Filter Tube (Fisher Scientific). Genome libraries were prepared using a modified paired-end tag protocol supplied by Illumina. We sequenced 20 Gbp of each library using an Illumina HiSeq 2500 system. We removed the adaptor sequences from both ends of the sequences with Cutadapt and trimmed the sequences using Trimmomatic (version 0.32) [27] to remove areas where the average CV (coefficient variation) was <20. We selected sequences that were >70 bp long and mapped them onto the amberjack reference genome sequence using the BWA software [31] to detect mutations. The sequences in regions where mismatches were high were realigned using the GATK software (Genome analysis tool kit, https://www.intel.co.jp/content/www/jp/ja/healthcare-it/solutions/genomicscode-gatk.html) to improve mapping accuracy.

2.5. Detection of Structural Variants by BreakDancer. We used BreakDancer-1.1 under GPLv3 [32], which provides genome-wide detection of structural variants from next generation paired-end sequencing reads, with the default parameters to detect insertions/deletions, inversion, and translocations in the pair-end resequenced data of the 20 greater amberjacks. The whole-genome resequenced assembly files were saved in Bam file format. These files were analyzed by BreakDancer to detect structural variations in each resequenced genome. In addition, we combined the resequenced data of the eight greater amberjacks captured off the Kochi Coast and combined the resequenced of the 12 greater amberjacks captured off Chinese Hainan Islands Coast and analyzed each of these data sets by BreakDancer. The structural variation data files were merged in a Circus plot (http://circos.ca/intro/genomic_data/) to visualize the structural variations on the genomes.

2.6. Transcriptome Sequencing. We bought a single alive female adult amberjack that was fished in the Pacific Ocean near Mie Prefecture for tissue sampling. Total RNAs were extracted from 12 tissues (muscle, brain, eye, heart, liver, intestine, kidney, spleen, gonad, gill, fin, and bladder) using RNAiso Plus (Takara, Shiga, Japan). After purification of

TABLE 1: Summary statistics of the whole-genome sequence assembly of greater amberjack.

	Hiseq 2500	Hiseq + PacBio
Total bases	662,587,481 bp	677,669,644 bp
Number of scaffolds	34,824	11,655
Number of gaps	32,742	9742
Mean of scaffolds	19,026 bp	19,554 bp
Longest bases	22,167,742 bp	24,919,768 bp
N50	4,989,656 bp	5,812,906 bp
Number of >2Kb	724	707
Total bases (>2Kb)	655,539,910 bp	670,698,073 bp

The HiSeq 2500 sequence assembly was compared with the HiSeq 2500 sequence data mapped onto the PacBio RSII sequence data.

poly(A) + RNA, first-strand cDNA synthesis was primed with an N6 randomized primer using a ScriptSeq RNA-Seq Library Prep Kit (Illumina). After hydroxyl apatite chromatography, the single-stranded cDNA was amplified by seven cycles of PCR. One sequencing run was performed on an Illumina HiSeq 2500 paired-end sequence platform using Illumina reagents and protocols. We removed the 3′ adaptor sequences from the read sequences with Cutadapt (http://cutadapt.readthedocs.io/en/stable/guide.html) and trimmed the read sequences using Trimmomatic (version 0.32) [27]. We mapped the trimmed sequences onto the amberjack reference genome using cufflinks (http://cole-trapnell-lab.github.io/cufflinks/) to identify transcribed regions of the genome. We then carried out BLASTX homology searches against the amino acid sequence data in the NCBI RefSeq Vertebrate Other dataset using Blast2GO [55].

3. Results and Discussion

3.1. Whole-Genome Sequence of Greater Amberjack. We prepared genomic DNA from the sperm of one male greater amberjack that had been cultured in the aquarium at the Komame Station of the National Research Institute of Aquaculture (Kochi, Japan). We sequenced three paired-end libraries and four mate pair libraries on an Illumina HiSeq 2500 platform (Supplementary Figure S1) and obtained 133.1 Gbp of sequence data with 200-fold coverage. Using the short-read de nova assembler SOAPdenovo2 [28], we obtained a greater amberjack genome assembly that was 663 Mbp long, with 724 scaffolds (>2 kbp) and N50 of 4.99 Mbp (Table 1). This high-quality assembly was obtained because the terminal sequences in the mate pair library integrated DNA fragments longer than 20 kbp, which was useful in building a more complete genome. We also sequenced 10 kbp libraries on a PacBio RSII platform with 10-fold coverage and obtained 7.27 Gbp of polymerase reads and 7.26 Gbp of subreads. Then, we mapped the assembled PacBio RSII sequence data to the assembled HiSeq 2500 sequence data, which allowed us to reduce the number of gaps in the scaffolds by one-third (Table 1). The final genome assembly was 678 Mbp long, with N50 of 5.8 Mbp (Table 1; DDBJ: BDQW01000001–BDQW01034655).

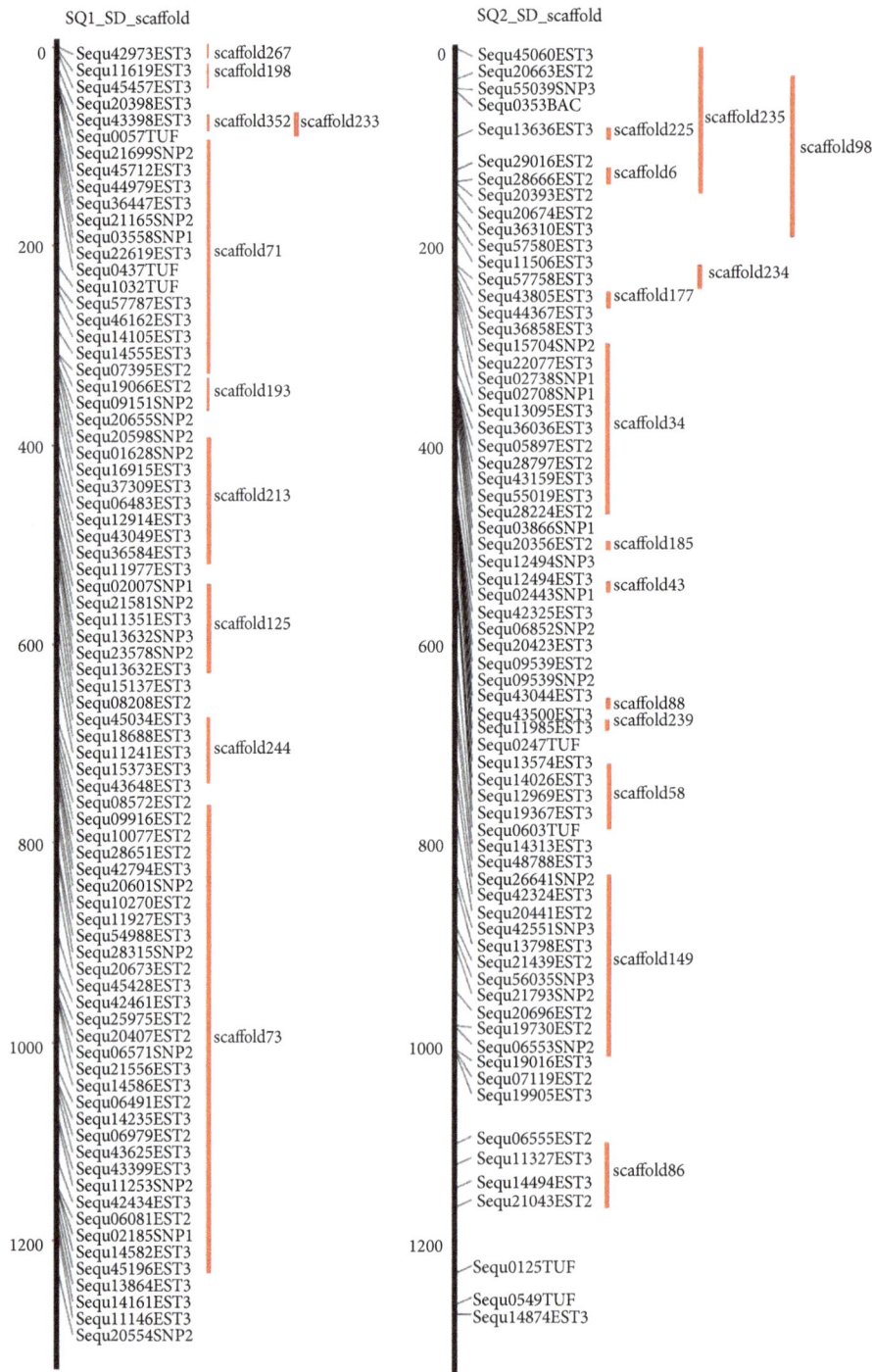

FIGURE 1: Greater amberjack scaffolds aligned onto two linkage groups of the yellowtail radiation hybrid physical map. A representative part of the yellowtail radiation hybrid (RH) physical map is shown with the greater amberjack scaffolds aligned. Numbers on the left indicate distance (cR) from the top of the RH map. Black lines indicate chromosomes. Red lines on the left indicate scaffold lengths. Seq numbers indicate mapped sequence number. Scaffold numbers identify the aligned scaffolds.

The Atlantic salmon (*Salmo salar*) reference genome assembly is 2.97 Gbp long with scaffold N50 of 2.97 Mbp [33]. For the common carp (*Cyprinus carpio*) reference genome assembly, the scaffold N50 is 1.0 Mbp [7]; for the channel catfish (*Ictalurus punctatus*) reference genome assembly, it is 7.73 Mbp [34]; for zebrafish (*Danio rerio*),

it is 1.55 Mbp [35]; for stickleback (*Gasterosteus aculeatus*), it is 10.8 Mbp [36]; and for medaka (*Oryzias latipes*), it is 5.1 Mbp [37]. Therefore, the quality of the greater amberjack reference genome assembly is equal to or better than that of the other fish reference genome assemblies that have been published so far.

TABLE 2: Number of mapped scaffolds and total length of the scaffolds of greater amberjack mapped to the 24 linkage groups onto yellowtail RH physical map.

LGNo	The number of mapped scaffold	Total length (bp) of mapped scaffolds on each LG (bp)
1	10	32,041,590
2	13	28,988,571
3	9	30,680,262
4	9	31,420,922
5	3	22,220,271
6	4	28,673,657
7	2	23,029,943
8	9	27,651,179
9	6	35,439,400
10	9	26,409,569
11	6	10,977,986
12	12	25,517,532
13	9	31,109,094
14	14	21,191,548
15	9	26,971,016
16	12	25,150,760
17	20	19,289,833
18	6	26,805,652
19	6	26,928,810
20	4	28,110,268
21	12	23,144,198
22	15	15,779,806
23	2	25,211,550
24	14	30,054,902
Total	215	622,798,319

LGNo indicates the linkage group number; number of mapped scaffold indicates the number of scaffolds mapped onto each linkage group; and total length of mapped scaffolds on each LG indicates the total length (bp) of the scaffold sequences mapped onto each linkage group.

TABLE 3: Summary of nonredundant SNPs mapped to the 24 linkage groups by resequencing 20 greater amberjack genomes.

LGNo	The number of mutations in each LG	The number of mapped SNPs onto each LG
1	373,859	7831
2	375,132	7968
3	360,266	7933
4	334,990	7725
5	269,110	7413
6	311,326	7640
7	262,037	7763
8	309,899	7802
9	383,874	7980
10	310,587	7791
11	132,260	7172
12	285,525	7602
13	323,954	7722
14	227,417	7583
15	259,349	7763
16	267,369	7882
17	253,519	7833
18	280,457	7785
19	300,795	7878
20	312,122	7870
21	273,678	7831
22	220,279	7863
23	300,003	7800
24	331,264	7829
Total	7,059,071	186,259

LGNo indicates the linkage group number; the number of mutations in each LG indicates the number of mutations found in each linkage group; and the number of mapped SNPs onto linkage group indicates the number of SNPs ordered onto each linkage group.

3.2. Alignment of the Assembled Amberjack Genome Sequence to the Yellowtail Radiation Hybrid Physical Map. Linkage maps indicate the genetic distances of mapped genes, whereas physical maps, where the distance between mapped genes reflects their physical distance on a genome, are more advantageous to accurately align scaffolds. However, to produce a physical map of greater amberjack using radiation hybrid (RH) panels will take a long time, because greater amberjack epidermal cells take 72 hours to divide, and a lot of time is needed to build greater amberjack RH panels. In a previous study [22], we reported RH physical and linkage maps of yellowtail (*Seriola quinqueradiata*) and mapped 300 to 600 bp long expressed sequence tags (ESTs) onto the yellowtail RH map. Greater amberjack is closely related to yellowtail evolutionarily, so we assumed that the genome sequences of greater amberjack and yellowtail will be highly homologous. We tried to map the assembled amberjack genome sequence onto the yellowtail RH map by BLAST searches using BWA software [31]. Then, we used the yellowtail ESTs for BLAST searches against the

scaffolds in the amberjack genome assembly. As a result, 215 of the longest amberjack sequences (total length 622.8 Mbp making up 92% of the total length of the genome scaffolds) were mapped onto the yellowtail RH map (Figure 1, Table 2, Supplementary Figure S2). Therefore, we considered that the yellowtail RH physical map was useful for lining up the assembled reference genome sequences of greater amberjack. This result suggested that the chromosome construction of greater amberjack and yellowtail may be conserved.

3.3. Resequencing of Amberjack Genomes and SNP Detection. We resequenced 20 amberjack genomes on an Illumina HiSeq 2500 platform and obtained an average of 18 Gbp of sequence data with about 27-fold coverage for each genome. The proportion of high-quality bases (≥Q30) to the total number of bases in each genome sequence was an average of 91%, and mean quality scores of the reads was an average of 38 (Supplementary Table S1). We mapped 99% of the resequenced data onto the amberjack reference genome using BWA software (Supplementary Table S2), which

(a)

(b)

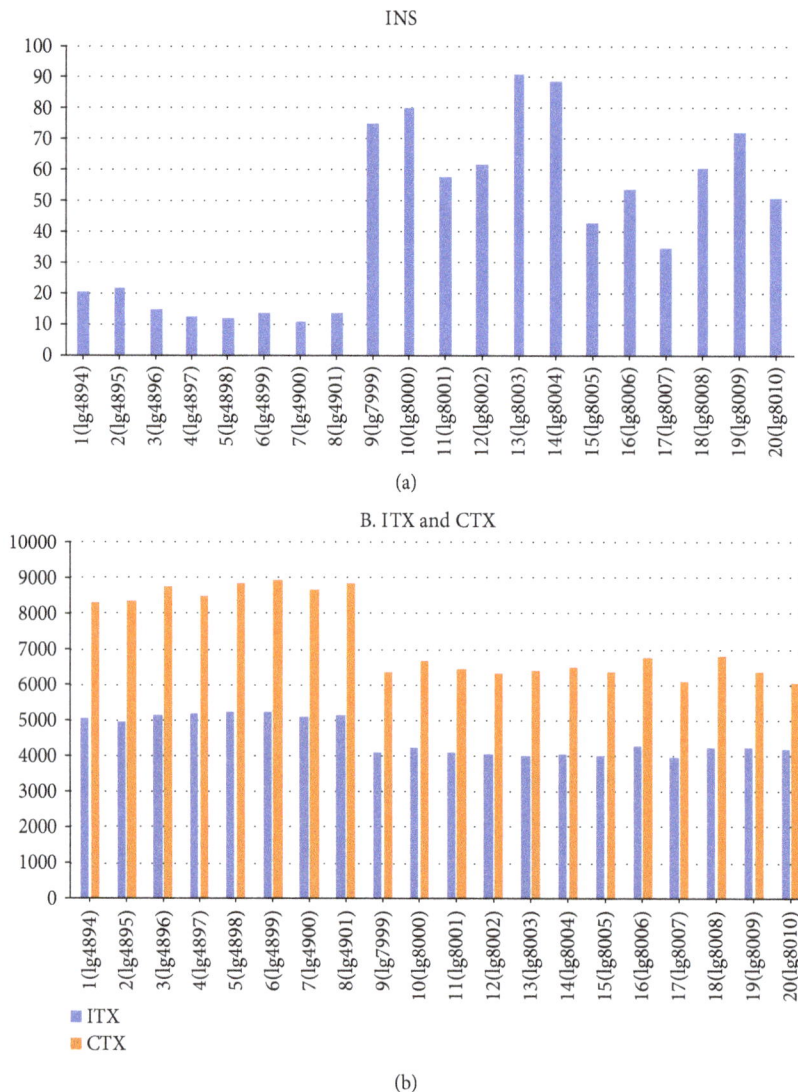

FIGURE 2: Comparison of structural variations in two greater amberjack populations. Greater amberjack samples 1–8 were captured in the sea near Kochi Prefecture, Japan. Greater amberjack samples 9–20 were captured in the sea near the Chinese Hainan Islands. (a) Number of insertions (INS) and (b) number of intra- (ITX) and inter- (CTX) chromosomal translocations detected in the 20 genomes. The vertical axis shows the number of structural variations.

resulted in 99.9% total coverage of the reference genome (Supplementary Table S3). We detected 7,059,071 mutations in the resequenced data mapped on the reference genome sequence by realigning the regions where mismatches were accumulated using Smatools and BCFtools (http://www.htslib.org/download/) and removing PCR duplicates using Picard (https://broadinstitute.github.io/picard/). We selected 186,259 nonredundant SNPs that showed polymorphisms at 10% or more in the 20 individuals and ordered them onto the reference genome sequence aligned on each linkage group (Table 3). There were 24 linkage groups corresponding to the 24 chromosomes.

Recently, SNP identification of Atlantic salmon, channel catfish, and common carp has been reported [11, 38, 39]. SNP discovery in Atlantic salmon was performed using extensive deep reduced representation sequencing, restriction site-associated DNA, and mRNA libraries derived from

farmed and wild Atlantic salmon samples, resulting in the discovery of >400 K putative SNPs, 132,023 of which were selected for an Affymetrix Axiom SNP array [38]. SNP identification in the common carp genome was performed separately for three strains by resequence, and a total of 24,272,905 nonredundant SNPs were detected, and 223,274 of them were selected for a carp SNP array system [39]. SNP identification in the catfish genome was performed within five strains by resequence, and 237,655 significant SNPs were detected and assigned to 29 tentative chromosomes based on the catfish linkage map [40]. Putative 2.12 million SNPs were identified in the rainbow trout genome within 12 fish by resequence, and 49,468 SNPs were finally selected for an Affymetrix SNP array [41]. We detected 7,059,071 mutations within the 20 greater amberjack genomes and selected 186,259 nonredundant SNPs for a SNP array. Thus, our SNP identification analysis of the

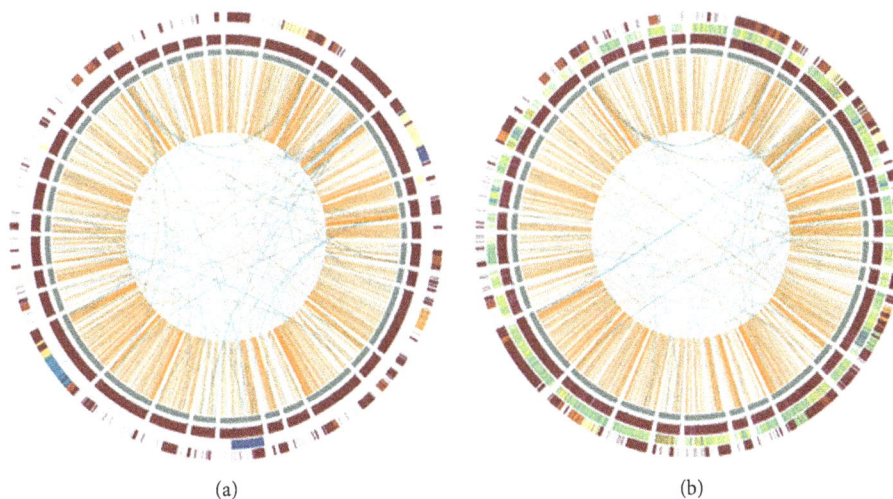

(a)   (b)

FIGURE 3: Genome-wide landscape of structural variations of greater amberjack. We linked the BreakDancer data to Circus plots to visualize regions of the genome that contained structural variations. We combined the resequenced data of (a) eight greater amberjacks captured off the Kochi coast and (b) 12 greater amberjacks captured off the Chinese Hainan Islands coast and analyzed these data by BreakDancer. Gene density of each contig is visualized by dark lines. The outermost circle shows inversion, next circle shows insertion, and third circle shows deletion. Orange lines show intrachromosomal translocations, and blue lines show interchromosomal translocations.

TABLE 4: Structural variations detected in the genomes of 20 greater amberjack.

Sample number	DEL	INS	INV	ITX	CTX
1(Ig4894)	6921	21	737	5042	8293
2(Ig4895)	7447	22	730	4954	8346
3(Ig4896)	6458	15	760	5147	8745
4(Ig4897)	7690	13	731	5213	8504
5(Ig4898)	6821	12	716	5262	8852
6(Ig4899)	7146	14	742	5243	8946
7(Ig4900)	7383	11	771	5118	8672
8(Ig4901)	6789	14	751	5159	8863
9(Ig7999)	5948	75	653	4107	6347
10(Ig8000)	6157	80	693	4236	6699
11(Ig8001)	6085	58	636	4133	6456
12(Ig8002)	5989	62	665	4061	6343
13(Ig8003)	6015	91	657	4027	6403
14(Ig8004)	6112	89	701	4056	6503
15(Ig8005)	5879	43	654	4025	6370
16(Ig8006)	6269	54	680	4290	6759
17(Ig8007)	6034	35	661	3990	6108
18(Ig8008)	6180	61	677	4261	6822
19(Ig8009)	5974	72	686	4230	6351
20(Ig8010)	6300	51	650	4201	6068

Structural variations were detected by BreakDancer using pair-end resequenced data for 20 greater amberjack genomes. Sample numbers 1–8 represent greater amberjacks captured near the Kochi coast, and sample numbers 9–20 represent greater amberjacks captured near Chinese Hainan Islands. The Ig numbers are the resequenced data analysis numbers. DEL: deletion; INS: insertion; INV: inversion; ITX: intrachromosomal translocation; and CTX: interchromosomal translocation.

greater amberjack genome produced SNP numbers that were about equal to those of the other aquaculture fish species.

To obtain molecular markers of shared genomic loci among individuals, many high-throughput technologies have been used to probe whole-genome polymorphisms, for example, SNP microarrays [10, 12], digital PCR [42], mass spectrometer SNP genotyping [43], and next generation sequencing [44]. The SNP information on the greater amberjack reference genome sequence that we obtained may be very useful for breeding amberjack by marker-assisted, gene-assisted, or genomic selection and for identifying functional genes for biological traits [45].

3.4. Detection of Structural Variants by BreakDancer. Recent developments in the analytical capacity of DNA sequencers have meant that massively parallel sequencing has been carried out in some species [46–48], making multiple data and methods available for the detection of structural variations. It has been suggested that small insertions/deletions and large structural variations may be major contributors to genetic diversity and traits [49–53].

We carried out BreakDancer [32] analysis to detect structural variants in the greater amberjack genome using resequenced data of the 20 greater amberjacks mapped onto the reference genome sequences, and the results are shown in Figures 2 and 3 and Table 4. BreakDancer Max can predict five types of structural variants (insertions, deletions, inversions, and intra- and interchromosomal translocations) from next generation short paired-end sequencing reads using read pairs that are mapped with unexpected separation distances or orientations [32]. Greater amberjack samples 1–8 (Figure 2, Table 4) were captured off the sea of Kochi Prefecture, Japan (Pacific Ocean side of Japan) and cultured in the Komame Station, National Research Institute of Aquaculture, and samples 9–20 (Figure 2, Table 4) were captured off the sea around the Chinese Hainan Islands and cultivated

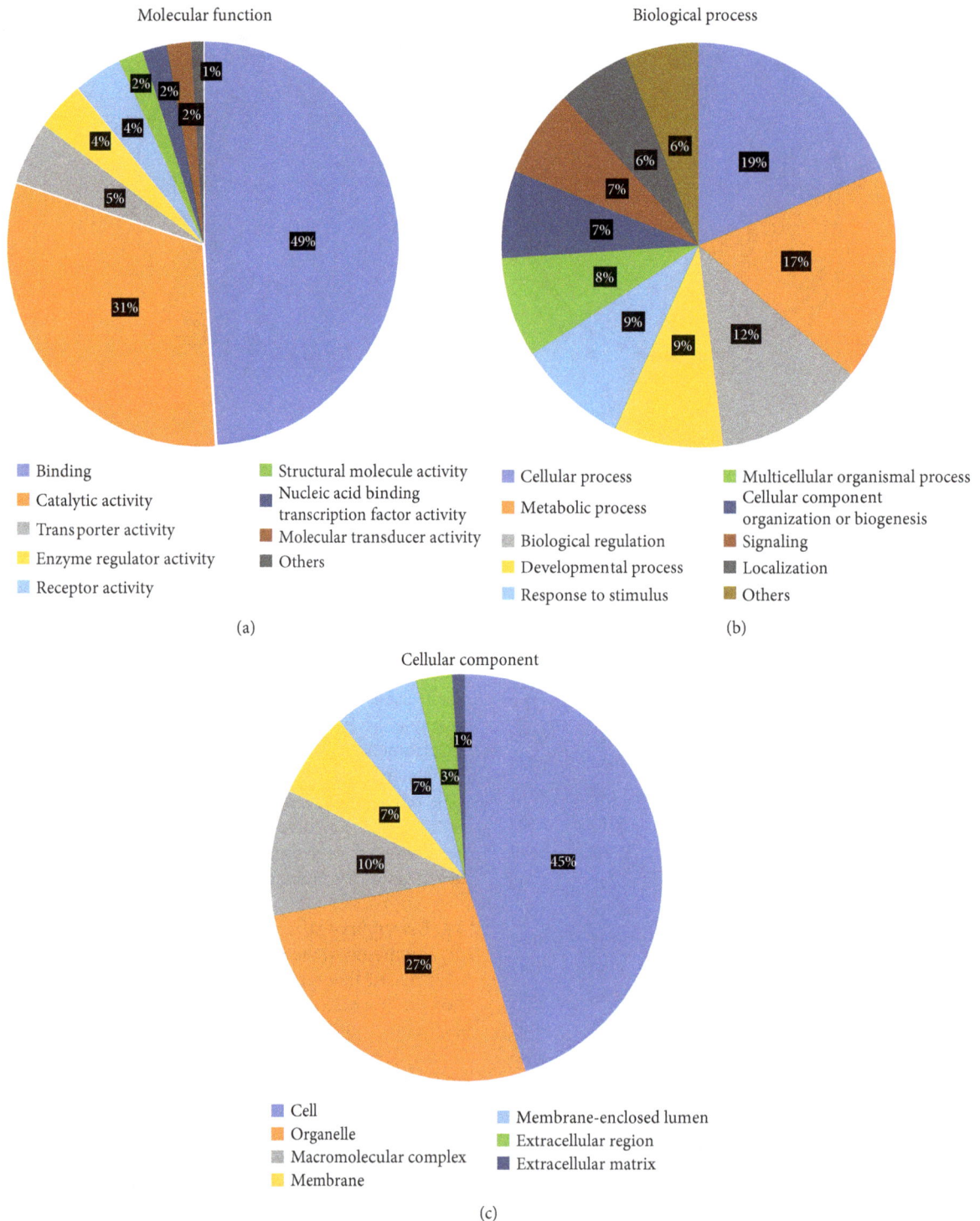

FIGURE 4: Gene ontology terms assigned to the assembled cDNA transcripts of greater amberjack. The gene ontology annotations under the three main categories: (a) molecular function, (b) biological process, and (c) cellular component.

in the marine crop of Owasebussan Co. Ltd. in Mie, Japan. We found an average of 19,083 structural variations per genome. The structural variations in the genomes of the Kochi amberjacks were different from those in the genomes of the Chinese amberjacks. More short insertions were detected in the Chinese amberjack genomes than in the Kochi amberjack genomes (Figure 2(a), Table 4), whereas

more intra- and interchromosome translocations were detected in the Kochi amberjack genomes than in Chinese amberjack genomes (Figure 2(b), Table 4). The total number of structural variations (average 21,634) in the Kochi amber-jack genomes were more than those (average 17,381) in the Chinese amberjack genomes. It is unclear why the greater amberjack captured off Kochi (Pacific Ocean side of Japan) had more chromosome translocations than the Chinese greater amberjack, and the greater amberjacks captured off China had many insertions in their genomes.

Greater amberjack is found in subtropical regions throughout the globe. In the Indo-West Pacific, this species has been reported from South Africa, the Persian Gulf, southern Japan, and the Hawaiian Islands, south to New Caledonia, and the Mariana and Caroline Islands in Micronesia. In the western Atlantic Ocean, greater amberjack is found off Nova Scotia, Canada south to Brazil including Bermuda, the Gulf of Mexico, and the Caribbean Sea. Gold and Richardson [54] analyzed the population structures of greater amberjack from the Gulf of Mexico and from the western Atlantic Ocean using variations of mitochondria DNA and suggested that one subpopulation existed in the northern Gulf of Mexico and a second subpopulation existed along the southeast Atlantic coast of America. Similarly, the results of our structural variation analysis may demonstrate that greater amberjack captured off the Hainan Islands in China, and Kochi in Japan could be classified as two subpopulations. If this is the case, the genome structural variations in the Chinese and Kochi greater amberjack subpopulations may have developed in different ways depending on the environment or population size because Hainan Islands in the South China Sea and Kochi on the Pacific side of Japan are geographically separated.

We combined the resequenced data of the eight Kochi amberjacks and combined the resequence data of the 12 Chinese amberjacks and analyzed separately their structure variations using BreakDancer. Then, we analyzed the distribution of the structural variations in the amberjack genomes using a Circus plot linked with the data obtained using BreakDancer (Figure 3). We found that the short insertions occurred genome wide in the Chinese amberjack (Figure 3(b)) and intra- and interchromosomal translocations occurred genome wide between narrow regions in the Kochi and Chinese amberjack populations (Figures 3(a) and 3(b)). There were some chromosome areas where intrachromosome translocations occurred in some individuals and interchromosome translocations occurred in other individuals (Figure 3(a), 3(b)).

This study is the first to analyze genome structural variations in fish species. However, we analyzed only 8 and 12 individuals in two populations. In future studies, we plan to resequence the genomes of more individuals to detect more accurately whole-genome structural variations in the greater amberjack genome.

3.5. Transcriptome Analysis. We analyzed the RNA sequences from 12 greater amberjack tissues (muscle, brain, eye, heart, liver, intestine, kidney, spleen, gonad, gill, fin, and bladder) by Illumina HiSeq 2500 sequencing. The trimmed reads were assembled by Cufflinks using the amberjack reference genome with BWA software [31]. We successfully mapped all the RNA sequences onto the reference genome sequence. After assembly, we obtained a total of 45,109 transcripts with N50 of 2624 bp; the longest contig

was 13,141 bp, and the average contig length was 1241 bp. BLASTX searches (E value $<1e-5$) were performed against the RefSeq vertebrate other sequences in NCBI using Blast2GO [55]. We identified 35,456 transcripts that shared high homology with RefSeq sequences (DDBJ: IACO01000001-IACO01045109), and each transcript was assigned to at least one gene ontology (GO) term under one of the three main categories: biological process, molecular function, and cellular component. The majority of assigned GO terms were under molecular function (41%), followed by biological process (35%), and cellular component (24%) (Figure 4). Under molecular function, binding and catalytic activity represented about 80% of the GO terms (Figure 4(a)); under biological process, cellular and metabolic processes (36%) and biological regulation (12%) were highly represented (Figure 4(b)); and under cellular component, cell and organelle represented about 72% of the GO terms (Figure 4(c)).

In a previous study, we reported the transcriptome assembly and GO analysis of yellowtail [22]. The GO analyses of the amberjack and yellowtail transcripts were similar and also similar to the results obtain for other fish species [56–58].

4. Conclusions

We carried out de novo genome sequence analysis and resequencing of 20 greater amberjacks to detect SNPs and to order the SNPs in aligned scaffolds. We obtained a high-quality genome assembly of 678 Mbp with N50 of 5.8 Mbp, and 215 scaffolds with 187,00 SNPs were ordered using the yellowtail RH physical map and homology between the yellowtail and greater amberjack genome sequences. Further, we analyzed structural variations in greater amberjack genomes using resequence data and found differences in the structural variations between two populations. We also analyzed the greater amberjack transcriptome, mapped the annotated sequences onto the reference genome sequence, and identified 35,456 transcripts that shared high homology with RefSeq sequences.

Conflicts of Interest

All authors declare that they have no competing interest.

Acknowledgments

The authors thank Owasebussan Cop. for providing the 12 greater amberjacks. The authors thank Margaret Biswas, PhD, from Edanz Group (http://www.edanzediting.com/ac) for editing a draft of this manuscript. This project was funded by a grant from the Ministry of Agriculture, Forestry and Fisheries of Japan (Research Project for Development of low-cost and stable production technology for promoting the export of cultured yellowtails).

Supplementary Materials

Supplementary 1. Figure S1: distribution of the lengths of the short read sequences obtained by HiSeq 2500 sequencing. Paired-end and mate pair libraries was constructed and sequenced using an Illumina HiSeq 2500 system. The vertical axis shows the percentage of each read length, and the horizontal axis shows the sequence length.

Supplementary 2. Figure S2: greater amberjack scaffolds aligned onto the linkage groups of the yellowtail radiation hybrid physical map. The yellowtail radiation hybrid (RH) physical map is shown with the greater amberjack scaffolds aligned. Numbers on the left indicate distance (cR) from the top of the RH map. Black line indicates chromosomes. Red lines on the left indicate scaffold lengths. Seq numbers indicate mapped sequence number. Scaffold numbers identify the aligned scaffold.

Supplementary 3. Table S1: summary of the quality of the 20 amberjack resequenced genomes. The 20 greater amberjack genomes were resequenced in parallel, and an average of 18 Gbp was obtained for each genome.

Supplementary 4. Table S2: summary of the results of mapping the resequenced reads on the greater amberjack reference genome. Overall, 99% of the resequenced reads in all 20 genomes were successfully mapped onto the reference genome sequence.

Supplementary 5. Table S3: summary of sequence coverage information. The coverage information was obtained by filtering the resequencing data and mapping it onto the greater amberjack reference genome sequence.

References

[1] W. S. Davidson, B. F. Koop, S. J. M. Jones et al., "Sequencing the genome of the Atlantic salmon (*Salmo salar*)," *Genome Biology*, vol. 11, p. 403, 2010.

[2] B. Star, A. J. Nederbragt, S. Jentoft et al., "The genome sequence of Atlantic cod reveals a unique immune system," *Nature*, vol. 477, no. 7363, pp. 207–210, 2011.

[3] C. Berthelot, F. Brunet, D. Chalopin et al., "The rainbow trout genome provides novel insights into evolution after whole-genome duplication in vertebrates," *Nature Communication*, vol. 5, article 3657, 2014.

[4] C. Shao, B. Bao, Z. Xie et al., "The genome and transcriptome of Japanese flounder provide insights into flatfish asymmetry," *Nature Genetics*, vol. 49, no. 1, pp. 119–124, 2017.

[5] S. Chen, G. Zhang, C. Shao et al., "Whole-genome sequence of a flatfish provides insights into ZW sex chromosome evolution and adaptation to a benthic lifestyle," *Nature Genetics*, vol. 46, no. 3, pp. 253–260, 2014.

[6] M. Schartl, R. B. Walter, Y. Shen et al., "The genome of the platyfish, *Xiphophorus maculatus*, provides insights into evolutionary adaptation and several complex traits," *Nature Genetics*, vol. 45, no. 5, pp. 567–572, 2013.

[7] P. Xu, X. Zhang, X. Wang et al., "Genome sequence and genetic diversity of the common carp, *Cyprinus carpio*," *Nature Genetics*, vol. 46, no. 11, pp. 1212–1219, 2014.

[8] X. Chen, L. Zhong, C. Bian et al., "High-quality genome assembly of channel catfish, *Ictalurus punctatus*," *GigaScience*, vol. 5, no. 1, pp. 1–4, 2016.

[9] J. E. Seeb, G. Carvalho, L. Hauser, K. Naish, S. Roberts, and L. W. Seeb, "Single-nucleotide polymorphism (SNP) discovery and applications of SNP genotyping in nonmodel organisms," *Molecular Ecology Resources*, vol. 11, no. s1, pp. 1–8, 2014.

[10] R. J. Lipshutz, S. P. Fodor, T. R. Gingeras, and D. J. Lockhart, "High density synthetic oligonucleotide arrays," *Nature Genetics*, vol. 21, pp. 20–24, 1999.

[11] Y. Hsin, D. Robledo, N. R. Lowe et al., "Construction and annotation of a high density SNP linkage map of the atlantic salmon (*Salmo salar*) genome," *Genes Genomics Genetics*, vol. 6, pp. 2173–2179, 2016.

[12] Q. Zeng, Q. Fu, Y. Li et al., "Development of a 690K SNP array in catfish and its application for genetic mapping and validation of the reference genome sequence," *Scientific Reports*, vol. 7, article 40347, pp. 1–14, 2017.

[13] Z. Liao, Q. Wan, X. Shang, and J. Su, "Large-scale SNP screenings identify markers linked with GCRV resistant traits through transcriptomes of individuals and cell lines in *Ctenopharygodon idella*," *Scientific Reports*, vol. 7, article 1184, 2017.

[14] W. Peng, J. Xu, Y. Zhang et al., "An ultra-high density linkage map and QTL mapping for sex and growth-related traits of common carp (*Cyprinus carpio*)," *Scientific Reports*, vol. 6, no. 1, article 26693, 2016.

[15] T. LaFramboise, "Single nucleotide polymorphism arrays: a decade of biological, computational and technological advance," *Nucleic Acids Research*, vol. 37, no. 13, pp. 4181–4193, 2009.

[16] M. D. Donato, S. O. Peters, S. E. Mitchell, T. Hussain, and L. G. Imumorin, "Genotyping-by-sequencing (GBS): a novel, efficient and cost-effective genotyping method for cattle using next-generation sequencing," *PLoS One*, vol. 8, no. 5, article e62137, 2013.

[17] P. M. P. van Poecke, M. Maccaferri, J. Tang et al., "Sequence-based SNP genotyping in durum wheat," *Plant Biotechnology Journal*, vol. 11, pp. 809–817, 2013.

[18] S. Das, H. D. Upadhyaya, D. Bajaj et al., "Deploying QTL-seq for rapid delineation of a potential candidate gene underlying major trait-associated QTL in chickpea," *DNA Research*, vol. 22, no. 3, pp. 193–203, 2015.

[19] N. Campbell, S. A. Harmon, R. Shawn, and S. R. Narum, "Genotyping-in-thousands by sequencing (GT-seq): a cost effective SNP genotyping method based on custom amplicon sequencing," *Molecular Ecology Resources*, vol. 15, no. 4, pp. 855–867, 2015.

[20] P. Muir, S. Li, S. Lou et al., "The real cost of sequencing: scaling computation to keep pace with data generation," *Genome Biology*, vol. 17, no. 53, pp. 1–9, 2016.

[21] J. Aoki, W. Kai, Y. Kawabata et al., "Construction of a radiation hybrid panel and the first yellowtail (*Seriola quinqueradiata*) radiation hybrid map using a nanofluidic dynamic array," *BMC Genomics*, vol. 15, p. 165, 2014.

[22] J. Aoki, W. Kai, Y. Kawabata et al., "Second generation physical and linkage maps of yellowtail (*Seriola quinqueradiata*)

and comparison of synteny with four model fish," *BMC Genomics*, vol. 16, article 406, 2015.

[23] Z. J. Liu and J. F. Cordes, "DNA marker technologies and their applications in aquaculture genetics," *Aquaculture*, vol. 238, no. 1-4, pp. 1-37, 2004.

[24] L. Coulibaly, K. Gharbi, R. Danzmann, J. Yao, and C. E. Rexroad, "Characterization and comparison of microsatellites derived from repeat-enriched libraries and expressed sequence tags," *Animal Genetics*, vol. 36, no. 4, pp. 309-315, 2005.

[25] A. P. Gutierrez, J. M. Yanez, S. Fukui, B. Swift, and W. S. Davidson, "Genome-wide association study (GWAS) for growth rate and age at sexual maturation in Atlantic salmon (*Salmo salar*)," *PLoS One*, vol. 10, no. 3, article e0119730, 2015.

[26] G.-F. Dianely, G. Guangtu, B. Matthew et al., "Genome-wide association study for identifying loci that affect fillet yield, carcass, and body weight traits in rainbow trout (*Oncorhynchus mykiss*)," *Frontiers Genetics*, vol. 22, no. 7, p. 203, 2016.

[27] A. M. Bolger, M. Lohse, and B. Usadel, "Trimmomatic: a flexible trimmer for Illumina sequence data," *Bioinformatics*, vol. 30, no. 15, pp. 2114-2120, 2014.

[28] R. Li, H. Zhu, J. Ruan et al., "De novo assembly of human genomes with massively parallel short read sequencing," *Genome Research*, vol. 20, no. 2, pp. 265-272, 2010.

[29] R. Kajitani, K. Toshimoto, H. Noguchi et al., "Efficient de novo assembly of highly heterozygous genomes from whole-genome shotgun short reads," *Genome Research*, vol. 24, no. 8, pp. 1384-1395, 2014.

[30] A. C. English, S. Richards, Y. Han et al., "Mind the gap: upgrading genomes with pacific biosciences RS long-read sequencing technology," *PLoS One*, vol. 7, no. 11, article e47768, 2012.

[31] H. Li and R. Durbin, "Fast and accurate short read alignment with burrows–wheeler transform," *Bioinformatics*, vol. 25, no. 14, pp. 1754-1760, 2009.

[32] K. Chen, J. M. Wallis, M. D. McLellan et al., "BreakDancer: an algorithm for high-resolution mapping of genomic structural variation," *Nature Methods*, vol. 6, no. 9, pp. 677-681, 2009.

[33] S. Lien, B. F. Koop, S. R. Sandve et al., "The Atlantic salmon genome provides insights into rediploidization," *Nature*, vol. 533, no. 7602, pp. 200-205, 2016.

[34] Z. Li, S. Liu, J. Yan et al., "The channel catfish genome sequence provides insights into the evolution of scale formation in teleosts," *Nature Communications*, vol. 7, article 11757, 2016.

[35] K. Howe, M. D. Clark, C. F. Torroja et al., "The zebrafish reference genome sequence and its relationship to the human genome," *Nature*, vol. 496, no. 7446, pp. 498-503, 2013.

[36] F. C. Jones, M. G. Grabherr, Y. F. Chan et al., "The genomic basis of adaptive evolution in threespine sticklebacks," *Nature*, vol. 484, no. 7392, pp. 55-61, 2012.

[37] H. Takeda, "Draft genome of the medaka fish: a comprehensive resource for medaka developmental genetics and vertebrate evolutionary biology," *Development, Growth & Differentiation*, vol. 50, no. s1, pp. S157-S166, 2008.

[38] R. D. Houston, J. B. Taggart, T. Cézard et al., "Development and validation of a high density SNP genotyping array for Atlantic salmon (*Salmo salar*)," *BMC Genomics*, vol. 15, no. 1, p. 90, 2014.

[39] J. Xu, Z. Zhao, X. Zhang et al., "Development and evaluation of the first high-throughput SNP array for common carp (*Cyprinus carpio*)," *BMC Genomics*, vol. 15, no. 1, p. 307, 2014.

[40] L. Sun, S. Llu, R. Wang et al., "Identification and analysis of genome-wide SNPs provide insight into signatures of selection and domestication in channel catfish (*Ictalurus punctatus*)," *PLoS One*, vol. 9, no. 10, article e109666, 2014.

[41] Y. Palti, G. Gao, S. Liu et al., "The development and characterization of a 57K single nucleotide polymorphism array for rainbow trout," *Molecular Ecology Resources*, vol. 15, no. 3, pp. 662-672, 2015.

[42] A. S. Whale, C. A. Bushell, P. R. Grant et al., "Detection of rare drug resistance mutations by digital PCR in a human influenza a virus model system and clinical samples," *Journal of Clinical Microbiology*, vol. 54, no. 2, pp. 392-400, 2016.

[43] X. Sun and B. Guo, "Genotyping single-nucleotide polymorphisms by matrix-assisted laser desorption/ionization time-of-flight-based mini-sequencing," *Methods in Molecular Medicine*, vol. 128, pp. 225-230, 2006.

[44] R. Nielsen, J. S. Paul, A. Albrechtsen, and Y. S. Song, "Genotype and SNP calling from next-generation sequencing data," *Nature Reviews Genetics*, vol. 12, no. 6, pp. 443-451, 2011.

[45] J. Seeb, G. Carvalho, L. Hauser, K. Naish, S. Roberts, and L. Seeb, "Single-nucleotide polymorphism (SNP) discovery and applications of SNP genotyping in nonmodel organisms," *Molecular Ecology Resources*, vol. 11, pp. 1-8, 2011.

[46] S. B. Ng, D. A. Nickerson, M. J. Bamshad, and B. J. Shendure, "Massively parallel sequencing and rare disease," *Human Molecular Genetics*, vol. 19, pp. R119-R124, 2010.

[47] J. P. Hamilton and C. R. Buell, "Advances in plant genome sequencing," *The Plant Journal*, vol. 70, no. 1, pp. 177-190, 2012.

[48] C. Drogemuller, J. Tetens, S. Sigurdsson et al., "Identification of the bovine Arachnomelia mutation by massively parallel sequencing implicates sulfite oxidase (SUOX) in bone development," *PLoS Genetics*, vol. 6, no. 8, article e1001079, 2010.

[49] H. Bai, Y. Cao, J. Quan et al., "Identifying the genome-wide sequence variations and developing new molecular markers for genetics research by re-sequencing a landrace cultivar of Foxtail millet," *PLoS One*, vol. 10, no. 9, article e73514, 2013.

[50] M. Boussaha, D. Esquerre, J. Barbieiri et al., "Genome-wide study of structural variants in Bovine Holstein, Montbéliarde and normande dairy breeds," *PLoS One*, vol. 10, no. 8, article e0135931, 2015.

[51] B. Yalcin, K. Wong, A. Agam et al., "Sequence-based characterization of structural variation in the mouse genome," *Nature*, vol. 477, no. 7364, pp. 326-329, 2011.

[52] A. C. English, W. J. Salerno, O. Hampton et al., "Assessing structural variation in a personal genome–towards a human reference diploid genome," *BMC Genomics*, vol. 16, no. 1, article 286, pp. 1-15, 2015.

[53] W. P. Kloosterman, L. C. Francioli, F. Hormozdiari et al., "Characteristics of de novo structural changes in the human genome," *Genome Research*, vol. 25, no. 6, pp. 792-801, 2015.

[54] J. R. Gold and L. R. Richardson, "Population structure in greater amberjack, *Seriola dumerili*, from the Gulf of Mexico and the Western Atlantic Ocean," *Fish Bull*, vol. 96, pp. 767-778, 1998.

[55] A. Conesa, S. Gotz, J. Miguel et al., "Blast2GO: a universal tool for annotation, visualization and analysis in functional genomics research," *Bioinformatics*, vol. 21, no. 18, pp. 3674-3676, 2005.

[56] A. Coppe, J. M. Pujolar, G. E. Maes et al., "Sequencing, *de novo* annotation and analysis of the first *Anguilla anguilla* transcriptome: EelBase opens new perspectives for the study of the critically endangered European eel," *BMC Genomics*, vol. 11, no. 1, article 635, 2010.

[57] M. Salem, C. E. Rexcoad, J. Wang, H. G. Thorgaard, and J. Yao, "Characterization of the rainbow trout transcriptome using Sanger and 454-pyrosequencing approaches," *BMC Genomics*, vol. 11, article 564, pp. 1–10, 2010.

[58] P. Pereiro, P. Balseiro, A. Romero et al., "High-throughput sequence analysis of turbot (*Scophthalmus maximus*) transcriptome using 454-pyrosequencing for the discovery of antiviral immune genes," *PLoS One*, vol. 7, no. 5, article e35369, 2012.

13

Comparative Genomics of the First and Complete Genome of "*Actinobacillus porcitonsillarum*" Supports the Novel Species Hypothesis

Valentina Donà and Vincent Perreten (iD)

Institute of Veterinary Bacteriology, Vetsuisse Faculty, University of Bern, Bern, Switzerland

Correspondence should be addressed to Vincent Perreten; vincent.perreten@vetsuisse.unibe.ch

Academic Editor: Marco Gerdol

"*Actinobacillus porcitonsillarum*" is considered a nonpathogenic member of the *Pasteurellaceae* family, which phenotypically resembles the pathogen *Actinobacillus pleuropneumoniae*. Previous studies suggested that "*A. porcitonsillarum*" may represent a new species closely related to *Actinobacillus minor*, yet no full genome has been sequenced so far. We implemented the Oxford Nanopore and Illumina sequencing technologies to obtain the highly accurate and complete genome sequence of the "*A. porcitonsillarum*" strain 9953L55. After validating our *de novo* assembly strategy by comparing the *A. pleuropneumoniae* S4074$^\mathrm{T}$ genome sequence obtained by Oxford Nanopore Technology combined with Illumina reads with a PacBio-sequenced S4074$^\mathrm{T}$ genome from the NCBI database, we performed comparative analyses of the 9953L55 genome with the *A. minor* type strain NM305$^\mathrm{T}$, *A. minor* strain 202, and *A. pleuropneumoniae* S4074$^\mathrm{T}$. The 2,263,191 bp circular genome of 9953L55 consisted of 2168 and 2033 predicted genes and proteins, respectively. The lipopolysaccharide cluster resembled the genetic organization of *A. pleuropneumoniae* serotypes 1, 9, and 11, possibly explaining the positive reactions observed previously in serotyping tests. In contrast to NM305$^\mathrm{T}$, we confirmed the presence of a complete *apxIICABD* operon in 9953L55 and 202 accounting for their hemolytic phenotype and Christie-Atkins-Munch-Petersen (CAMP) reaction positivity. Orthologous gene cluster analysis provided insight into the differential ability of strains of the *A. minor*/"*porcitonsillarum*" complex and *A. pleuropneumoniae* to ferment lactose, raffinose, trehalose, and mannitol. The four strains showed distinct and shared transposable elements, CRISPR/Cas systems, and integrated prophages. Genome comparisons based on average nucleotide identity and *in silico* DNA-DNA hybridization confirmed the close relationship among strains belonging to the *A. minor*/"*porcitonsillarum*" complex compared to other *Actinobacillus* spp., but also suggested that 9953L55 and 202 belong to the same novel species closely related to *A. minor*, namely, "*A. porcitonsillarum*." Recognition of the taxon as a separate species would improve diagnostics and control strategies of pig pleuropneumonia.

1. Introduction

"*Actinobacillus porcitonsillarum*" is a Gram-negative rod belonging to the *Pasteurellaceae* family, which is regularly isolated from the tonsils of healthy pigs and phenotypically resembles *Actinobacillus pleuropneumoniae*, the causative agent of porcine pleuropneumonia, which is associated with high economic burdens in the pig industry worldwide [1, 2]. "*A. porcitonsillarum*" mimics the major antigenic factors of *A. pleuropneumoniae* causing cross-reactivity in serological tests [1], which negatively affects serological diagnosis of *A. pleuropneumoniae*, potentially leading to the unnecessary depopulation and/or antimicrobial treatment of pig herds.

The "*A. porcitonsillarum*" strain 9953L55 (CCUG 46996) was firstly isolated from the tonsils of a healthy pig belonging to a high-health status herd considered to be free from *A. pleuropneumoniae*, in which regular serological testing suddenly evidenced a low number of pigs showing weak positive reactions for *A. pleuropneumoniae* serogroups 1, 9, and 11 [3]. Subsequent phenotypic and biochemical analyses indicated that this strain appeared to be identical to *A.*

pleuropneumoniae, including the hemolytic growth on blood agar plates and the Christie-Atkins-Munch-Petersen (CAMP) activity, i.e., a cohemolytic effect observed on blood agar plates in the presence of a sphingomyelinase (β-hemolysin)-producing *Staphylococcus aureus*, but with the exception that it did not ferment mannitol [1]. Serotyping by three different methods showed a positive reaction with antiserum raised against serotype 1 *A. pleuropneumoniae* S4074[T] [1]. Nevertheless, three *A. pleuropneumoniae*-specific PCRs were negative, indicating also the absence of the *apxIV* gene, which was previously proven to be species-specific for *A. pleuropneumoniae* [1, 4]. Toxin gene typing PCR for the major RTX toxins (ApxI, ApxII, and ApxIII) additionally revealed that *apxII*, but not *apxI* or *apxIII* genes were present [1].

Phylogenetic analysis of the 16S rRNA gene sequence indicated that "*A. porcitonsillarum*" was most closely related to *Actinobacillus minor* strain 202 (formerly named "*Haemophilus* strain 202", but subsequently classified as a borderline *A. minor* strain [5]), and to the *A. minor* type strain NM305[T], although it distinguished itself phenotypically from the latter by the hemolysis and CAMP activity [1]. Interestingly, a later study provided evidence that *A. minor* 202 also produced the ApxII toxin and appeared to be genetically rather more related to "*A. porcitonsillarum*" than to *A. minor* NM305[T] [6].

Despite these previous observations suggesting that "*A. porcitonsillarum*" may represent a new species, it has not been recognized as a distinct species so far, mainly due to the absence of sufficient phenotypic markers to distinguish it from *A. minor* [7]. However, a clear differentiation of the commensal "*A. porcitonsillarum*" from the pathogen *A. pleuropneumoniae* would be essential in diagnostics and, particularly, in eradication programs.

To corroborate these previous observations at a genomic level, we used the Oxford Nanopore and Illumina sequencing technologies to sequence the "*A. porcitonsillarum*" strain 9953L55, which was proposed as a type strain for "*A. porcitonsillarum*." After the validation of our *de novo* assembly approach by obtaining the genome of the *A. pleuropneumoniae* strain S4074[T] with the Oxford Nanopore Technology combined with Illumina reads and comparing it with the PacBio-sequenced genome of the same strain found in the NCBI database, we implemented this method to obtain the highly accurate circular genome sequence of strain 9953L55, which was further used for comparative analyses with the genome sequences of *A. minor* 202, *A. minor* NM305[T], and *A. pleuropneumoniae* S4074[T].

2. Materials and Methods

2.1. Bacterial Strains, Growth Conditions, and Sugar Fermentation Test. "*A. porcitonsillarum*" 9953L55 (CCUG 46996), *A. minor* NM305[T] (CCUG 38923[T]), and *A. pleuropneumoniae* S4074[T] (ATCC 27088[T]) were grown on chocolate agar plates supplemented with Polyvitex (BioMérieux) at 37°C with 5% CO_2. Lactose, raffinose, and trehalose fermentation was assessed, using S4074[T] and NM305[T] as control

strains, in PPLO broth (Difco) supplemented with 40 μg/ml NAD, as described previously [1, 8].

2.2. DNA Isolation and Sequencing. DNA was isolated with a modified phenol/chloroform extraction method, treated for 30 min with 0.5 μl RNase (20 mg/ml) (Qiagen), and purified with 0.8X Agencourt AMPure beads (Beckman Coulter) [9]. The purified DNA was subsequently sheared to 8–10 kb fragments with a g-TUBE (Covaris), and library preparation was performed with the SQK-LSK108 1D ligation sequencing kit (Oxford Nanopore), as per the manufacturer's instructions. The sequencing library was sequenced on a R9.4 SpotON flow cell (Oxford Nanopore) with the MinION Mk 1B sequencing device (Oxford Nanopore) for 24 hours. In parallel, the DNA was also submitted to GATC, Constance, Germany, for 2 × 150 paired-end sequencing on an Illumina HiSeq (Illumina) platform.

2.3. Genome Assembly. Base calling and quality filtering of the Oxford Nanopore Technology (ONT) reads were performed with Albacore v2.0.1. Pairing, trimming, and quality filtering of the Illumina reads were performed with Trimmomatic v0.33. ONT reads were assembled with Canu v1.3 with default parameters and the option corOutCoverage = 100 [10]. Paired-end Illumina reads were mapped to the Canu-generated scaffold with BWA-MEM v0.7.13 and polished with Pilon 1.22 twice [11]. A third mapping of the Illumina reads was performed with BWA-MEM for the final inspection and curation of the polished sequence with the Geneious software v10.2.3 (Biomatters). In case of repetitive regions leading to unbalanced (low) read coverage, these regions were extracted to locally repeat read mapping with BWA-MEM. The final circular genome sequences of strains 9953L55 and S4074[T] were first annotated with Prokka v1.12 for primary sequence analysis and subsequently with the NCBI prokaryotic genome annotation pipeline [12]. Paired-end Illumina reads were used to run plasmidSPAdes v3.9.0 with default parameters [13].

2.4. Genome Analysis and Comparison. The whole-genome shotgun and complete genome sequences, which were retrieved from the NCBI database for the genome comparisons, are deposited under the following GenBank accession numbers: *A. minor* NM305[T] (ACQL01000001-ACQL01000197), *A. minor* 202 (ACFT01000001-ACFT01000154), *A. pleuropneumoniae* S4074[T] (PacBio, CP029003; Roche 454, ADOD01000001-ADOD01000044), *A. equuli* 19392[T] (CP007715), *A. succinogenes* 130Z[T] (NC_009655), *A. suis* ATCC 33415[T] (NZ_CP009159), *A. ureae* ATCC 25976[T] (AEVG01000001-AEVG01000183), *A. capsulatus* DSM 19761[T] (ARFN01000001-ARFN01000049), and *A. seminis* ATCC 15768[T] (NLFK01000001-NLFK01000022). Genome alignments were performed with progressiveMauve v2.3.1 [14]. OrthoVenn was used to identify orthologous genes [15]. Online available platforms were used to characterize the presence of known resistance genes (ResFinder) [16], plasmids (PlasmidFinder) [17], insertion sequences (IS, ISfinder) [18], clustered regularly interspaced short palindromic repeat (CRISPR) arrays and CRISPR-

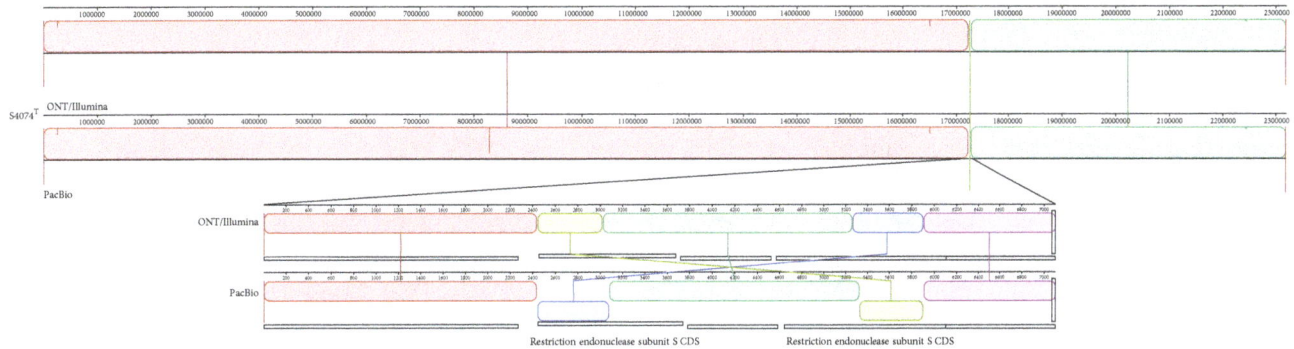

FIGURE 1: MAUVE alignment of the genome sequence of *A. pleuropneumoniae* S4074[T] obtained by Oxford Nanopore Technology (ONT)/ Illumina (top) and PacBio (bottom) sequencing. The same color boxes, i.e., locally collinear blocks (LCB), represent homologous regions of sequence without rearrangement. The inset underneath magnifies the only rearrangement found between the two sequences.

associated gene (CRISPR/Cas) systems (CRISPRone) [19], and phage sequences (PHASTER) [20]. The circular map of 9953L55 including the BLAST-based comparison with the genome sequences of *A. minor* NM305[T], *A. minor* 202, and *A. pleuropneumoniae* S4074[T] was generated with the BLAST Ring Image Generator (BRIG) [21]. Comparisons of average nucleotide identity (ANI) based on BLAST and MUMmer pairwise sequence alignments (ANIb and ANIm, respectively) were obtained with JSpeciesWS [21, 22]. The distance matrix representing the ANI divergence (defined as 100% − ANI) was used to compute a complete linkage hierarchical clustering with the hclust function in R v3.0.1, as done previously [23]. *In silico* DNA-DNA hybridization (*is*DDH) based on genome BLAST distance phylogeny was performed with GGDC 2.1 [24]. Only results based on formula 2 were used for analysis, since it estimates *is*DDH values independently of genome length and is therefore recommended for incomplete genomes [25, 26].

2.5. Nucleotide Sequence GenBank Accession Number. The complete nucleotide sequences of the "*A. porcitonsillarum*" strain 9953L55 and of the *A. pleuropneumoniae* strain S4074[T] were deposited in DDBJ/EMBL/GenBank under the accession numbers CP029206 and CP030753, respectively.

3. Results and Discussion

After base calling and quality filtering, 580,932 1D pass ONT reads corresponding to 4.08 Gbp (>1800X coverage) and 10,478,015 paired-end Illumina reads were obtained for the *A. pleuropneumoniae* strain S4074[T]. Assembly of the ONT reads generated a single 2.32 Mbp contig with overlapping ends, which was circularized and polished with paired-end Illumina reads.

Alignment of the obtained S4074[T] genome with the complete genome sequence of the same strain previously sequenced with PacBio technology, which we retrieved from the NCBI database (accession number CP029003), indicated a very high sequence homology (Figure 1). Only one rearrangement was identified between the two genome sequences, which mapped to the 5′-end region of two genes in opposite orientation both encoding a restriction

endonuclease subunit S (Figure 1). High sequence divergence was observed solely in a 5 kbp region comprising 5 genes encoding a D-alanine–D-alanine ligase, cell division proteins (*ftsQ*, *ftsA*, and *ftsZ*), and the UDP-3-O-[3-hydroxymyristoyl] N-acetylglucosamine deacetylase (*lpxC*). All 5 genes are annotated as frameshifted and contain internal stop codons in the PacBio-generated S4074[T] genome sequence. In contrast, all genes were intact in the genome obtained by ONT/Illumina sequencing, and a comparison with a previous whole-genome shotgun assembly of the same strain obtained by Roche 454 sequencing technology, which we retrieved from the NCBI database (accession number ADOD01000001-ADOD01000044), showed 100% identity. A 5 bp indel was found in a gene encoding a methyltransferase annotated as incomplete in the PacBio-generated genome sequence, restoring the completeness of the gene in the genome obtained by ONT/Illumina sequencing. Only 2 additional single nucleotide polymorphisms (SNPs) and one indel (in an intergenic region) were identified between the two genomes.

Taken together, these results confirmed that our approach, i.e., *de novo* assembly of ONT reads combined with Illumina polishing, can be successfully applied to generate a complete and highly accurate bacterial genome sequence. Therefore, this strategy was further used to obtain the full genome sequence of the "*A. porcitonsillarum*" strain 9953L55.

After base calling and quality filtering, 721,267 1D pass ONT reads corresponding to 4.82 Gbp (>2000X coverage) and 5,367,150 paired-end Illumina reads were obtained for strain 9953L55. Assembly of the ONT reads generated a single 2.26 Mbp contig with overlapping ends, which was circularized and polished with paired-end Illumina reads to obtain the complete circular sequence of the "*A. porcitonsillarum*" 9953L55 chromosome, as done for S4074[T]. *In silico* analysis with PlasmidFinder and plasmidSPAdes using paired-end Illumina reads suggested the absence of plasmids.

The circular genome of 9953L55 consisted of 2,263,191 bp with an average 39.7% GC content and displayed, as expected, a high nucleotide sequence similarity with the genome sequences of *A. minor* NM305[T] and 202, while *A. pleuropneumoniae* S4074[T] was more dissimilar

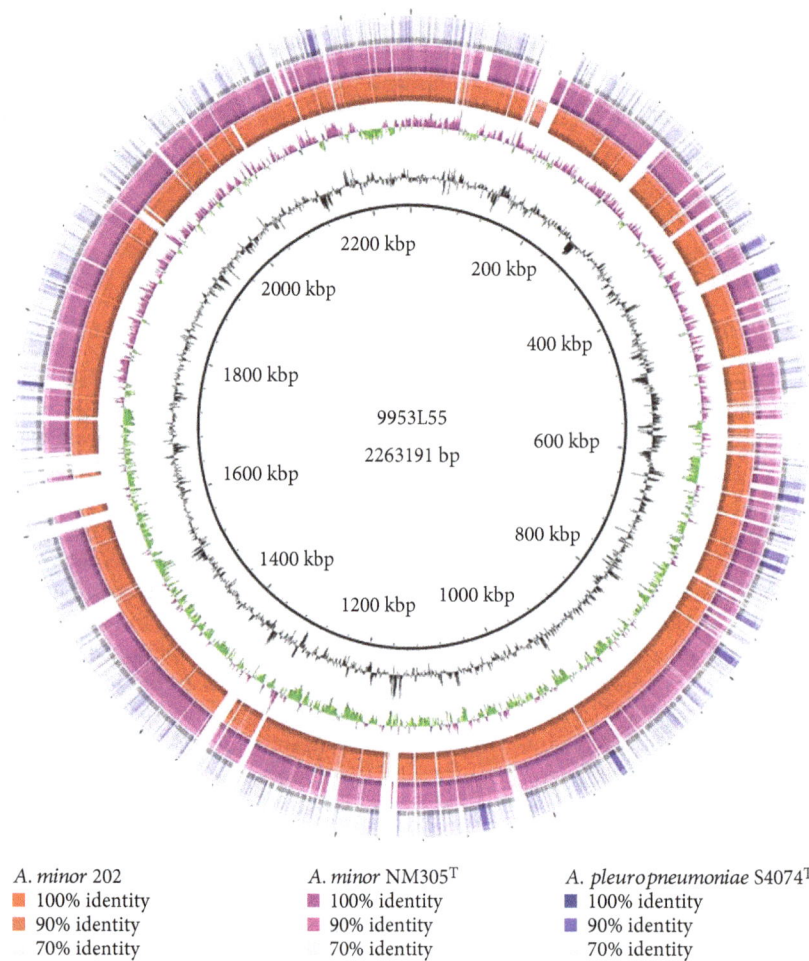

Figure 2: Circular map of the "*A. porcitonsillarum*" strain 9953L55. The scale ring shows the coordinates in kilobase pairs. The second ring represents the average GC content. The third ring represents the GC skew. The colored outer rings display regions of homology based on BLASTn. First outer ring (red): *A. minor* 202, second outer ring (pink): *A. minor* NM305[T], and third outer ring (blue): *A. pleuropneumoniae* S4074[T].

(Figure 2). In total, 2168 genes, including six copies of the *rrn* operon encoding the 16S, 23S, and 5S rRNA, as well as 2087 CDS, of which 2033 encode proteins and 54 are pseudogenes, were predicted.

Analysis with OrthoVenn showed that the four strains display 1523 common clusters of orthologous genes (COGs), of which 1507 were single-copy clusters, indicating few duplication events before speciation (Figure 3). The two *A. minor* strains and "*A. porcitonsillarum*" 9953L55 shared 130 additional COGs, reflecting their closer phylogenetic relationship compared to *A. pleuropneumoniae*, as suggested previously [1]. Interestingly, most of the COGs present only in 9953L55 and S4074[T] were genes belonging to the lipopolysaccharide (LPS) cluster, which was located (as in the other three strains) between the *erpA* and *rpsU* genes and closely resembled the genetic organization of the *A. pleuropneumoniae* serotype 1, 9, and 11 LPS cluster. This may explain the previously observed cross-reactivity with antiserum against S4074[T] and, particularly, the positive reaction in the dot-ELISA test with a monoclonal antibody recognizing a common *O*-chain LPS epitope of *A. pleuropneumoniae* serotypes 1, 9, and 11 [1].

The further analysis of the COGs shed some light on the different phenotypes observed in the biochemical tests [1, 5, 27].

RTX toxins (ApxI, ApxII, and ApxIII) in *A. pleuropneumoniae* are responsible for its hemolytic activity and CAMP positivity [28, 29]. Orthologs for the *apxIICA* genes were identified in all strains but not in *A. minor* NM305[T]. In fact, an intact and a complete *apxIICABD* operon was located between the *aspC* and *folC* genes in both "*A. porcitonsillarum*" 9953L55 and *A. minor* 202, but no *apxI*, *apxIII*, or *apxIVA* genes were found, consistent with previous observations [1, 7, 30]. This *apxIICABD* operon was shown to be responsible and sufficient for RTX toxin ApxII expression and secretion and, consequently, for their hemolytic phenotype [7, 30].

Regarding the main differences in sugar utilization, orthologs encoding the *β*-galactosidase were present in all four strains accounting for their positive reaction with the o-nitrophenyl-*β*-D-galactopyranoside (ONPG) test [1, 5]. However, a full *lac* operon, i.e., including genes encoding the transcriptional regulator (*lacI*), the lactose permease (*lacY*), and the *α*-galactosidase (*melA*), was identified only

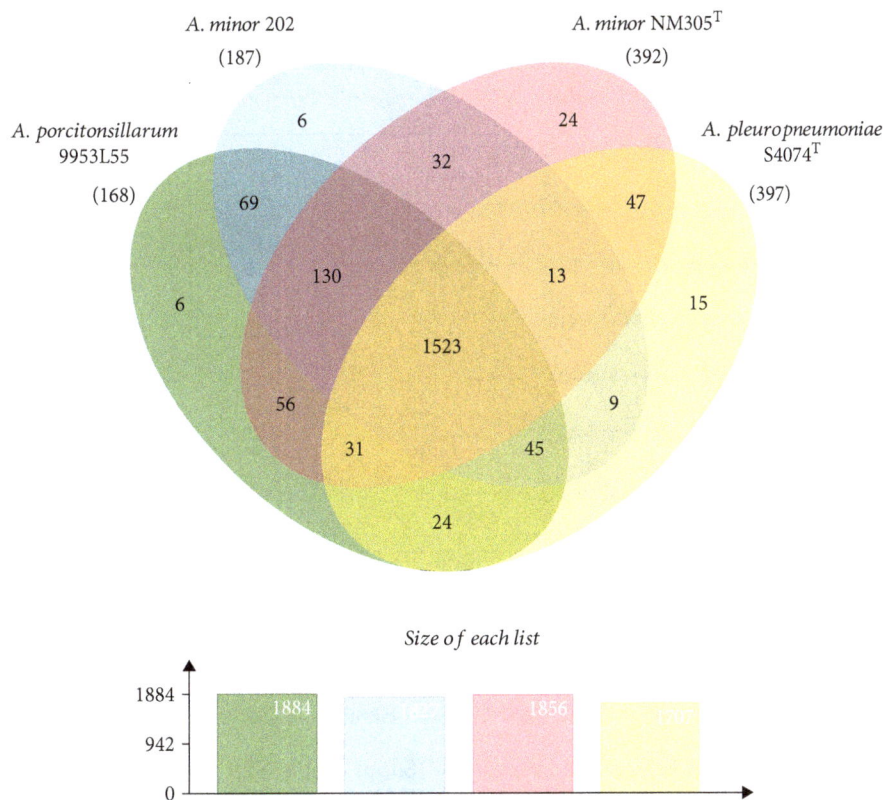

FIGURE 3: Venn diagram depicting clusters of orthologous genes (COGs) in "*A. porcitonsillarum*" 9953L55, *A. minor* 202, *A. minor* NM305[T], and *A. pleuropneumoniae* S4074[T]. The number of singletons for each strain is shown in brackets. The total number of COGs for each strain is displayed in the graph underneath.

in the "*A. porcitonsillarum*" and in both *A. minor* strains. The absence of the lactose permease and the α-galactosidase provides an explanation for the inability of *A. pleuropneumoniae* to ferment lactose and raffinose, respectively, in contrast to most *A. minor* strains [5, 27]. Consistently, we confirmed by testing lactose and raffinose fermentation that "*A. porcitonsillarum*" 9953L55 also produces acid from both sugars.

While *A. pleuropneumoniae* does not ferment trehalose, most *A. minor* strains are trehalose fermenters [5]. We identified a full *tre* operon with genes encoding the HTH transcriptional regulator (*treR*), the PTS trehalose transporter (*treP*), and the trehalose-6-phosphate hydrolase (*treA*) only in "*A. porcitonsillarum*" 9953L55 and *A. minor* NM305[T], suggesting that both strains are able to import and ferment trehalose. As expected, when testing their ability to utilize trehalose, acid production was observed for both 9953L55 and NM305[T], but not for S4074[T].

On the other hand, we found no orthologs in "*A. porcitonsillarum*" 9953L55 and in both *A. minor* strains for the *mtlD* and *mtlA* genes, which code for the PTS mannitol transporter and the mannitol-1-phosphate-5-dehydrogenase in *A. pleuropneumoniae*, respectively, providing an explanation for their inability to assimilate and/or ferment mannitol [1, 5].

Most COGs shared only by "*A. porcitonsillarum*" 9953L55 and *A. minor* 202 were genes involved in different metabolic pathways, iron transport, response to stimuli, and quorum sensing. However, we also identified many

orthologs for genes related to the CRISPR/Cas system, which represents the bacterial adaptive immune system against phages. Further analysis showed that both 9953L55 and 202 possess a subtype I-C CRISPR/Cas system, including a CRISPR array containing 37 repeat units in the "*A. porcitonsillarum*" strain. In contrast, complete subtypes II-C and I-F (yet in a particular genetic rearrangement) were identified in *A. minor* NM305[T] and *A. pleuropneumoniae* S4074[T], respectively.

Regarding phages, only one intact HP2-related *Haemophilus* prophage and an incomplete prophage region of 6.2 kb were identified in "*A. porcitonsillarum*" 9953L55 between positions 1,610,930–1,645,906 and 1,650,425–1,656,579, respectively. Interestingly, different intact *Haemophilus* as well as enterobacterial prophages were found in *A. minor* NM305[T] and *A. pleuropneumoniae* S4074[T], but not in *A. minor* 202 (data not shown).

Among the COGs identified exclusively in "*A. porcitonsillarum*" 9953L55 and *A. minor* NM305[T], the IS*Apl1* was the most abundant with 10 copies present in the genome of the "*A. porcitonsillarum*" strain. This IS is typically found in *Actinobacillus* spp. and has been recently associated with the widespread of the colistin-resistance gene *mcr-1* in different genetic backgrounds [31, 32]. However, in 9953L55, the IS*Apl1* did not flank any known antibiotic resistance genes.

Nevertheless, both "*A. porcitonsillarum*" 9953L55 and *A. minor* NM305[T] strains possessed a tetracycline resistance operon containing *tet*(B), which was located on a Tn*10*

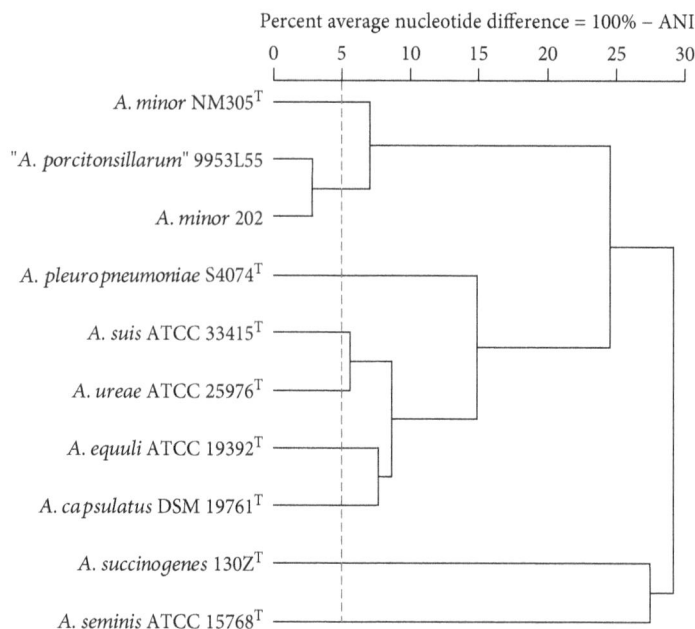

FIGURE 4: Cluster analysis of average nucleotide identity (ANI) values obtained by BLAST-based pairwise comparisons of 10 *Actinobacillus* spp. genome sequences. The distance matrix representing the ANI divergence (defined as 100% – ANI) was used for the complete linkage hierarchical clustering. The vertical dashed line represents the 95% species cutoff value.

mobile element flanked by two IS*Vsa5* in opposite orientation [33]. This mobile element, which is widely disseminated among different bacterial species, was also found on integrative conjugative elements in *A. pleuropneumoniae* (ICE*Apl1*) and other *Pasteurellaceae*, such as *Haemophilus parainfluenzae* (ICE*Hpa*T3T1) [34, 35].

Regarding antimicrobial resistance, we note that in "*A. porcitonsillarum*" 9953L55, *tet*(B) was the only resistance gene identified by the *in silico* analysis with ResFinder. However, we additionally identified in this strain, as well as in *A. minor* 202, an ortholog encoding a major facilitator superfamily (MFS) transporter (LmrB) potentially associated with lincomycin resistance.

Pairwise comparisons of the genome sequences of these four strains and the type strains of six other *Actinobacillus* spp. based on ANI confirmed the close relationship between "*A. porcitonsillarum*" and *A. minor* (Figure 4). However, ANIb values for "*A. porcitonsillarum*" 9953L55 and *A. minor* 202 were above the 95% species criteria between each other, but were below for both strains when compared with *A. minor* NM305^T (Suppl. Table S1) [24, 36], indicating that 9953L55 and 202 may belong to a distinct new species closely related to *A. minor*, as suggested previously [1, 7]. Since ANIm may be more robust for genomes sharing >90% sequence similarity [37], we also implemented ANIm for pairwise comparisons of "*A. porcitonsillarum*" 9953L55 and both *A. minor* strains. The ANIm values for 9953L55 and 202 correlated well with the ANIb values; that is, both were 97.3% between each other but 93.6% when compared with NM305^T, supporting once more the novel species hypothesis.

The same conclusions were drawn also from *is*DDH based on genome BLAST distance phylogeny, with only "*A. porcitonsillarum*" 9953L55 and *A. minor* 202 exhibiting *is*DDH values > 70%, i.e., above the same-species threshold

(Suppl. Table S2) [25]. Intriguingly, in a previous study, DNA-DNA relatedness assessed by a classic DNA-DNA hybridization (DDH) method showed borderline species-level values for 202 compared with other *A. minor* strains and, in particular, DDH values < 70% and a melting temperature difference > 5°C for 202 compared with NM305^T, already indicating that these two strains may not belong to the same species [5, 38]. Of note, it has been suggested previously that ANI values ≥ 96% and *is*DDH values > 70% (at the upper 95% confidence interval) are good predictors of the same-species genomes in *Aeromonas* spp. [39].

4. Conclusions

In conclusion, we implement herein the ONT and Illumina sequencing technologies to obtain the first, complete, and highly accurate genome sequence of an "*A. porcitonsillarum*" strain and highlight its main features and differences compared to those of *A. pleuropneumoniae* and *A. minor*. Pairwise genome comparisons of 9953L55 with *A. minor* 202 and NM305^T based on both ANI and *is*DDH support previous observations that "*A. porcitonsillarum*" should be recognized as a new species closely related to *A. minor*, to which strain 202 also belongs. This would be essential to clearly differentiate this nonpathogenic species from the pathogenic *A. pleuropneumoniae* in diagnostic settings and, consequently, in eradication programs.

Conflicts of Interest

The authors declare that there is no conflict of interest regarding the publication of this paper.

Acknowledgments

This work was supported by CTI grant number 25291.2 PFLS-LS. We thank Dr. Marcelo Gottschalk for kindly donating the "*A. porcitonsillarum*" strain 9953L55 and Dr. Joachim Frey for advice.

References

[1] M. Gottschalk, A. Broes, K. R. Mittal et al., "Non-pathogenic *Actinobacillus* isolates antigenically and biochemically similar to *Actinobacillus pleuropneumoniae*: a novel species?," *Veterinary Microbiology*, vol. 92, no. 1-2, pp. 87–101, 2003.

[2] E. L. Sassu, J. T. Bossé, T. J. Tobias, M. Gottschalk, P. R. Langford, and I. Hennig-Pauka, "Update on *Actinobacillus pleuropneumoniae*—knowledge, gaps and challenges," *Transboundary and Emerging Diseases*, vol. 65, no. S1, pp. 72–90, 2018.

[3] M. Gottschalk, E. Altman, N. Charland, F. De Lasalle, and J. D. Dubreuil, "Evaluation of a saline boiled extract, capsular polysaccharides and long-chain lipopolysaccharides of *Actinobacillus pleuropneumoniae* serotype 1 as antigens for the serodiagnosis of swine pleuropneumonia," *Veterinary Microbiology*, vol. 42, no. 2-3, pp. 91–104, 1994.

[4] A. Schaller, S. P. Djordjevic, G. J. Eamens et al., "Identification and detection of *Actinobacillus pleuropneumoniae* by PCR based on the gene *apx*IVA," *Veterinary Microbiology*, vol. 79, no. 1, pp. 47–62, 2001.

[5] K. Møller, V. Fussing, P. A. D. Grimont, B. J. Paster, F. E. Dewhirst, and M. Kilian, "*Actinobacillus minor* sp. nov., *Actinobacillus porcinus* sp. nov., and *Actinobacillus indolicus* sp. nov., three new V factor-dependent species from the respiratory tract of pigs," *International Journal of Systematic Bacteriology*, vol. 46, no. 4, pp. 951–956, 1996.

[6] G. Arya and D. F. Niven, "Production of haemolysins by strains of the *Actinobacillus minor*/"*porcitonsillarum*" complex," *Veterinary Microbiology*, vol. 141, no. 3-4, pp. 332–341, 2010.

[7] P. Kuhnert and H. Christensen, "International Committee on Systematics of Prokaryotes. Subcommittee on the taxonomy of *Pasteurellaceae*: minutes of the meetings, 6 August 2008, Istanbul, Turkey," *International Journal of Systematic and Evolutionary Microbiology*, vol. 59, no. 1, pp. 202-203, 2009.

[8] P. Kielstein, H. Wuthe, O. Angen, R. Mutters, and P. Ahrens, "Phenotypic and genetic characterization of NAD-dependent *Pasteurellaceae* from the respiratory tract of pigs and their possible pathogenetic importance," *Veterinary Microbiology*, vol. 81, no. 3, pp. 243–255, 2001.

[9] J. Sambrook and D. W. Russell, "Purification of nucleic acids by extraction with phenol:chloroform," *Cold Spring Harbor Protocols*, vol. 2006, no. 1, 2006.

[10] S. Koren, B. P. Walenz, K. Berlin, J. R. Miller, and A. M. Phillippy, "Canu: scalable and accurate long-read assembly via adaptive *k*-mer weighting and repeat separation," *Genome Research*, vol. 27, no. 5, pp. 722–736, 2017.

[11] B. J. Walker, T. Abeel, T. Shea et al., "Pilon: an integrated tool for comprehensive microbial variant detection and genome assembly improvement," *PLoS One*, vol. 9, no. 11, article e112963, 2014.

[12] T. Seemann, "Prokka: rapid prokaryotic genome annotation," *Bioinformatics*, vol. 30, no. 14, pp. 2068-2069, 2014.

[13] D. Antipov, N. Hartwick, M. Shen, M. Raiko, A. Lapidus, and P. A. Pevzner, "plasmidSPAdes: assembling plasmids from whole genome sequencing data," *Bioinformatics*, vol. 32, no. 22, pp. 3380–3387, 2016.

[14] A. E. Darling, B. Mau, and N. T. Perna, "progressiveMauve: multiple genome alignment with gene gain, loss and rearrangement," *PLoS One*, vol. 5, no. 6, article e11147, 2010.

[15] Y. Wang, D. Coleman-Derr, G. Chen, and Y. Q. Gu, "OrthoVenn: a web server for genome wide comparison and annotation of orthologous clusters across multiple species," *Nucleic Acids Research*, vol. 43, no. W1, pp. W78–W84, 2015.

[16] E. Zankari, H. Hasman, S. Cosentino et al., "Identification of acquired antimicrobial resistance genes," *The Journal of Antimicrobial Chemotherapy*, vol. 67, no. 11, pp. 2640–2644, 2012.

[17] A. Carattoli, E. Zankari, A. García-Fernández et al., "*In silico* detection and typing of plasmids using PlasmidFinder and plasmid multilocus sequence typing," *Antimicrobial Agents and Chemotherapy*, vol. 58, no. 7, pp. 3895–3903, 2014.

[18] P. Siguier, J. Perochon, L. Lestrade, J. Mahillon, and M. Chandler, "ISfinder: the reference centre for bacterial insertion sequences," *Nucleic Acids Research*, vol. 34, no. 90001, pp. D32–D36, 2006.

[19] Q. Zhang and Y. Ye, "Not all predicted CRISPR-Cas systems are equal: isolated cas genes and classes of CRISPR like elements," *BMC Bioinformatics*, vol. 18, no. 1, p. 92, 2017.

[20] D. Arndt, J. R. Grant, A. Marcu et al., "PHASTER: a better, faster version of the PHAST phage search tool," *Nucleic Acids Research*, vol. 44, no. W1, pp. W16–W21, 2016.

[21] N. F. Alikhan, N. K. Petty, N. L. Ben Zakour, and S. A. Beatson, "BLAST Ring Image Generator (BRIG): simple prokaryote genome comparisons," *BMC Genomics*, vol. 12, no. 1, p. 402, 2011.

[22] J. Goris, J. A. Klappenbach, P. Vandamme, T. Coenye, K. T. Konstantinidis, and J. M. Tiedje, "DNA–DNA hybridization values and their relationship to whole-genome sequence similarities," *International Journal of Systematic and Evolutionary Microbiology*, vol. 57, no. 1, pp. 81–91, 2007.

[23] M. Richter, R. Rosselló-Móra, F. Oliver Glöckner, and J. Peplies, "JSpeciesWS: a web server for prokaryotic species circumscription based on pairwise genome comparison," *Bioinformatics*, vol. 32, no. 6, pp. 929–931, 2016.

[24] C. O'Flynn, O. Deusch, A. E. Darling et al., "Comparative genomics of the genus *Porphyromonas* identifies adaptations for heme synthesis within the prevalent canine oral species *Porphyromonas cangingivalis*," *Genome Biology and Evolution*, vol. 7, no. 12, pp. 3397–3413, 2015.

[25] J. P. Meier-Kolthoff, A. F. Auch, H.-P. Klenk, and M. Göker, "Genome sequence-based species delimitation with confidence intervals and improved distance functions," *BMC Bioinformatics*, vol. 14, no. 1, p. 60, 2013.

[26] A. F. Auch, M. von Jan, H. P. Klenk, and M. Göker, "Digital DNA-DNA hybridization for microbial species delineation by means of genome-to-genome sequence comparison," *Standards in Genomic Sciences*, vol. 2, no. 1, pp. 117–134, 2010.

[27] V. J. Rapp, R. F. Ross, and T. F. Young, "Characterization of *Haemophilus* spp. isolated from healthy swine and evaluation of cross-reactivity of complement-fixing antibodies to *Haemophilus pleuropneumoniae* and *Haemophilus* taxon "minor group"," *Journal of Clinical Microbiology*, vol. 22, no. 6, pp. 945–950, 1985.

[28] R. Jansen, J. Briaire, E. M. Kamp, A. L. Gielkens, and M. A. Smits, "The CAMP effect of *Actinobacillus pleuropneumoniae* is caused by Apx toxins," *FEMS Microbiology Letters*, vol. 126, no. 2, pp. 139–143, 1995.

[29] J. Frey, "Virulence in *Actinobacillus pleuropneumoniae* and RTX toxins," *Trends in Microbiology*, vol. 3, no. 7, pp. 257–261, 1995.

[30] P. Kuhnert, Y. Schlatter, and J. Frey, "Characterization of the type I secretion system of the RTX toxin ApxII in "*Actinobacillus porcitonsillarum*"," *Veterinary Microbiology*, vol. 107, no. 3-4, pp. 225–232, 2005.

[31] L. Poirel, N. Kieffer, and P. Nordmann, "*In vitro* study of ISA*pl1*-mediated mobilization of the colistin resistance gene *mcr-1*," *Antimicrobial Agents and Chemotherapy*, vol. 61, no. 7, 2017.

[32] V. Donà, O. J. Bernasconi, J. Pires et al., "Heterogeneous genetic location of *mcr-1* in colistin-resistant *Escherichia coli* isolates from humans and retail chicken meat in Switzerland: emergence of *mcr-1*-carrying IncK2 plasmids," *Antimicrobial Agents and Chemotherapy*, vol. 61, no. 11, 2017.

[33] T. D. Lawley, V. Burland, and D. E. Taylor, "Analysis of the complete nucleotide sequence of the tetracycline-resistance transposon Tn*10*," *Plasmid*, vol. 43, no. 3, pp. 235–239, 2000.

[34] J. T. Bossé, Y. Li, R. F. Crespo et al., "ICE*Apl1*, an integrative conjugative element related to ICE*Hin1056*, identified in the pig pathogen *Actinobacillus pleuropneumoniae*," *Frontiers in Microbiology*, vol. 7, p. 810, 2016.

[35] M. Juhas, P. M. Power, R. M. Harding et al., "Sequence and functional analyses of *Haemophilus* spp. genomic islands," *Genome Biology*, vol. 8, no. 11, article R237, 2007.

[36] J. Z. Chan, M. R. Halachev, N. J. Loman, C. Constantinidou, and M. J. Pallen, "Defining bacterial species in the genomic era: insights from the genus *Acinetobacter*," *BMC Microbiology*, vol. 12, no. 1, p. 302, 2012.

[37] M. Richter and R. Rosselló-Móra, "Shifting the genomic gold standard for the prokaryotic species definition," *Proceedings of the National Academy of Sciences of the United States of America*, vol. 106, no. 45, pp. 19126–19131, 2009.

[38] L. G. Wayne, "International Committee on Systematic Bacteriology: announcement of the report of the ad hoc Committee on Reconciliation of Approaches to Bacterial Systematics," *Zentralblatt für Bakteriologie, Mikrobiologie und Hygiene. Series A: Medical Microbiology, Infectious Diseases, Virology, Parasitology*, vol. 268, no. 4, pp. 433-434, 1988.

[39] S. M. Colston, M. S. Fullmer, L. Beka, B. Lamy, J. P. Gogarten, and J. Graf, "Bioinformatic genome comparisons for taxonomic and phylogenetic assignments using *Aeromonas* as a test case," *MBio*, vol. 5, no. 6, article e02136, 2014.

14

Genome-Wide Expression Profiles of Hemp (*Cannabis sativa* L.) in Response to Drought Stress

Chunsheng Gao (iD), Chaohua Cheng, Lining Zhao, Yongting Yu (iD), Qing Tang, Pengfei Xin, Touming Liu, Zhun Yan, Yuan Guo, and Gonggu Zang (iD)

Institute of Bast Fiber Crops, Chinese Academy of Agricultural Sciences/Key Laboratory of the Biology and Process of Bast Fiber Crops, Ministry of Agriculture, Changsha 410205, China

Correspondence should be addressed to Chunsheng Gao; gaochunsheng@caas.cn and Gonggu Zang; ibfczgg@sohu.com

Academic Editor: Gunvant B. Patil

Drought is the main environmental factor impairing hemp growth and yield. In order to decipher the molecular responses of hemp to drought stress, transcriptome changes of drought-stressed hemp (DS1 and DS2), compared to well-watered control hemp (CK1 and CK2), were studied with RNA-Seq technology. RNA-Seq generated 9.83, 11.30, 11.66, and 11.31 M clean reads in the CK1, CK2, DS1, and DS2 libraries, respectively. A total of 1292 differentially expressed genes (DEGs), including 409 (31.66%) upregulated and 883 (68.34%) downregulated genes, were identified. The expression patterns of 12 selected genes were validated by qRT-PCR, and the results were accordant with Illumina analysis. Gene Ontology (GO) and KEGG analysis illuminated particular important biological processes and pathways, which enriched many candidate genes such as NAC, B3, peroxidase, expansin, and inositol oxygenase that may play important roles in hemp tolerance to drought. Eleven KEGG pathways were significantly influenced, the most influenced being the plant hormone signal transduction pathway with 15 differentially expressed genes. A similar expression pattern of genes involved in the abscisic acid (ABA) pathway under drought, and ABA induction, suggested that ABA is important in the drought stress response of hemp. These findings provide useful insights into the drought stress regulatory mechanism in hemp.

1. Introduction

Abiotic stresses, such as drought, high salt, and extremes of temperature, are major environmental factors that can limit plant growth and development. Among various abiotic stressors, drought has the greatest impact on crop culture and world agriculture [1]. Currently, global warming is increasing the frequency and severity of extreme weather events, including drought, worldwide. It is therefore important to improve plant drought tolerance and further understand the relationship between drought stress and water use for plant growth.

Drought stress or water deficit induces a series of morphological, physiological, biochemical, and molecular changes that influence plant growth, development, and productivity. During their long-term evolution, plants have developed three main mechanisms to adapt to drought stress, including drought escape, drought avoidance, and drought tolerance. It is important to understand the genetic basis of these mechanisms of plants encountering a water deficit [2]. Many plant genes responding to drought stress have been identified by molecular and genomic analyses of *Arabidopsis*, rice, and other plants. These genes are classified into two groups according to their putative functional modes. One group contains proteins that are likely involved in abiotic stress tolerance, such as chaperones, late embryogenesis abundant (LEA) proteins, osmotin, mRNA-binding proteins, key enzymes for osmolyte biosynthesis, water channel proteins, sugar and proline transporters, detoxification enzymes, and various proteases. Another group contains regulatory proteins involved in signal transduction and stress-responsive gene expression, including various transcription factors (TFs), protein kinases, protein phosphatases, enzymes involved in phospholipid metabolism, and other signaling molecules, such as calmodulin-binding proteins [3]. Moreover, some of these proteins (LEA, osmotin, zinc finger protein, NAC, WRKY, bZIP, AP2/ERF, MYB, etc.) have been overexpressed in transgenetic plants, and conferred transgenetic lines that enhanced drought tolerance [4]. All these data verify the contribution of these genes to drought stress response.

Plant cell signaling and molecular regulation networks during drought stress have also been investigated and highlighted. Two important pathways of transcriptional networks were found in *Arabidopsis* and rice, the two model plants, when they were grown under drought conditions: an abscisic acid- (ABA-) dependent signaling pathway and an ABA-independent regulatory network mediated by dehydration-responsive element-binding- (DREB-) type TFs [4, 5]. In the first type, the ABA-responsive element (ABRE) is the major *cis*-element and TFs are the master regulators of drought-responsive gene expression, as they control gene expression in an ABA-dependent manner. Additionally, ABA receptors (PYLs), group A 2C-type protein phosphatases (PP2Cs), and SNF1-related protein kinases 2 (SnPK2) were core components and controlled the ABA signaling pathway. In the ABA-independent regulatory network, NAC TFs were also involved in drought stress response, along with DREB TFs [6].

Hemp (*Cannabis sativa* L.) was one of the earliest domesticated crops and is used today in multiple industrial applications, including the production of fiber, foods, and oils [7, 8]. Hemp has been cultivated to produce textile for more than 6000 years in China, and China is currently the largest producer of hemp seed and textiles for domestic use and exports [9]. Drought stress is the main environmental factor that influences hemp production, limiting growth and reducing fiber quality and yield [10–12]. For instance, more than 10 days of consecutive drought increases the incidence of hemp tip dieback over a 1-month period after sowing [13]. Moreover, as the largest developing country with the largest population worldwide, China has begun to face serious water scarcity issues [14]. Therefore, investigating the mechanisms that regulate drought tolerance in hemp is important for drought-tolerant cultivar development in breeding programs. The complete genome and transcriptome sequences of hemp have been reported [15], along with a diversity analysis based on the large-scale development of expressed sequence tag- (EST-) derived simple sequence repeat (SSR) markers [16]. However, the responsive genes and stress regulatory mechanisms of hemp subjected to drought stress remain elusive.

High-throughput sequencing has become a powerful tool in many research fields due to its cost-efficiency and rapidness [17–23]. The data yielded facilitates the development of genetic analyses and functional genomics studies among species, especially for many nonmodel plants. In particular, this technology has been widely used to understand drought stress response in various plants, such as rice [24], wheat [25], maize [26], soybean [27], cotton [28, 29], potato [30], and ramie [18]. In the present study, the transcriptome changes of hemp to drought stress were investigated and potential responsive genes were identified using the Illumina HiSeq™ 2000 platform. As a result, this study presents the first genome-wide expression profile of hemp responding to drought stress. The findings of this study are expected to provide a foundation for a comprehensive understanding of the mechanisms of hemp subjected to drought stress, along with identifying potential drought resistance genes, which can be used to improve the drought tolerance ability of hemp in breeding programs.

2. Materials and Methods

2.1. Plant Material, Stress Treatment, and RNA Extraction. Hemp cultivar Yunma 1, which is sensitive to drought [31], was used in this study. The hemp seeds were sowed in a pot (31 cm deep × 34 cm diameter) filled with 16 kg soil, 15 seeds per pot. The pots were kept at $26°C \pm 1°C$, 75 ± 1 RH, and a photoperiod of 14 : 10 (L : D) in the greenhouse of the Institute of Bast Fiber Crops, Chinese Academy of Agricultural Sciences. After seeds germinated, seedlings were thinned out to keep 10 seedlings per pot. Six potted hemp plants were used as drought stress plants (DS), whereas the other six potted plants were used as control plants (CK). Two replicates of CK and DS were designated, which were named CK1 and CK2 and DS1 and DS2, respectively. Watering was withheld at 30 days after sowing in the soil. CK plants were grown under well-watered conditions, where DS plants were treated with drought stress by controlling the relative water content of soil at no more than 20% (severe drought stress) by natural drying [31]. After 7 days, the CK and DS plants were uprooted, and the leaves, roots, stem bast, and stem shoots were collected separately. For abscisic acid (ABA) treatment, plants were collected at 3 and 6 hours after being sprayed 3 times with 100 μM ABA [32]. The sampled tissues were immediately frozen in liquid nitrogen and stored at −80°C until use. Total RNA was extracted from the tissue of CK1, CK2, DS1, and DS2 using TRIzol reagent (Invitrogen, USA) and treated with DNase I (Fermentas, USA) according to the manufacturer's protocol. RNA quality, purity, and integrity were determined by a NanoDrop 2000 spectrophotometer and an Agilent 100 Bioanalyzer.

2.2. cDNA Library Construction and Sequencing. Equal amounts of total RNA from each DS and CK sample were pooled together. Then, the poly (A)$^+$ RNAs were purified from 20 μg total RNA by oligo(dT) Dynabeads. To avoid priming bias when synthesizing cDNA, the purified mRNA was first fragmented into small pieces. First-strand cDNAs were synthesized using random hexamer primers, and second-strand cDNAs were synthesized using dNTPs, RNase H, and DNA polymerase I. Double-stranded cDNAs were purified by a QIAquick PCR Purification Kit and repaired, and an adenine base was added to the 3$'$ end. Two different adapters were ligated to the 5$'$ and 3$'$ ends, respectively. The ligated fragments were separated on gel and purified. After amplification by PCR, the fragments were separated using electrophoresis and purified. Paired-end sequencing was performed by the Illumina sequencing platform (HiSeq™ 2000) at Biomarker Technologies Co., Ltd, Beijing, China (http://www.biomarker.com.cn/) according to the manufacturer's instructions (Illumina, San Diego, CA).

2.3. Data Processing and Mapping to Reference Genome.
Sequencing errors usually bring difficulties to subsequent analyses. Therefore, first all reads with adaptor contamination were discarded. Then, the low-quality reads with more than 10% ambiguous "N" sequences were ruled out. Finally, the low-quality reads ($Qvalue < 20$) were removed. Q30 is equivalent to the probability of an incorrect base call 1 in 1000 times. Clean reads were submitted to Gene Expression Omnibus in NCBI with the GEO accession number GSE56964. RNA-Seq data was assessed by saturation, duplicate reads, and gene coverage analysis, using RSeQC software (http://rseqc.sourceforge.net/). High-quality reads were mapped to the *C. sativa* (strain Purple Kush) genome (GenBank accession no. AGQN00000000) [15] using TopHat2 (http://ccb.jhu.edu/software/tophat/index.shtml). After aligning reads to the genome, Cufflinks (http://cole-trapnell-lab.github.io/cufflinks/) was used to assemble aligned reads into transcripts.

2.4. Identification of Differentially Expressed Genes (DEGs).
Differentially expressed genes were identified by comparing the abundance of the same transcript in CK and DS samples. The hemp genome sequences described above were used as a reference. The number of clean reads mapped to the reference genome was used to calculate gene expression levels via RSEM (http://deweylab.github.io/RSEM/). The FPKM (fragments per kilobase of exon per million mapped reads) value of each transcript was used to represent the gene expression level. Differentially expressed genes (DEGs) were identified using DESeq2 [33]. All statistical tests were corrected for multiple testing with the Benjamini–Hochberg false discovery rate (FDR < 0.01). DEGs were defined as being significantly differentially expressed along with an FDR < 0.01 and a more than two-fold change (>1 or <−1 in log 2 ratio value).

2.5. Functional Classification and Annotation for DEGs.
Unigenes were annotated using BLASTx alignment ($Evalue < 10^{-5}$) against five databases, including NCBI nonredundant protein database (Nr, http://www.ncbi.nlm.nih.gov), Swiss-Prot (http://www.expasy.ch/sprot), Gene Ontology (GO, http://wego.genomics.org.cn/cgi-bin/wego/index.pl), Clusters of Orthologous Groups (COG, https://www.ncbi.nlm.nih.gov/COG/), and the Kyoto Encyclopedia of Genes and Genomes (KEGG) pathway database [34]. Nr and GO annotation was performed using the Blast2GO program [35]. GO classification was performed using WEGO software to view the distribution of gene functions.

2.6. Pathway Enrichment of DEG.
Pathway enrichment analysis based on the KEGG pathway database (http://www.genome.jp/kegg) was used to identify markedly enriched metabolic pathways or signal transduction pathways in differentially expressed genes, comparing the whole genomic background. The following formula was used for the calculation of

$$p = 1 - \sum_{i=0}^{m-1} \frac{\binom{M}{i}\binom{N-M}{n-i}}{\binom{N}{n}}, \quad (1)$$

where N is the number of all genes with a KEGG annotation, n is the number of DEGs in N, M is the number of all genes annotated to specific pathways, and m is the number of DEGs in M.

2.7. Quantitative Real-Time PCR Analyses.
For quantitative real-time PCR (qRT-PCR) analysis, total RNA was extracted from the tissue of CK and DS hemp using TRIzol reagent (Invitrogen, USA) and treated with DNase I (Fermentas, USA) according to the manufacturer's protocol. Subsequently, 2 μg total RNA was reverse-transcribed into cDNA. qRT-PCR was performed in an iQ5 multicolor real-time PCR detection system (Bio-Rad, USA) with a 20 μL reaction system containing 10 μL iQ™ SYBR Green Supermix (Bio-Rad, USA), 10 pmol each of the forward and reverse gene-specific primers, and 5 μL diluted cDNA (1 : 50). Gene-specific primers (Table S1) were designed using Primer Premier 5 software. The hemp actin gene was selected as an internal control to normalize the total amount of cDNA present in each reaction. In brief, following a denaturation step at 95°C for 30 s, the amplification was carried out with 40 cycles at a melting temperature of 95°C for 10 s and an annealing temperature of 55°C for 30 s. The qRT-PCR experiments were performed in triplicate.

3. Results

3.1. Sequencing and Mapping Reads to the Hemp Reference Genome.
Four samples (two control samples, CK1 and CK2, and two drought stress samples, DS1 and DS2) were collected separately. After filtering, a total of 44.10 M clean reads were obtained in the CK and DS libraries (Table 1). The number of clean reads for CK1, CK2, DS1, and DS2 was 9.83 M, 11.30 M, 11.66 M, and 11.31 M, respectively. The GC contents of the four libraries ranged from 44.18% to 44.30%. In addition, the Q30 for all libraries were all over 90%, indicating the high accuracy and quality of the sequencing data. To annotate the function of the reads, Bowtie was used to align the reads with the published hemp genome. The mapped reads from the four hemp samples were 8.61 M (87.61%), 9.94 M (87.94%), 10.25 M (87.96%), and 9.92 M (87.65%), respectively. These results indicate that reliable sequence alignment results were obtained for gene analysis. Saturation of the library was determined by checking the number of detected genes. When the sequencing level reached 6 M or more, only a few new genes were detected in the four libraries (Figure 1), indicating all four libraries were sequenced to saturation.

TABLE 1: Quality of Illumina sequencing and the statistics of reads.

Sample	Total reads (M)	Total base (GB)	GC percent (%)	Q30 percent (%)	Mapped reads (%)	Uniquely mapped reads (%)
CK1	9.83	1.99	44.20	90.03	8.61 (87.61%)	7.08 (82.21%)
CK2	11.30	2.28	44.30	90.06	9.94 (87.94%)	8.18 (82.33%)
DS1	11.66	2.35	44.22	90.07	10.25 (87.96%)	8.41 (82.03%)
DS2	11.31	2.29	44.18	90.01	9.92 (87.65%)	8.08 (81.51%)
Total	44.10	8.91				

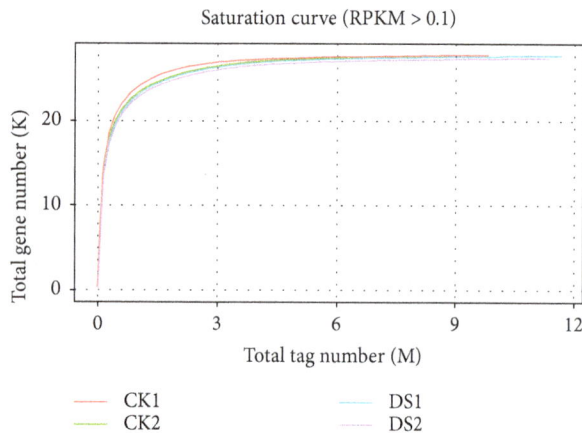

FIGURE 1: Saturation evaluations of CK and DS hemp.

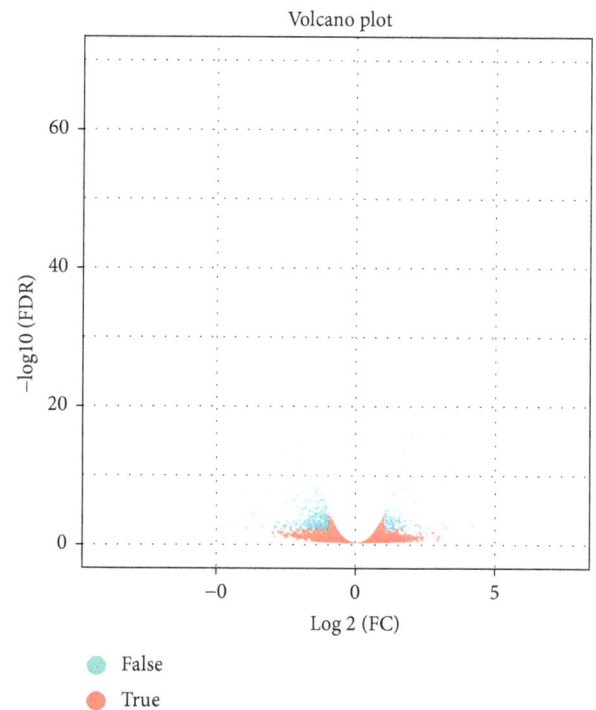

FIGURE 2: Volcano analysis of differentially expressed genes (DEGs) between CK and DS hemp. In the volcano plot, statistical significance ($\log_{10}$ of P value; y-axis) has been plotted against $\log_2$-fold change (x-axis).

3.2. Comparison of the Gene Expression Level between CK and DS Libraries. For analysis of gene expression levels, the number of unambiguous clean reads for each gene was calculated and normalized to RPKM. Distribution and density distribution of RPKM showed there to be few different genes among the samples (Figure S1). Additionally, the RPKM values for all genes were compared between both the two DS replicates and the two CK replicates. There were significant correlations between the two CK replicates and the two DS replicates, with Pearson's correlation coefficients of the CK and DS groups of 0.82 and 0.99, respectively. Scatter diagrams were created, in which the logarithmic RPKM values of each gene in the two replicates of each gene, in the two replicates of each treatment, were assigned as coordinate values of two axes, showed that all data points were distributed in the region of the diagonal (Figure S2). These results suggested that the abundances and expression levels of genes in the two DS libraries and in the two CK libraries were similar.

The expression levels of DEGs between the two libraries were evaluated by detecting the sequence frequencies. Following data analysis, only transcripts with FDR < 0.01 and fold change > 2 were selected to represent differentially expressed genes. Based on these criteria, a total of 1292 DEGs were identified, including 409 (31.66%) upregulated genes and 883 (68.34%) downregulated genes (Figure 2, Table S2). Of the total DEGs, 126 showed more than 20-fold ($|\log 2FC| > 4.32$) the expression differences, including 43 upregulated genes and 83 downregulated genes.

3.3. Functional Annotation and Classification of DEGs. To investigate the function of the 1292 DEGs, five databases were used to screen for sequence similarities. These databases included the NCBI Nr database, the Swiss-Prot protein database, the GO database, the KEGG database, and the COG database. The results indicated that 1255 (97.14%), 1067 (82.59%), 1105 (85.53%), 158 (12.23%), and 493 (38.16%) DEGs exhibited significant similarity to known genes in the five databases, respectively (Figure 3). Overall, 1258 (97.37%) DEGs, including 394 (31.32%) upregulated and 864 (68.68%) downregulated DEGs, exhibited similarities to known genes in the five databases (Table S2). This information provided a good reference for gene function analysis.

We used the COG database for genome-scale analysis of protein functions and evolution. A total of 416 DEGs were assigned to the 22 COG classifications, of which 6 COG classifications contained more than 50 DEGs (Figure 4). The 6

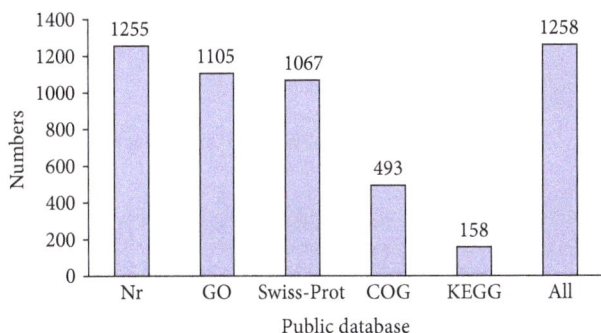

FIGURE 3: Numbers of differentially expressed genes (DEGs) annotated in five public databases.

largest COG categories were "general function prediction only" (R, 133 DEGs, 31.197%), "carbohydrate transport and metabolism" (G, 60 DEGs, 14.92%), "replication recombination and repair" (L, 57 DEGs, 13.70%), "signal transduction mechanisms" (T, 57 DEGs, 13.70%), "transcription" (K, 56 DEGs, 13.46%), and "secondary metabolite biosynthesis, transport and catabolism" (Q, 55 DEGs, 13.22%).

3.4. GO and KEGG Enrichment of DEGs.

The GO database is a tool for the unification of biology, providing structured, controlled vocabularies and classifications that cover several domains of molecular and cellular biology. Using GO analysis, a total of 1106 DEGs matching known genes were assigned to 53 functional terms of GO for biological processes, cellular components, and molecular function categories (Figure 5). The three GO categories enriched 1011, 997, and 933 DEGs, respectively (Table S3). Among these GO terms, "cellular process" (898 DEGs, 69.50%), "metabolic process" (855 DEGs, 66.18%), "response to stimulus" (744 DEGs, 57.59%), "biological regulation" (639 DEGs, 48.46%), and "developmental process" (514 DEGs, 39.78%) were the dominant biological process terms; "cell part" (937 DEGs, 72.52%), "cell" (908 DEGs, 70.28%), "organelle" (806 DEGs, 62.38%), and "membrane" (577 DEGs, 44.66%) were the dominant cellular component terms; and "binding" (680 DEGs, 52.63%) and "catalytic activity" (638 DEGs, 49.38%) were the most abundant molecular function terms. Additionally, the "transporter activity" and "receptor activity" GO terms enriched 104 (2.55%) and 33 (8.05%) DEGs, respectively.

Biological pathways play a key role in advanced genomics studies. The influence of drought stress on biological pathways in hemp was analyzed by enrichment analysis of DEGs. A total of 77 pathways (210 DEGs) were possibly influenced, out of which 11 pathways were significantly influenced (P value < 0.05) (Table 2). Among the 11 pathways, the top five most significantly influenced pathways were plant hormone signal transduction (ko04075), phenylpropanoid biosynthesis (ko00940), cyanoamino acid metabolism (ko00460), carbon fixation in photosynthetic organisms (ko00710), and plant hormone signal transduction (ko00475). The plant hormone signal transduction pathway enriched the most DEGs (15), distributed in the

abscisic acid (ABA) (8), auxin (5), jasmonic acid (1), and salicylic acid (1) metabolic pathways (Table S4). Of the 15 DEGs that were differentially regulated by drought stress, 7 DEGs were upregulated and 8 DEGs were downregulated. Interestingly, all the DEGs in the auxin, jasmonic acid, and salicylic acid metabolic pathways were significantly downregulated. In contrast, the DEGs in the ABA metabolic pathway were significantly upregulated. The photosynthesis pathway was the most significantly enriched pathway, with the lowest P value. All DEGs enriched by the photosynthesis pathway (ko00195), and photosynthesis–antenna proteins (ko00196), were downregulated, indicating that photosynthesis in drought-stressed hemp was reduced. All DEGs enriched in cyanoamino acid metabolism, and most of the DEGs enriched by phenylpropanoid biosynthesis and carbon fixation in photosynthetic organisms, were downregulated, with ratios of 12/14 and 8/9, respectively. These results suggest the three pathways were significantly suppressed.

3.5. Identification of Genes Responding to Drought Stress.

To identify drought stress-responding genes in hemp, DEGs with the following characteristics were analyzed: (1) involvement in significantly influenced GO terms and KEGG pathways, (2) connection to a biological process related to water deprivation and desiccation and stomatal movement, and (3) exhibition of a higher fold change (>20) in differential expression. Of the significantly influenced biological processes (P value < 0.05) related to water stress, response to water deprivation (GO: 0009414), response to desiccation (GO: 0009269), and regulation of stomatal movement (GO: 0010119) enriched 94, 22, and 29 DEGs, respectively. Moreover, the DEGs with a high ratio or higher fold change in expression included peroxidase (POD), glycine-rich protein (GRP), expansin, nitrate transporter, β-amylase 1 (BAM), laccase, protein phosphatase 2C (PP2C), and a variety of transcription factors (TFs) (Table S5). Among the significantly enriched pathways, DEGs with a high ratio or higher fold change in expression included POD, inositol oxygenase, β-glucosidase, and (R)-mandelonitrile lyase (Table S4).

A total of 51 transcription factors (TFs), including 20 upregulated and 31 downregulated TFs, were identified to be significantly differentially expressed under drought stress (Table 3). Among these TFs, all 3 NAC genes, 4/7 MYB genes, and 3/5 HD-Zip genes were upregulated, while all 5 B3 (B3 domain-containing transcription factor), 2 KNOX genes, 2 C2H2L genes, and 4/6 bHLH genes were downregulated (Table S6).

3.6. Analysis of DEGs by qRT-PCR.

To validate DEGs identified by RNA-Seq, the expression pattern of 12 genes in drought-stressed hemp was studied by qRT-PCR. Among these genes, 8 (PYL4, PP2C-1 to PP2C-6, and SAPK3) were ABA metabolism-related genes, and 4 (X15-1, X15-2, IAA-1, and IAA-2,) were auxin metabolism-related genes. The results showed that 7 genes were upregulated and 5 genes were downregulated, and this trend in the change of gene expression was consistent with that detected by RNA-Seq

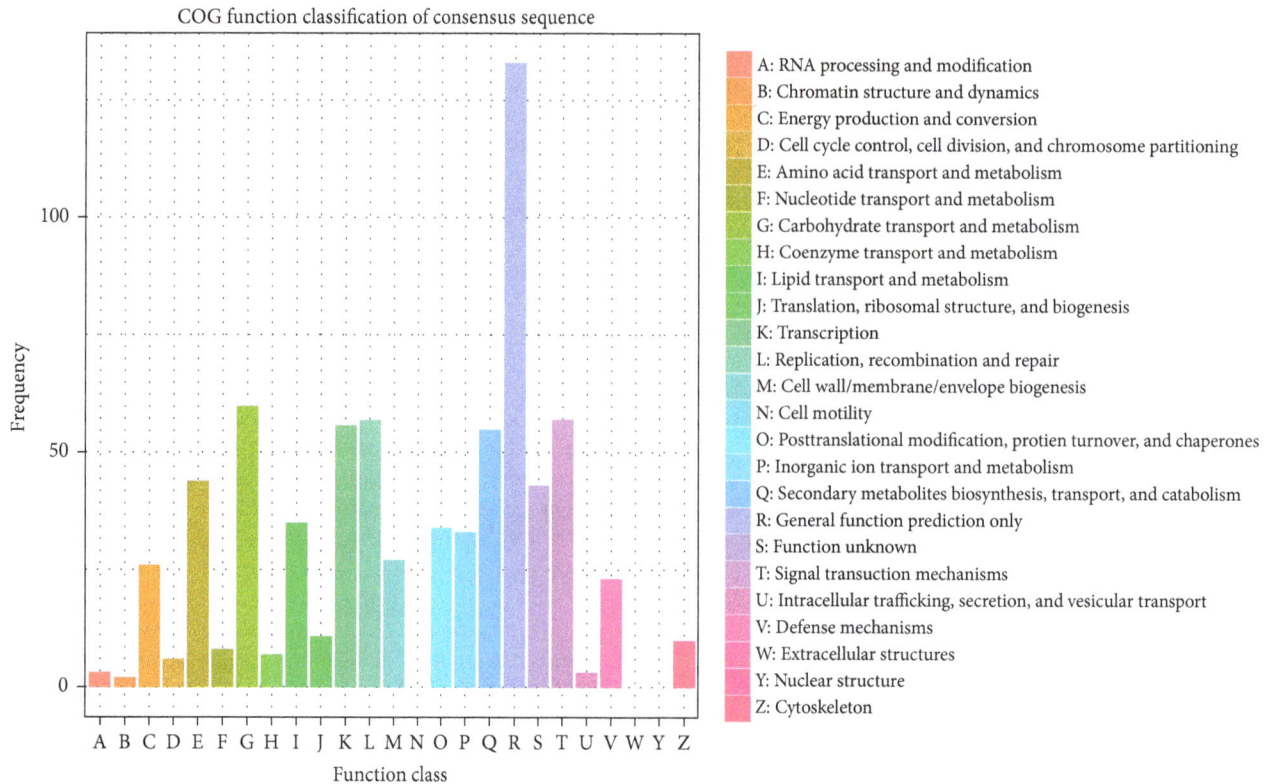

FIGURE 4: Clusters of orthologous group (COG) classification of differentially expressed genes (DEGs) in drought-stressed hemp.

(Figure 6). In addition, there was a strong correlation (Pearson correlation coefficient of $R = 0.777$) between the qRT-PCR and RNA-Seq data. To investigate the relevance of the 8 ABA metabolism-related genes to drought stress, expression profiles of these genes at 3 h and 6 h after ABA treatment in hemp were also studied by RT-PCR. As shown in Figure 7, the expression of 7 genes of *PP2C* and *SAPK3* were upregulated, and a *PYL4* gene was suppressed in ABA-treated hemp plants (Figure 7), which was in accordance with analysis obtained from drought-stressed hemp. These results signify that ABA may play an important role in regulating the response of hemp to drought stress.

4. Discussion

4.1. Identification of 1292 Genes Responded to Drought Stress in Hemp. Drought stress is one of the most important environmental factors limiting hemp growth; however, the mechanism of hemp tolerance to drought stress remains unclear. China, as the country with the largest population worldwide, needs more irrigable farmland to grow grain crops to ensure food security; thus, hemp has been mainly planted on un-irrigable dry land and hill slopes. It is vital to study the drought stress response mechanisms of hemp to know how cultivars have adapted to thrive under adverse conditions. However, few studies have focused on identifying the drought response genes and regulatory mechanism of hemp. In this study, a total of 1292 potential drought stress-responsive genes were

identified in hemp using RNA-Seq technology. Out of these genes, 1258 (97.37%) were annotated by the five widely used databases (Table S2). These potential drought stress-responsive genes are expected to be useful for investigating the molecular mechanisms of hemp drought tolerance.

4.2. Dramatic Changes of Hemp in Response to Drought Stress. Drought stress induces a range of physiological and biochemical responses in plants. These responses include the repression of cell growth and photosynthesis and the activation of respiration [36]. Similar responses occurred in hemp. For example, most DEGs enriched in significantly influenced biological processes involved in cell growth were downregulated (cell wall modification involved in multidimensional cell growth, GO:0042547; multidimensional cell growth, GO:0009825), photosynthesis (photosynthesis, light reaction, GO:0019684; photosynthesis, GO:0015979; photosynthesis, light harvesting, GO:0009765; regulation of photosynthesis, GO:0010109) (Table S2). Additionally, almost all the genes related to photosynthesis and the photosynthesis-related pathways were downregulated. Photosynthesis is a process used by plants and other organisms to convert light energy into chemical energy, which provides plants with the food and energy they need to grow. Our results indicated that drought noticeably altered energy metabolism to avoid damaging hemp. This phenomenon has also been documented in other higher plants [37]. It is shown that some plant growth-related processes were suppressed and a series of responses were activated to facilitate the survival of hemp under drought stress.

(a)

(b)

(c)

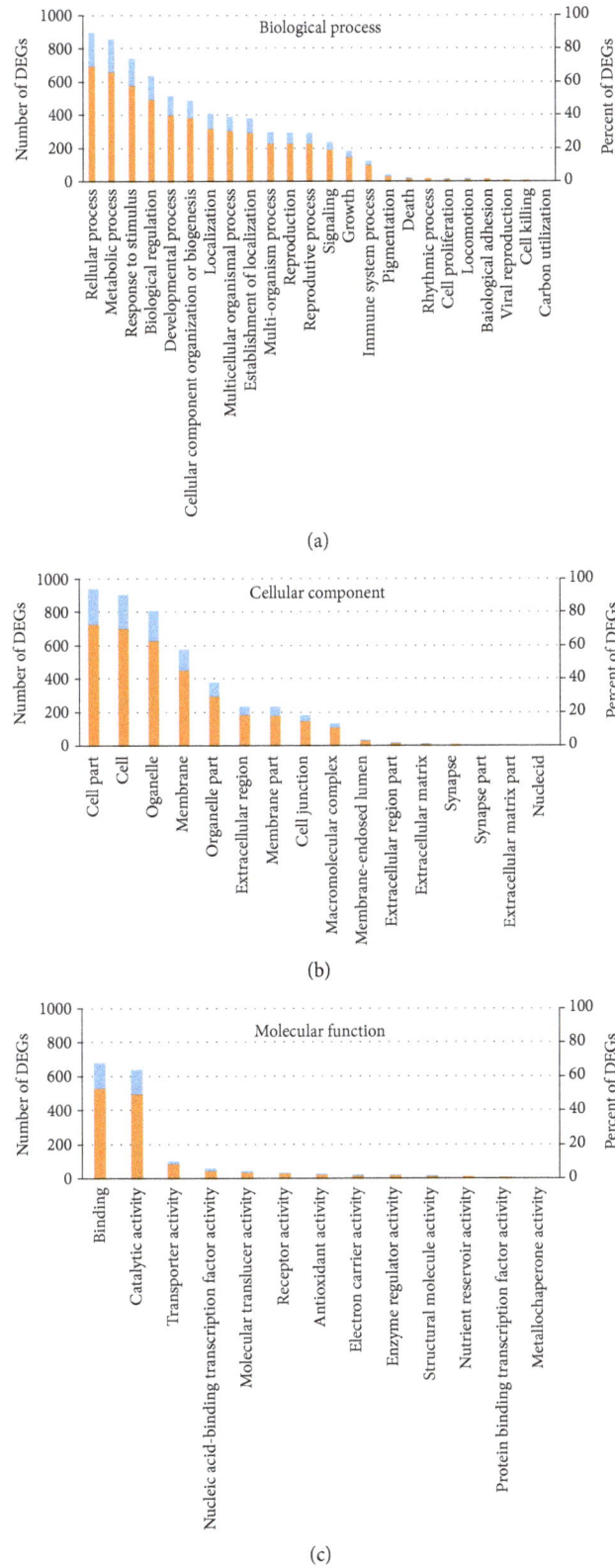

FIGURE 5: Gene Ontology (GO) enrichment of differentially expressed genes (DEGs) in drought-stressed hemp. The results are summarized in three main categories: biological processes (a), cellular components (b), and molecular functions (c). The blue (corresponding to the left y-axis) and orange columns (corresponding to the right y-axis) represent the number and percent of DEGs in that category, respectively.

TABLE 2: List of pathways significantly enriched in DEGs ($P < 0.05$).

KEGG pathway	Number of genes	Number of DEGs	Up	Down	P value	Pathway ID
Photosynthesis	62	14	0	14	$1.47E - 8$	ko00195
Phenylpropanoid biosynthesis	132	14	2	12	$1.67E - 4$	ko00940
Cyanoamino acid metabolism	55	8	0	8	$5.45E - 4$	ko00460
Carbon fixation in photosynthetic organisms	84	9	1	8	$2.35E - 3$	ko00710
Plant hormone signal transduction	199	15	7	8	$3.57E - 3$	ko04075
Photosynthesis—antenna proteins	25	4	0	4	$1.01E - 2$	ko00196
Ascorbate and aldarate metabolism	45	5	4	1	$1.90E - 2$	ko00053
Nitrogen metabolism	66	6	0	6	$2.63E - 2$	ko00910
Phenylalanine metabolism	108	8	1	7	$3.34E - 2$	ko00360
Pentose and glucuronate interconversions	54	5	0	5	$3.85E - 2$	ko00040
Inositol phosphate metabolism	74	6	5	1	$4.28E - 2$	ko00562
Total number	904	94	20	74		

Most DEGs involved in biosynthesis of secondary metabolites (phenylpropanoid biosynthesis) and amino acid metabolism (phenylalanine metabolism and cyanoamino acid metabolism) were significantly downregulated, suggesting that these pathways were also suppressed. Among these DEGs, 16 of 22 were peroxidase homologs of hemp. POD is an important antioxidant enzyme in ROS metabolism. ROS is always enhanced when plants suffer drought stress [38]. A decrease of POD activities in drought-stressed hemp implied the decrease in scavengers of ROS, thus resulting in the steady state of cellular ROS breaks and an increase of ROS in hemp.

Interestingly, in the "ascorbate and aldarate metabolism" and "inositol phosphate metabolism" pathways, more DEGs (5 and 4, resp.) were upregulated and limited DEGs (1 and 1, resp.) were downregulated. The upregulated genes contained 3 inositol oxygenase homologs and an inositol-3-phosphate synthase homolog of hemp (Table S5). Among the two pathways, the same upregulated pathway was D-glucuronate synthesis. D-glucuronate is directly synthesized from inositol and is used in the production of cell wall components, glycoproteins, gums, and mucilage. Inositol itself and these methylated derivatives increase in some animal and plant cells, in association with high external NaCl concentrations and dehydration [4, 39]. OsMIOX, a myo-inositol oxygenase gene, has been shown to improve drought tolerance of rice [40]. Thus, D-glucuronate synthesis and inositol may be important in the hemp drought response.

4.3. Changes of Genes Involved in Sucrose/Starch Synthesis and Cell Wall Plasticity.
It has been reported that genes involved in sucrose\starch metabolism are always affected by water stress during grain filling in some crops [41, 42]. In the drought-stressed hemp, the starch and sucrose metabolism (ko00500) pathway (enriched 3 upregulated and 7 downregulated DEGs) was not significantly influenced (P value > 0.05). No starch synthesis-related enzyme, such as sucrose transporter, sucrose synthase, starch synthase, and branching enzyme of hemp, were significantly differentially

expressed in drought-stressed hemp. However, three β-amylase 1 homologs of hemp were upregulated. During osmotic stress, starch can be degraded by stress-activated β-amylase 1 to release sugar and sugar-derived osmolytes [43]. Simply put, under drought stress, starch degradation was increased in hemp plants.

Plant cell walls are complex structures composed of cellulose, hemicellulose, pectin, protein, lignin, and various inorganic compounds. Cell wall plasticity has been reported to be related to activities of xylosyltransferase and cellulose synthesis inhibitors [44]. The content of xyloglucan in a cell well affects the mechanical properties of a plant [45]. GO analysis showed that most of the DEGs involved in xyloglucan and cellulose biosynthetic processes, and metabolic processes (GO:0030243, GO:0030244, GO:0010411, GO:0009969, GO:0030244, and GO:0052541), were downregulated. In addition, a zeatin O-xylosyltransferase and a xylosyltransferase 1 homolog of hemp were down- and upregulated, respectively, under drought stress, while the fold change of the former was higher than that of the latter. These data suggest that synthesis and metabolism of xyloglucan and cellulose may be reduced, and the cell wall plasticity was weakened.

4.4. Transcription Factors Responding to Drought Stress.
Transcription factors are master regulators that control gene clusters. Recent studies demonstrated that many TFs, such as AP2/ERF, NAC, HD-Zip, and WRKY, have important roles in response to abiotic stresses in plants [46, 47]. In the drought-stressed hemp, 51 DEGs of families including bHLH, MYB, NAC, WRKY, and AP2/ERF were identified. Among these TFs, all 5 B3 domain-containing transcription factors were significantly downregulated (>20-fold). Although there was still no substantial evidence that B3 TFs were involved in plant drought tolerance, they may play some role in drought adaptation. Abundant studies have reported that overexpression of genes of the NAC family can enhance drought tolerance of plants [48–50]. The observation that all 3 NAC TFs were upregulated indicated that these 3 NAC homologs of hemp were likely involved in drought resistance. Further

TABLE 3: Summary of DEGs annotated as transcription factor.

Gene family	Total	Number of genes Downregulated	Upregulated
AP2/ERF	2	1	1
B3	5	5	0
bHLH	6	4	2
bZIP	3	2	1
C2H2L	2	2	0
Dof	1	1	0
HD-Zip	5	2	3
KNOX	2	2	0
LHY	1	0	1
MYB	7	3	4
MYB1R1	2	1	1
NAC	3	0	3
NF-YA	2	1	1
SPB	1	0	1
Trihelix	1	0	1
Wox	1	1	0
WRKY	2	1	1
YABBY	1	1	0
ZF (CO-like)	3	2	1
ZF-HD	1	1	0
Total	51	30	21

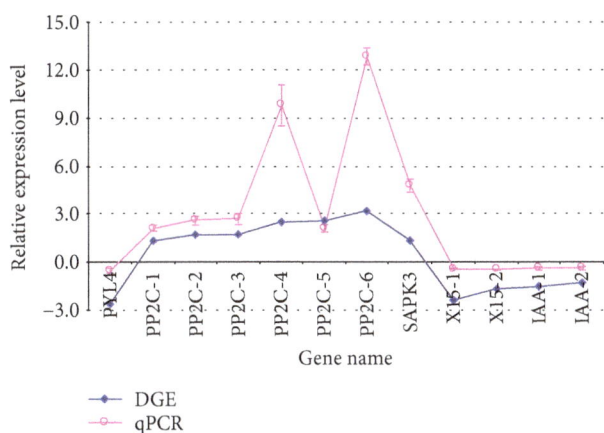

FIGURE 6: Validation of the expression of drought stress-induced genes by qRT-PCR. Data from qRT-PCR are means of three replicates, and bars represent standard error. PYL4: abscisic acid receptor PYL4 (JP449530); PP2C-1: protein phosphatase 2C (JP474397); PP2C-2: protein phosphatase 2C (JP473042); PP2C-3: protein phosphatase 2C (JP478284); PP2C-4: protein phosphatase 2C (JP477418); PP2C-5: protein phosphatase 2C (JP469562); PP2C-6: protein phosphatase 2C (JP455237); X15-1: auxin-induced protein X15 (JP473847); X15-2: auxin-induced protein X15 (JP472037); IAA-1: indole-3-acetic acid amido synthetase (JP480028); IAA-2: indole-3-acetic acid-induced protein ARG7 (JP461329); SAPK3: serine/threonine-protein kinase (JP477935). Pearson correlation coefficient ($R = 0.777$) analysis showed strong correlation between the qRT-PCR and RNA-Seq data.

analyses of these transcription factors may provide new insight into the complex regulatory gene networks in response to drought stress in hemp.

4.5. ABA May Be a Key Regulation Factor in Hemp Responding to Drought. Plant hormones are important for regulating developmental processes and signaling networks in plant responses to biotic and abiotic stresses, including drought [51, 52]. ABA is critical to osmotic stress regulation among plant hormones; in fact, it is sometimes defined as a stress hormone because of its rapid accumulation and mediation in plant survival when subjected to various stresses. In our study, eight genes in ABA biosynthesis were noticeably influenced by drought stress, including PYL4, PP2C, and SAPK3 (Figure 6). The Illumina sequencing results were also confirmed by qRT-PCR (Figure 6) and were in accordance with those in ABA-treated hemp plants (Figure 7). PYL-PP2C-SnRK2 families function as the core components of ABA signaling, PYLs are ABA receptors; PP2Cs and SnPKs are important negative and positive regulators of ABA signaling, respectively [53, 54]; and SAPK3 belongs to the SnRK2 family [55]. Overexpression of some PYL, PP2C, and SAPK2 genes significantly increases or decreases drought tolerance in transgenic plants [56–58]. Moreover, the ABA receptor PYL5 can inhibit the activity of clade A PP2Cs [59], and PP2CA together with ABI1 inhibits SnRK2.4 activity and regulates plant responses to salinity [60]. These data clarify the function and nature of the cascade relationship of these genes in the stress response of plants. In this study, the expression of PYL4 was downregulated, while PP2Cs and SAPK3 were both upregulated; PP2Cs and SAPK3 displayed the same expression trends in both drought-stressed and ABA-treated hemp plants. Similar findings were also reported in drought- or salinity-stressed tomato and jute plants [23, 61] and ABA-treated tomato plants [62]. Additionally, a gene encoding PYL4 was also downregulated in drought-stressed potato plants [30]; *CsPYL3* was downregulated, although *CsPYL1*, *CsPYL2*, *CsPP2C2*, and *CsSnKR2.2* were upregulated in roots and stems of cucumber seedlings under drought conditions [63]. These data indicated that different numbers of PYLs, PP2Cs, and SnKR2s display different responses to ABA or drought stress. ABA is also important for the regulation of stomatal movement and closure in response to drought [64, 65]. Taken together, ABA may play an important role in regulating stomata closure, reducing water evaporation, and launching the resistance response in hemp.

The plant hormone auxin is essential to many aspects of plant growth and development. In our study, transcripts in auxin biosynthesis pathways clearly decreased in hemp under drought stress (Figure 6). The GH3 gene encodes IAA-amido synthetase, which acts as an auxin-responsive gene, and can help to maintain auxin homeostasis by conjugating excess IAA to amino acids [66]. A rice GH3 gene, OsGH3-2, is involved in negatively regulating ABA levels and drought tolerance [67]. The GH3 homolog of hemp was significantly downregulated under drought stress. These results demonstrate that decreased IAA expression is responsible for repressed cell enlargement, photosynthesis, and plant growth during water stress in hemp.

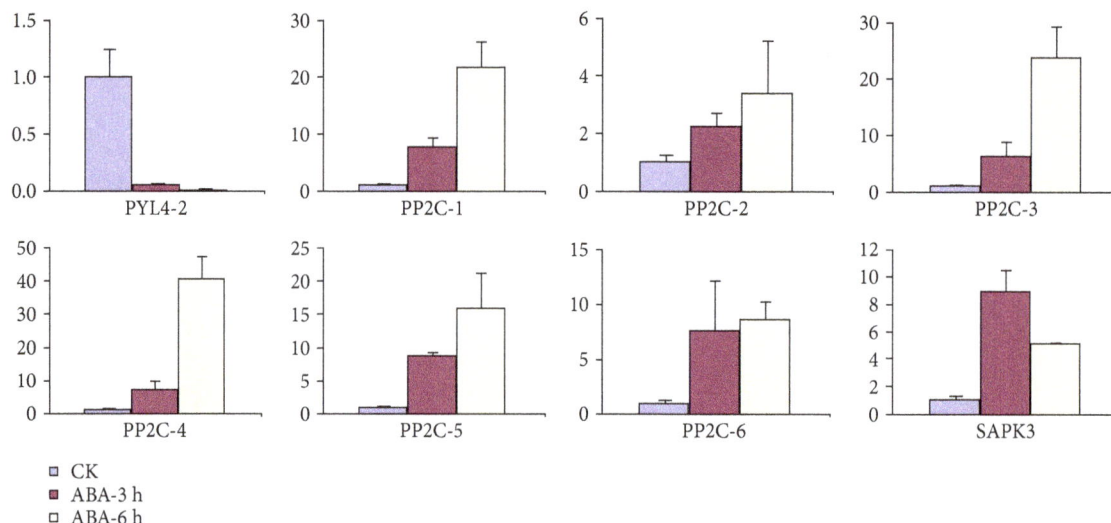

FIGURE 7: Expression profiles of drought stress-induced genes after ABA treatment determined by qRT-PCR. Data from qRT-PCR are means of three replicates, and bars represent standard error. PYL4: abscisic acid receptor PYL4 (JP449530); PP2C-1: protein phosphatase 2C (JP474397); PP2C-2: protein phosphatase 2C (JP473042); PP2C-3: protein phosphatase 2C (JP478284); PP2C-4: protein phosphatase 2C (JP477418); PP2C-5: protein phosphatase 2C (JP469562); PP2C-6: protein phosphatase 2C (JP455237); SAPK3: serine/threonine-protein kinase (JP477935).

5. Conclusions

This study presents, for the first time, the characterization of genome-wide expression profiling of hemp in response to drought stress. A total of 1292 DEGs were identified, including 409 upregulated genes and 883 downregulated genes. Some genes, including POD, expansin, NAC, and B3 TFs, were shown to be significantly enriched by GO and KEGG analyses, and they may play important roles in hemp adaptation or tolerance to the osmotic and oxidative stresses caused by drought stress. These genes may contribute to further study of the drought stress tolerance of hemp. In addition, we demonstrated that ABA and auxin are crucial in the response of hemp to drought stress. These results are expected to help improve our understanding of the drought stress regulatory mechanism of hemp and improve the drought tolerance ability of the crops.

Conflicts of Interest

The authors declare that there are no competing interests.

Authors' Contributions

Gonggu Zang and Chunsheng Gao designed the research. Chunsheng Gao, Chaohua Cheng, Lining Zhao, and Pengfei Xin performed the research. Chunsheng Gao, Lining Zhao, Qing Tang, and Pengfei Xin analyzed the data. Yongting Yu, Yuan Guo, Zhun Yan, and Touming Liu assisted with the experiments and reviewed the manuscript. All authors have read and approved the final manuscript.

Acknowledgments

This work was financially supported by the China Agricultural Science and Technology Innovation project (ASTIP-IBFC), the National Modern Agro-industry Technology Research System (nycytx-19-E04), and the National Natural Science Foundation of China (no. 31301376). The authors would like to thank Enpub (http://www.enpub.cn) for English language editing.

Supplementary Materials

Supplementary 1. Figure S1: RPKM distribution and RPKM density distribution of different genes in the DS and CK hemp. A: box-plot of RPKM. *x*-axis means sample names, and *y*-axis means log10(RPKM). B: RPKM distribution. *x*-axis represent log10(RPKM), and *y*-axis represents gene density. E1, E2, E3, and E4 represent DS1, DS2, CK1 and CK2, respectively. Figure S2: correlations between two repeats of the CK and DS samples. E1 and E2 represent DS1 and DS2, and E3 and E4 represent CK1 and CK2, respectively. The logarithmic values of (FPKM) for each gene in the two replicates were assigned as coordinate values of the two axes.

Supplementary 2. Table S1: the sequence of primers used for qRT-PCR analysis.

Supplementary 3. Table S2: summary of DEGs annotated by 5 databases.

Supplementary 4. Table S3: summary of GO-enriched DEGs.

Supplementary 5. Table S4: details of 11 significantly enriched KEGG pathways in DEGs.

Supplementary 6. Table S5: GO terms responding to drought stress.

Supplementary 7. Table S6: details of DEGs annotated as transcription factor.

References

[1] B. Vinocur and A. Altman, "Recent advances in engineering plant tolerance to abiotic stress: achievements and limitations," *Current Opinion in Biotechnology*, vol. 16, no. 2, pp. 123–132, 2005.

[2] Q. Zhang, "Strategies for developing Green Super Rice," *Proceedings of the National Academy of Sciences of the United States of America*, vol. 104, no. 42, pp. 16402–16409, 2007.

[3] K. Yamaguchi-Shinozaki and K. Shinozaki, "Transcriptional regulatory networks in cellular responses and tolerance to dehydration and cold stresses," *Annual Review of Plant Biology*, vol. 57, no. 1, pp. 781–803, 2006.

[4] D. Todaka, K. Shinozaki, and K. Yamaguchi-Shinozaki, "Recent advances in the dissection of drought-stress regulatory networks and strategies for development of drought-tolerant transgenic rice plants," *Frontiers in Plant Science*, vol. 6, p. 84, 2015.

[5] G. T. Huang, S. L. Ma, L. P. Bai et al., "Signal transduction during cold, salt, and drought stresses in plants," *Molecular Biology Reports*, vol. 39, no. 2, pp. 969–987, 2012.

[6] K. Nakashima, K. Yamaguchi-Shinozaki, and K. Shinozaki, "The transcriptional regulatory network in the drought response and its crosstalk in abiotic stress responses including drought, cold, and heat," *Frontiers in Plant Science*, vol. 5, p. 170, 2014.

[7] A. Abot, C. Bonnafous, F. Touchard et al., "Effects of cultural conditions on the hemp (*Cannabis sativa*) phloem fibres: Biological development and mechanical properties," *Journal of Composite Materials*, vol. 47, no. 8, pp. 1067–1077, 2012.

[8] J. D. House, J. Neufeld, and G. Leson, "Evaluating the quality of protein from hemp seed (*Cannabis sativa* L.) products through the use of the protein digestibility-corrected amino acid score method," *Journal of Agricultural and Food Chemistry*, vol. 58, no. 22, pp. 11801–11807, 2010.

[9] P. Bouloc, S. Allegret, and L. Arnaud, *Hemp: Industrial Production and Uses*, CABI, Wallingford, UK, 2012.

[10] T. Schäfer and B. Honermeier, "Effect of sowing date and plant density on the cell morphology of hemp (*Cannabis sativa* L.)," *Industrial Crops and Products*, vol. 23, no. 1, pp. 88–98, 2006.

[11] S. Amaducci, A. Zatta, F. Pelatti, and G. Venturi, "Influence of agronomic factors on yield and quality of hemp (*Cannabis sativa* L.) fibre and implication for an innovative production system," *Field Crops Research*, vol. 107, no. 2, pp. 161–169, 2008.

[12] M. Mihoc, G. Pop, E. Alexa, and I. Radulov, "Nutritive quality of Romanian hemp varieties (*Cannabis sativa* L.) with special focus on oil and metal contents of seeds," *Chemistry Central Journal*, vol. 6, no. 1, p. 122, 2012.

[13] C. M. Deng, J. Li, T. Sun, and C. J. Li, "Occurrence regularity and control measurements of main diseases and insect pests in hemp," *Yunnan Nongye Keji*, vol. 4, pp. 48-49, 2007.

[14] Y. Jiang, "China's water scarcity," *Journal of Environmental Management*, vol. 90, no. 11, pp. 3185–3196, 2009.

[15] H. van Bakel, J. M. Stout, A. G. Cote et al., "The draft genome and transcriptome of *Cannabis sativa*," *Genome Biology*, vol. 12, no. 10, article R102, 2011.

[16] C. Gao, P. Xin, C. Cheng et al., "Diversity analysis in *Cannabis sativa* based on large-scale development of expressed sequence tag-derived simple sequence repeat markers," *PLoS One*, vol. 9, no. 10, article e110638, 2014.

[17] Y. Hong, W. Zhang, and X. Wang, "Phospholipase D and phosphatidic acid signalling in plant response to drought and salinity," *Plant, Cell & Environment*, vol. 33, no. 4, pp. 627–635, 2010.

[18] T. Liu, S. Zhu, Q. Tang, Y. Yu, and S. Tang, "Identification of drought stress-responsive transcription factors in ramie (*Boehmeria nivea* L. Gaud)," *BMC Plant Biology*, vol. 13, no. 1, p. 130, 2013.

[19] Y. Yu, L. Zeng, Z. Yan et al., "Identification of ramie genes in response to *Pratylenchus coffeae* infection challenge by digital gene expression analysis," *International Journal of Molecular Sciences*, vol. 16, no. 9, pp. 21989–22007, 2015.

[20] L. Zeng, A. Shen, J. Chen et al., "Transcriptome analysis of ramie (*Boehmeria nivea* L. Gaud.) in response to ramie moth (*Cocytodes coerulea* Guenée) infestation," *BioMed Research International*, vol. 2016, Article ID 3702789, 10 pages, 2016.

[21] Y. Guo, C. Qiu, S. Long et al., "Digital gene expression profiling of flax (*Linum usitatissimum* L.) stem peel identifies genes enriched in fiber-bearing phloem tissue," *Gene*, vol. 626, pp. 32–40, 2017.

[22] H. Li, D. Li, A. Chen, H. Tang, J. Li, and S. Huang, "Characterization of the Kenaf (*Hibiscus cannabinus*) global transcriptome using Illumina Paired-End sequencing and development of EST-SSR markers," *PLoS One*, vol. 11, no. 3, article e0150548, 2016.

[23] Z. Yang, A. Yan, R. Lu et al., "De novo transcriptome sequencing of two cultivated jute species under salinity stress," *PLoS One*, vol. 12, no. 10, article e0185863, 2017.

[24] L. Huang, F. Zhang, F. Zhang et al., "Comparative transcriptome sequencing of tolerant rice introgression line and its parents in response to drought stress," *BMC Genomics*, vol. 15, no. 1, p. 1026, 2014.

[25] Z. Liu, M. Xin, J. Qin et al., "Temporal transcriptome profiling reveals expression partitioning of homeologous genes contributing to heat and drought acclimation in wheat (*Triticum aestivum* L.)," *BMC Plant Biology*, vol. 15, no. 1, p. 152, 2015.

[26] J. Xu, Y. Yuan, Y. Xu et al., "Identification of candidate genes for drought tolerance by whole-genome resequencing in maize," *BMC Plant Biology*, vol. 14, no. 1, p. 83, 2014.

[27] X. D. Fan, J. Q. Wang, N. Yang et al., "Gene expression profiling of soybean leaves and roots under salt, saline–alkali and drought stress by high-throughput Illumina sequencing," *Gene*, vol. 512, no. 2, pp. 392–402, 2013.

[28] M. J. Bowman, W. Park, P. J. Bauer et al., "RNA-Seq transcriptome profiling of upland cotton (*Gossypium hirsutum* L.) root tissue under water-deficit stress," *PLoS One*, vol. 8, no. 12, article e82634, 2013.

[29] F. Xie, Q. Wang, R. Sun, and B. Zhang, "Deep sequencing reveals important roles of microRNAs in response to drought and salinity stress in cotton," *Journal of Experimental Botany*, vol. 66, no. 3, pp. 789–804, 2015.

[30] N. Zhang, B. Liu, C. Ma et al., "Transcriptome characterization and sequencing-based identification of drought-responsive

genes in potato," *Molecular Biology Reports*, vol. 41, no. 1, pp. 505–517, 2014.

[31] G. Yuan, W. Yu-Fu, Q. Cai-Sheng, L. Song-Hua, D. Xin, and H. Dong-Mei, "Preliminary study on effects of drought stress on physiological characteristics and growth of different hemp cultivars (*Cannabis Sativa* L.)," *Plant Fiber Sciences in China*, vol. 33, no. 5, pp. 235–239, 2011.

[32] H. Hu, M. Dai, J. Yao et al., "Overexpressing a NAM, ATAF, and CUC (NAC) transcription factor enhances drought resistance and salt tolerance in rice," *Proceedings of the National Academy of Sciences of the United States of America*, vol. 103, no. 35, pp. 12987–12992, 2006.

[33] M. I. Love, W. Huber, and S. Anders, "Moderated estimation of fold change and dispersion for RNA-seq data with DESeq2," *Genome Biology*, vol. 15, no. 12, p. 550, 2014.

[34] M. Kanehisa, Y. Sato, M. Kawashima, M. Furumichi, and M. Tanabe, "KEGG as a reference resource for gene and protein annotation," *Nucleic Acids Research*, vol. 44, no. D1, pp. D457–D462, 2016.

[35] A. Conesa, S. Gotz, J. M. Garcia-Gomez, J. Terol, M. Talon, and M. Robles, "Blast2GO: a universal tool for annotation, visualization and analysis in functional genomics research," *Bioinformatics*, vol. 21, no. 18, pp. 3674–3676, 2005.

[36] K. Shinozaki and K. Yamaguchi-Shinozaki, "Gene networks involved in drought stress response and tolerance," *Journal of Experimental Botany*, vol. 58, no. 2, pp. 221–227, 2007.

[37] A. R. Reddy, K. V. Chaitanya, and M. Vivekanandan, "Drought-induced responses of photosynthesis and antioxidant metabolism in higher plants," *Journal of Plant Physiology*, vol. 161, no. 11, pp. 1189–1202, 2004.

[38] M. H. Cruz de Carvalho, "Drought stress and reactive oxygen species: production, scavenging and signaling," *Plant Signaling & Behavior*, vol. 3, no. 3, pp. 156–165, 2014.

[39] H. Zhai, F. Wang, Z. Si et al., "A *myo*-inositol-1-phosphate synthase gene, *IbMIPS1*, enhances salt and drought tolerance and stem nematode resistance in transgenic sweet potato," *Plant Biotechnology Journal*, vol. 14, no. 2, pp. 592–602, 2016.

[40] J. Duan, M. Zhang, H. Zhang et al., "*OsMIOX*, a *myo*-inositol oxygenase gene, improves drought tolerance through scavenging of reactive oxygen species in rice (*Oryza sativa* L.)," *Plant Science*, vol. 196, pp. 143–151, 2012.

[41] J. Yang, J. Zhang, Z. Wang, G. Xu, and Q. Zhu, "Activities of key enzymes in sucrose-to-starch conversion in wheat grains subjected to water deficit during grain filling," *Plant Physiology*, vol. 135, no. 3, pp. 1621–1629, 2004.

[42] Z. Wang, Y. Xu, T. Chen, H. Zhang, J. Yang, and J. Zhang, "Abscisic acid and the key enzymes and genes in sucrose-to-starch conversion in rice spikelets in response to soil drying during grain filling," *Planta*, vol. 241, no. 5, pp. 1091–1107, 2015.

[43] M. Thalmann, D. Pazmino, D. Seung et al., "Regulation of leaf starch degradation by abscisic acid is important for osmotic stress tolerance in plants," *The Plant Cell*, vol. 28, no. 8, pp. 1860–1878, 2016.

[44] K. J. D. Lee, S. E. Marcus, and J. P. Knox, "Cell wall biology: perspectives from cell wall imaging," *Molecular Plant*, vol. 4, no. 2, pp. 212–219, 2011.

[45] D. M. Cavalier, O. Lerouxel, L. Neumetzler et al., "Disrupting two *Arabidopsis thaliana* xylosyltransferase genes results in plants deficient in xyloglucan, a major primary cell wall component," *The Plant Cell*, vol. 20, no. 6, pp. 1519–1537, 2008.

[46] K. Nakashima, Y. Ito, and K. Yamaguchi-Shinozaki, "Transcriptional regulatory networks in response to abiotic stresses in Arabidopsis and grasses," *Plant Physiology*, vol. 149, no. 1, pp. 88–95, 2009.

[47] Y. Fujita, M. Fujita, K. Shinozaki, and K. Yamaguchi-Shinozaki, "ABA-mediated transcriptional regulation in response to osmotic stress in plants," *Journal of Plant Research*, vol. 124, no. 4, pp. 509–525, 2011.

[48] Y. Hong, H. Zhang, L. Huang, D. Li, and F. Song, "Overexpression of a stress-responsive NAC transcription factor gene *ONAC022* improves drought and salt tolerance in rice," *Frontiers in Plant Science*, vol. 7, p. 4, 2016.

[49] V. P. Thirumalaikumar, V. Devkar, N. Mehterov et al., "NAC transcription factor JUNGBRUNNEN1 enhances drought tolerance in tomato," *Plant Biotechnology Journal*, vol. 16, no. 2, pp. 354–366, 2018.

[50] D. Chen, S. Chai, C. L. McIntyre, and G. P. Xue, "Overexpression of a predominantly root-expressed NAC transcription factor in wheat roots enhances root length, biomass and drought tolerance," *Plant Cell Reports*, vol. 37, no. 2, pp. 225–237, 2018.

[51] J. K. Zhu, "Salt and drought stress signal transduction in plants," *Annual Review of Plant Biology*, vol. 53, no. 1, pp. 247–273, 2002.

[52] R. Bari and J. D. G. Jones, "Role of plant hormones in plant defence responses," *Plant Molecular Biology*, vol. 69, no. 4, pp. 473–488, 2009.

[53] T. Umezawa, K. Nakashima, T. Miyakawa et al., "Molecular basis of the core regulatory network in ABA responses: sensing, signaling and transport," *Plant & Cell Physiology*, vol. 51, no. 11, pp. 1821–1839, 2010.

[54] S. Y. Park, P. Fung, N. Nishimura et al., "Abscisic acid inhibits type 2C protein phosphatases via the PYR/PYL family of START proteins," *Science*, vol. 324, no. 5930, pp. 1068–1071, 2009.

[55] A. Kulik, I. Wawer, E. Krzywińska, M. Bucholc, and G. Dobrowolska, "SnRK2 protein kinases—key regulators of plant response to abiotic stresses," *OMICS: A Journal of Integrative Biology*, vol. 15, no. 12, pp. 859–872, 2011.

[56] H. Kim, K. Lee, H. Hwang et al., "Overexpression of *PYL5* in rice enhances drought tolerance, inhibits growth, and modulates gene expression," *Journal of Experimental Botany*, vol. 65, no. 2, pp. 453–464, 2014.

[57] L. Liu, X. Hu, J. Song, X. Zong, D. Li, and D. Li, "Overexpression of a *Zea mays* L. protein phosphatase 2C gene (*ZmPP2C*) in *Arabidopsis thaliana* decreases tolerance to salt and drought," *Journal of Plant Physiology*, vol. 166, no. 5, pp. 531–542, 2009.

[58] T. T. Phan, B. Sun, J. Q. Niu et al., "Overexpression of sugarcane gene *SoSnRK2.1* confers drought tolerance in transgenic tobacco," *Plant Cell Reports*, vol. 35, no. 9, pp. 1891–1905, 2016.

[59] J. Santiago, A. Rodrigues, A. Saez et al., "Modulation of drought resistance by the abscisic acid receptor PYL5 through inhibition of clade A PP2Cs," *The Plant Journal*, vol. 60, no. 4, pp. 575–588, 2009.

[60] E. Krzywińska, A. Kulik, M. Bucholc, M. A. Fernandez, P. L. Rodriguez, and G. Dobrowolska, "Protein phosphatase type 2C PP2CA together with ABI1 inhibits SnRK2.4 activity

and regulates plant responses to salinity," *Plant Signaling & Behavior*, vol. 11, no. 12, article e1253647, 2016.

[61] L. Sun, Y. P. Wang, P. Chen et al., "Transcriptional regulation of *SlPYL*, *SlPP2C*, and *SlSnRK2* gene families encoding ABA signal core components during tomato fruit development and drought stress," *Journal of Experimental Botany*, vol. 62, no. 15, pp. 5659–5669, 2011.

[62] P. Chen, Y. F. Sun, W. B. Kai et al., "Interactions of ABA signaling core components (SlPYLs, SlPP2Cs, and SlSnRK2s) in tomato (*Solanum lycopersicon*)," *Journal of Plant Physiology*, vol. 205, pp. 67–74, 2016.

[63] Y. Wang, Y. Wu, C. Duan et al., "The expression profiling of the *CsPYL*, *CsPP2C* and *CsSnRK2* gene families during fruit development and drought stress in cucumber," *Journal of Plant Physiology*, vol. 169, no. 18, pp. 1874–1882, 2012.

[64] M. R. Blatt, "Cellular signaling and volume control in stomatal movements in plants," *Annual Review of Cell and Developmental Biology*, vol. 16, no. 1, pp. 221–241, 2000.

[65] L. Yang, W. Ji, P. Gao et al., "GsAPK, an ABA-activated and Calcium-independent SnRK2-type kinase from *G. soja*, mediates the regulation of plant tolerance to salinity and ABA stress," *PLos One*, vol. 7, no. 3, article e33838, 2012.

[66] P. E. Staswick, B. Serban, M. Rowe et al., "Characterization of an Arabidopsis enzyme family that conjugates amino acids to indole-3-acetic acid," *The Plant Cell*, vol. 17, no. 2, pp. 616–627, 2005.

[67] H. Du, N. Wu, J. Fu et al., "A GH3 family member, OsGH3-2, modulates auxin and abscisic acid levels and differentially affects drought and cold tolerance in rice," *Journal of Experimental Botany*, vol. 63, no. 18, pp. 6467–6480, 2012.

Identification of Wheat Inflorescence Development-Related Genes using a Comparative Transcriptomics Approach

Lingjie Ma, Sheng-Wei Ma, Qingyan Deng, Yang Yuan, Zhaoyan Wei, Haiyan Jia⑩, and Zhengqiang Ma

The Applied Plant Genomics Laboratory of Crop Genomics and Bioinformatics Centre, Nanjing Agricultural University, Jiangsu 210095, China

Correspondence should be addressed to Haiyan Jia; hyjia@njau.edu.cn

Academic Editor: Marco Gerdol

Inflorescence represents the highly specialized plant tissue producing the grains. Although key genes regulating flower initiation and development are conserved, the mechanism regulating fertility is still not well explained. To identify genes and gene network underlying inflorescence morphology and fertility of bread wheat, expressed sequence tags (ESTs) from different tissues were analyzed using a comparative transcriptomics approach. Based on statistical comparison of EST frequencies of individual genes in EST pools representing different tissues and verification with RT-PCR and RNA-seq data, 170 genes of 59 gene sets predominantly expressed in the inflorescence were obtained. Nearly one-third of the gene sets displayed differentiated expression profiles in terms of their subgenome orthologs. The identified genes, most of which were predominantly expressed in anthers, encode proteins involved in wheat floral identity determination, anther and pollen development, pollen-pistil interaction, and others. Particularly, 25 annotated gene sets are associated with pollen wall formation, of which 18 encode enzymes or proteins participating in lipid metabolic pathway, including fatty acid ω-hydroxylation, alkane and fatty alcohol biosynthesis, and glycerophospholipid metabolism. We showed that the comparative transcriptomics approach was effective in identifying genes for reproductive development and found that lipid metabolism was particularly active in wheat anthers.

1. Introduction

Inflorescences are the reproductive architectures of plants, composed of stems, stalks, bracts, and flowers. *Poaceae* (also called *Gramineae*) is one of the largest families in the monocotyledonous flowering plants, including the major cereal crops, such as maize (*Zea mays* L.), rice (*Oryza sativa* L.), and wheat (*Triticum aestivum* L.). Inflorescences of this family are characterized by their panicle or spike shapes [1], complex branches, and unique spikelets, as well as inconspicuous and anemophilous flowers without obvious petals and sepals [2, 3]. Species in the genus *Triticum* takes the shape of the spike with the spikelets, each containing one or more florets, attached to rachis nodes. The wheat floret consists of a carpel with its ovary, style, and stigma, with three anthers attached to the base through slender filaments, which are enclosed by bract-like organs called lemma and palea.

Inflorescence development and regulation have attracted great attention from plant biologists and crop breeders since they are crucial for reproduction of flowering plants and for production of food grains in cereal crops. Although the key genes regulating the flower initiation and development are conserved in higher plants [4, 5], the diversity of reproductive structure and behaviors is still not well explained. Transcriptomes reflect the complete set of RNA transcripts expressed by the genome under developmentally or physiologically distinct states; therefore, its comparison allows identification of genes under common regulation. High-throughput methods, such as serial analysis of gene expression (SAGE) [6], microarray technology [7], and next-generation sequencing [8] have enabled transcriptome studies in an unprecedented scale in many plant species, especially in those with full genome sequence information available. This has led to the discovery of a number of genes involved in flower development. These genes display tempo-spatial expression patterns not only in transcriptome level [9–12] but also at the proteome level, such as those in pollen development of tomatoes [13], indicating their strictly regulated functions. Anther tissues were used in most of these studies because they are easy to isolate and specific in biological roles. The availability of these data has shed some light on the gene networks that contribute to the formation of unique reproductive structures and flower development, although it might still not be enough because of the particularity of different plant species and highly specific flower components.

In bread wheat (*T. aestivum* L.), an allohexaploid species ($2n = 42$) with three closely related subgenomes, that is, A, B, and D [14], large-scale transcriptomic investigations have been conducted in some tissue types. Crismani et al. performed microarray-based expression analysis of anthers across various stages of meiosis in an attempt to build the link between the model and wheat and identify early meiotic genes [15]. McIntosh et al. investigated the transcriptome in wheat caryopsis development using data generated by SAGE [16]. Yang et al. compared the RNA-seq tags of pistillody stamen and pistil from a pistillody wheat mutant and stamen from the wild-type control to identify differentially expressed genes [17]. In these studies, the annotation of the identified transcripts was based on short tags or sequence reads which is impossible to ensure accuracy for species such as wheat that has a complex genome and does not have detailed sequence information so far. Moreover, gene identity determination according to short sequence fragments could be problematic due to functional diversification of the homoeologous genes in polyploidy species.

Expressed sequence tags (ESTs) are single-pass sequence reads by sequencing cDNA libraries and usually have >400-base read length. They represent part of the transcriptome in a given tissue and/or at a given developmental stage. As of January 1, 2013, over one million ESTs are available for wheat in GenBank of the National Center for Biotechnology Information (NCBI) and are valuable for identifying genes involved in biotic and abiotic stress [18–22], kernel development and quality [23, 24], and development [21, 25, 26].

In this paper, we reported mining of 170 genes predominantly expressed in floral organs through comparative transcriptomic profiling using ESTs of 67 wheat cDNA libraries deposited in GenBank genes and the identification of a few metabolic pathways involved in anther development.

2. Materials and Methods

2.1. Wheat ESTs and Contigs.

ESTs were downloaded from the NCBI dbEST database (ftp://ncbi.nlm.nih.gov/repository/dbEST). The ESTs used in this study were produced with 67 cDNA libraries prepared using inflorescences (spike at flowering or before flowering, anther, pistil, ovary, palea, and lemma), roots, stems, leaves (including seedlings and crown tissues), and seeds (matured or immature embryos) from normally grown seedlings or plants and represented 434,658 cloning events (Table S1). Libraries subjected to enrichment or normalization treatment and those with less than 1000 ESTs were not included in the expression profiling analysis. The 163rd release of unique wheat transcripts, including 77,657 contigs, was downloaded from the PUT database (http://www.plantgdb.org).

2.2. Gene Mining.

To identify putative genes predominantly expressed in wheat flowers, BLAST search was conducted using the PUT contig sequences as queries against the EST database consisting of the 67 cDNA libraries. ESTs matching each PUT contig with ≥90% homology were recorded and classified according to their library origins from the five aforementioned tissue types. The contigs with matched ESTs solely from inflorescences and those with significantly more matched ESTs from inflorescence tissues than from any of the other four types of tissues were considered to be the putative genes predominantly expressed in inflorescences. The probability to achieve the ESTs matching to a contig in the noninflorescence libraries was estimated based on the random sampling principle using the equation: $P = C_N^n f^n (1-f)^{N-n}$, as described by Ding et al. [27], where f is the EST frequency of a contig in the inflorescence libraries estimated by dividing the number of matched ESTs by the total number of ESTs in the libraries and n and N are the number of matched ESTs and the total number of ESTs in libraries of other tissues types, respectively. A sampling probability ≤ 0.0001 was considered an indication of significant difference in expression levels between the inflorescence tissues and noninflorescence tissues.

Chromosome-specific genomic DNA sequences corresponding to the retained contigs were obtained by retrieving the chromosome-assigned homologous scaffolds in the TGAC database [28] (http://www.tgac.ac.uk/grassroots-genomics) that showed 95% homology to the contig sequences. Coding DNA sequences (CDS) corresponding to the contigs in the scaffolds were predicted using the gene-finding program FGENESH [29] (http://linux1.softberry.com) and verified by alignment with the homologous ESTs. Scaffolds that did not contain the full-length target CDS were extended via in silico walking, using a parameter of 100% match, with the Roche 454 sequence reads of Chinese Spring [30] (http://www.cerealsdb.uk.net).

To confirm the expression predominance of the identified genes in wheat flowers, ESTs with ≥99% homology to the individual CDS identified from the sixty-seven libraries were subjected to further analysis. The sampling probability to achieve the EST frequency of a CDS in the noninflorescence libraries was again calculated. A sampling probability ≤ 0.01 was used as the threshold for declaration of significant difference. Expression specificity of the candidate genes was estimated similarly.

The identified genes were coded in numerical order with the prefix "IDG" (inflorescence development-related gene). Orthologous genes were given a common name but marked with a chromosome assignment suffix. Multiple copies derived from duplication of an ancestral gene at the same chromosome were differentiated by a numerical suffix.

2.3. RNA-Seq Data Analysis. RNA-data sets used in the analysis included the developmental time course series in five tissues (spike or inflorescence, root, leaf, grain, and stem) each with three different developmental stages. The stages of inflorescence included Z32, two nodes stage; Z39, meiosis; Z65, anthesis of C.S. [31] (http://wheat.pw.usda.gov/WheatExp), those for C.S. pistil and stamen by Yang et al. [17], and those for anther at meiosis data from the URGI public database (https://wheat-urgi.versailles.inra.fr/Seq-Repository/Expression). After trimmed adapters and removing vectors and low quality reads with Adapter-Removal in a setting of quality base = 33 [32], the RNA-seq reads were mapped to the sequence database consisting of genomic DNA sequences of each identified candidate gene using HISAT2 [33]. FPKM of a gene was estimated with reads matching to each gene with 100% identity, which were counted using featureCounts with both readExtension5 and readExtension3 set at 70 [34].

To estimate the relative expression specificity (RES) of a gene in a certain tissue (A), the tissue (B) in which the gene was most highly expressed among all but tissue A was identified. RES was then measured by dividing the difference of the normalized expression values between A and B by the expression value in A. The higher the RES value is, the more specific the gene expression is in A. Significance of the expression difference between two tissues was examined via χ^2 test. A gene was considered to be differentially expressed when the difference reached significance at $P = 0.05$.

To reflect the relative expression abundance of each gene in each tissue among the identified genes, a heat map was drawn with the quotients obtained from dividing the FPKM value (the maximal one if two or more developmental stages of a tissue were involved) of a specific gene in a tissue by the FPKM means of all genes. A log-transformation was applied to facilitate the mapping drawing.

2.4. RT-PCR. Root, node, internode, flag leaf, glume, lemma, palea, lodicule, stamen, pistil, and rachis tissues of the common wheat landrace "Wangshuibai," grown in a field during the normal growing season, were collected at the heading stage, and developing kernels were collected at the 9th day postanthesis for RNA extraction. RNA was extracted using TRIzol reagent (Invitrogen, CA) following the manufacturer's protocol and quantified with an Ultrospec 2100 Pro spectrometer (Amersham Pharmacia, UK). To eliminate DNA contamination, RNA samples were treated with RNase-free DNase I (Fermentas, Canada), following the product manual. First-strand cDNA was synthesized using oligo(dT) primer with 3 μg of total RNA using M-MLV reverse transcriptase (Life Technologies, CA) according to the manufacturer's instructions.

Semiquantitative RT-PCR (sqRT-PCR) was performed in a 25 μl total reaction volume supplemented with 10–20 ng of first-strand cDNA as the template, 5 pmol concentration of each primer, 5 nmol dNTPs for each, 1 U rTaq DNA polymerase (Takara, Japan), and 1x PCR buffer supplied together with the enzyme. The wheat α-tubulin gene was used in calibrating cDNA templates. The following thermal cycle profile was observed: 94°C for 3 min; 26–32 cycles of 94°C

for 20 s, 56–62°C (depending on the primer sets) for 25 s, and 72°C for 30 s, and a final extension step of 72°C for 5 min. PCR products were resolved in 2.0% (w/v) agarose gels and visualized with ethidium bromide staining. RT-PCR reactions were independently repeated three times or more to ensure reproducibility.

Quantitative real-time RT-PCR (qRT-PCR) amplifications in 20 μL volumes, containing 10 μL SYBRÒ Green qRCR Mix (Toyoba), 20 ng template, and 8 pmol of each primer, were performed using the StepOneTM Real-Time PCR instrument (Applied Biosystems), following the protocol described in [35]. The reactions were conducted in triplicate. The cycle threshold values for each target gene were normalized based on values obtained in corresponding reactions for the wheat α-tubulin gene. The relative expression was estimated by employing the $2^{-\Delta\Delta CT}$ method [36].

The RT-PCR primers used in the present study were designed with MacVector 11 (MacVector, NC) and are listed in Table S2. The product length ranged from 150–300 bp.

2.5. Gene Annotation and Pathway Assignment. Genes were functionally annotated through homologous search of the NCBI nonredundant (Nr) protein, KEGG [37] (http://www.kegg.jp/blastkoala), and Pfam [38] (http://pfam.xfam.org) databases. Subcellular locations of the proteins were predicted using the TargetP 1.1 server [39] (http://www.cbs.dtu.dk/services/TargetP). Pathway assignments were based on KEGG pathway mapping (http://www.kegg.jp/kegg/tool/map_pathway1.html) and keyword search of the plant metabolic pathway databases (http://www.plantcyc.org).

3. Results

3.1. Identification of Genes Preferentially Expressed in Inflorescence. Of the 67 cDNA libraries in line with the screening conditions, 29 libraries containing 140,092 sequences were from seed tissues; only three cDNA libraries including 17,732 sequences were from stem tissues (Table S1). The number of ESTs in each library ranged from 1000 to more than 10,000. For identification of the inflorescence development-related genes, these libraries were classified into five types according to the tissues used in library preparation, including seedling-stage leaf and stem, seedling to tillering-stage root, seed (from DPA3 to mature), and inflorescence (including premeiotic anthers, anthers at meiosis, pistil and ovary, immature inflorescence, lemma and palea, spike before flowering, and spike at flowering).

Alignment of the 77,657 wheat EST contigs, downloaded from the PlantGDB-assembled unique transcript (PUT) database, with ESTs of the abovementioned cDNA libraries led to the identification of 335 contigs that matched (95% homology) significantly more ESTs from inflorescence tissues, including nearly one-third with ESTs only from inflorescence tissue libraries. Using these contig sequences as queries in search of the TGAC database [28] with 90% as the homology threshold, 318 nonabundant genomic DNA sequences were obtained, 294 of which yield the expected open reading frames (ORF) with EST

TABLE 1: Chromosome distribution of the identified wheat inflorescence development-related genes.

Gene_ID	Chromosomes			Gene_ID	Chromosomes		
IDG001	1A	1B	1D	IDG031	c	c	3D
IDG002	1A	1B	1D	IDG032	4A	b	5D
IDG003	1A	1B	1D	IDG033	4A	4B	4D
IDG004	1A	1B	1D	IDG034	c	4B	4D
IDG005	1A	1B	1D	IDG035	5A (2)	4B	4D (2)
IDG006	1A	1B	1D	IDG036	5A	5B	5D
IDG007	1A	1B	1D	IDG037	5A	5B	5D
IDG008	1A	1B	1D	IDG038	5A	5B	5D
IDG009	1A	1B	1D	IDG039	5A	5B	5D
IDG010	1A	1B	1D	IDG040	c	5B (2)	5D
IDG011	1A	c^1	a	IDG041	b	5B	5D
IDG012	c	1B	1D	IDG042	6A (8)	6B (3)	6D (3)
IDG013	b	1B	b	IDG043	b	6B	6D (2)
IDG014	2A	2B	2D	IDG044	6A	6B	6D
IDG015	2A	a	2D	IDG045	6A	6B	6D
IDG016	2A	2B	2D	IDG046	6A	c	b
IDG017	2A	2B	b	IDG047	6A	6B	6D
IDG018	a	2B	2D	IDG048	6A	6B	6D
IDG019	3A	3B	3D	IDG049	6A (2)	6B	6D
IDG020	3A	3B	3D	IDG050.1	6A	a	6D
IDG021	3A	3B	3D	IDG050.2	6A	b	6D
IDG022	3A	3B	3D	IDG051	b	6B	6D
IDG023	3A	3B	3D	IDG052	c	6B (2)	c
IDG024	3A	3B	3D	IDG053	a	6B	a
IDG025	3A	3B	3D	IDG054	7A	7B	7D
IDG026	3A	3B	3D	IDG055	7A	7B	7D
IDG027	3A	c	3D	IDG056	7A	7B	7D
IDG022	3A	3B	3D	IDG057	7A	a	7D
IDG029	3A (2)	3B (2)	3D (2)	IDG058	b	7B	b
IDG030	3A	3B	3D	IDG059	c	c	7D

a: corresponding gene was not found; b: genomic DNA is available but not supported by ESTs; c: the gene was not inflorescence predominantly expressed. Copy number is shown in parenthesis.

support in gene prediction with the gene-finding program FGENESH [29]. With 99% homology as the cutoff, 187 genes matched significantly more ESTs from the inflorescence tissues than from any of the other four types of tissues ($P = 0.01$). We disregarded 17 of these genes that were neither more abundant in the inflorescence RNA-seq datasets nor in the anther and pistil data sets. Interestingly, most ESTs matching to 14 of the 17 genes came from Ogihara's unpublished cDNA libraries Wh_FL or Wh_f, which were constructed with spikelets or spikes at flowering stage. Probably, these genes are expressed at a developmental inflorescence stage not included in the tissues for the RNA sequencing. The remaining 170 genes represented 59 inflorescence development-related nonredundant gene sets, since wheat is an allohexaploid species and the majority of the genes have three orthologous copies (Table 1 and Table S3).

In terms of genomic distribution, the identified genes distributed to chromosomes of homoeologous groups 1, 3, and 6 accounted for 68.2%, those to group 4 chromosomes accounted for only 5.3%. Majority of the nonredundant gene sets have homologs in all three homoeologous chromosomes; however, not all of them showed inflorescence development-related preferential expression, suggesting functional differentiation have occurred among them. Of the 59 nonredundant gene sets, 13 had inflorescence development-related homologs in only two of three subgenomes, eight were solely identified in a single subgenome. Eight nonredundant gene sets had intrachromosomal duplications, of which five were distributed to the homoeologous group 6 chromosomes. A notable example was IDG042 that had duplications in all three group 6 chromosomes, with a total of 14 copies showing inflorescence development-related preferential expression.

3.2. Expression Specificity of the Identified Genes. In mining genes predominantly expressed in inflorescence, we considered all tissues from inflorescence as a whole. It was noted, in the EST analysis, that some genes were solely expressed in spikes at or before anthesis, some were solely in stamen or in pistil, apart from those expressed more abundantly in spikes than in other tissues. To validate the expression patterns of these genes, we tested the significance of expression difference and estimated the expression specificity (RES) of inflorescence (including the stages of Z32, Z39, and Z65), stamen, and pistil relative to the individual vegetative tissues (including kernel) using the RNA-seq data. Based on the expression profiles, we were able to classify the 170 genes into three groups, each with two subgroups (Figure 1 and Table S3). Within each subgroup, the expression profiles were similar between genes, but the relative abundances were not identical even between orthologous genes.

The first group, G1, consisted of 42 genes from 10 gene sets. Overall, they had a low expression level and were expressed more abundantly and specifically in majority of the cases, in the inflorescence as a whole. A few genes, such as *IDG035.2-5A*, *IDG035.1-4D*, *IDG035.2-4D*, *IDG042.1-6A*, *IDG042.3-6A*, *IDG042.4-6A*, and *IDG042.8-6A*, were supported by ESTs but were matched to a negligible number of reads in the RNA-seq datasets. They were classified together with their orthologous or paralogous homologs, since their expression profiles in EST analysis were similar. Genes in subgroup G1-1 also showed enhanced expression, even though less abundantly, in stamen or pistil. Different from those in G1-1, genes in subgroup G1-2 had negligible fragments per kilobase of cDNA model per million mapped reads (FPKM) values from tissues other than the inflorescence.

The second group, G2, consisted of 20 gene sets and 48 genes. This group was basically characterized by expression in one or more of the vegetative tissues and an even higher level of expression in the stamen and/or pistil. *IDG011-1A* was the only exception, which was predominantly expressed in the inflorescence as a whole in spite of a significantly higher expression in the stamen and pistil relative to the vegetative tissues. Genes in subgroup G2-1 all had a significantly higher level of expression in the pistil relative to the vegetative tissues, and except for *IDG044-6B* and *IDG044-6D*, a significantly higher level of expression in the stamen as well. A few genes in this subgroup were expressed more abundantly in the inflorescence (*IDG011-1A*) or in the stamen (*IDG001-1A*, *IDG001-1D*, and *IDG045*). Genes in subgroup G2-2 were expressed more abundantly in the stamen relative to the vegetative tissues, but their expression levels in the pistil were not different from or even lower than those in at least one of the vegetative tissues.

Group G3 was the largest group, including 80 genes from 29 gene sets, characterized by a predominant expression in stamen and a negligible level of expression in the vegetative tissues. The RES values (stamen versus vegetative tissues) were high, ranged from 0.82–1.0. G3-1 consisted of 39 genes, all with a negligible FPKM value from the inflorescence. The remaining genes were different from genes in G3-1 by a significantly enhanced expression in the inflorescence as well

relative to the vegetative tissues, although the expression level was much lower in most cases.

To verify the expression specificity experimentally, sqRT-PCR was performed with tissues of root, node, internode, flag leaf, glume, lemma, palea, lodicule, stamen, pistil, rachis, and kernel 9th day postanthesis, using 27 pairs of primers that corresponds to 64 members of the identified genes (Table S2). All PCR reactions revealed a pattern of predominant expression in at least one of the floral tissues or organs but not in kernels and vegetative organs (Figure 2), which, by and large, were in agreement with results from the EST and RNA-seq data analysis. The qRT-PCR of a selected set of genes, including those with a relatively lower expression level and those expressed in multiple tissues, further confirmed these results (Figure S1).

Generally speaking, the tissue expression profiles between orthologous genes inferred based on the EST data were similar to those from the RNA-seq analysis. But a few exceptions were noted. In most of these cases, a low expression level of certain orthologous members was likely the cause of the discrepancy, for instance, some members of *IDG035* and *IDG042*.

3.3. Functional Annotation of the Identified Genes. Forty-nine of the identified gene sets were annotated via homology search and classified into five categories according to their putative biological functions (Table 2). It has to be mentioned that some genes could functionally fall into multiple categories.

The first category included only two gene sets; both encode allergenic proteins. The biological functions of this class of proteins in floral development have not been well characterized. *IDG035* encodes group 3 grass pollen allergens, which have sequence similarity to expansins that promote plant cell wall enlargement and thereby serve as cell wall-loosening agents [40].

The second category included 11 gene sets, most of which belonged to the G2 expression type. In this category, five gene sets code for proteins related to JA, ET, and GA signaling, three for MADS-box transcription factors and three for H2A and H2B proteins. The MADS-box transcription factor proteins encoded by *IDG001* and *IDG021* have 87% similarity. *IDG001* is orthologous to the rice *OsMADS4*. According to the ABCDE model for floral organ identity specification [41], *IDG001* and *IDG021* belong to class B MADS-box genes. *IDG044* encodes an AGL6-like MADS transcription factor and is functionally similar to class E genes [42].

The third category included 28 gene sets, accounting for 57% of the annotated. They were all predominantly or specifically expressed in stamen. Most of them were associated with substance production, transportation, and assembly for anther and pollen development. Of this category, 18 gene sets code for proteins associated with fatty acid and lipid metabolism. Homologs in other plants of most of these genes have been associated with the process of pollen wall development, such as suberin biosynthesis [43, 44], cutin biosynthesis [45–47], pollen sporopollenin biosynthesis [48], and pollen exine formation [49–51]. Other genes in this category have also been associated

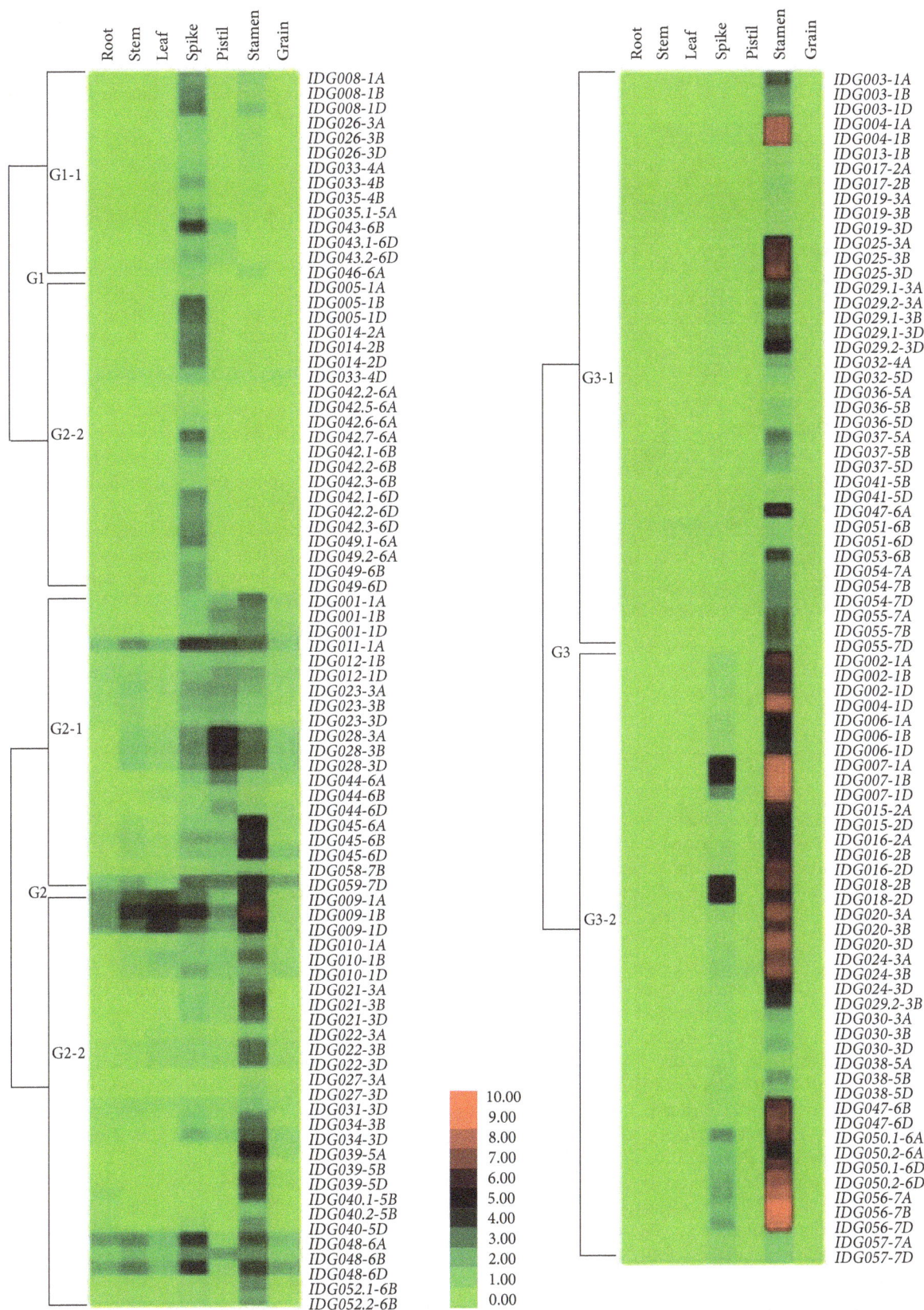

FIGURE 1: Relative expression abundance of the identified genes in different tissues based on RNA-seq data. Apart from seven genes with a negligible number of reads, the remaining 163 genes were divided according to their expression profiles into three groups, each of which had two subgroups.

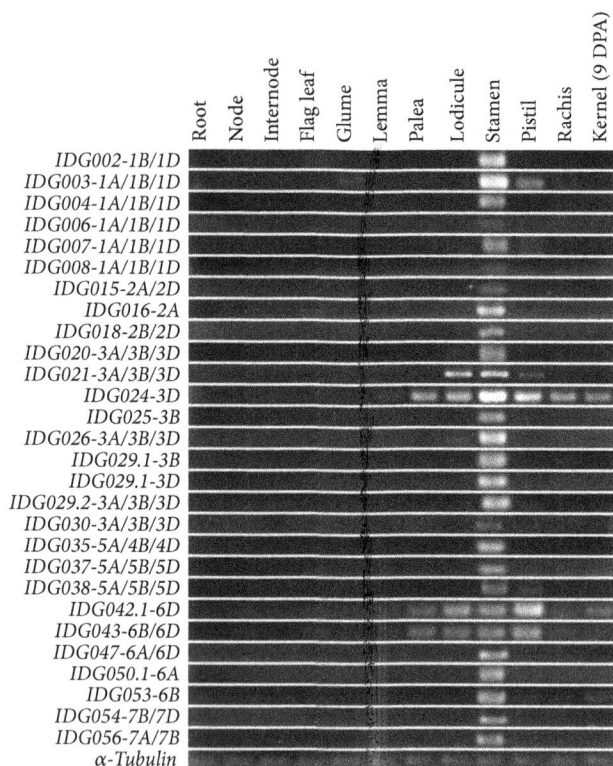

FIGURE 2: Expression of a selected set of identified genes in different tissues. The tissues used included kernel 9th day postanthesis, root, node, internode, flag leaf, glume, lemma, palea, lodicule, stamen, pistil, and rachis at the heading stage of common wheat landrace "Wangshuibai."

with anther and pollen development. *IDG038* encodes an acidic peroxidase that might participate in the synthesis of phenylpropanoids present in sporopollenin [52]. The products encoded by *IDG007*, *IDG018*, *IDG050*, *IDG055*, and *IDG056* were related to pollen wall formation [53–55]. Both *IDG004* and *IDG025* code for anther-specific RTS-like proteins, required for male fertility and affecting tapetal development [56]. *IDG006* encodes a galactosyltransferase, which is implicated in the accumulation control of glycosylated flavonols in pollen [57]. In *Arabidopsis*, a type II β-(1,3)-galactosyltransferase is required for pollen exine development [58]. *IDG020* codes for a late cornified envelope- (LCE-) like proline-rich protein. LCE proteins are involved in the cornified cell envelope assembly of skins and associated with ROS detoxification [59].

The fourth category only has four gene sets and is associated with pollination and pollen-stigma interactions. For instance, the subtilisin-like protease encoded by *IDG002* was related to anther dehiscence [60]; the products of *IDG008* were related to pollen tube growth [61]; the chemocyanins encoded by *IDG040* could be involved in the pollination process [62] and induce pollen tube chemotropism as a diffusible chemotropic factor [63]. *IDG028* codes for a protodermal factor 1-like protein mainly in the pistil. This protein is related to reactive oxygen species (ROS) homeostasis [64]. ROS are involved in pollen tube growth

and rupture [65, 66], implying the role of modulating ROS levels in male reproductive development [67].

The fifth category, the "other" in Table 2, had only four annotated gene sets. Their specific functions in reproductive development still require clarification.

4. Discussion

Inflorescence represents a highly specialized plant tissue producing seeds for propagation. Deciphering genes involved in its development is the first step to understand the essence of reproduction and of great importance for seed production manipulation. In this study, we identified 59 nonredundant wheat gene sets that were differentially expressed in wheat inflorescence and encode proteins with diverse functions. Majority of the identified genes were associated with metabolic activities and wall assembly required for the specialized process of pollen maturation and pollination, while few showed predominance or specificity in macrosporogenesis. On one hand, this could be attributed to the fact that much more ESTs from libraries made with anthers were used in the analysis, which had limitation of development stage and tissue-type coverage; on the other hand, it was probably due to the specific structure of pollen grains whose formation requires expression of a specific set of genes or gene network that made the related genes easily recognized through the differential analysis. Our results complemented well with previous studies involved in wheat floral development. The microarray-based transcriptomic analysis of anthers by Crismani et al. was mainly focused on identification of early meiotic genes [15]. In the RNA-seq data comparison of pistillody stamen versus pistil, pistillody stamen versus stamen, and pistil versus stamen, Yang et al. identified 206 genes highly correlated with stamen and pistil development [17]. Among them, however, only a few were functionally annotated as identically as the genes presented in this paper. It was noted that nearly one-third of the identified gene sets in the present study displayed differentiated expression profiles in terms of their subgenome orthologs, implying functional diversification in polyploidy wheat for the inflorescence development.

The whole process of inflorescence development is under regulatory control. A set of MADS transcription factors regulate floral organ identity specification [41, 68, 69]. The three MADS-box genes we identified, one for E-class MADS proteins and two for B-class MADS proteins, differed in their expression profiles (Figure 1), even though both *IDG001* and *IDG021* were mainly expressed in stamen, suggesting they act in concert in determining the anther and pistil identity. The pistillody in alloplasmic wheat was related to expression pattern alteration of class B genes [68]. The identification of genes related to JA, ET, and GA signaling added support for the important roles of JA, ET, and GA signaling cross-talks playing in stamen development [70, 71]. ET signaling is involved in multiple aspects of floral organ development, for instance, nectar secretion, accumulation of stigmatic exudate, and development of the self-incompatible response [72], floral organ

TABLE 2: Functional classifications of the annotated gene sets.

Gene set	Annotation	Molecular or biological function	Reference
1. Allergen			
IDG035	Group 3 grass pollen allergen		[93]
IDG039	Peamaclein-like		
2. Expression regulation			
IDG001	MADS-box transcription factor WM14	Organ identity specification	[68]
IDG003	Gibberellin-regulated protein II-like	Gibberellin signaling	[70]
IDG005	Cysteine protease-like protein	Ethylene signaling	[73]
IDG011	Histone H2A	Chromosome modeling	
IDG012	Histone H2A	Chromosome modeling	
IDG021	MADS-box transcription factor	Organ identity specification	[68]
IDG031	Glyoxysomal fatty acid beta-oxidation multifunctional protein MFP-a	JA signaling	[94]
IDG044	Transcription factor AGL6-like	Organ identity specification	[69]
IDG046	1-aminocyclopropane-1-carboxylate oxidase, EC 1.14.17.4	Ethylene signaling	[72, 95]
IDG048	Histone H2B	Chromosome modeling	
IDG052	1-aminocyclopropane-1-carboxylate oxidase, EC 1.14.17.4	Ethylene signaling	[72, 95]
3. Anther development and pollen wall formation			
IDG004	Anther-specific protein RTS-like	Tapetal development	[56]
IDG006	Galactosyltransferase, EC 2.4.1.-	Cell wall assembly	[81]
IDG007	Protease inhibitor/seed storage/LTP family protein-like	Pollen wall formation	[53]
IDG009	Phosphoethanolamine N-methyltransferase EC 2.1.1.103	Lipid metabolism	[92]
IDG010	Long chain acyl-CoA synthetase, EC 6.2.1.3	Lipid metabolism	[92]
IDG013	Triacylglycerol lipase, EC 3.1.1.3	Lipid metabolism	
IDG017	Cytochrome P450 86B1-like, EC 1.14.15.-	Lipid metabolism	[83]
IDG018	Putative RAFTIN1/BURP domain-containing protein	Pollen wall formation	[53]
IDG019	Cytochrome P450 94A1-like	Lipid metabolism	[45]
IDG020	Late cornified envelope-like proline-rich protein	ROS detoxification	[59]
IDG022	Glycerol-3-phosphate acyltransferase 6, EC 2.3.1.198	Lipid metabolism	[46]
IDG025	Anther-specific protein RTS-like	Tapetal development	[56]
IDG027	Nonspecific phospholipase C2, EC 3.1.4.3	Lipid metabolism	
IDG030	Cytochrome P450 86A2-like	Lipid metabolism	[83]
IDG032	Acyl-[acyl-carrier-protein] desaturase, EC 1.14.19.2	Lipid metabolism	
IDG033	Fatty acyl-CoA reductase 2. EC 1.2.1.42	Lipid metabolism	[48]
IDG036	Cytochrome P450 86B1, fatty acid ω-hydroxylase, EC 1.14.15.3	Lipid metabolism	[83]
IDG037	Acyl-[acyl-carrier-protein] desaturase, EC 1.14.19.2	Lipid metabolism	
IDG038	Acidic peroxidase, EC 1.11.1.7	Phenylpropanoid biosynthesis	[52]
IDG041	Cytochrome P450 86A1-like, fatty acid ω-hydroxylase	Lipid metabolism	[83]
IDG045	Fatty aldehyde decarbonylase, EC 4.1.99.5	Lipid metabolism	[50, 51]
IDG050	Nonspecific lipid-transfer protein C6-like	Pollen wall formation	[53]
IDG051	Acyltransferase-like protein, EC 2.3.1.-	Lipid metabolism	
IDG054	Triacylglycerol lipase 2-like, EC 3.1.1.13	Lipid metabolism	
IDG055	Bidirectional sugar cr SWEET	Pollen wall formation	[54, 55]
IDG056	RAFTIN1 protein/BURP domain protein-like	Pollen wall formation	[53]
IDG057	Acyltransferase-like protein, EC 2.3.1.-	Lipid metabolism	
IDG059	Acyl carrier protein	Lipid metabolism	
4. Pollination and pollen-stigma interactions			
IDG002	Subtilisin-like protease	Anther dehiscence	[60]
IDG008	Chalcone synthase-like	Biosynthesis of flavonols	[61]
IDG028	Protodermal factor 1-like	ROS homeostasis	[64]

TABLE 2: Continued.

Gene set	Annotation	Molecular or biological function	Reference
IDG040	Chemocyanin	Pollen tube attraction	[62]
5. Others			
IDG014	Nitrate-induced NOI protein	Plant defense	[96]
IDG034	Nucleoside diphosphate kinase, EC 2.7.4.6	Nucleotide triphosphate generation	
IDG053	Zinc transporter-like	Early reproductive development	
IDG058	Heat shock protein	Thermotolerance	

senescing [73], pollen thermotolerance [74], and timing of anther dehiscence [71]. Cheng et al. demonstrated that GA regulates stamen development through JA signaling [75]. Mutations of genes encoding JA-biosynthetic enzymes result in failure of filament elongation, delayed anther dehiscence, and unviable pollens [70, 76, 77].

A few identified genes were associated with pollen-stigma interactions. All but *IDG028* had transcripts in spikes at meiosis and anthesis stages as well as in anthers but the abundance varied considerably. The subtilase-encoding genes related to anther dehiscence (*IDG002*) were overwhelmingly expressed in anthers, especially at the tetrad stage [60]. In male-sterile lines, their expression was downregulated [77]. Transcripts of pollen allergen-encoding *IDG035* and ACO-encoding *IDG046* were also present in spikes at this stage. Accumulating evidence indicates the involvement of ET signaling in fertilization [78, 79]. Valdiva et al. showed that disruption of a maize group-I allergen affected pollen-pollen competition for access to the ovules [80]. In addition, the galactosyltransferase-encoding *IDG006*, predominantly expressed in stamen, was related to pollen tube elongation [81].

Most of the genes showing anther-specific or predominant expression are related to tapetal and pollen developments (Table 2). Particularly worth mentioning are the gene set-encoding enzymes or proteins participating in lipid metabolism. Lipid metabolism is important to pollen development because the distinct pollen wall structure is mainly made of fatty (lipid) substances produced in the tapetum of anthers [82]. Among the lipid metabolism-related gene sets, five (*IDG017*, *IDG019*, *IDG030*, *IDG036*, and *IDG041*) encode members of 86A, 86B, and 94A subfamilies of cytochrome P450 proteins that are related to fatty acid ω-hydroxylation in primary fatty alcohols and suberin monomer biosynthesis for formation of anther cuticle and pollen sporopollenin in monocots and dicots [43–45, 47, 83–86]. These cytochrome P450 proteins might function in different subcellular locations, since *IDG017*, *IDG019*, and *IDG030* have secretory signal peptides, while *IDG036* and *IDG041* do not.

Pathways emerging from the lipid metabolism-related genes included those for alkane and fatty alcohol production and glycerophospholipid metabolism. Genes encoding long-chain acyl-CoA synthetase (*IDG010*) and fatty aldehyde decarboxylase (*CER1*, *IDG045*), the two enzymes involved in the plant alkane-forming pathway [87], were coexpressed in the wheat stamen. This *CER1* gene was

downregulated in pistil or pistillody stamen [17], suggesting its specificity to stamen development. Very long-chain (VLC) alkanes are major components of the tryphine layer covering pollen grains and are needed for proper pollen-pistil signaling and fertility [50]. Mutation of *CER1* in both *Arabidopsis* and rice caused defective pollens [50, 51]. In *Arabidopsis*, the acyl-CoA synthetase gene *ACOS5* was upregulated in the tapetal cells. Its mutation led to failure in pollen production and pollen wall formation [88].

Fatty alcohols are components of surface lipid barriers such as anther cuticle and pollen wall [89]. Fatty acyl-CoA reductase, encoded by *IDG033* in wheat and *Ms2* in *Arabidopsis*, is the key enzyme for the production of fatty alcohols in plastids [48]. Mutation of *Ms2* led to abnormal pollen wall development and reduced pollen fertility [90, 91]. Like Ms2, the *IDG033*-encoded proteins have plastidic localization signal peptides. The encoded products of *IDG059*, *IDG032*, and *IDG037*, probably involved in the production of fatty acyl-CoA reductase substrates, all have plastidic localization signals. The diacylglycerol acyltransferase- (DGAT-) like protein encoded by *IDG057* could also carry a plastidic peptide. The *Arabidopsis DGAT1* contributes to triacylglycerol biosynthesis and its function loss causes critical defects in normal pollen and embryo development [84]. However, information about its link to the plastidial fatty alcohol pathway is still lacking. The expression profiles of these genes were different, although all expressed in stamen. *IDG059*, *IDG032*, and *IDG037* appeared to be specifically expressed in this organ.

Among the lipid metabolism-related genes, five (*IDG009*, *IDG013*, *IDG022*, *IDG027*, and *IDG054*) encode proteins putatively associated with glycerophospholipid metabolism, in agreement with the findings of Yang et al. [17]. The *IDG009*-encoded phosphoethanolamine *N*-methyltransferase (PEAMT) is the committing enzyme for choline biosynthesis. In *Arabidopsis*, silencing the PEAMT gene resulted in temperature-sensitive male sterility and salt hypersensitivity [92]; knockdown of *GPAT6*, the homolog of *IDG022*, caused defective pollen grains [46]. PEAMT and GPAT6 also affected pollen tube growth [46, 92]. Moreover, *IDG051* encode proteins predicted with lysophosphatidylethanolamine acyltransferase activities, which probably participates in the phospholipid metabolism in mitochondria.

Only 59 inflorescence development-related nonredundant gene sets were identified in this study. This could be much less than the actual number of genes differentially expressed in inflorescence tissues. We reasoned that the EST libraries used in gene mining, which had limitations in volume size and representation of tissue and developmental stages, and the strict standard used in gene mining were the main causes. A sampling probability ≤ 0.0001 has very likely increase type II error; however, it could minimize false positives, as shown in RNA-seq data analysis and RT-PCR validation, which is beneficial for correct data interpretation.

5. Conclusions

In this study, we identified 170 wheat genes for floral identity determination, anther and pollen development, pollen-pistil interaction, and others using the comparative transcriptomics approach. The potential importance of the identified genes to wheat inflorescence development was manifested in the enhanced or specific expression in the floral tissues. We noted that nearly one-third of the gene sets have undergone subgenome differentiation. Of the identified genes, those coding for enzymes or proteins participating in lipid metabolic pathway accounted for the largest category, implying the particularity and important roles of lipid metabolism in wheat reproductive development. This study is useful for understanding the gene network underlying wheat inflorescence morphology and fertility, which eventually will allow us to purposely manipulate fertility in breeding.

Conflicts of Interest

The authors declare that there are no conflicts of interest regarding the publication of this article.

Acknowledgments

This study was partially supported by Natural Science Foundation of China (31430064 and 30025030), Ministry of Science and Technology of China (2016YFD0101004 and 2016ZX08002003), Jiangsu collaborative innovation initiative for modern crop production (JCIC-MCP), "111" project B08025, and fund from the innovation team program for Jiangsu universities (2014).

Supplementary Materials

Supplementary 1. Figure S1: expression of a selected set of identified genes in different tissues, examined using qRT-PCR. The expression levels were estimated relative to that in stamen. The tissues used included kernel 9th day post-anthesis, root, node, internode, flag leaf, glume, lemma, palea, lodicule, stamen, pistil, and rachis at the heading stage of common wheat landrace "Wangshuibai."

Supplementary 2. Table S1: cDNA libraries used in this study.

Supplementary 3. Table S2: primers used in RT-PCR.

Supplementary 4. Table S3: sequence, annotation, and expression specificity of 170 identified genes. * and ** significantly different at $P = 0.05$ and 0.01, respectively.

References

[1] J. Kyozuka, "Grass inflorescence: basic structure and diversity," *Advances in Botanical Research*, vol. 72, pp. 191–219, 2014.

[2] H. T. Clifford, *Spikelet and Floral Morphology*, Washington, D.C, USA, Smithsonian, 1987.

[3] L. G. Clark and R. W. Pohl, *Agnes Chase's First Book of Grasses: The Structure of Grasses Explained for Beginners*, Washington D.C, USA, Smithsonian, 1996.

[4] G. Theissen and H. Saedler, "Plant biology: floral quartets," *Nature*, vol. 409, no. 6819, pp. 469–471, 2001.

[5] D. B. Zhang and Z. Yuan, "Molecular control of grass inflorescence development," *Annual Review of Plant Biology*, vol. 65, no. 1, pp. 553–578, 2014.

[6] J. Y. Lee and D. H. Lee, "Use of serial analysis of gene expression technology to reveal changes in gene expression in Arabidopsis pollen undergoing cold stress," *Plant Physiology*, vol. 132, no. 2, pp. 517–529, 2003.

[7] T. H. Lee, Y. K. Kim, T. T. Pham et al., "RiceArrayNet: a database for correlating gene expression from transcriptome profiling, and its application to the analysis of coexpressed genes in rice," *Plant Physiology*, vol. 151, no. 1, pp. 16–33, 2009.

[8] H. P. Buermans and J. T. den Dunnen, "Next generation sequencing technology: advances and applications," *Biochimica et Biophysica Acta (BBA) - Molecular Basis of Disease*, vol. 1842, no. 10, pp. 1932–1941, 2014.

[9] J. A. Schrauwen, P. F. de Groot, M. M. van Herpen et al., "Stage-related expression of mRNAs during pollen development in lily and tobacco," *Planta*, vol. 182, no. 2, pp. 298–304, 1990.

[10] J. Ma, D. S. Skibbe, J. Fernandes, and V. Walbot, "Male reproductive development: gene expression profiling of maize anther and pollen ontogeny," *Genome Biology*, vol. 9, no. 12, article R181, 2008.

[11] P. Deveshwar, W. D. Bovill, R. Sharma, J. A. Able, and S. Kapoor, "Analysis of anther transcriptomes to identify genes contributing to meiosis and male gametophyte development in rice," *BMC Plant Biology*, vol. 11, no. 1, pp. 78–78, 2011.

[12] N. Rutley and D. Twell, "A decade of pollen transcriptomics," *Plant Reproduction*, vol. 28, no. 2, pp. 73–89, 2015.

[13] P. Chaturvedi, T. Ischebeck, V. Egelhofer, I. Lichtscheidl, and W. Weckwerth, "Cell-specific analysis of the tomato pollen proteome from pollen mother cell to mature pollen provides evidence for developmental priming," *Journal of Proteome Research*, vol. 12, no. 11, pp. 4892–4903, 2013.

[14] T. Marcussen, S. R. Sandve, L. Heier et al., "Ancient hybridizations among the ancestral genomes of bread wheat," *Science*, vol. 345, no. 6194, article 1250092, 2014.

[15] W. Crismani, U. Baumann, T. Sutton et al., "Microarray expression analysis of meiosis and microsporogenesis in hexaploid bread wheat," *BMC Genomics*, vol. 7, no. 1, p. 267, 2006.

[16] S. McIntosh, L. Watson, P. Bundock et al., "SAGE of the developing wheat caryopsis," *Plant Biotechnology Journal*, vol. 5, no. 1, pp. 69–83, 2007.

[17] Z. J. Yang, Z. S. Peng, S. H. Wei, M. L. Liao, Y. Yu, and Z. Y. Jang, "Pistillody mutant reveals key insights into stamen and pistil development in wheat (*Triticum aestivum* L.)," *BMC Genomics*, vol. 16, no. 1, p. 211, 2015.

[18] M. Houde, M. Belcaid, F. Ouellet et al., "Wheat EST resources for functional genomics of abiotic stress," *BMC Genomics*, vol. 7, no. 1, p. 149, 2006.

[19] N. Z. Ergen and H. Budak, "Sequencing over 13,000 expressed sequence tags from six subtractive cDNA libraries of wild and modern wheats following slow drought stress," *Plant, Cell & Environment*, vol. 32, no. 3, pp. 220–236, 2009.

[20] A. Manickavelu, K. Kawaura, K. Oishi et al., "Comparative gene expression analysis of susceptible and resistant near-isogenic lines in common wheat infected by *Puccinia triticina*," *DNA Research*, vol. 17, no. 4, pp. 211–222, 2010.

[21] A. Manickavelu, K. Kawaura, K. Oishi et al., "Comprehensive functional analyses of expressed sequence tags in common wheat (*Triticum aestivum*)," *DNA Research*, vol. 19, no. 2, pp. 165–177, 2012.

[22] M. Song, W. Xu, Y. Xiang, H. Jia, L. Zhang, and Z. Ma, "Association of jacalin-related lectins with wheat responses to stresses revealed by transcriptional profiling," *Plant Molecular Biology*, vol. 84, no. 1-2, pp. 95–110, 2014.

[23] O. D. Anderson, N. Huo, and Y. Q. Gu, "The gene space in wheat: the complete -gliadin gene family from the wheat cultivar Chinese Spring," *Functional & Integrative Genomics*, vol. 13, no. 2, pp. 261–273, 2013.

[24] K. Rikiishi and M. Maekawa, "Seed maturation regulators are related to the control of seed dormancy in wheat (*Triticum aestivum* L.)," *PLoS One*, vol. 9, no. 9, article e107618, 2014.

[25] M. Domoki, A. Szucs, K. Jager, S. Bottka, B. Barnabas, and A. Feher, "Identification of genes preferentially expressed in wheat egg cells and zygotes," *Plant Cell Reports*, vol. 32, no. 3, pp. 339–348, 2013.

[26] Z. Y. Chen, X. J. Guo, Z. X. Chen et al., "Genome-wide characterization of developmental stage- and tissue-specific transcription factors in wheat," *BMC Genomics*, vol. 16, no. 1, p. 125, 2015.

[27] L. N. Ding, H. B. Xu, H. Y. Yi et al., "Resistance to hemibiotrophic *F.graminearum* infection is associated with coordinated and ordered expression of diverse defense signaling pathways," *PLoS One*, vol. 6, no. 4, article e19008, 2011.

[28] P. A. Wilkinson, M. O. Winfield, G. L. Barker et al., "CerealsDB 3.0: expansion of resources and data integration," *BMC Bioinformatics*, vol. 17, no. 1, p. 256, 2016.

[29] V. Solovyev, P. Kosarev, I. Seledsov, and D. Vorobyev, "Automatic annotation of eukaryotic genes, pseudogenes and promoters," *Genome Biology*, vol. 7, article S10, Supplement 1, 2006.

[30] R. Brenchley, M. Spannagl, M. Pfeifer et al., "Analysis of the bread wheat genome using whole-genome shotgun sequencing," *Nature*, vol. 491, no. 7426, pp. 705–710, 2012.

[31] F. Choulet, A. Alberti, S. Theil et al., "Structural and functional partitioning of bread wheat chromosome 3B," *Science*, vol. 345, no. 6194, article 1249721, 2014.

[32] S. Lindgreen, "AdapterRemoval: easy cleaning of next-generation sequencing reads," *BMC Research Notes*, vol. 5, no. 1, p. 337, 2012.

[33] D. Kim, B. Landmead, and S. L. Salzberg, "HISAT: a fast spliced aligner with low memory requirements," *Nature Methods*, vol. 12, no. 4, pp. 357–360, 2015.

[34] Y. Liao, G. K. Smyth, and W. Shi, "featureCounts: an efficient general purpose program for assigning sequence reads to genomic features," *Bioinformatics*, vol. 30, no. 7, pp. 923–930, 2014.

[35] M. Song, W. Q. Xu, Y. Xiang, H. Y. Jia, L. X. Zhang, and Z. Q. Ma, "Association of jacalin-related lectins with wheat responses to stresses revealed by transcriptional profiling," *Plant Molecular Biology*, vol. 84, no. 1-2, pp. 95–110, 2014.

[36] K. J. Livak and T. D. Schmittgen, "Analysis of relative gene expression data using real-time quantitative PCR and the $2^{-\Delta\Delta C}{}_T$ method," *Methods*, vol. 25, no. 4, pp. 402–408, 2001.

[37] M. Kanehisa, Y. Sato, and K. Morishima, "BlastKOALA and GhostKOALA: KEGG tools for functional characterization of genome and metagenome sequences," *Journal of Molecular Biology*, vol. 428, no. 4, pp. 726–731, 2016.

[38] R. D. Finn, P. Coggill, R. Y. Eberhardt et al., "The Pfam protein families database: towards a more sustainable future," *Nucleic Acids Research*, vol. 44, no. D1, pp. D279–D285, 2016.

[39] O. Emanuelsson, H. Nielsen, S. Brunak, and G. von Heijne, "Predicting subcellular localization of proteins based on their N-terminal amino acid sequence," *Journal of Molecular Biology*, vol. 300, no. 4, pp. 1005–1016, 2000.

[40] D. J. Cosgrove, P. Bedinger, and D. M. Durachko, "Group I allergens of grass pollen as cell wall-loosening agents," *Proceedings of the National Academy of Sciences of the United States of America*, vol. 94, no. 12, pp. 6559–6564, 1997.

[41] L. M. Zahn, J. H. Leebens-Mack, J. M. Arrington et al., "Conservation and divergence in the *AGAMOUS* subfamily of MADS-box genes: evidence of independent sub- and neofunctionalization events," *Evolution & Development*, vol. 8, no. 1, pp. 30–45, 2006.

[42] H. F. Li, W. Q. Liang, R. D. Jia et al., "The *AGL6*-like gene *OsMADS6* regulates floral organ and meristem identities in rice," *Cell Research*, vol. 20, no. 3, pp. 299–313, 2010.

[43] R. Hofer, I. Briesen, M. Beck, F. Pinot, L. Schreiber, and R. Franke, "The *Arabidopsis* cytochrome P450 *CYP86A1* encodes a fatty acid ω-hydroxylase involved in suberin monomer biosynthesis," *Journal of Experimental Botany*, vol. 59, no. 9, pp. 2347–2360, 2008.

[44] V. Compagnon, P. Diehl, I. Benveniste et al., "CYP86B1 is required for very long chain ω-hydroxyacid and α,ω-dicarboxylic acid synthesis in root and seed suberin polyester," *Plant Physiology*, vol. 150, no. 4, pp. 1831–1843, 2009.

[45] N. Tijet, C. Helvig, F. Pinot et al., "Functional expression in yeast and characterization of a clofibrate-inducible plant cytochrome P-450 (CYP94A1) involved in cutin monomers synthesis," *Biochemical Journal*, vol. 332, no. 2, pp. 583–589, 1998.

[46] X. C. Li, J. Zhu, J. Yang et al., "Glycerol-3-phosphate acyltransferase 6 (GPAT6) is important for tapetum development in *Arabidopsis* and plays multiple roles in plant fertility," *Molecular Plant*, vol. 5, no. 1, pp. 131–142, 2012.

[47] S. G. Rupasinghe, H. Duan, and M. A. Schuler, "Molecular definitions of fatty acid hydroxylases in *Arabidopsis thaliana*," *Proteins*, vol. 68, no. 1, pp. 279–293, 2007.

[48] W. Chen, X. H. Yu, K. Zhang et al., "*Male Sterile2* encodes a plastid-localized fatty acyl carrier protein reductase required for pollen exine development in Arabidopsis," *Plant Physiology*, vol. 157, no. 2, pp. 842–853, 2011.

[49] A. A. Dobritsa, Z. T. Lei, S. Nishikawa et al., "*LAP5* and *LAP6* encode anther-specific proteins with similarity to calcone synthase essential for pollen exine development in *Arabidopsis thaliana*," *Plant Physiology*, vol. 153, no. 3, pp. 937–955, 2010.

[50] M. G. Aarts, C. J. Keijzer, W. J. Stiekema, and A. Pereira, "Molecular characterization of the CER1 gene of arabidopsis

involved in epicuticular wax biosynthesis and pollen fertility," *The Plant Cell*, vol. 7, no. 12, pp. 2115–2127, 1995.

[51] K. H. Jung, M. J. Han, D. Y. Lee et al., "*Wax-deficient anther1* is involved in cuticle and wax production in rice anther walls and is required for pollen development," *The Plant Cell*, vol. 18, no. 11, pp. 3015–3032, 2006.

[52] A. Skirycz, S. Jozefczuk, M. Stobiecki et al., "Transcription factor AtDOF4;2 affects phenylpropanoid metabolism in *Arabidopsis thaliana*," *New Phytologist*, vol. 175, no. 3, pp. 425–438, 2007.

[53] A. M. Wang, Q. Xia, W. S. Xie, R. Datla, and G. Selvaraj, "The classical Ubisch bodies carry a sporophytically produced structural protein (RAFTIN) that is essential for pollen development," *Proceedings of the National Academy of Sciences of the United States of America*, vol. 100, no. 24, pp. 14487–14492, 2003.

[54] B. Yang, A. Sugio, and F. F. White, "*Os8N3* is a host disease-susceptibility gene for bacterial blight of rice," *Proceedings of the National Academy of Sciences of the United States of America*, vol. 103, no. 27, pp. 10503–10508, 2006.

[55] Y. F. Guan, X. Y. Huang, J. Zhu, J. F. Gao, H. X. Zhang, and Z. N. Yang, "*RUPTURED POLLEN GRAIN1*, a member of the MtN3/saliva gene family, is crucial for exine pattern formation and cell integrity of microspores in *Arabidopsis*," *Plant Physiology*, vol. 147, no. 2, pp. 852–863, 2008.

[56] H. Luo, J. Y. Lee, Q. Hu et al., "*RTS*, a rice anther-specific gene is required for male fertility and its promoter sequence directs tissue-specific gene expression in different plant species," *Plant Molecular Biology*, vol. 62, no. 3, pp. 397–408, 2006.

[57] L. P. Taylor and K. D. Miller, "The use of a photoactivatable kaempferol analogue to probe the role of flavonol 3-O-galactosyltransferase in pollen germination," *Advances in Experimental Medicine and Biology*, vol. 505, pp. 41–50, 2002.

[58] T. Suzuki, J. O. Narciso, W. Zeng et al., "KNS4/UPEX1: a type II arabinogalactan β-(1,3)-galactosyltransferase required for pollen exine development," *Plant Physiology*, vol. 173, no. 1, pp. 183–205, 2017.

[59] W. P. Vermeij, A. Alia, and C. Backendorf, "ROS quenching potential of the epidermal cornified cell envelope," *Journal of Investigative Dermatology*, vol. 131, no. 7, pp. 1435–1441, 2011.

[60] A. A. Taylor, A. Horsch, A. Rzepczyk, C. A. Hasenkampf, and C. D. Riggs, "Maturation and secretion of a serine proteinase is associated with events of late microsporogenesis," *The Plant Journal*, vol. 12, no. 6, pp. 1261–1271, 1997.

[61] G. J. vanEldik, W. H. Reijnen, R. K. Ruiter, M. M. A. vanHerpen, J. A. M. Schrauwen, and G. J. Wullems, "Regulation of flavonol biosynthesis during anther and pistil development, and during pollen tube growth in *Solanum tuberosum*," *The Plant Journal*, vol. 11, no. 1, pp. 105–113, 1997.

[62] J. Dong, S. T. Kim, and E. M. Lord, "Plantacyanin plays a role in reproduction in Arabidopsis," *Plant Physiology*, vol. 138, no. 2, pp. 778–789, 2005.

[63] S. Kim, J. C. Mollet, J. Dong, K. L. Zhang, S. Y. Park, and E. M. Lord, "Chemocyanin, a small basic protein from the lily stigma, induces pollen tube chemotropism," *Proceedings of the National Academy of Sciences of the United States of America*, vol. 100, no. 26, pp. 16125–16130, 2003.

[64] F. L. Deng, L. L. Tu, J. F. Tan, Y. Li, Y. C. Nie, and X. L. Zhang, "GbPDF1 is involved in cotton fiber initiation via the core cis-element HDZIP2ATATHB2," *Plant Physiology*, vol. 158, no. 2, pp. 890–904, 2012.

[65] M. Potocky, M. A. Jones, R. Bezvoda, N. Smirnoff, and V. Zarsky, "Reactive oxygen species produced by NADPH oxidase are involved in pollen tube growth," *New Phytologist*, vol. 174, no. 4, pp. 742–751, 2007.

[66] Q. H. Duan, D. Kita, E. A. Johnson et al., "Reactive oxygen species mediate pollen tube rupture to release sperm for fertilization in *Arabidopsis*," *Nature Communications*, vol. 5, article 3129, 2014.

[67] L. F. Hu, W. Q. Liang, C. S. Yin et al., "Rice MADS3 regulates ROS homeostasis during late anther development," *The Plant Cell*, vol. 23, no. 2, pp. 515–533, 2011.

[68] E. Hama, S. Takumi, Y. Ogihara, and K. Murai, "Pistillody is caused by alterations to the class-B MADS-box gene expression pattern in alloplasmic wheats," *Planta*, vol. 218, no. 5, pp. 712–720, 2004.

[69] T. Zhao, Z. F. Ni, Y. Dai, Y. Y. Yao, X. L. Nie, and Q. X. Sun, "Characterization and expression of 42 MADS-box genes in wheat (*Triticum aestivum* L.)," *Molecular Genetics and Genomics*, vol. 276, no. 4, pp. 334–350, 2006.

[70] Z. Y. Peng, Z. Xhou, L. C. Li et al., "Arabidopsis hormone database: a comprehensive genetic and phenotypic information database for plant hormone research in Arabidopsis," *Nucleic Acids Research*, vol. 37, Supplement 1, pp. D975–D982, 2009.

[71] I. Rieu, M. Wolters-Arts, J. Derksen, C. Mariani, and K. Weterings, "Ethylene regulates the timing of anther dehiscence in tobacco," *Planta*, vol. 217, no. 1, pp. 131–137, 2003.

[72] X. Tang, A. Gomes, A. Bhatia, and W. R. Woodson, "Pistil-specific and ethylene-regulated expression of 1-aminocyclopropane-1-carboxylate oxidase genes in petunia flowers," *The Plant Cell*, vol. 6, no. 9, pp. 1227–1239, 1994.

[73] M. L. Jones, P. B. Larsen, and W. R. Woodson, "Ethylene-regulated expression of a carnation cysteine proteinase during flower petal senescence," *Plant Molecular Biology*, vol. 28, no. 3, pp. 505–512, 1995.

[74] N. Firon, E. Pressman, S. Meir, R. Khoury, and L. Altahan, "Ethylene is involved in maintaining tomato (*Solanum lycopersicum*) pollen quality under heat-stress conditions," *Aob Plants*, vol. 2012, no. 0, article pls024, 2012.

[75] H. Cheng, S. S. Song, L. T. Xiao et al., "Gibberellin acts through jasmonate to control the expression of *MYB21*, *MYB24*, and *MYB57* to promote stamen filament growth in *Arabidopsis*," *PLoS Genetics*, vol. 5, no. 3, article e1000440, 2009.

[76] S. Ishiguro, A. Kawai-Oda, J. Ueda, I. Nishida, and K. Okada, "*The defective in anther dehiscence1* gene encodes a novel phospholipase A1 catalyzing the initial step of jasmonic acid biosynthesis, which synchronizes pollen maturation, anther dehiscence, and flower opening in *Arabidopsis*," *The Plant Cell*, vol. 13, no. 10, pp. 2191–2209, 2001.

[77] P. Lou, J. G. Kang, G. Y. Zhang, G. Bonnema, Z. Y. Fang, and X. W. Wang, "Transcript profiling of a dominant male sterile mutant (*Ms-cd1*) in cabbage during flower bud development," *Plant Science*, vol. 172, no. 1, pp. 111–119, 2007.

[78] S. Bhattacharya and I. T. Baldwin, "The post-pollination ethylene burst and the continuation of floral advertisement are harbingers of non-random mate selection in *Nicotiana attenuata*," *The Plant Journal*, vol. 71, no. 4, pp. 587–601, 2012.

[79] R. Volz, J. Heydlauff, D. Ripper, L. von Lyncker, and R. Gross-Hardt, "Ethylene signaling is required for synergid

degeneration and the establishment of a pollen tube block," *Developmental Cell*, vol. 25, no. 3, pp. 310–316, 2013.

[80] E. R. Valdivia, Y. Wu, L. C. Li, D. J. Cosgrove, and A. G. Stephenson, "A group-1 grass pollen allergen influences the outcome of pollen competition in maize," *PLoS One*, vol. 2, no. 1, article e154, 2007.

[81] S. J. Roy, T. L. Holdaway-Clarke, G. R. Hackett, J. G. Kunkel, E. M. Lord, and P. K. Hepler, "Uncoupling secretion and tip growth in lily pollen tubes: evidence for the role of calcium in exocytosis," *The Plant Journal*, vol. 19, no. 4, pp. 379–386, 1999.

[82] S. Blackmore, A. H. Wortley, J. J. Skvarla, and J. R. Rowley, "Pollen wall development in flowering plants," *New Phytologist*, vol. 174, no. 3, pp. 483–498, 2007.

[83] D. S. Zhang, W. Q. Liang, Z. Yuan et al., "Tapetum degeneration retardation is critical for aliphatic metabolism and gene regulation during rice pollen development," *Molecular Plant*, vol. 1, no. 4, pp. 599–610, 2008.

[84] D. B. Zhang and Z. A. Wilson, "Stamen specification and anther development in rice," *Chinese Science Bulletin*, vol. 54, no. 14, pp. 2342–2353, 2009.

[85] D. B. Zhang, X. Luo, and L. Zhu, "Cytological analysis and genetic control of rice anther development," *Journal of Genetics and Genomics*, vol. 38, no. 9, pp. 379–390, 2011.

[86] D. S. Zhang, W. Q. Liang, C. S. Yin, J. Zong, F. W. Gu, and D. B. Zhang, "*OsC6*, encoding a lipid transfer protein, is required for postmeiotic anther development in rice," *Plant Physiology*, vol. 154, no. 1, pp. 149–162, 2010.

[87] A. Bernard, F. Domergue, S. Pascal et al., "Reconstitution of plant alkane biosynthesis in yeast demonstrates that *Arabidopsis eceriferum1* and *eceriferum3* are core components of a very-long-chain alkane synthesis complex," *The Plant Cell*, vol. 24, no. 7, pp. 3106–3118, 2012.

[88] V. F. Souza, M. S. Pagliarini, C. B. Valle, N. C. Bione, M. U. Menon, and A. B. Mendes-Bonato, "Meiotic behavior of *Brachiaria decumbens* hybrids," *Genetics and Molecular Research*, vol. 14, no. 4, pp. 12855–12865, 2015.

[89] O. Rowland and F. Domergue, "Plant fatty acyl reductases: enzymes generating fatty alcohols for protective layers with potential for industrial applications," *Plant Science*, vol. 193-194, pp. 28–38, 2012.

[90] M. G. Aarts, R. Hodge, K. Kalantidis et al., "The *Arabidopsis male sterility 2* protein shares similarity with reductases in elongation/condensation complexes," *The Plant Journal*, vol. 12, no. 3, pp. 615–623, 1997.

[91] A. A. Dobritsa, S. I. Nishikawa, D. Preuss et al., "*LAP3*, a novel plant protein required for pollen development, is essential for proper exine formation," *Sexual Plant Reproduction*, vol. 22, no. 3, pp. 167–177, 2009.

[92] Z. L. Mou, X. Q. Wang, Z. M. Fu et al., "Silencing of phosphoethanolamine *N*-methyltransferase results in temperature-sensitive male sterility and salt hypersensitivity in Arabidopsis," *The Plant Cell*, vol. 14, no. 9, pp. 2031–2043, 2002.

[93] M. A. Zaidi, S. O'Leary, S. B. Wu et al., "A molecular and proteomic investigation of proteins rapidly released from triticale pollen upon hydration," *Plant Molecular Biology*, vol. 79, no. 1-2, pp. 101–121, 2012.

[94] A. Baker, I. A. Graham, M. Holdsworth, S. M. Smith, and F. L. Theodoulou, "Chewing the fat: β-oxidation in signalling and development," *Trends in Plant Science*, vol. 11, no. 3, pp. 124–132, 2006.

[95] S. F. Yang and N. E. Hoffman, "Ethylene biosynthesis and its regulation in higher plants," *Annual Review of Plant Physiology*, vol. 35, no. 1, pp. 155–189, 1984.

[96] A. J. Afzal, J. H. Kim, and D. Mackey, "The role of NOI-domain containing proteins in plant immune signaling," *BMC Genomics*, vol. 14, no. 1, p. 327, 2013.

Relationship of SNP rs2645429 in Farnesyl-Diphosphate Farnesyltransferase 1 Gene Promoter with Susceptibility to Lung Cancer

Mehdi Dehghani,[1] **Zahra Samani,**[2] **Hassan Abidi,**[2] **Leila Manzouri,**[3] **Reza Mahmoudi,**[2] **Saeed Hosseini Teshnizi ⓘ,**[4] **and Mohsen Nikseresht ⓘ**[2]

[1]*Hematology and Medical Oncology Department, Hematology Research Center, Shiraz University of Medical Sciences, Shiraz, Iran*
[2]*Cellular and Molecular Research Center, Yasuj University of Medical Sciences, Yasuj, Iran*
[3]*Social Determinant of Health Research Center, Yasuj University of Medical Sciences, Yasuj, Iran*
[4]*Biostatistican, Molecular Medicine Research Center, Hormozgan University of Medical Sciences, Bandar Abbas, Iran*

Correspondence should be addressed to Mohsen Nikseresht; nikmohsen65@gmail.com

Academic Editor: Mohamed Salem

Background and Purpose. The mevalonate pathway is one of the major metabolic pathways that use acetyl-CoA to produce sterols and isoprenoids. These compounds can be effective in the growth and development of tumors. One of the enzymes involved in the mevalonate pathway is FDFT1. Different variants of this gene are involved in the risk of suffering various diseases. The present study examined the relationship between FDFT1 rs2645429 polymorphism and the risk of nonsmall cell lung cancer (NSCLC) in a population from southern Iran. *Method.* The genotypes of rs2645429 polymorphism of FDFT1 gene were examined in 95 samples: 34 patients with NSCLC and 61 healthy individuals by RFLP method. *Results.* The results of this study indicated that C allele of this polymorphism was effectively associated with the risk of NSCLC in the Iranian population (p value = 0.023; OR = 2.71; 95% CI = 1.12–6.59) and CC genotype has significant relation with susceptibility to NSCLC (p value = 0.029; OR = 3.02; 95% CI = 1.09–8.39). This polymorphism is located in the promoter region FDFT1 gene, and CC genotype may increase the activity of this promoter. This study also found a significant relationship between C allele and metastatic status. C allele was more common in NSCLC patients. (p = 0.04). *Conclusion.* C allele of FDFT1 rs2645429 polymorphism gene can be a risk factor for NSCLC, whereas T allele probably has a low protective role.

1. Introduction

Nonsmall cell lung cancer (NSCLC) is lethal cancer in which approximately half of the patients show metastasis [1, 2]. Lung cancer in the western hemisphere is the second common cancer and is more common among men than women after breast cancer [3, 4]. Lung cancer mortality is higher in developed countries compared to less developed ones and is higher in men than in women. In the world, the incidence of lung cancer in North America is higher than that in the rest of the world, and East Asia ranks fifth worldwide [5]. In Iran, lung cancer is among the five common cancers [6]. The difference in the incidence of lung cancer in different parts of the world can be due to environmental factors and genetic diversity. Genetic variations in different genes can affect the risk of lung cancer [7–14].

One of the hallmarks in different kinds of human cancers is alteration in metabolic pathways such as carbohydrate, lipid, and protein metabolism. The mevalonate pathway is one of these pathways that biosynthesize sterols and isoprenoids [15]. The first reaction of this pathway is the forming of 3-hydroxy-3-methylgrutaryl coenzyme A (HMG-CoA) from the condensation of acetyl-CoA with acetoacetyl-CoA. Then, HMG-CoA was reduced to mevalonate by the HMG-CoA reductase (HMG-CoAR) enzyme. This enzyme is the rate-limiting enzyme for the pathway. This pathway has been

Chromosome 8

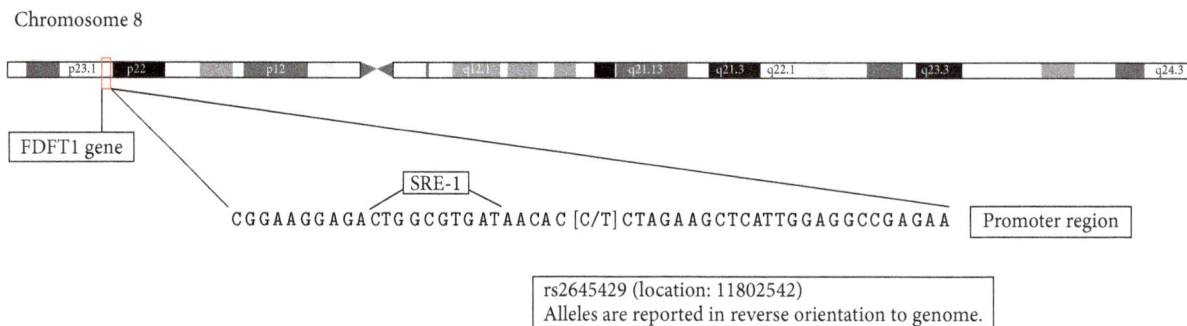

FIGURE 1: Location of the rs2645429 promoter SNP in the FDFT1 gene.

studied in various cancers, including breast cancer, multiple myeloma, myeloid leukemia, pancreatic cancer, and liver cancer, and the role of various genes of this pathway has been demonstrated in the molecular mechanism of the mentioned cancers [16–22]. Recent studies showed that the inhibitors of HMG-CoAR that called statins could contribute to the treatment of cancers such as ovarian, breast, and lung cancer [23–25]. The mevalonate pathway can be effective in various cellular processes. Two intermediate products of this route are farnesyl-diphosphate (FPP) and geranylgeranyl diphosphate (GGPP) that participate in the protein prenylation. This is a posttranslational modification of proteins and can affect cell functions such as growth, differentiation, and tumor formation. Many small guanosine triphosphate hydrolases (GTPases) that participate in tumorigenesis, such as RAS and RHO proteins, are isoprenylated, and the inhibition of this pathway can reduce the isoprenylation of these small GTPases. This process can induce the death of some cancer cells [26–31].

FPP is converted into squalene by squalene synthase (SQS). The gene encoding SQS enzyme is farnesyl-diphosphate farnesyltransferase 1 (FDFT1) with the chromosomal position of 8p23. The product of this gene weighs 47 kDa and plays an important role in the production of mevalonate-pathway processes, either from sterol or from the nonsterol route. This gene is expressed in all body tissues, especially in the liver and hypothalamus. This gene has several isoforms; the most common of which has 8 exons, and the promoter of this gene contains sterol regulatory element- (SRE-) like regions [32, 33].

The genetic diversity of FDFT1 gene has been studied in a number of diseases, including cancers. In investigating the SNP rs2645424 polymorphism of this gene in hepatitis C patients, it was found that this polymorphism was associated with advanced fibrosis in patients with a nonfatty liver [34]. In prostate cancer, the association of the rs2645429 polymorphism of FDFT1 gene has also been observed with progression and invasive phenotypes of this cancer. FDFT1 rs2645429 polymorphism, where nucleotide T replaces C (rs2645429 alleles are reported in reverse orientation to the genome), is located on the promoter of FDFT1 gene (Figure 1). FDFT1 rs2645429 is located at 6 bp upstream of a putative SRE-1, and it has been suggested that the promoter activity of this gene is affected by the replacement of one base in this polymorphism [35]. Given the mentioned points and

the importance of the mevalonate pathway in the onset and development of cancer, the present study was designed to determine the relationship between NSCLC and FDFT1 rs2645429 polymorphism gene for the first time.

2. Materials and Methods

Thirty-four patients with NSCLC and 61 healthy subjects as the controls were enrolled. The control group was matched to the case group in terms of age and gender. The control group was the people whose pathological tests were negative for lung cancer and who had no history of other cancers, cardiovascular disease, or cholesterol-lowering drugs. The case group was those whose lung cancer, not secondary cancer, was confirmed to be NSCLC by an oncologist, and in fact, results showed no metastases of other nonlung cancers, and they did not have any other chronic disease at the same time. Sampling from the case was done with informed consent and voluntarily. For this purpose, 3 ml of the whole blood was taken from each case to extract DNA. Clinical information of the patients, such as metastasis, stage of disease, and response to treatment was recorded.

DNA was extracted using QIAamp DNA Mini (QIAgen Inc., Santa Clarita, CA). In summary, $100 \mu l$ of the whole blood was transferred to a 1.5 ml microtube to extract DNA. Then, $400 \mu l$ of lysis buffer was added and vortexed for 20 seconds. Then, $300 \mu l$ of precipitant solution was added to the microtubes containing blood and lysis buffer, and vertex was performed for 5 seconds. All of the above solution ($800 \mu l$) was transferred to the separator column, and then, the column was centrifuged for 12 minutes at $12000 g$ for one minute, and after centrifugation and washing with solutions 1 and 2 and centrifugation in $12000 g$, DNA was extracted by means of an elution buffer and in accordance with kit instructions.

The PCR-RFLP method was used to study genotypes. Primers needed for PCR were designed by the NCBI primer design and Primer3. The fragment containing SNP rs2645429 polymorphism was amplified using forward $5'$-GCTGGACCTGTGGAGTAGGT-$3'$ and reverse $5'$-CTCCTGCGCATCCTAAGC-$3'$ primers. The restriction endonuclease enzyme suitable for the rs2645429 polymorphism was identified by Geneious Basic 4.8.5 software; XbaI (Thermo Fisher Scientific Co., USA) was purchased for this purpose.

TABLE 1: Demographic data for healthy controls and NSCLC patients.

Demographic data	Healthy controls Number (%)	NSCLC patients Number (%)	p value
Samples	61	34	—
Age (years)			
Male	65.44 ± 7.91	64.25 ± 12.05	0.052^a
Female	61.59 ± 13.55	55.07 ± 12.43	
Gender			
Male	39 (64)	15 (44)	0.06^b
Female	22 (36)	19 (56)	
Smoking status			
Smoker	25 (65.9)	13 (34.1)	0.79^b
Nonsmoker	36 (63.2)	21 (36.8)	

[a]Based on independent sample *t*-test. [b]Based on chi-square test. Statistically, there was no difference between age and sex of the healthy and NSCLC groups ($p > 0.05$), but the number of smokers of the healthy control group was significantly higher than that in the NSCLC patient group ($p = 0.005$).

After DNA extraction using the primers designed, the PCR test was conducted for all the samples. The steps of performing PCR in brief are 3 minutes at 95°C and 45 cycles for 30 seconds at 95, 58, and 72°C, respectively, and 3 minutes at 72°C. The product of this PCR was 376 bp long and was visible by electrophoresis of 1.5% agarose gel. The samples were then incubated for 16 hours at 37°C by the XbaI restriction enzyme until digested. After incubation, the products were electrophoresed on 2% agarose gel.

Statistical analyses were performed using SPSS 23. *t*-test and chi-square tests were used to compare the means and to examine the relationship between the desired polymorphism genotypes and lung cancer, respectively. In all results, the significance level was considered $p < 0.05$.

3. Results

3.1. Study Characteristics. The demographic characteristics of the controls and NSCLC cases are presented in Table 1.

The results of this study showed that the variable age followed the normal distribution ($p = 0.052$). There were no significant differences in the frequency distribution of gender in the patient and healthy groups. Although NSCLC was more common in women, this difference was not significant ($p = 0.06$).

3.2. Genotypes and NSCLC Risk. In determining the genotype of the FDFT1 gene, the existence of only one 376 bp band shows homozygous CC genotypes, three bands with lengths of 376, 224, and 152 represent heterozygous CT, and the presence of two bands with lengths of 224 and 152 represents the homozygous TT genotype (Figure 2).

The rs2645429 genotype distributions in the control group were compatible with the Hardy-Weinberg equilibrium ($p = 0.76$).

The average efficiency of the extracted DNA was $5.5\,\mu g/100\,\mu l$, and the optical absorption ratio 260/80 was equal to 1.68–1.92.

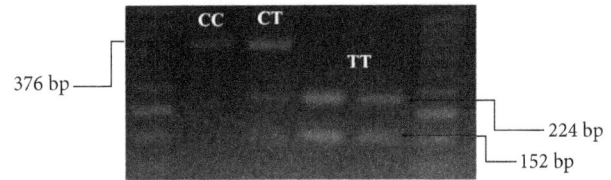

FIGURE 2: PCR product agarose gel electrophoresis after digestion with XbaI restriction enzyme.

In the present study, 6 (6.3%) subjects had TT genotype, 24 (25.3%) subjects had CT genotype, and 65 (68.4%) subjects had CC genotype (Table 2; Figure 3).

The frequency distribution of the CC genotype was significantly different in the control and case groups ($p = 0.029$). In the case group, 28 cases were CC, five cases were CT, and one case was TT genotypes, and in the control group, five persons were TT, nine persons were CT, and 37 persons were CC genotypes.

There was no significance associated with disease stages and all genotypes ($p = 0.10$). However, those with the CC genotype were at stage IV of cancer. The frequency distribution of C allele was significantly correlated with metastatic status ($p = 0.04$) Table 3.

4. Discussion and Conclusion

In the present study, for the first time, we examined the SNP rs2645429 genotype from the FDFT1 gene in patients with NSCLC. The study revealed that C allele could be a risk factor for this cancer.

One of the prominent features of cancer is a change in the metabolic structure of the cells. To support cell proliferation, tumor cells are able to alter the metabolism of carbohydrates, lipids, and proteins. Although most studies have investigated the disorder in regulating carbohydrate metabolism in cancer, recent studies have shown that lipid metabolism can also be associated with the progression of cancer [36–38]. For example, an increase in LDL cholesterol in ovarian cancer patients has been associated with their prognosis [39].

Lung cancer is a global health problem with more than 1.6 million cases a year. It is the second most common cancer and one of the main causes of death from cancer in men and women. Most patients are diagnosed at the advanced level of the disease (III, IV), where the average survival is about 8 to 12 months. There are three types of this cancer, 85% of which are nonsmall cell, 10 to 15% of small cell type, and less than 5% of lung carcinoid tumor type [4, 40].

There are many studies in relation to cholesterol and the risk of lung cancer and its implications; for example, Kucharska-Newton et al. found that the incidence of this cancer was higher in smokers compared to nonsmokers with low levels of HDL cholesterol [41]. In a study by Li et al. [42], it was shown that measuring cholesterol levels before treatment was used as a new independent predictor of NSCLC. In the study of Wu et al., the relationship between serum cholesterol and drug resistance in patients with pulmonary adenocarcinoma has been observed [43].

TABLE 2: Genotypic and allelic frequency of FDFT1 polymorphisms in NSCLC patients and control group.

	Control (N = 61)	NSCLC (N = 34)	OR (95% CI)	p value
Age (years)	62.32 ± 4.12	61.08 ± 7.93		
Gender				
Male	39 (64)	15 (44)	0.71 (0.28–1.77)	0.46[a]
Female	22 (36)	12 (56)	Reference	
Smoking status				
Smoker	25 (65.9)	13 (34.1)	0.89 (0.38–2.11)	0.79[a]
Nonsmoker	36 (63.2)	21 (36.8)	Reference	
Codominant				
CC, n (%)	37 (60.65)	28 (82.35)		0.09[a]
CT, n (%)	19 (31.15)	5 (14.7)		
TT, n (%)	5 (8.2)	1 (2.95)		
Dominant				
CC, n (%)	37 (60.65)	28 (82.35)	3.02 (1.09–8.39)	0.029[a]
CT + TT, n (%)	24 (39.35)	6 (17.65)		
Recessive				
TT, n (%)	5 (8.2)	1 (2.95)	0.33 (0.03–3.03)	0.31[a]
CT + CC, n (%)	56 (91.8)	33 (97. 05)		
Allele				
C, n (%)	93 (76.23)	61 (89.7)	2.71 (1.12–6.59)	0.023[a]
T, n (%)	29 (23.77)	7 (10.3)	Reference	

[a]Based on chi-square test.

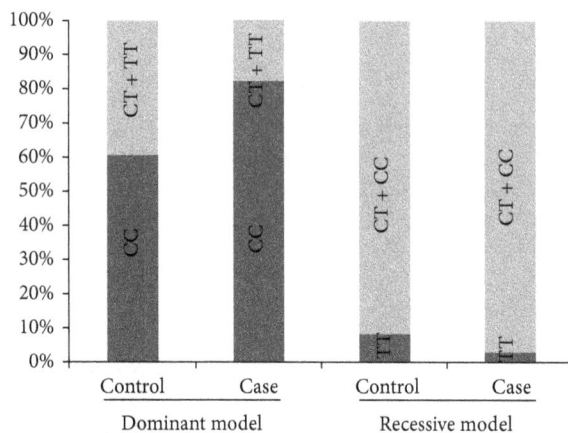

FIGURE 3: Genotype distribution in case and controls for rs2645429 site. CC genotype frequency was significantly higher in the NSCLC case than the control group under dominant model (p value = 0.029; OR = 3.02; 95% CI = 1.09–8.39). CC + CT versus TT was not significant (p value = 0.31).

Given the above data, the importance of the mevalonate pathway in different types of cancers, including lung cancer, was shown. Moreover, it is important to study the role of enzymes and other factors involved in this pathway at different genetic and protein levels. In this study, one of the SNP polymorphisms of FDFT1 (rs2645429) was studied. Stättermayer et al. worked on two rs2645424 polymorphisms of

FDFT1 gene and rs738409 of adiponutrin gene (PNPLA3) in patients with chronic hepatitis C, and the association of FDFT1 polymorphism with advanced fibrosis was observed in this disease, which probably resulted from the metabolic changes that this polymorphism can produce in the mevalonate pathway [34]. In the present study, a significant correlation was found between the presence of the CC genotype of FDFT1 gene and the risk of NSCLC with a chance ratio of 3.02 (p = 0.029). This finding may indicate C allele being a risk factor for NSCLC.

Schneider et al. studied GSTM1, GSTP1, or GSTT1 gene polymorphism in Caucasian smokers with lung cancer, where there was no significant association between the risk of this cancer and these polymorphisms. Thus, different genotypes of these genes do not play a decisive role in smokers suffering lung cancer [44]; while in the present study, the polymorphisms in FDFT1 gene could be associated with the risk of NSCLC.

Yang et al. [45] conducted a meta-analysis study on 23 studies with a population of 9815 people with lung cancer. There were no significant relationships between the risk of lung cancer and myeloperoxidase- (MPO-) 463G>A [45]. This is contrary to the present study that suggests that the presence of C alleles from FDFT1 gene polymorphism in individuals may increase the risk of lung cancer.

In this study, 19 patients, whose lung cancer stages had been determined based on pathologic findings and the approval of an oncologist, were examined for association with FDFT1 gene rs2645429 polymorphism genotypes. It was observed that individuals who had the C allele of this

TABLE 3: Association between FDFT1 gene polymorphism and clinicopathological characteristics of Non-small Cell lung cancer.

Clinical characteristics	Genotype ($N = 19$)			Allele ($N = 38$)	
Clinical staging	CC	CT	TT	T	C
Stage IIIA, IIIB	3 (15.79)	0 (0.0)	1 (5.26)	2 (5.26)	6 (15.79)
Stage IV	12 (63.16)	3 (15.79)	0 (0.0)	3 (7.89)	27 (71.05)
Fisher's exact test		$p = 0.1$			$p = 0.26$	
Metastasis					
Positive	8 (42.11)	0 (0.0)	0 (0.0)	0 (0.0)	16 (42.1)
Negative	7 (36.84)	3 (15.79)	1 (5.26)	5 (13.16)	17 (44.74)
Fisher's exact test		$p = 0.15$			$p = 0.04^*$	
Treatment					
Response	8 (42.11)	2 (10.53)	1 (5.26)	4 (10.53)	18 (47.37)
Nonresponse	7 (36.84)	1 (5.26)	0 (0.0)	1 (2.63)	15 (39.47)
Fisher's exact test		$p = 0.62$			$p = 0.28$	
NSCLC types					
Adenocarcinoma	11 (57.89)	3 (15.79)	1 (5.26)	5 (13.16)	25 (65.79)
Squamous cell carcinoma	4 (21.05)	0 (0.0)	0 (0.0)	0 (0.0)	8 (21.05)
Fisher's exact test		$p = 0.5$			$p = 0.21$	

*There is a significant correlation between metastatic status and C allele distribution ($p = 0.04$).

gene were at more advanced stages of cancer but there was no significance ($p = 0.26$), and this may indicate that C allele increases not only the risk of lung cancer but also the chance of progression. However, due to the small number of samples, this can be checked out more. In other studies, the importance of the polymorphism of some genes in the progression of cancer has been studied; for example, Javid et al. [46] observed that CASP3 (-1337C>G) polymorphism is likely to play a role in the development of NSCLC.

There are several studies which have also shown that SNP polymorphisms can be effective in the advancing and development of other cancers [47, 48] and the present study. These studies indicate the importance of examining the association of the polymorphism of some genes with the progression of cancer.

The results of this study showed that the rs2645429 polymorphism of FDFT1 gene could be associated with the risk of NSCLC, and C allele was considered a risk factor for lung cancer in the patient group. C allele of this gene is also associated with more advanced stages of this cancer, which may indicate that C allele increases not only the risk of lung cancer but also the chance of progression.

Although smoking is a major contributor to lung cancer, due to an increase in the incidence of lung cancer in nonsmokers, it is very important to study other risk factors. As previously mentioned, the alterations in lipid metabolism pathways were shown in different kinds of cancer, such as lung cancer. The cholesterol biosynthesis pathway in relation to cancer has been very much considered. The present study is the first study about one of the other risk factors in the cholesterol biosynthesis pathway in relation with NSCLC, and it is a preliminary study, and to understand how this polymorphism affects lung cancer, we should study more at the function of all genotypes of FDFT1 rs2645429 polymorphism.

Disclosure

This article was extracted from the MS thesis at the Yasuj University of Medical Sciences.

Conflicts of Interest

The authors declare that there is no conflict of interests regarding the publication of this paper.

Acknowledgments

The authors would like to thank the Deputy of Research Affairs at the University for funding this project and Farzanegan Clinical Laboratory, Shiraz, Iran.

References

[1] R. L. Siegel, K. D. Miller, and A. Jemal, "Cancer statistics, 2015," *CA: A Cancer Journal for Clinicians*, vol. 65, no. 1, pp. 5–29, 2015.

[2] D. J. Raz, J. A. Zell, S.-H. Ignatius Ou, D. R. Gandara, H. Anton-Culver, and D. M. Jablons, "Natural history of stage I non-small cell lung cancer: implications for early detection," *Chest Journal*, vol. 132, no. 1, pp. 193–199, 2007.

[3] D. Behera, "Epidemiology of lung cancer-global and Indian perspective," *Journal Indian Academy of Clinical Medicine*, vol. 13, pp. 131–137, 2012.

[4] R. L. Siegel, K. D. Miller, and A. Jemal, "Cancer statistics, 2016," *CA: A Cancer Journal for Clinicians*, vol. 66, no. 1, pp. 7–30, 2016.

[5] C. S. Dela Cruz, L. T. Tanoue, and R. A. Matthay, "Lung

cancer: epidemiology, etiology, and prevention," *Clinics in Chest Medicine*, vol. 32, no. 4, pp. 605–644, 2011.

[6] F. Moradpour and Z. Fatemi, "Estimation of the projections of the incidence rates, mortality and prevalence due to common cancer site in Isfahan, Iran," *Asian Pacific Journal of Cancer Prevention*, vol. 14, no. 6, pp. 3581–3585, 2013.

[7] M. C. Turner, A. Cohen, M. Jerrett et al., "Interactions between cigarette smoking and fine particulate matter in the risk of lung cancer mortality in cancer prevention study II," *American Journal of Epidemiology*, vol. 180, no. 12, pp. 1145–1149, 2014.

[8] C. Steinmaus, C. Ferreccio, Y. Yuan et al., "Elevated lung cancer in younger adults and low concentrations of arsenic in water," *American Journal of Epidemiology*, vol. 180, no. 11, pp. 1082–1087, 2014.

[9] Y. Liu, K. Steenland, Y. Rong et al., "Exposure-response analysis and risk assessment for lung cancer in relationship to silica exposure: a 44-year cohort study of 34,018 workers," *American Journal of Epidemiology*, vol. 178, no. 9, pp. 1424–1433, 2013.

[10] P. J. Villeneuve, M. Jerrett, D. Brenner, J. Su, H. Chen, and J. R. McLaughlin, "A case-control study of long-term exposure to ambient volatile organic compounds and lung cancer in Toronto, Ontario, Canada," *American Journal of Epidemiology*, vol. 179, no. 4, pp. 443–451, 2014.

[11] A. G. Schwartz, J. E. Bailey-Wilson, and C. I. Amos, "6 genetic susceptibility to lung cancer," *Iaslc Thoracic Oncology E-Book*, vol. 46, 2017.

[12] M. S. Nørskov, M. Dahl, and A. Tybjærg-Hansen, "Genetic variation in *GSTP1*, lung function, risk of lung cancer, and mortality," *Journal of Thoracic Oncology*, vol. 12, no. 11, pp. 1664–1672, 2017.

[13] Y. Yamamoto, C. Kiyohara, S. Ogata-Suetsugu, N. Hamada, and Y. Nakanishi, "Association between genetic polymorphisms involved in the hypoxia-inducible factor pathway and lung cancer risk: a case–control study in Japan," *Asia-Pacific Journal of Clinical Oncology*, vol. 13, no. 3, pp. 234–242, 2017.

[14] Y. Zhang, D. C. Wang, L. Shi, B. Zhu, Z. Min, and J. Jin, "Genome analyses identify the genetic modification of lung cancer subtypes," *Seminars in Cancer Biology*, vol. 42, pp. 20–30, 2017.

[15] P. J. Mullen, R. Yu, J. Longo, M. C. Archer, and L. Z. Penn, "The interplay between cell signalling and the mevalonate pathway in cancer," *Nature Reviews Cancer*, vol. 16, no. 11, pp. 718–731, 2016.

[16] C. A. Goard, M. Chan-Seng-Yue, P. J. Mullen et al., "Identifying molecular features that distinguish fluvastatin-sensitive breast tumor cells," *Breast Cancer Research and Treatment*, vol. 143, no. 2, pp. 301–312, 2014.

[17] O. Bjarnadottir, Q. Romero, P.-O. Bendahl et al., "Targeting HMG-CoA reductase with statins in a window-of-opportunity breast cancer trial," *Breast Cancer Research and Treatment*, vol. 138, no. 2, pp. 499–508, 2013.

[18] E. R. Garwood, A. S. Kumar, F. L. Baehner et al., "Fluvastatin reduces proliferation and increases apoptosis in women with high grade breast cancer," *Breast Cancer Research and Treatment*, vol. 119, no. 1, pp. 137–144, 2010.

[19] J. W. Clendening, A. Pandyra, Z. Li et al., "Exploiting the mevalonate pathway to distinguish statin-sensitive multiple myeloma," *Blood*, vol. 115, no. 23, pp. 4787–4797, 2010.

[20] N. E. Martin, T. B. Brunner, K. D. Kiel et al., "A phase I trial of the dual farnesyltransferase and geranylgeranyltransferase

inhibitor L-778,123 and radiotherapy for locally advanced pancreatic cancer," *Clinical Cancer Research*, vol. 10, no. 16, pp. 5447–5454, 2004.

[21] J. Dimitroulakos, D. Nohynek, K. L. Backway et al., "Increased sensitivity of acute myeloid leukemias to lovastatin-induced apoptosis: a potential therapeutic approach," *Blood*, vol. 93, no. 4, pp. 1308–1318, 1999.

[22] H. Graf, C. Jüngst, G. Straub et al., "Chemoembolization combined with pravastatin improves survival in patients with hepatocellular carcinoma," *Digestion*, vol. 78, no. 1, pp. 34–38, 2008.

[23] U. Walther, K. Emmrich, R. Ramer, N. Mittag, and B. Hinz, "Lovastatin lactone elicits human lung cancer cell apoptosis via a COX-2/PPARγ-dependent pathway," *Oncotarget*, vol. 7, no. 9, pp. 10345–10362, 2016.

[24] S. F. Nielsen, B. G. Nordestgaard, and S. E. Bojesen, "Statin use and reduced cancer-related mortality," *The New England Journal of Medicine*, vol. 368, no. 6, pp. 576–577, 2013.

[25] Y. Ling, L. Yang, H. Huang et al., "Prognostic significance of statin use in colorectal cancer: a systematic review and meta-analysis," *Medicine*, vol. 94, no. 25, p. e908, 2015.

[26] B. M. Willumsen, A. Christensen, N. L. Hubbert, A. G. Papageorge, and D. R. Lowy, "The p21 ras C-terminus is required for transformation and membrane association," *Nature*, vol. 310, no. 5978, pp. 583–586, 1984.

[27] K. C. Hart and D. J. Donoghue, "Derivatives of activated H-ras lacking C-terminal lipid modifications retain transforming ability if targeted to the correct subcellular location," *Oncogene*, vol. 14, no. 8, pp. 945–953, 1997.

[28] S. L. Moores, M. D. Schaber, S. D. Mosser et al., "Sequence dependence of protein isoprenylation," *Journal of Biological Chemistry*, vol. 266, no. 22, pp. 14603–14610, 1991.

[29] P. J. Casey and M. C. Seabra, "Protein prenyltransferases," *Journal of Biological Chemistry*, vol. 271, no. 10, pp. 5289–5292, 1996.

[30] Z. Xia, M. M. Tan, W. Wei-Lynn Wong, J. Dimitroulakos, M. D. Minden, and L. Z. Penn, "Blocking protein geranylgeranylation is essential for lovastatin-induced apoptosis of human acute myeloid leukemia cells," *Leukemia*, vol. 15, no. 9, pp. 1398–1407, 2001.

[31] S. Ghavami, B. Yeganeh, G. L. Stelmack et al., "Apoptosis, autophagy and ER stress in mevalonate cascade inhibition-induced cell death of human atrial fibroblasts," *Cell Death & Disease*, vol. 3, no. 6, article e330, 2012.

[32] W. Likus, K. Siemianowicz, K. Bieńk et al., "Could drugs inhibiting the mevalonate pathway also target cancer stem cells?," *Drug Resistance Updates*, vol. 25, pp. 13–25, 2016.

[33] R. Do, R. Kiss, D. Gaudet, and J. Engert, "Squalene synthase: a critical enzyme in the cholesterol biosynthesis pathway," *Clinical Genetics*, vol. 75, no. 1, pp. 19–29, 2009.

[34] A. F. Stättermayer, K. Rutter, S. Beinhardt et al., "Role of FDFT1 polymorphism for fibrosis progression in patients with chronic hepatitis C," *Liver International*, vol. 34, no. 3, pp. 388–395, 2014.

[35] Y. Fukuma, H. Matsui, H. Koike et al., "Role of squalene synthase in prostate cancer risk and the biological aggressiveness of human prostate cancer," *Prostate Cancer and Prostatic Diseases*, vol. 15, no. 4, pp. 339–345, 2012.

[36] S. Riedel, S. Abel, S. Swanevelder, and W. C. A. Gelderblom, "Induction of an altered lipid phenotype by two cancer pro-

moting treatments in rat liver," *Food and Chemical Toxicology*, vol. 78, pp. 96–104, 2015.

[37] C. Huang and C. Freter, "Lipid metabolism, apoptosis and cancer therapy," *International Journal of Molecular Sciences*, vol. 16, no. 12, pp. 924–949, 2015.

[38] S. Hashmi, Y. Wang, D. S. Suman et al., "Human cancer: is it linked to dysfunctional lipid metabolism?," *Biochimica et Biophysica Acta (BBA) - General Subjects*, vol. 1850, no. 2, pp. 352–364, 2015.

[39] A. J. Li, R. Geoffrey Elmore, I. Y.-d. Chen, and B. Y. Karlan, "Serum low-density lipoprotein levels correlate with survival in advanced stage epithelial ovarian cancers," *Gynecologic Oncology*, vol. 116, no. 1, pp. 78–81, 2010.

[40] M. M. Salvador, M. G. de Cedrón, J. M. Rubio et al., "Lipid metabolism and lung cancer," *Critical Reviews in Oncology/Hematology*, vol. 112, pp. 31–40, 2017.

[41] A. M. Kucharska-Newton, W. D. Rosamond, J. C. Schroeder, A. M. McNeill, J. Coresh, and A. R. Folsom, "HDL-cholesterol and the incidence of lung cancer in the atherosclerosis risk in communities (ARIC) study," *Lung Cancer*, vol. 61, no. 3, pp. 292–300, 2008.

[42] J.-R. Li, Y. Zhang, and J.-L. Zheng, "Decreased pretreatment serum cholesterol level is related with poor prognosis in resectable non-small cell lung cancer," *International Journal of Clinical and Experimental Pathology*, vol. 8, no. 9, pp. 11877–11883, 2015.

[43] Y. Wu, R. Si, H. Tang et al., "Cholesterol reduces the sensitivity to platinum-based chemotherapy via upregulating ABCG2 in lung adenocarcinoma," *Biochemical and Biophysical Research Communications*, vol. 457, no. 4, pp. 614–620, 2015.

[44] J. Schneider, U. Bernges, M. Philipp, and H.-J. Woitowitz, "*GSTM1*, *GSTT1*, and *GSTP1* polymorphism and lung cancer risk in relation to tobacco smoking," *Cancer Letters*, vol. 208, no. 1, pp. 65–74, 2004.

[45] W.-J. Yang, M.-Y. Wang, F.-Z. Pan, C. Shi, and H. Cen, "Association between MPO-463G > A polymorphism and cancer risk: evidence from 60 case-control studies," *World Journal of Surgical Oncology*, vol. 15, no. 1, p. 144, 2017.

[46] J. Javid, R. Mir, and A. Saxena, "Involvement of CASP3 promoter polymorphism (− 1337 C> G) in the development and progression of non-small cell lung cancer," *Tumor Biology*, vol. 37, no. 7, pp. 9255–9262, 2016.

[47] P. B. Meka, S. Jarjapu, S. K. Vishwakarma et al., "Influence of BCL2-938 C>A promoter polymorphism and BCL2 gene expression on the progression of breast cancer," *Tumor Biology*, vol. 37, no. 5, pp. 6905–6912, 2016.

[48] S. Matsusaka, S. Cao, D. L. Hanna et al., "CXCR4 polymorphism predicts progression-free survival in metastatic colorectal cancer patients treated with first-line bevacizumab-based chemotherapy," *The Pharmacogenomics Journal*, vol. 17, no. 6, pp. 543–550, 2016.

Marker-Assisted Introgression of *Saltol* QTL Enhances Seedling Stage Salt Tolerance in the Rice Variety "Pusa Basmati 1"

Vivek Kumar Singh,[1] Brahma Deo Singh,[2] Amit Kumar,[1] Sadhna Maurya,[3] Subbaiyan Gopala Krishnan,[1] Kunnummal Kurungara Vinod ⓘ,[4] Madan Pal Singh,[3] Ranjith Kumar Ellur,[1] Prolay Kumar Bhowmick,[1] and Ashok Kumar Singh ⓘ[1]

[1]ICAR-Indian Agricultural Research Institute, Division of Genetics, New Delhi 110012, India
[2]Banaras Hindu University, School of Biotechnology, Varanasi 221005, Uttar Pradesh, India
[3]ICAR-Indian Agricultural Research Institute, Division of Plant Physiology, New Delhi 110012, India
[4]Rice Breeding and Genetics Research Centre, ICAR-Indian Agricultural Research Institute, Aduthurai 612 101, India

Correspondence should be addressed to Ashok Kumar Singh; aks_gene@yahoo.com

Academic Editor: Gunvant B. Patil

Marker-assisted selection is an unequivocal translational research tool for crop improvement in the genomics era. Pusa Basmati 1 (PB1) is an elite Indian Basmati rice cultivar sensitive to salinity. Here, we report enhanced seedling stage salt tolerance in improved PB1 genotypes developed through marker-assisted transfer of a major QTL, *Saltol*. A highly salt tolerant line, FL478, was used as the *Saltol* donor. Parental polymorphism survey using 456 microsatellite (SSR)/QTL-linked markers revealed 14.3% polymorphism between PB1 and FL478. Foreground selection was carried out using three *Saltol*-linked polymorphic SSR markers RM8094, RM493, and RM10793 and background selection by 62 genome-wide polymorphic SSR markers. In every backcross generation, foreground selection was restricted to the triple heterozygotes of foreground markers, which was followed by phenotypic and background selections. Twenty-four near isogenic lines (NILs), with recurrent parent genome recovery of 96.0–98.4%, were selected after two backcrosses followed by three selfing generations. NILs exhibited agronomic traits similar to those of PB1 and additional improvement in the seedling stage salt tolerance. They are being tested for per se performance under salt-affected locations for release as commercial varieties. These NILs appear promising for enhancing rice production in salinity-affected pockets of Basmati Geographical Indication (GI) areas of India.

1. Introduction

Rice plants suffer severe salt injury in both seedling and reproductive stages; the most common damages are attributed to osmotic imbalance, membrane destabilisation, and failure of photosynthetic machinery [1]. The damage due to salt stress is often cumulative as the seedling stage sensitivity leads to poor crop establishment, and reproductive stage sensitivity results in reduced yields [2]; the combined effect of damages at both the stages may lead to total crop loss. Nevertheless, seedling stage tolerance can sustain crop production in salinity prone areas by promoting good initial establishment leading to healthy vegetative growth that can augment crop yield [3]. There are some saline ecosystem-adapted traditional rice landraces such as Pokkali and Nona Bokra that are known to be salt tolerant. Salt tolerance in rice is manifested through morphological, physiological, and metabolic responses that includes stomatal changes, sodium exclusion, tissue tolerance, apoplastic salt compartmentalization, salt sequestration into older tissues, and regulation of the antioxidants [2–5]. Apart from the understanding of physiological and metabolic responses to salt stress, quantitative trait loci (QTLs) and genes governing salt tolerance have also been reported in rice. These include a major QTL, *Saltol* identified on chromosome 1 of Pokkali, and *SKC1* (*OsHKT1;5*), a gene located within the *Saltol* region identified from Nona Bokra. The QTL *Saltol* imparts salt tolerance by regulating Na^+/K^+ homeostasis under salt stress [6–9].

In India, of the estimated area of 7.0 million ha (mha) occupied by saline soils, a sizeable fraction occurs in the Indo-Gangetic plains covering the states of Haryana, Punjab, Uttar Pradesh, Rajasthan, and Bihar [10]. Basmati rice is exclusively grown in an area of over 1.68 mha spanning the Indo-Gangetic plains; this region is recognised as its Geographical Indication (GI) area [11–14]. In recent times, soil salinity has become a major problem affecting Basmati rice cultivation, especially in the state of Haryana [11]. Haryana has about 1.0 mha under Basmati rice, majority of which is threatened by inland salinity resulting from the continuous use of brackish irrigation water [12]. None of the popular Basmati cultivars is reported to be tolerant to salt stress.

Basmati rice is preferred globally for its aromatic grains with unparalleled cooking qualities [11] such as extra-long slender grains, rich aroma, white kernels, translucent endosperm, high cooking elongation, fluffy cooked kernels, good palatability, and medium amylose content. Commercially released in 1989 by ICAR-Indian Agricultural Research Institute (ICAR-IARI), Pusa Basmati 1 (PB1) is the first semidwarf and high-yielding Basmati variety in the world. The release of PB1 revolutionized Basmati rice production in India, because of several advantages over the traditional Basmati cultivars: (a) It had an average yield of more than 4.5 t/ha, as against the low average yield of 2.5 t/ha for the traditional cultivars; (b) PB1 was shorter with robust plant stature, and (c) PB1 matured faster than the late and photosensitive traditional Basmati cultivars [15, 16]. Soon after the release, PB1 got established as a premium cultivar and was extensively cultivated. Twenty-eight years after its commercial release, even today, PB1 is cultivated in about 0.16 mha (~10% of the total Basmati area) in India. It is used extensively in Basmati rice improvement programmes as donor for quality traits as well as high yield. However, PB1 is sensitive to several biotic stresses such as diseases (bacterial blight, blast, sheath blight, and bakanae) and pests (brown plant hopper) and also to abiotic stresses, such as soil salinity and drought. PB1 has been improved for resistance to bacterial blight [17], blast [18], and sheath blight [19] using molecular markers as indirect selection tools, but improvement of salinity tolerance of PB1 is yet to be achieved.

In recent times, marker-aided selection (MAS) has been widely acclaimed as the most effective method of transferring desirable traits [1, 8, 20–22] in rice, including salinity tolerance. The conventional breeding efforts for salinity tolerance in rice had limited success, possibly due to their long turnover time, cumbersome screening procedures, and complex genetic control of the trait [1]. For transferring seedling stage salt tolerance, *Saltol* QTL is the only best-known target locus that is amenable to MAS. As the donor for *Saltol*, FL478 (IR 66946-3R-178-1-1), a highly salt tolerant RIL derived from the cross IR29/Pokkali, has been successfully deployed in breeding programmes in many countries. The SSR markers RM3412, AP3206, and RM8094 are used for transfer of *Saltol* [11, 23–26]. Since grain quality traits are of paramount

importance, MAS in Basmati rice needs special attention, especially when the transferred gene(s)/QTL(s) are sourced from non-Basmati donors [27, 28]. The *Saltol* donor, FL478, is a non-Basmati line that has grain characteristics such as medium bold shape, red pericarp, chalky endosperm, no aroma, high amylose content, and low gel consistency. The recovery of grain quality is achieved by integrating phenotypic selection for these traits, in every MAS stage [11].

In this paper, we report marker-aided introgression of the QTL *Saltol* from FL478 into PB1 and the resulting improvement in seedling stage salt tolerance of the PB1 near isogenic lines (NILs). Other agronomic features and grain quality of the NILs were comparable to those of the recurrent parent, PB1.

2. Materials and Methods

2.1. Plant Materials. The parents used in the present study were (a) PB1, as the recurrent parent (RP), and (b) FL478, as the donor parent for *Saltol*. FL478 is a breeding line with very high level of seedling stage salt tolerance; it can endure salt solutions with electrical conductivity (EC) of up to 15 dSm^{-1} for more than a fortnight. Both the parents were first evaluated for tolerance to 100 mM NaCl solution (EC of 11.6 dSm^{-1}) at seedling stage to validate their salt tolerance levels before initiating the crossing programme. Salt tolerance was scored using the standard evaluation system (SES) for rice developed by the International Rice Research Institute, Manila, Philippines [29, 30]. In the pre-screening, the recurrent parent, PB1, was found highly sensitive to salt stress and recorded a score of nine, while the donor parent, FL478, was tolerant and recorded a score of one (Supplementary Figure 1). Crosses were made at IARI-Rice Breeding and Genetics Research Centre, Aduthurai, Tamil Nadu (IARI-RBGRC), and subsequent generations were shuttled between the ICAR-IARI, New Delhi, during *Kharif* season and IARI-RBGRC during off-season.

2.2. Breeding Strategy. PB1 was crossed as the female parent with FL478, and the hybridity of the F_1 plants was confirmed using the SSR marker, RM493. The confirmed F_1s were backcrossed to PB1 (always used as the female parent in backcrosses) to generate the BC_1F_1 seeds. The plant selected in F_1 was designated as Pusa 1822; the lines derived from the backcross programme (Figure 1) carried the designation as the prefix, for example, Pusa 1822-6-14-9. The parental lines were screened for polymorphism at the target QTL locus using twenty-one *Saltol*-linked SSR markers, of which three markers RM8094, RM493, and RM10793 were found to be polymorphic (Supplementary Figure 2); all the three markers were used for foreground selection. Further, the genome-wide polymorphism between the parents was tested using 435 SSR markers, which identified 62 polymorphic markers that were employed for background selection (Supplementary Table 1). In the BC_1F_1 generation, the plants that tested positive for the three markers used for foreground selection were screened with the 62 SSR markers and subjected to phenotypic evaluation for agronomic traits, including grain characteristics. The plant with the highest recurrent

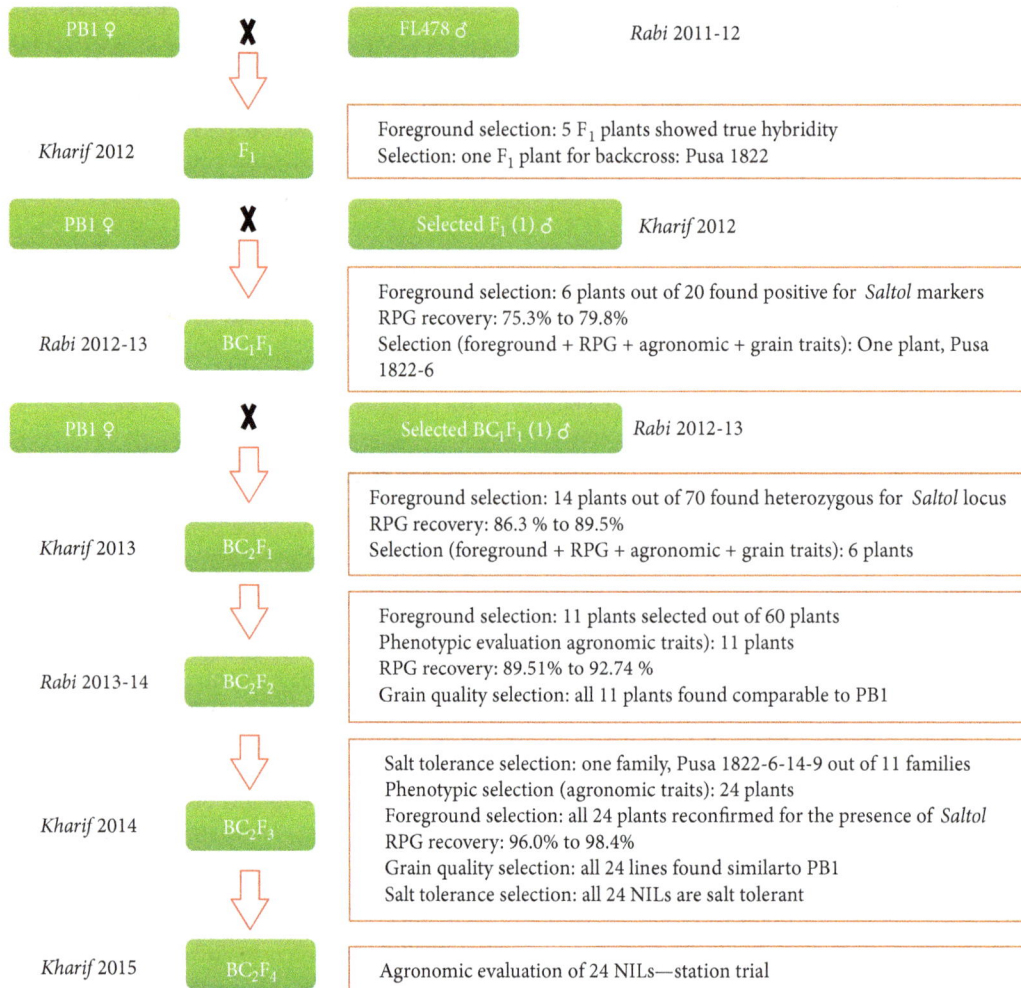

FIGURE 1: Breeding scheme used in the marker-assisted backcross programme for the transfer of *Saltol* locus in the background of the elite rice variety, Pusa Basmati 1.

parent genome (RPG) recovery, and having most similarity to the RP, was backcrossed to PB1 to produce the BC_2F_1 seeds. The BC_2F_1 plants were handled in the same manner, except for using only those markers for background selection that were heterozygous in BC_1F_1. The selected BC_2F_1 plants were selfed to produce BC_2F_2 generation. Each BC_2F_2 plant was subjected to foreground selection to identify plants homozygous for all the three foreground markers. The selected plants were subjected to background and phenotypic selections, the former to assess the recovery of RPG, using markers that were heterozygous in BC_2F_1, and the later to determine the recovery of Basmati quality traits. The selected BC_2F_2 plants were selfed to raise BC_2F_3 families, which were screened for seedling stage salt tolerance. The family showing the highest level of salt tolerance was transplanted in the field and evaluated for agronomic performance and grain quality. Agronomically superior members of the tolerant family were subjected to foreground screening to confirm the presence of *Saltol* alleles in homozygous state and background selection based on markers that were heterozygous in BC_2F_2 family to assess further increase in the

RPG recovery. The salt tolerant lines were advanced to BC_2F_4 generation.

2.3. Molecular Analyses. Genomic DNA was isolated from young leaves of the test lines when they were about 40 days old using the standard Cetyl Trimethyl Ammonium Bromide protocol [31]. Polymerase chain reaction- (PCR-) based amplification of the target genomic fragments by the primer pairs for each selected marker was performed in a $10\,\mu l$ reaction mix constituted by adding 25–30 ng genomic DNA, 5 pmol each of the two primers, 0.05 mM each of the four dNTPs, and PCR buffer (10x) containing 10 mM Tris (pH 8.4), 50 mM KCl, and 1.8 mM $MgCl_2$. To this mix, 0.5 U of Taq DNA polymerase was added, and the volume made up to $10\,\mu l$ using nuclease free water. The PCR was run for 35 cycles comprising of denaturation for one minute at 94°C, followed by annealing for one minute at 55°C, and primer elongation for two minutes at 72°C, sandwiched between an initial denaturation for five minutes at 94°C and the final extension for seven minutes at 72°C. The amplified products were electrophoresed in 3.5% agarose gel, and the products were visualized using a gel documentation system. The marker segregation data was graphically compiled in each generation using Graphical GenoTypes (GGT) version 2.0 software [32].

2.4. Marker-Aided Selection. Details of three *Saltol*-linked SSR markers, RM8094, RM493, and RM10793, used for foreground selection such as their physical position on chromosome 1, primer nucleotide sequences, and physical locations within the *Saltol* QTL are given in Supplementary Table 2. The details of all the genome-wide polymorphic SSR markers used for assessing the background polymorphism were sourced from the rice marker database at Gramene (http://www.gramene.org). In each backcross generation, background selection was done after foreground and phenotypic selections. In the background selection process using 62 polymorphic markers, the number of plants with homozygous alleles similar to PB1 and heterozygotes was counted sepa-

TABLE 1: Genetic diversity between the recurrent parent PB1 and the *Saltol* donor FL478. The foreground survey was limited to *Saltol* region alone, whereas the background survey included all chromosomes, including the *Saltol* carrier chromosome 1.

Class of markers	Markers surveyed	Polymorphic markers	Polymorphism (%)
Foreground	21	3	14.29
Background*	435	62	14.75
Chromosome 1§	63	7	11.11

* includes markers on chromosome 1, excluding *Saltol*-linked markers; §Based on all markers used including *Saltol*-linked markers.

rately for each marker. A reductionist strategy was followed for background selection; markers that became homozygous for the PB1 allele in a given generation were not included in the assay for the subsequent generations. RPG recovery was computed using following formula:

$$RPG\ recovery\% = \left[\frac{\text{number of marker homozygous for RP alleles} + (0.5 \times \text{number of heterozygotes markers})}{\text{total number of polymorphic markers}}\right] \times 100. \quad (1)$$

To ensure maximum recovery of the carrier chromosome of the *Saltol* QTL, the chromosome 1 was surveyed with 42 evenly distributed SSR markers together with 21 markers linked to the *Saltol* region. Complete recovery of the chromosome 1 together with *Saltol* was specifically targeted, while exercising selections for the background genome.

2.5. Screening for Seedling Stage Salt Tolerance. The PB1 NILs homozygous for the *Saltol* QTL along with the two parents were screened for seedling stage salt tolerance at the National Phytotron Facility, ICAR-IARI, New Delhi. Average day/night temperature of approximately 32/25°C and relative humidity of 70–80% were maintained at the screen house throughout the study period. Polystyrene floats with a 14×8 matrix of holes lined with a nylon net at the bottom side, and suspended in plastic crates filled with 10 litres of Yoshida nutrient solution [1, 33], were used for screening. The experiment was set up according to a randomized complete block design with two treatments (0.0 mM as control and 100 mM (EC of 11.6 dSm^{-1}) NaCl for salt stress) and three replications. Each replication comprised two plastic crates, one crate having six plants each of 12 NILs and the parents PB1 and FL478. The parents served as susceptible and salt tolerant checks. Four-day-old pregerminated seeds were surface sterilized using 70% ethanol and 5% sodium hypochlorite for five minutes each, transferred into the holes in the polystyrene floats, and allowed to germinate over the nutrient solution. The seedlings were subjected to salinity stress after 14 days, starting with an EC of 3 dSm^{-1} by adding 26 mM NaCl concentration in the nutrient solution, and subsequently elevating to 11.6 dSm^{-1} (100 mM NaCl) three days after. The same volume of deionized water was added in the control set. The nutrient solution was replaced once a week,

and its pH was maintained daily at 5.8 (adjusted by adding either 1 N NaOH or HCl). The EC of the nutrient solution was recorded daily. Sixteen days after imposing the full salt stress, the symptoms were scored as per SES for rice [30]. The genotypes showing score of 1–3 were classified as tolerant, those with a score of 5 were moderately tolerant, and those with scores of 7–9 were rated as susceptible.

2.6. Agronomic and Grain Quality Assessment. Agronomic evaluation of the BC_2F_4 NILs along with both the parents was carried out during *Kharif* 2015 at the research farm of the Division of Genetics, ICAR-IARI, New Delhi, in a field trial laid out in a randomized complete block design with two replications and plot size of 5m^2. Twenty-five-day-old seedlings were transplanted at a spacing of 20 cm × 15 cm, and the trial was maintained adopting recommended agronomic practices. From each replication, data on various agronomic traits, namely, days to 50% flowering (DFF), plant height (PH), effective tillers per plant (ETP), panicle length (PL), spikelet fertility (SF), weight of 1000 grains (TW), and grain yield per plant (YLD), were recorded from five random plants selected from each entry. The harvested grains from the NILs and their parents were assessed for quality traits like hulling recovery (HUL), milling recovery (MIL), and cooking-related characters, such as kernel length before and after cooking (KLBC and KLAC, resp.), kernel elongation on cooking (ER), alkali spreading value (ASV), and aroma (AROM) as described earlier [34].

2.7. Na$^+$ and K$^+$ Contents in Shoots and Roots. Since *Saltol* acts by balancing the Na$^+$ and K$^+$ ions in the plant system to counter the salt stress, we estimated the cationic concentrations in shoots and roots from the salt-stressed and salt-

TABLE 2: Agronomic performance, salt tolerance, and recurrent parent genome recovery of *Saltol*-introgressed NILs of Pusa Basmati 1.

NILS	Agronomic traits								RPG recovery			
	DFF	PH	ETP	PL	SF	TW	YLD	STS	RP	HT	DP	RPG %
NIL1	101.0^{a-e}	97.7^a	20.7^{ab}	27.8^a	77.4^{ab}	19.4^a	43.6^a	1.0	60	1	1	97.58
NIL2	98.5^{c-e}	93.3^a	17.7^{ab}	28.4^a	70.8^{ab}	19.4^a	43.4^a	1.0	60	1	1	97.58
NIL3	97.0^{de}	86.7^a	14.6^{ab}	25.3^a	71.0^{ab}	16.5^a	38.7^a	1.0	59	2	1	96.77
NIL4	106.0^a	86.0^a	14.1^{ab}	27.2^a	81.7^a	19.9^a	38.3^a	1.0	60	1	1	97.58
NIL5	105.0^{a-c}	88.9^a	15.1^{ab}	26.5^a	74.1^{ab}	17.8^a	39.1^a	1.0	59	2	1	96.77
NIL6	106.0^a	95.6^a	14.9^{ab}	27.8^a	78.7^{ab}	19.8^a	38.5^a	1.0	59	2	1	96.77
NIL7	106.0^a	91.5^a	12.8^b	26.9^a	69.5^{ab}	19.9^a	39.0^a	1.0	60	1	1	97.58
NIL8	98.5^{c-e}	93.7^a	18.9^{ab}	29.0^a	76.8^{ab}	18.9^a	40.2^a	1.0	60	1	1	97.58
NIL9	98.5^{c-e}	93.8^a	14.3^{ab}	26.9^a	81.3^a	18.0^a	39.7^a	1.0	60	1	1	97.58
NIL10	99.0^{b-e}	90.3^a	16.7^{ab}	25.2^a	79.2^a	19.1^a	38.9^a	1.0	59	2	1	96.77
NIL11	103.0^{a-d}	95.0^a	17.6^{ab}	28.7^a	82.7^a	18.9^a	44.9^a	1.0	61	0	1	98.39
NIL12	103.0^{a-d}	88.0^a	15.4^{ab}	27.5^a	78.0^{ab}	19.4^a	42.9^a	1.0	59	2	1	96.77
NIL13	99.0^{b-e}	91.2^a	12.7^b	27.5^a	79.3^a	18.1^a	42.6^a	1.0	59	2	1	96.77
NIL14	105.0^{a-c}	97.6^a	13.4^{ab}	27.7^a	76.5^{ab}	19.9^a	40.0^a	1.0	60	1	1	97.58
NIL15	105.5^{ab}	95.4^a	15.2^{ab}	27.2^a	77.6^{ab}	17.6^a	41.3^a	1.0	59	2	1	96.77
NIL16	100.5^{a-e}	92.3^a	17.6^{ab}	25.1^a	75.8^{ab}	16.9^a	39.0^a	1.0	60	1	1	97.58
NIL17	102.0^{a-d}	94.0^a	17.9^{ab}	25.8^a	72.9^{ab}	16.7^a	41.6^a	1.0	60	1	1	97.58
NIL18	99.0^{b-e}	91.4^a	18.0^{ab}	27.0^a	78.8^a	17.5^a	41.2^a	1.0	60	1	1	97.58
NIL19	95.0^e	93.8^a	14.3^{ab}	26.0^a	78.3^{ab}	19.6^a	42.5^a	1.0	60	1	1	97.58
NIL20	98.5^{c-e}	96.2^a	16.6^{ab}	27.1^a	80.2^a	18.0^a	40.3^a	1.0	61	0	1	98.39
NIL21	102.0^{a-d}	100.6^a	22.5^a	27.4^a	73.3^{ab}	19.4^a	42.2^a	1.0	59	2	1	96.77
NIL22	106.0^a	99.7^a	17.7^{ab}	26.0^a	58.4^b	16.6^a	37.2^a	1.0	59	2	1	96.77
NIL23	100.0^{a-e}	89.6^a	17.5^{ab}	27.0^a	78.6^{ab}	17.6^a	40.4^a	1.0	58	3	1	95.97
NIL24	104.0^{a-c}	101.5^a	14.7^{ab}	28.1^a	76.2^{ab}	19.9^a	41.6^a	1.0	60	1	1	97.58
PB1	103.0^{a-d}	98.9^a	18.4^{ab}	27.9^a	76.8^{ab}	19.5^a	43.3^a	9.0	100	0	0	—
FL478	83.5	97.0	13.0	25.0	85.5	26.2	45.5	1.0	0	0	100	—
CV (%)	1.56	5.52	13.80	4.00	6.50	6.85	5.36	—	—	—	—	—
SE	1.59	5.17	2.26	1.08	4.95	1.27	2.19	—	—	—	—	—

Means followed by same letters are statistically not different ($p < 0.05$), by Tukey's honest significance test. DFF: days to 50% flowering; PH: plant height in cm; ETP: effective tillers per plant; PL: panicle length in cm; TW: weight of 1000 grains in grams; YLD: yield in g per hill; STS: salt tolerance score (IRRI, 2013); RP: number of recurrent parent homozygotes; DP: number of donor parent homozygotes; HT: heterozygotes; RPG: recurrent parent genome recovery; CV: coefficient of variation; SE: standard error.

unstressed plants of the NILs and the parents [35]. The plant samples were prepared by carefully cleaning the shoots and roots and then drying them at 80°C for 24 h. The dried samples were ground to fine powder in a rotary mill. 500 mg of the powder was then digested in 10 ml of diacid digestion mixture (HNO_3 and $HClO_4$, 9 : 4). The digest was cooled and washed into a volumetric flask, and the volume made up to 50 ml. The mixture was filtered with Whatman number 42 filter paper and analysed for Na^+ and K^+ using Systronics Type 128 flame photometer (Systronics India).

2.8. Statistical Analyses. The data were analysed for standard statistical tests using the software package Statistical Tools for Agricultural Research STAR 2.0.1 [36].

3. Results

3.1. Polymorphism between the Parents. Of the 21 *Saltol*-linked markers tested, three markers, RM8094, RM493 and RM10793, were found polymorphic between the parents. Further, four markers were found polymorphic among the 42 tested on the flanking regions of the *Saltol*, resulting in a cumulative polymorphism of 11.1% on chromosome 1. Genome-wide polymorphism survey using 435 SSR markers (this included 42 markers tested on chromosome 1) identified a total of 62 polymorphic markers between PB1 and FL478, ranging from 4–7 markers spanned on each chromosome, resulting in an overall polymorphism of 14.7% between the parents (Table 1).

3.2. Development of Near Isogenic Lines by Marker-Assisted Selection. Five out of the 15 F_1 plants from the cross PB1/FL478 were found to be true F_1s as they were heterozygous for the *Saltol*-linked marker RM493. One of the true F_1 plants was backcrossed with PB1 to produce 20 BC_1F_1 plants. Foreground analysis using the three *Saltol*-linked markers identified six of the 20 plants to be heterozygous for all the three

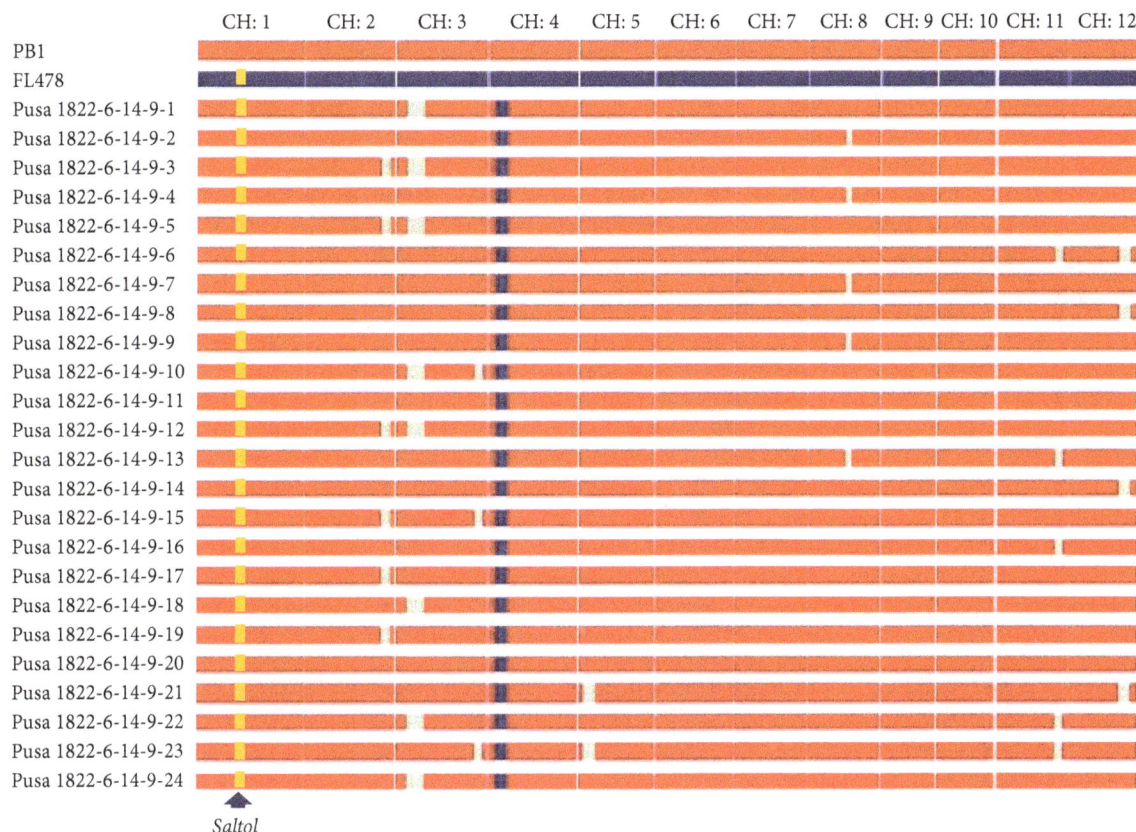

FIGURE 2: Graphical representation of the genotypes of 24 *Saltol*-introgressed NILs of PB1. The recurrent parent genome recovery ranged between 96.0 and 98.4%. All the NILs had maximum recovery on the carrier chromosome 1. CH: chromosome.

markers. These six BC_1F_1 plants were phenotypically closer to PB1 than the other plants. Background analysis of these six plants using the 62 SSR markers indicated an average RPG recovery of 75.3% (range, 72.6 to 79.8%). One progeny with the highest RPG recovery (79.8%), Pusa 1822-6, was backcrossed with PB1 to generate 70 BC_2F_1 seeds. In the BC_2F_1 generation, 14 plants were heterozygous for all three *Saltol*-linked markers; these plants were subjected to phenotypic selection to identify six plants that were phenotypically closer to PB1 for agro-morphological and grain quality traits than the remaining plants. Background analysis of these six plants, using 21 unfixed markers in Pusa 1822-6, indicated RPG recovery from 86.3% to 89.5%, with an average of 87.8%. All the six plants were selfed to generate six BC_2F_2 families. A total of 60 BC_2F_2 plants, 10 plants each from a family, were subjected to foreground selection. Eleven BC_2F_2 plants were found to be homozygous for the three *Saltol*-linked markers. Background analysis of these plants showed RPG recovery ranging from 89.9 to 92.7% with an average of 91.2%. These plants were further characterised for morphological and grain quality traits. Screening of eleven BC_2F_3 families raised by selfing of the selected BC_2F_2 plants for seedling stage salinity tolerance, identified one family, Pusa 1822-6-14-9, with a salt tolerance level comparable to that of FL478. All the plants of this family from the screening system were field transplanted to raise BC_2F_3 population.

The BC_2F_3 plants from Pusa 1822-6-14-9 were evaluated for both agro-morphological and grain quality traits, and 24 plants were selected for closer similarity with PB1. Background analysis of these plants, using six unfixed markers in the previous generation, indicated a cumulative RPG recovery of 96.0 to 98.4% (Table 2). All the 24 plants were advanced to BC_2F_4 generation by selfing. No selection was done beyond BC_2F_4 as all the *Saltol*-introgressed PB1 NILs had an average RPG recovery of more than 97%. Some residual donor segments were observed in chromosomes 2, 3, 5, 8, 11, and 12, whereas complete recovery was achieved in chromosomes 1, 6, 7, 9, and 10 (Figure 2). These lines are sequentially identified as NIL1 (Pusa 1822-6-14-9-1) to NIL24 (Pusa 1822-6-14-9-24).

3.3. *Seedling Stage Salinity Tolerance.* All the 24 NILs showed good seedling stage salinity tolerance (score of 1) comparable to that of FL478 under a salt stress of 11.6 dSm^{-1} (100 mM of NaCl) for sixteen days. In contrast, PB1 showed a highly sensitive reaction (score of 9) (Table 2; Supplementary Figure 1 [b]). The concentrations of major cations, Na$^+$ and K$^+$, that influence the salt response in rice seedlings are presented in Table 3. Broadly, there was significant variation among the NILs and their parents for cation contents and their ratios in both shoots and roots under both stressed as well as unstressed conditions. Under unstressed conditions, the root and shoot cation contents of the parental lines were

TABLE 3: Cation (Na^+ and K^+) content in the *Saltol*-introgressed PB1 lines and the donor and recipient parents under salt-stressed and salt-unstressed treatments.

| NILs | Unstressed | | | | | | Salt stressed | | | | | |
| | Shoot | | | Root | | | Shoot | | | Root | | |
	Na^+	K^+	Na^+/K^+	Na^+	K^+	Na^+/K^+	Na^+	K^+	Na^+/K^+	Na^+	K^+	Na^+/K^+
NIL1	2.8^{a-f*}	26.5^{b-d}	0.11^{a-f}	3.3^{ij}	27.2^{a-c}	0.12^{j}	15.2^{i-l}	23.6^{a-d}	0.65^{b}	20.7^{d-f}	23.9^{b-e}	0.86^{e-i}
NIL2	2.7^{a-f}	29.0^{a-d}	0.09^{b-f}	3.6^{e-j}	17.7^{k}	0.21^{a-d}	13.8^{j-l}	23.2^{a-d}	0.60^{b}	20.1^{e-g}	20.8^{g-i}	0.97^{c-f}
NIL3	2.4^{c-f}	28.9^{a-d}	0.08^{c-f}	5.1^{a}	24.1^{c-f}	0.21^{a-c}	12.2^{l}	22.0^{a-e}	0.56^{b}	19.5^{e-h}	23.6^{b-f}	0.82^{e-j}
NIL4	3.4^{a-e}	28.7^{a-d}	0.12^{a-f}	3.6^{f-j}	18.0^{k}	0.2^{a-f}	24.5^{b-d}	21.2^{a-e}	1.16^{b}	14.6^{m-o}	15.1^{lm}	0.98^{b-e}
NIL5	2.1^{ef}	32.4^{a-d}	0.07^{ef}	3.7^{e-j}	21.9^{f-i}	0.17^{e-i}	23.0^{c-f}	25.7^{a-c}	0.89^{b}	19.1^{e-h}	26.3^{b}	0.73^{f-j}
NIL6	2.5^{b-f}	28.7^{a-d}	0.09^{b-f}	3.4^{g-j}	23.2^{e-g}	0.15^{ij}	22.4^{d-f}	23.3^{a-d}	0.96^{b}	19.2^{e-h}	23.0^{c-g}	0.83^{e-j}
NIL7	3.8^{a-e}	26.8^{b-d}	0.14^{a-e}	4.7^{a-d}	27.9^{ab}	0.17^{e-i}	27.0^{bc}	25.0^{a-d}	1.10^{b}	20.9^{de}	18.6^{i-k}	1.13^{b-d}
NIL8	2.7^{a-f}	30.6^{a-d}	0.09^{b-f}	5.0^{ab}	29.0^{a}	0.17^{d-i}	21.2^{d-h}	22.5^{a-d}	0.94^{b}	11.0^{q}	16.3^{k-m}	0.68^{g-j}
NIL9	3.5^{a-e}	31.2^{a-d}	0.11^{a-f}	4.3^{b-e}	18.2^{jk}	0.23^{a}	18.7^{f-i}	21.6^{a-e}	0.86^{b}	24.0^{c}	19.6^{h-j}	1.22^{b}
NIL10	4.3^{a-c}	25.2^{c-e}	0.18^{a}	4.8^{a-c}	27.2^{a-c}	0.17^{c-i}	12.4^{kl}	12.2^{e}	1.21^{b}	18.2^{g-j}	20.6^{g-i}	0.88^{d-i}
NIL11	4.5^{ab}	30.3^{a-d}	0.15^{a-d}	4.1^{c-g}	26.6^{a-c}	0.16^{g-j}	13.2^{kl}	28.1^{ab}	0.47^{b}	14.0^{no}	21.1^{f-i}	0.66^{h-j}
NIL12	2.4^{c-f}	31.1^{a-d}	0.08^{d-f}	3.3^{h-j}	19.0^{i-k}	0.18^{c-i}	13.9^{j-l}	17.9^{c-e}	0.78^{b}	11.3^{pq}	15.3^{lm}	0.74^{e-j}
NIL13	2.6^{b-f}	17.7^{e}	0.15^{a-d}	4.7^{a-d}	25.3^{b-e}	0.19^{b-h}	16.2^{i-l}	21.3^{a-e}	0.76^{b}	11.3^{pq}	16.7^{k-m}	0.68^{g-j}
NIL14	2.6^{b-f}	26.7^{b-d}	0.10^{b-f}	4.0^{d-h}	21.4^{f-i}	0.19^{b-h}	25.6^{b-d}	24.7^{a-d}	1.05^{b}	13.4^{op}	14.5^{m}	0.92^{d-g}
NIL15	2.8^{a-f}	27.0^{b-d}	0.11^{a-f}	5.1^{a}	21.9^{f-i}	0.23^{a}	27.1^{bc}	24.6^{a-d}	1.15^{b}	26.7^{b}	22.6^{d-g}	1.18^{bc}
NIL16	4.1^{a-e}	28.5^{a-d}	0.14^{a-e}	3.0^{j}	19.6^{h-k}	0.16^{g-j}	17.8^{g-j}	15.5^{de}	1.22^{b}	13.8^{no}	20.6^{g-i}	0.67^{h-j}
NIL17	3.1^{a-f}	34.9^{a}	0.09^{b-f}	4.2^{c-f}	22.9^{e-g}	0.18^{b-i}	21.5^{d-g}	24.8^{a-d}	0.87^{b}	16.2^{j-m}	22.9^{d-g}	0.71^{g-j}
NIL18	2.9^{a-f}	32.9^{a-c}	0.09^{b-f}	4.2^{c-f}	21.2^{f-j}	0.2^{a-e}	23.5^{b-e}	22.9^{a-d}	1.03^{b}	18.7^{f-i}	21.8^{e-h}	0.86^{e-i}
NIL19	4.4^{ab}	32.0^{a-d}	0.14^{a-e}	4.2^{c-f}	19.6^{h-k}	0.22^{ab}	24.9^{b-d}	24.7^{a-d}	1.01^{b}	15.9^{k-n}	24.9^{b-d}	0.64^{ij}
NIL20	1.3^{f}	31.8^{a-d}	0.04^{f}	5.1^{a}	26.8^{a-c}	0.19^{b-g}	16.9^{h-k}	18.1^{c-e}	0.94^{b}	13.6^{o}	17.5^{j-l}	0.78^{e-j}
NIL21	2.2^{d-f}	33.8^{ab}	0.07^{ef}	3.9^{e-i}	18.9^{i-k}	0.20^{a-e}	22.6^{c-f}	22.7^{a-d}	0.99^{b}	19.1^{e-h}	21.2^{f-i}	0.90^{d-h}
NIL22	4.1^{a-d}	24.9^{de}	0.17^{ab}	4.2^{c-f}	23.4^{d-f}	0.18^{c-i}	22.1^{d-g}	18.6^{b-e}	1.20^{b}	17.0^{i-l}	19.8^{h-j}	0.86^{e-i}
NIL23	4.6^{a}	29.0^{a-d}	0.16^{a-c}	4.1^{c-g}	26.5^{a-d}	0.15^{h-j}	22.9^{c-f}	26.9^{a-c}	0.85^{b}	15.4^{l-o}	25.6^{bc}	0.60^{j}
NIL24	4.3^{a-c}	32.8^{a-c}	0.13^{a-e}	3.5^{g-j}	21.3^{f-j}	0.16^{f-i}	19.4^{e-i}	23.4^{a-d}	0.83^{b}	17.8^{h-k}	20.4^{g-i}	0.87^{e-i}
PB1	3.2^{a-f}	25.5^{cd}	0.13^{a-e}	3.8^{e-i}	20.1^{g-k}	0.19^{b-h}	34.4^{a}	12.2^{e}	2.83^{a}	22.3^{cd}	8.2^{n}	2.72^{a}
FL478	3.6^{a-e}	28.6^{a-d}	0.13^{a-e}	5.2^{a}	22.3^{e-h}	0.23^{a}	27.9^{b}	29.4^{a}	0.95^{b}	29.3^{a}	30.4^{a}	0.96^{c-f}
CV (%)	15.25	6.48	17.33	4.12	3.33	4.82	5.30	10.84	19.34	2.88	3.17	6.57
SE	0.49	1.88	0.02	0.17	0.76	0.01	1.10	2.40	0.19	0.51	0.65	0.06

*Means followed by the same letter are statistically not different at $p < 0.05$, by Tukey's honest significance test. CV: coefficient of variation; SE: standard error.

TABLE 4: Interrelationships of cation content and their proportions in root and shoots under unstressed (lower diagonal) and salt-stressed (upper diagonal) conditions. Cross correlations between unstressed and stressed conditions are given as diagonal elements.

Parameters[†]	St: Na^+	St: K^+	St: Na^+/K^+	Rt: Na^+	Rt: K^+	Rt: Na^+/K^+	STS
St: Na^+	0.072	0.101	0.658*	0.380*	−0.130	0.545*	0.494*
St: K^+	−0.063	0.171	−0.625*	0.186	0.567*	−0.389*	−0.437*
St: Na^+/K^+	0.864*	−0.534*	0.107	0.197	−0.519*	0.801*	0.833*
Rt: Na^+	−0.105	−0.161	0.013	0.227	0.405*	0.465*	0.201
Rt: K^+	0.052	−0.250	0.176	0.480*	0.115	−0.537*	−0.551*
Rt: Na^+/K^+	−0.135	0.086	−0.148	0.521*	−0.490*	0.244	0.908*

[†]St: Shoot; Rt: Root; *Correlation coefficients are significant at $p < 0.01$ level; STS: salt tolerance score.

comparable, but there were significant differences between some NILs, and some of them differed significantly from the parents as well. However, under stressed conditions, PB1 and FL478 had significantly distinct cation concentration both in shoots and roots; while the K^+ levels in the shoots and roots of FL478 were much higher than those in PB1, the Na^+ content in the shoots of FL478 was significantly lower than PB1, whereas Na^+ content in the roots of PB1 was lower than FL478. All the NILs showed shoot and root K^+ levels closer to those of FL478 than to PB1. The root Na^+

concentration of NILs were closer to that of PB1, but shoot Na^+ content was comparable or marginally lower than that of FL478. Further, there were several NILs that showed Na^+/K^+ ratio lower than that of FL478.

The correlations between cation content in shoots and roots (Table 4) under stressed and nonstressed conditions were insignificant. Under salt stress, salt tolerance score was found to have a significant positive association with shoot Na^+ content, while shoot K^+ level showed a negative association. Similar trend was observed for root ion concentrations under stress, except for root Na^+ content, which exhibited nonsignificant correlation. The ionic proportions had shown very high negative association with salt tolerance score in both shoots and roots.

Correlations among the cation contents in shoots and roots under salt stress, indicated several significant associations such as a positive trend between shoot Na^+ content and root Na^+ content (0.38), as well as between shoot K^+ and root K^+ contents (0.57). There were no associations between shoot Na^+ and root K^+ levels and vice versa. Na^+ content showed a major positive association with Na^+/K^+ ratio in both shoots and roots (0.66 and 0.47, resp.), while the K^+ content showed significant negative association with the Na^+/K^+ ratio (−0.63 and −0.54, resp.). The cation ratios between shoots and roots also showed a positive trend (0.80). Further, cross associations were also noticed for shoot ion concentrations with root cation ratios (0.55 and −0.39, resp., for shoot Na^+ and K^+ contents), while root K^+ showed a negative association with shoot Na^+/K^+ ratio (−0.52), but no such association was found with root Na^+ content. Further, root Na^+ and K^+ contents showed a positive association (0.41).

3.4. Agronomic Performance. Mean performance of each of the 24 PB1 NILs for yield and yield-related traits is presented in Table 2. The NILs were essentially comparable to the recurrent parent, PB1, for agronomic traits, such as plant height, panicle length, weight of 1000 grains, and yield per plant. The days to 50% flowering ranged from 95.0 days (NIL19) to 106 days (NIL4, 6, 7, and 22): 23 NILs were at par with PB1 (103 days), while NIL19 was significantly flowering earlier than PB1.

3.5. Grain and Cooking Quality. The mean grain and cooking quality parameters of the NILs are presented in Table 5. Hulling and milling percentages for all the NILs were similar to those of the recurrent parent, PB1. Further, all the NILs possessed extra-long slender grain type (Figure 3) with strong aroma and with low gelatinization temperature as indicated by the alkali spreading value of 7.0, which is the same as that of PB1. Some of the NILs had significantly longer grain length before/after cooking than the RP while few had them significantly shorter, but these differences were rather small (0.15 mm or less).

4. Discussion

Growing demand for Basmati rice has resulted in its increased cultivation in the north-western areas of India

TABLE 5: Grain and cooking quality of *Saltol*-introgressed NILs of Pusa Basmati 1 (Pusa 1822).

NILS	HUL	MIL	KLBC	KLAC	ER	ASV	AROM
NIL1	75.5^a	66.9^a	7.29^a	13.32^a	1.822^{de}	7.0	2.0
NIL2	73.7^a	65.3^a	7.21^{d-g}	13.21^{b-d}	1.832^{c-e}	7.0	2.0
NIL3	74.7^a	68.7^a	7.27^{a-c}	13.24^{bc}	1.819^{ef}	7.0	2.0
NIL4	72.8^a	64.7^a	7.20^{e-h}	13.21^{b-d}	1.837^{a-c}	7.0	2.0
NIL5	72.7^a	64.6^a	7.22^{d-f}	13.21^{b-d}	1.825^{de}	7.0	2.0
NIL6	74.8^a	66.0^a	7.19^{f-i}	13.12^{fg}	1.822^{de}	7.0	2.0
NIL7	69.7^a	62.9^a	7.23^{c-f}	13.20^{cd}	1.827^{c-e}	7.0	2.0
NIL8	74.0^a	66.7^a	7.29^{ab}	13.26^b	1.819^{ef}	7.0	2.0
NIL9	74.2^a	66.5^a	7.25^{b-d}	13.22^{bc}	1.822^{de}	7.0	2.0
NIL10	73.5^a	66.5^a	7.19^{f-i}	13.20^{cd}	1.836^{a-c}	7.0	2.0
NIL11	75.6^a	67.5^a	7.29^{ab}	13.31^a	1.826^{c-e}	7.0	2.0
NIL12	74.7^a	66.2^a	7.28^{ab}	13.07^g	1.795^g	7.0	2.0
NIL13	73.4^a	65.7^a	7.21^{d-g}	13.21^{b-d}	1.830^{c-e}	7.0	2.0
NIL14	76.4^a	67.9^a	7.20^{e-h}	13.19^{cd}	1.832^{c-e}	7.0	2.0
NIL15	73.7^a	65.6^a	7.16^{hi}	13.15^{ef}	1.837^{b-d}	7.0	2.0
NIL16	73.5^a	67.6^a	7.27^{a-c}	13.25^{bc}	1.823^{ef}	7.0	2.0
NIL17	73.5^a	67.7^a	7.24^{c-e}	13.00^h	1.796^g	7.0	2.0
NIL18	74.2^a	66.9^a	7.17^{g-i}	13.23^{bc}	1.845^a	7.0	2.0
NIL19	76.5^a	68.7^a	7.22^{d-f}	13.21^{b-d}	1.835^{c-e}	7.0	2.0
NIL20	73.6^a	66.5^a	7.21^{d-g}	13.18^{de}	1.828^{c-e}	7.0	2.0
NIL21	73.4^a	65.6^a	7.20^{e-h}	13.21^{b-d}	1.837^{a-c}	7.0	2.0
NIL22	71.9^a	64.9^a	7.20^{e-h}	13.20^{cd}	1.835^{c-e}	7.0	2.0
NIL23	74.3^a	65.1^a	7.15^i	13.20^{cd}	1.846^a	7.0	2.0
NIL24	73.4^a	71.6^a	7.15^i	13.20^{cd}	1.844^{ab}	7.0	2.0
PB1	73.9^a	66.0^a	7.30^a	13.21^{b-d}	1.810^{fg}	7.0	2.0
FL478	79.1	64.4	6.28	9.13	1.454	5.0	0.0
CV (%)	2.14	3.60	0.14	0.09	0.18	—	—
SE	1.58	2.39	0.01	0.01	0.00	—	—

Means followed by same letters are statistically not different at $p < 0.05$, by Tukey's honest significance test. HUL: hulling recovery in percentage; MIL: milling recovery in percentage; KLBC: kernel length before cooking in mm; KLAC: kernel length after cooking in mm; ASV: alkali spreading value; AROM: aroma score from panel test; CV: coefficient of variation; SE: standard error.

[11, 12]. However, soil salinization in these regions poses a major threat to cropping as salinity stress leads to poor crop establishment and survival resulting in significant yield losses. Therefore, it is important to develop salt stress tolerant Basmati cultivars for cultivation in these areas [37]. In the present study, marker-assisted backcross breeding based on the established step-wise selection approach, namely, foreground, phenotypic, and background selections in the given order, was successful in improving the salt tolerance of PB1 Basmati rice variety. Stringent phenotypic selection carried out after the foreground selection is reported to accelerate RP genome recovery process [5, 11, 38, 39] and is expected to reduce the cost of background selection by reducing the number of test plants. It is noteworthy that very high (~96–98%) RPG recovery was achieved with only two backcrosses, and the recovery of the Basmati grain and cooking

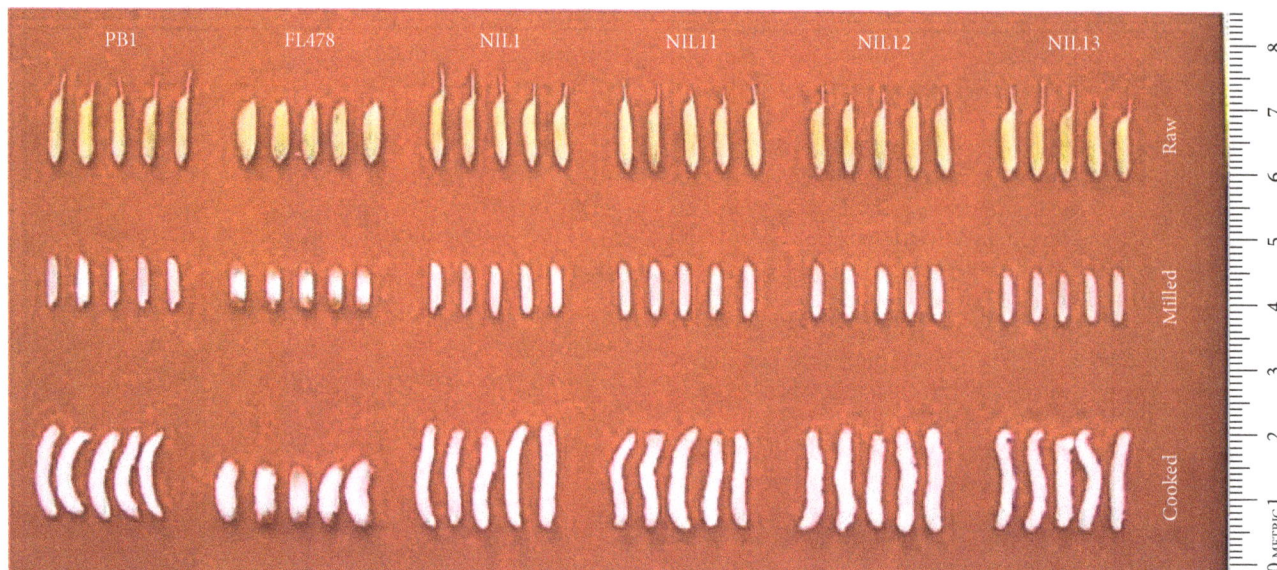

FIGURE 3: Grain and cooking quality of some of the NILs of Pusa Basmati 1 carrying *Saltol* locus.

quality traits was almost complete. Further, there was complete recovery of the carrier chromosome (chromosome 1) together with *Saltol*, the target QTL (Figure 2; Supplementary Figure 1; Supplementary Table 1); this might have been facilitated by the relatively low level of polymorphism (11.1%) for this chromosome. This indicates the effectiveness of the selection procedure used in the study.

It is pertinent here to mention that PB1 was reported to possess a *Saltol* haplotype that was different from other Basmati cultivars. The PB1 haplotype shared a close homology with the *Saltol* locus of FL478, by differing only for three markers RM8094, RM493, and RM10793 [37]. Among these, RM8094 was the only recognised *Saltol*-linked marker that has been used for marker-assisted breeding, while RM493 and RM10793 were centromeric distal markers [20]. This implied that PB1 *Saltol* locus was very similar to FL478 locus, except for the region proximal to RM8094 marker locus. Therefore, the contrasting salt stress response between PB1 and FL478 can be arbitrarily assigned to a segment within 10.8 to 11.4 Mbp on chromosome 1. The NILs showed seedling stage salt tolerance levels comparable to that of FL478. Although, there was up to 4% residual donor genome present in some of the NILs, there was little effect of the donor genome on the agronomic performance, except for days to 50% flowering that was significantly lower than PB1 in one of the NILs, Pusa 1822-6-14-9-19.

Inspite of huge strides made in genomics-assisted breeding, development of salt-tolerant rice cultivars continues to be a major challenge due to the complex nature of *Saltol* region. Although, all the selected eleven BC_2F_2 genotypes possessed the target marker alleles in homozygous condition, they exhibited differential tolerance response ranging from susceptibility to complete tolerant at 11.6 dSm^{-1} ECE level (Supplementary Table 3). The sensitive response of some BC_2F_2 lines suggests the possibility that some genomic regions of PB1 may harbour genes/QTLs that have inhibitory

effect on the *Saltol* QTL. Another possibility is cryptic intra-*Saltol* QTL recombination that could not be detected by the three markers used for the foreground selection. It is emphasised that *Saltol* region is fairly large (having a size of ~1.5Mbp) enough to accede intra-QTL recombination, as evident from its highly fragmented existence in the rice genome [40]. Further, introgression of additional hitherto unidentified QTLs from donor into the salt tolerant NILs cannot be ruled out.

The *Saltol* QTL region consists of several genes associated with salt response. These include transcription factors, signal transduction components, cell wall components, and membrane transporters [41, 42]. Specific genes, such as Na$^+$ transporter gene *OsHKT1;5* [20, 43], osmoprotection-associated *SalT* [44], cation-proton exchanger (*OsCHX11*), cyclic nucleotide-gated ion channel (*OsCNGC1*) [45], high affinity potassium transporter (*HKT1*), and ATP-binding cassette transporter (*ABC1*) [42, 46], have been recognized in this region. However, since *Saltol* QTL is associated with Na$^+$/K$^+$ balance in the shoot tissues, the implicit mechanism of tolerance is attributed to Na$^+$/K$^+$ homeostasis driven by *OsHKT1;5*. The *OsHKT1;5* gene, also known as *SKC1*, encodes for a xylem-expressed Na$^+$ transporter and acts by preferentially unloading Na$^+$ ions from xylem vessels while regulating K$^+$ homeostasis [19]. Current observation of absence of any relation between the cation content between stressed and unstressed conditions indicated that ion homeostasis mechanisms might be active only under salt stress. Further, under stress, the shoot cation content outweighed root cation status in determining the salt tolerance, among which Na$^+$ content was more deterministic of the level of tolerance than the K$^+$ content. This strongly suggested Na$^+$ transport as the major mechanism of salt tolerance in *Saltol*. The shift in Na$^+$/K$^+$ cation balance in shoot tissues of NILs towards the ratio in the donor parent FL478 tends to support this suggestion. Successful recovery of Basmati grain and cooking quality traits, together with pleasing aroma, were achieved in this study, as in several previous studies [1, 3, 4, 11, 39], in spite of the donor parent having poor grain and cooking quality traits. This was possible solely due to the marker-assisted selection strategy that combined a rigorous phenotypic selection in every generation.

5. Conclusions

In the present investigation, incorporation of seedling stage salinity tolerance in PB1 was achieved by introgression of the Saltol QTL using marker-assisted backcross breeding. The improved lines showed marked enhancement of salt tolerance in seedling stage. Since salt tolerance in Basmati cultivars is absent, the newly developed lines together with Saltol-introgressed NILs of Pusa Basmati 1121, another premium Basmati cultivar [11], will now offer choice of cultivars to be grown in salt-affected soils. Two of the improved NILs, Pusa 1822-6-14-9-11 and Pusa 1822-6-14-9-20 (Figure 3; Supplementary Figure 1 [c]), may be evaluated for their suitability for commercial cultivation and/or in breeding programmes for improving their reproductive stage salt tolerance, since the genetic controls of seedling and reproductive stage salt tolerance are different [47]. Additionally, a comprehensive evaluation of the NILs under salt-affected soil will reveal, other than agronomic performance, physiological improvements such as photosynthetic efficiency gained by incorporation of salt tolerance.

Conflicts of Interest

The authors declare that they have no competing interests.

Authors' Contributions

Ashok Kumar Singh conceptualised the project; Ashok Kumar Singh, Subbaiyan Gopala Krishnan, and Brahma Deo Singh led the experiment and did the evaluation and midcourse corrections; Vivek Kumar Singh, Amit Kumar, Ranjith Kumar Ellur, Subbaiyan Gopala Krishnan, and Prolay Kumar Bhowmick designed and conducted the field and phytotron experiments; Sadhna Maurya, Madan Pal Singh, and Vivek Kumar Singh did the physiological and biochemical evaluation; Vivek Kumar Singh, Kunnummal Kurungara Vinod, and Subbaiyan Gopala Krishnan did the data curation and analyses; Vivek Kumar Singh, Kunnummal Kurungara Vinod, Subbaiyan Gopala Krishnan, Brahma Deo Singh, and Ashok Kumar Singh wrote the paper. All the authors have read and approved the final manuscript.

Acknowledgments

The senior author acknowledges the help rendered by the technical and supporting staff of the rice section, Division of Genetics, for field work and Division of Soil Science, ICAR-IARI, for biochemical analysis of samples.

Supplementary Materials

Supplementary Table 1: chromosome wise list of polymorphic STMS markers between PB1 and FL478. Supplementary Table 2: information on Saltol-linked SSR markers on chromosome1 polymorphic between PB1 and FL478. Supplementary Table 3: agronomic performance, salt tolerance, and recurrent parent genome recovery of Saltol-positive homozygous BC_2F_2 plants. Supplementary Figure 1: phenotypic evaluation for salt tolerance between parents and introgressed lines with their field view. Supplementary Figure 2: screening of foreground markers in the parental lines. Three markers RM8094, RM493, and RM10793 showed polymorphism between PB1 and FL478. (Supplementary Materials)

References

[1] K. K. Vinod, S. G. Krishnan, N. N. Babu, M. Nagarajan, and A. K. Singh, "Improving salt tolerance in rice: looking beyond the conventional," in Salt Stress in Plants: Signalling, Omics and Adaptations, P. Ahmad, M. M. Azooz and M. N. V. Prasad, Eds., pp. 219–260, Springer, New York, 2013.

[2] F. Moradi and A. M. Ismail, "Responses of photosynthesis, chlorophyll fluorescence and ROS-scavenging systems to salt stress during seedling and reproductive stages in rice," Annals of Botany, vol. 99, no. 6, pp. 1161–1173, 2007.

[3] T. M. L. Hoang, T. T. Tran, T. K. T. Nguyen et al., "Improvement of salinity stress tolerance in rice: challenges and opportunities," Agronomy, vol. 6, no. 4, p. 54, 2016.

[4] A. R. Yeo and T. J. Flowers, "Salinity resistance in rice (Oryza sativa L.) and a pyramiding approach to breeding varieties for saline soils," Australian Journal of Plant Physiology, vol. 13, no. 1, pp. 161–174, 1986.

[5] A. M. Ismail, S. Heuer, M. J. Thomson, and M. Wissuwa, "Genetic and genomic approaches to develop rice germplasm for problem soils," Plant Molecular Biology, vol. 65, no. 4, pp. 547–570, 2007.

[6] H. X. Lin, M. Z. Zhu, M. Yano et al., "QTLs for Na+ and K+ uptake of the shoots and roots controlling rice salt tolerance," Theoretical and Applied Genetics, vol. 108, no. 2, pp. 253–260, 2004.

[7] Z. H. Ren, J. P. Gao, L. G. Li et al., "A rice quantitative trait locus for salt tolerance encodes a sodium transporter," Nature Genetics, vol. 37, no. 10, pp. 1141–1146, 2005.

[8] M. J. Thomson, M. Ocampo, J. Egdane et al., "Characterizing the Saltol quantitative trait locus for salinity tolerance in rice," Rice, vol. 3, no. 2-3, pp. 148–160, 2010.

[9] J. D. Platten, J. A. Egdane, and A. M. Ismail, "Salinity tolerance, Na+ exclusion and allele mining of HKT1; 5 in Oryza sativa and O. glaberrima: many sources, many genes, one mechanism?," BMC Plant Biology, vol. 13, no. 1, p. 32, 2013.

[10] B. B. Patel, B. B. Patel, and R. S. Dave, "Studies on infiltration of saline–alkali soils of several parts of Mehsana and Patan districts of north Gujarat," Journal of Applied Technology in Environmental Sanitation, vol. 1, pp. 87–92, 2011.

[11] N. N. Babu, S. G. Krishnan, K. K. Vinod et al., "Marker aided incorporation of Saltol, a major QTL associated with seedling stage salt tolerance, into Oryza sativa 'Pusa basmati 1121'," Frontiers in Plant Science, vol. 8, 2017.

[12] E. A. Siddiq, L. R. Vemireddy, and J. Nagaraju, "Basmati rices: genetics, breeding and trade," *Agricultural Research*, vol. 1, no. 1, pp. 25–36, 2012.

[13] R. K. Ellur, A. Khanna, A. Yadav et al., "Improvement of basmati rice varieties for resistance to blast and bacterial blight diseases using marker assisted backcross breeding," *Plant Science*, vol. 242, pp. 330–341, 2016.

[14] E. Bienabe and D. Marie-Vivien, "Institutionalizing geographical indications in southern countries: lessons learned from basmati and rooibos," *World Development*, vol. 98, pp. 58–67, 2015.

[15] V. P. Singh, "Basmati rice of India," in *Aromatic Rices*, R. K. Singh, U. S. Singh and G. S. Khush, Eds., pp. 135–154, Oxford & IBH Publishing Co. Pvt. Limited, New Delhi, 2000.

[16] S. Gopalakrishnan, R. K. Sharma, K. A. Rajkumar et al., "Integrating marker assisted background analysis with foreground selection for identification of superior bacterial blight resistant recombinants in basmati rice," *Plant Breeding*, vol. 127, no. 2, pp. 131–139, 2008.

[17] M. Joseph, S. Gopalakrishnan, R. K. Sharma et al., "Combining bacterial blight resistance and basmati quality characteristics by phenotypic and molecular marker assisted selection in rice," *Molecular Breeding*, vol. 13, no. 4, pp. 377–387, 2004.

[18] A. Khanna, V. Sharma, R. K. Ellur et al., "Development and evaluation of near isogenic lines for major blast resistance gene(s) in basmati rice," *Theoretical and Applied Genetics*, vol. 128, no. 7, pp. 1243–1259, 2015.

[19] A. Singh, V. K. Singh, S. P. Singh et al., "Molecular breeding for the development of multiple disease resistant basmati rice," *AoB Plants*, vol. 2012, article pls029, 2012.

[20] R. Aliyu, A. M. Adam, S. Muazu, S. O. Alonge, and G. B. Gregario, "Tagging and validation of SSR markers to salinity tolerance in rice," in *2010 International Conference on Biology, Environment and Chemistry (IPCBEE)*, vol. 1, pp. 328–332, Singapore, 2011.

[21] T. T. H. Vu, D. D. Le, A. M. Ismail, and H. H. Le, "Marker-assisted backcrossing (MA-C) for improved salinity tolerance in rice (*Oryza sativa* L.) to cope with climate change in Vietnam," *Australian Journal of Crop Science*, vol. 6, pp. 1649–1654, 2012.

[22] B. D. Singh and A. K. Singh, *Marker-Assisted Plant Breeding: Principles and Practices*, Springer, New Delhi, 2015.

[23] G. B. Gregorio, M. R. Islam, G. V. Vergara, and S. Thirumeni, "Recent advances in rice science to design salinity and other abiotic stress tolerant rice varieties," *SABRAO Journal of Breeding and Genetics*, vol. 45, pp. 31–41, 2013.

[24] V. T. Ho, M. J. Thomson, and A. M. Ismail, "Development of salt tolerant IR64 near isogenic lines through marker-assisted breeding," *Journal of Crop Science and Biotechnology*, vol. 19, no. 5, pp. 373–381, 2016.

[25] L. T. N. Huyen, L. M. Cuc, A. M. Ismail, and L. H. Ham, "Introgression the salinity tolerance QTLs *Saltol* into AS996, the elite rice variety of Vietnam," *American Journal of Plant Science*, vol. 3, no. 7, pp. 981–987, 2012.

[26] M. M. Hasan, M. Y. Rafii, M. R. Ismail et al., "Marker-assisted backcrossing: a useful method for rice improvement," *Biotechnology and Biotechnological Equipment*, vol. 29, no. 2, pp. 237–254, 2015.

[27] A. K. Singh, S. Gopalakrishnan, V. P. Singh et al., "Marker assisted selection: a paradigm shift in basmati breeding," *Indian Journal of Genetics and Plant Breeding*, vol. 71, pp. 120–128, 2011.

[28] A. K. Singh and S. G. Krishnan, "Genetic improvement of basmati rice—the journey from conventional to molecular breeding," in *Molecular Breeding for Sustainable Crop Improvement. Sustainable Development and Biodiversity*, V. Rajpal, S. Rao and S. Raina, Eds., vol. 11, Springer, Cham, 2016.

[29] G. B. Gregorio, D. Senadhira, and R. D. Mendoza, *Screening Rice for Salinity Tolerance*, IRRI Discussion Paper Series no 22, Los Baños, International Rice Research Institute, 1997.

[30] IRRI, *Standard Evaluation System (SES) for Rice*, Los Baños, International Rice Research Institute, 5th edition, 2013.

[31] V. K. Singh, V. K. Singh, R. K. Ellur, S. G. Krishnan, and A. K. Singh, "Validation of rapid DNA extraction protocol and their effectiveness in marker assisted selection in crop plants," *Indian Journal of Genetics and Plant Breeding*, vol. 75, no. 1, pp. 110–113, 2015.

[32] R. Van Berloo, "Computer note. GGT: software for display of graphical genotypes," *Journal of Heredity*, vol. 90, no. 2, pp. 328–329, 1999.

[33] S. Yoshida, D. A. Forno, J. H. Cock, and K. A. Gomez, *Laboratory Manual for Physiological Studies of Rice*, International Rice Research Institute, Los Baños, 3rd edition, 1976.

[34] S. H. Basavaraj, V. K. Singh, A. Singh et al., "Marker-assisted improvement of bacterial blight resistance in parental lines of Pusa RH10, a superfine grain aromatic rice hybrid," *Molecular Breeding*, vol. 26, no. 2, pp. 293–305, 2010.

[35] B. S. Bhargava and H. B. Raghupathi, "Analysis of plant materials for macro and micronutrients," in *Methods of Analysis of Soils, Plants, Water and Fertilizers*, H. L. S. Tandon, Ed., pp. 49–82, Fertilization Department Consultant Organization, New Delhi, 1993.

[36] IRRI, *STAR Version 2.0.1*, Biometrics and Breeding Informatics, PBGB Division, International Rice Research Institute, Los Baños, 2014.

[37] N. N. Babu, K. K. Vinod, S. G. Krishnan et al., "Marker based haplotype diversity of *Saltol QTL* in relation to seedling stage salinity tolerance in selected genotypes of rice," *Indian Journal of Genetics and Plant Breeding*, vol. 74, no. 1, pp. 16–25, 2014.

[38] V. K. Singh, A. Singh, S. P. Singh et al., "Marker assisted simultaneous but stepwise backcross breeding for pyramiding blast resistance genes *Piz5* and *Pi54* into an elite basmati rice restorer line 'PRR78'," *Plant Breeding*, vol. 132, pp. 486–495, 2013.

[39] R. K. Ellur, A. Khanna, S. G. Krishnan et al., "Marker-aided incorporation of *Xa38*, a novel bacterial blight resistance gene, in PB1121 and comparison of its resistance spectrum with *xa13+Xa21*," *Scientific Reports*, vol. 6, no. 1, 2016.

[40] N. N. Babu, K. K. Vinod, S. L. Krishnamurthy et al., "Microsatellite based linkage disequilibrium analyses reveal *Saltol* haplotype fragmentation and identify novel QTLs for seedling stage salinity tolerance in rice (*Oryza sativa* L.)," *Journal of Plant Biochemistry and Biotechnology*, vol. 26, no. 3, pp. 310–320, 2016.

[41] H. Walia, C. Wilson, P. Condamine et al., "Comparative transcriptional profiling of two contrasting rice genotypes under salinity stress during the vegetative growth stage," *Plant Physiology*, vol. 139, no. 2, pp. 822–835, 2005.

[42] H. Walia, C. Wilson, L. Zeng, A. M. Ismail, P. Condamine, and T. J. Close, "Genome-wide transcriptional analysis of salinity stressed *japonica* and *indica* rice genotypes during panicle

initiation stage," *Plant Molecular Biology*, vol. 63, no. 5, pp. 609–623, 2007.

[43] J. D. Platten, O. Cotsaftis, P. Berthomieu et al., "Nomenclature for *HKT* transporters, key determinants of plant salinity tolerance," *Trends in Plant Science*, vol. 11, no. 8, pp. 372–374, 2006.

[44] A. B. Garcia, J. D. Engler, S. Iyer, T. Gerats, M. Van Montagu, and A. B. Caplan, "Effects of osmoprotectants upon NaCl stress in rice," *Plant Physiology*, vol. 115, no. 1, pp. 159–169, 1997.

[45] P. Senadheera, R. K. Singh, and F. J. M. Maathuis, "Differentially expressed membrane transporters in rice roots may contribute to cultivar dependent salt tolerance," *Journal of Experimental Botany*, vol. 60, no. 9, pp. 2553–2563, 2009.

[46] T. B. De Leon, S. Linscombe, and P. K. Subudhi, "Molecular dissection of seedling salinity tolerance in rice (*Oryza sativa* L.) using a high-density GBS-based SNP linkage map," *Rice*, vol. 9, no. 1, article 52, 2016.

[47] R. Mohammadi, M. S. Mendioro, G. Q. Diaz, G. B. Gregorio, and R. K. Singh, "Genetic analysis of salt tolerance at seedling and reproductive stages in rice (*Oryza sativa*)," *Plant Breeding*, vol. 133, no. 5, pp. 548–559, 2014.

Enriching Genomic Resources and Transcriptional Profile Analysis of *Miscanthus sinensis* under Drought Stress based on RNA Sequencing

Gang Nie,[1] Linkai Huang,[1] Xiao Ma,[1] Zhongjie Ji,[2] Yajie Zhang,[1] Lu Tang,[1] and Xinquan Zhang[1]

[1]*Department of Grassland Science, Animal Science and Technology College, Sichuan Agricultural University, Chengdu, Sichuan 611130, China*
[2]*Department of Agronomy, Purdue University, West Lafayette, IN 47906, USA*

Correspondence should be addressed to Xinquan Zhang; zhangxq@sicau.edu.cn

Academic Editor: Graziano Pesole

Miscanthus × giganteus is wildly cultivated as a potential biofuel feedstock around the world; however, the narrow genetic basis and sterile characteristics have become a limitation for its utilization. As a progenitor of *M. × giganteus*, *M. sinensis* is widely distributed around East Asia providing well abiotic stress tolerance. To enrich the *M. sinensis* genomic databases and resources, we sequenced and annotated the transcriptome of *M. sinensis* by using an Illumina HiSeq 2000 platform. Approximately 316 million high-quality trimmed reads were generated from 349 million raw reads, and a total of 114,747 unigenes were obtained after de novo assembly. Furthermore, 95,897 (83.57%) unigenes were annotated to at least one database including NR, Swiss-Prot, KEGG, COG, GO, and NT, supporting that the sequences obtained were annotated properly. Differentially expressed gene analysis indicates that drought stress 15 days could be a critical period for *M. sinensis* response to drought stress. The high-throughput transcriptome sequencing of *M. sinensis* under drought stress has greatly enriched the current genomic available resources. The comparison of DEGs under different periods of drought stress identified a wealth of candidate genes involved in drought tolerance regulatory networks, which will facilitate further genetic improvement and molecular studies of the *M. sinensis*.

1. Introduction

The genus *Miscanthus* is a species of promising C4 perennial nonfood bioenergy grasses for cellulosic biofuel production [1]. Specifically, *Miscanthus × giganteus*, a hybrid generated from a cross between tetraploid *Miscanthus sacchariflorus* and diploid *Miscanthus sinensis*, has been intensively studied in Europe and North America as a biomass feedstock [2–7]. However, it is the only genotype currently available for use in most countries by its natural sterility and a narrow genetic base [8, 9]. Furthermore, it is highly risky and genetically difficult to improve *M. × giganteus* through breeding, posing limitations to its biomass productivity, abiotic stress tolerance, and climatic adaptation under some extreme conditions [10–12]. As a progenitor of *M. × giganteus*, *M. sinensis* was widely distributed around East Asia and it was shown that abundant wild *M. sinensis* resources were distributed in China providing a comparable yield and well abiotic stress tolerance in some places [12–16].

Drought is a common environmental stress which induces adverse impacts on almost all aspects of plant development, growth, reproduction, and yield in a temperate area, and plants must adapt to this stress to survive [17, 18]. Plant drought tolerance is a complex quantitative trait, involving multiple pathways, regulatory networks, and cellular compartments [19]. Many of drought-induced or drought-repressed genes with diverse functions had been identified by molecular and genomic analysis in model plants. In *Arabidopsis*, 299 drought-inducible genes were identified through 7000 full-length cDNA microarray [20] and were classified into two groups including function proteins and regulatory proteins [21]. In rice, 73 dependable

TABLE 1: The quality report of *M. sinensis* RNA sample sequencing and unigene assembling under drought stress.

Samples	TCRs	Q20 (%)	TNU	MLU (nt)	N50	DC	DS
M0	55,128,976	97.19	92,444	931	1609	42,705	49,739
M1	54,603,020	97.23	90,148	883	1570	39,303	50,845
M2	52,019,400	97.30	88,455	830	1421	38,735	49,720
M3	51,366,992	97.20	81,159	846	1473	34,424	46,735
M4	51,745,196	97.26	89,320	852	1507	38,541	50,779
M5	51,337,262	97.29	88,416	855	1480	38,862	49,554
Total	316,200,846	/	114,747	1288	1854	65,203	49,544

TCRs: number of total clean reads; Q20: percentage of bases whose quality is larger than 20 in clean reads (%); TNU: total number of unigene; MLU: mean length of unigene (nt); DC: distinct clusters, which means that there are several unigenes wherein similarity between them is more than 70% in one cluster; DS: distinct singletons, which means a single unigene comes from a single gene.

genes were confirmed which were induced by drought, high salinity, or cold stress [22]. In a comparative analysis of 73 genes with those identified in *Arabidopsis*, 51 of them performed a similar function and revealed a considerable degree of similarity of drought stress response at the molecular level. Specially, genes involved in antioxidative metabolic pathways play an important role in detoxifying reactive oxygen species that can accumulate under drought stress conditions. In addition, it is well known that the phytohormone abscisic acid (ABA) level is essential for drought stress responses. Several genes involved in ABA biosynthesis and catabolism pathway in the drought stress responses have been identified in model plant and crops. In *Arabidopsis*, the 9-*cis*-epoxycarotenoid dioxygenase (NCED) family gene *AtNCED3* transcripts are rapidly induced by drought stress, which proved that it plays a crucial role in drought stress-inducible ABA biosynthesis [23]. CYP707A3 is strongly induced by rehydration after dehydration condition, which is a major enzyme for ABA catabolism in the drought stress response [24, 25]. Significantly, the introduction of many stress-inducible genes via gene transfer resulted in improved plant stress tolerance [25–28]. In consequence, discovering differential expression genes and analyzing the functions of these genes are important to further our understanding of the molecular mechanisms of plant drought stress response and tolerance regulation and ultimately facilitate the enhancement of plant drought tolerance through genetic manipulation.

It has been well known that many wild plants show high tolerant phenotypes against abiotic stresses, such as salt, drought, and oxidative stresses [29–31]. Based on the previous tests of drought and cold tolerance of *M. sinensis* in Europe, a much broader range of adaptation than *M. × giganteus* was found in this diploid species [10, 32], indicating that *M. sinensis* is considered possible to breed varieties with higher tolerance for frost and drought than *M. × giganteus* [14]. However, gene identification and molecular mechanisms involving in drought tolerance of *M. sinensis* are not well understood. Recently, high-throughput RNA sequencing technology was proved a powerful tool for gene discovering, gene expression, and physiological and biochemical metabolism realization under abiotic stress [33, 34]. In this study, we explored differential expression gene and transcriptional profiles of *M. sinensis* under different drought stress stages based on transcriptome analysis, which contributes to well

understand the change of regulatory mechanism processes and provide molecular bases for further revealing of metabolic networks associated with drought tolerance. The drought tolerance-related genes identified in this study aimed to provide varied candidate genes for *M. sinensis* genetic improvement and crop breeding.

2. Results and Discussion

2.1. Sequencing Analysis and De Novo Assembly. A total of 6 RNA samples from *M. sinensis* drought tolerance genotype "M2010228" were sequenced using the Illumina HiSeq 2000 platform, and 349,393,396 raw reads were generated. After filtering and trimming the raw reads, a total of 316,200,846 high-quality clean reads were assembled into 114,747 unigenes using Trinity [35]. The length of unigenes ranged from 200 to 12,275 nt, the average length of unigenes was 1288 nt, and the total N_{50} was 1854 nt. Compared to previous de novo assembly of *M. × giganteus* by using ABySS and Phrap, the contigs obtaining longer than 200 bp were greater in this study, and a contig N_{50} length was longer than that of 1459 bp. All the unigenes were divided into two classes by gene family clustering, with a total of 65,203 distinct clusters identified which contained several similar unigene sequences (more than 70%) in each cluster and the other total of 49,544 distinct singletons generated with single unigene (Table 1).

The Illumina HiSeq 2000 platform, a short-read-based technology [36], was used leading to the generation of large-scale genomic and transcriptomic data for nonmodel crops. Furthermore, high-throughput RNA sequencing (RNA-Seq) technologies are more accurate and sensitive for detecting both low and high levels of gene expression [37]. Generally, Illumina sequencing platform is more cost-effective when compared to Roche 454 sequencing [38] and has been successfully used in transcriptome sequencing of many plant species [39–42]. In this study, an average of 52,700,141 clean reads was got from each sample, in which the number of clean reads was greatly larger than that of sequencing based on the 454 platform [43], although the average length of unigene was shorter. Therefore, the results of this study not only provide additional valuable genomic resources for *M. sinensis* but also construct reference for the comparison between two sequencing methods, which could

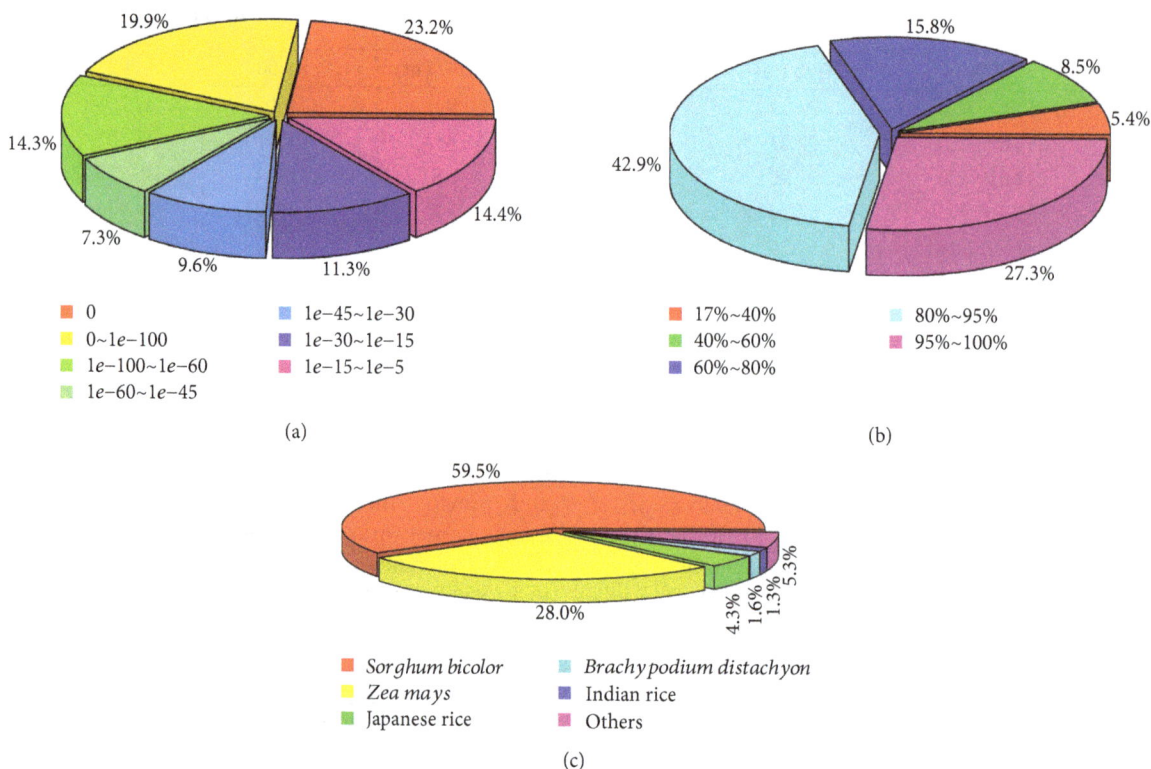

FIGURE 1: Unigene annotation results in the NR database. (a) displays the *e* value of the unigene annotation. (b) displays the identity of the similarity distribution. (c) displays the species distribution of annotated unigenes in NR database.

be used for further gene discovery and the identification of molecular markers.

2.2. Unigene Function Annotation and Classification Analysis. Unigene annotation and classification provide rich information about expression profiles and predict the potential functions of the assembly. Furthermore, various databases for annotation could shed light on intracellular metabolic pathways and biological behaviors of genes. In this study, the 114,747 obtained unigene sequences from six *M. sinensis* samples were aligned to protein databases NR, Swiss-Prot, KEGG, COG, and GO by BLASTx and nucleotide database NT by BLASTn (*e* value < 1*e*−5). Amongst them, 91,894 (80.08%) unigenes had significant hits in NT database; 84,456 (73.60%) in NR; 64,145 (55.90%) in GO; 55,807 (48.63%) in Swiss-Prot; 55,557 (48.42%) in KEGG; and 38,458 (33.52%) in COG. In total, 95,897 (83.57%) unigenes were annotated using at least one database.

Within NR annotation, 74.3% of *e* value was <1*e*−30 (Figure 1(a)) and 86% of similarity distribution was >60% (Figure 1(b)). For the species distribution of all unigenes identified from six samples, the most frequent and significant annotation hits in the databases were matched to two well-annotated Poaceae plant species, including 59.5% of them which were annotated to *Sorghum bicolor* and 28.0% to *Zea mays* (Figure 1(c)). It is no surprise that the unigenes were annotated to these two species. Included within the Andropogoneae are major crops such as maize (*Zea mays* L.), sorghum (*Sorghum bicolor* L. Moench), and sugarcane (*Saccharum officinarum* L.) and species in the genus

Miscanthus. Domesticated and wild grass species in the Andropogoneae tribe are important sources of food, feed, fiber, and fuel [44]. Previous studies showed that the high utility of sorghum as a reference genome sequence for Andropogoneae grasses was widely used for *M. sinensis* in genome-wide association analysis and QTL mapping [16, 44–46]. Swaminathan et al. [47] constructed a framework genetic map of *M. sinensis* using single-nucleotide variant (SNV) markers which were developed by deep RNA sequencing and comparison with the genomes of sorghum maize and rice (*Oryza sativa*). Ma et al. [48] created high-resolution genetic mapping of *M. sinensis* and revealed that sorghum has the closest phylogenetic relationship to *Miscanthus* by comparing the genome sequences to several grass species. Besides, the similarity of *Miscanthus* transcripts to the gene models and ESTs of sorghum, sugarcane, maize, rice, and *Brachypodium distachyon* was assessed by Barling et al. [49] and showed that a large portion of similarity was contributed to sugarcane ESTs and sorghum gene models with most matches sharing over 95% identity. In this study, the results of unigene annotation by *M. sinensis* RNA-seq showed that most of the gene annotations (59.5%) were aligned to the sorghum database, which is consistent with previous studies with the high utility of sorghum as a reference genome sequence for genus *Miscanthus*, supporting that the sequences obtained in our study were annotated properly.

The Clusters of Orthologous Groups (COGs) of proteins were delineated by comparing protein sequences encoded in complete genomes, representing major phylogenetic lineages. Each COG consists of individual proteins or groups of

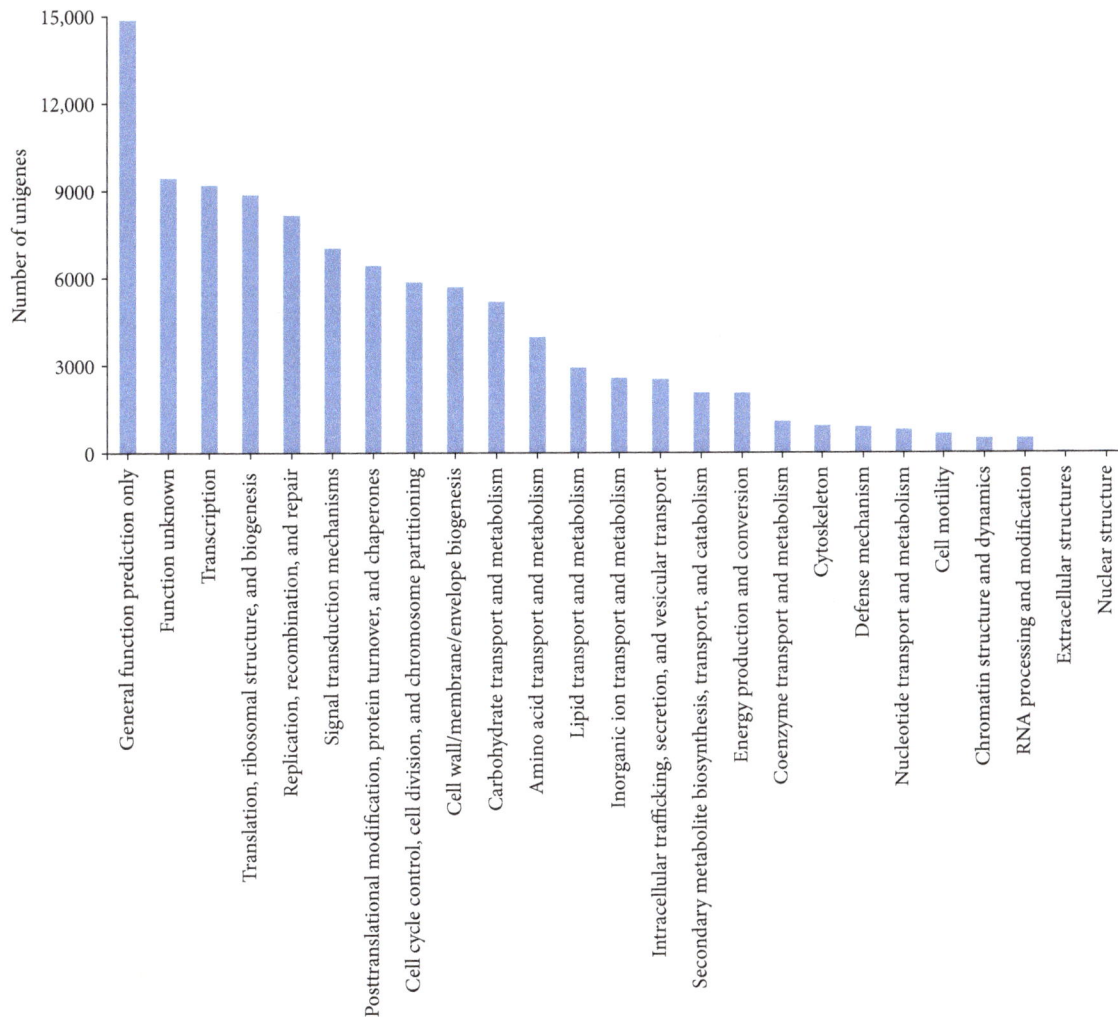

FIGURE 2: Function classification of all unigenes annotated in the COG database. A total of 38,458 unigenes were annotated under 25 function categories including (1) nuclear structure; (2) extracellular structures; (3) RNA processing and modification; (4) chromatin structure and dynamics; (5) cell motility; (6) nucleotide transport and metabolism; (7) defense mechanisms; (8) cytoskeleton; (9) coenzyme transport and metabolism; (10) energy production and conversion; (11) secondary metabolite biosynthesis, transport, and catabolism; (12) intracellular trafficking, secretion, and vesicular transport; (13) inorganic ion transport and metabolism; (14) lipid transport and metabolism; (15) amino acid transport and metabolism; (16) carbohydrate transport and metabolism; (17) cell wall/membrane/envelop biogenesis; (18) cell cycle control, cell division, and chromosome partitioning; (19) posttranslational modification, protein turnover, and chaperones; (20) signal transduction mechanisms; (21) replication, recombination, and repair; (22) translation, ribosomal structure, and biogenesis; (23) transcription; (24) function unknown; and (25) general function prediction only.

paralogs from at least 3 lineages and thus corresponds to an ancient conserved domain. The results of COG analysis indicated that 38,458 unigenes were annotated under 25 categories (Figure 2), among which were mainly classified into general function prediction (14,852 unigenes, 38.6%); transcription (9175 unigenes, 23.9%); translation, ribosomal structure, and biogenesis (8864 unigenes, 23.0%); replication, recombination, and repair (8162 unigenes, 21.2%); signal transduction mechanisms (6998 unigenes, 18.2%); and posttranslational modification, protein turnover, and chaperone function (6437 unigenes, 16.7%). In addition, there still have 9442 unigenes (24.6%) classified into unknown function category indicating that the unigenes identified from M. sinensis transcriptome under drought stresses were very different in biological functions involving in transport

and metabolism, cellular processes and signaling, and information storage and processing.

As an international standardized gene functional classification system, Gene Ontology (GO) offers a dynamic updated controlled vocabulary and a strictly defined concept to describe properties of genes and their products in any organism [50, 51]. In total, 64,145 (55.9%) unigenes were annotated to at least one of the three ontologies: molecular function, cellular component, and biological process in the GO database. In comparison, 58% of sorghum genes have GO annotation [52] as the most closely related plant to M. sinensis, indicating that our transcript assemblies afforded functional annotation of a comparable percentage of gene products to that of annotated plant species despite the current lack of a reference genome sequence.

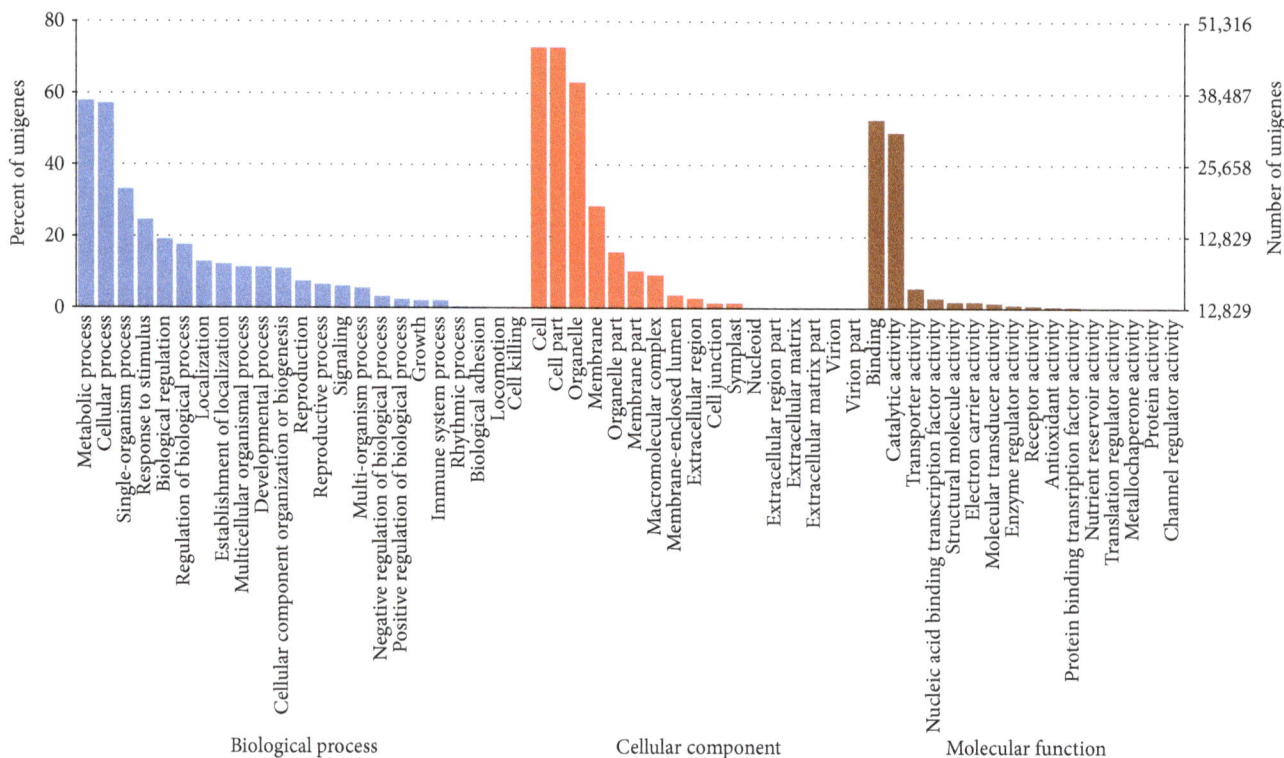

FIGURE 3: Function classification of all unigenes annotated in the GO database. A total of 64,145 unigenes were annotated under 23 different biological process categories, 17 cellular component categories, and 16 molecular function categories.

The distribution of all annotated unigenes in these three GO categories is shown in Figure 3. Among 23 different biological processes, metabolic process (57.9%), cellular process (57.0%), and single-organism process (33.2%) were the three most abundant GO categories responding to stimuli and biological regulation, suggesting active cellar and metabolic functions in *M. sinensis* leaves when exposed to drought stress. The frequent classes in cellar component were the cell part (72.9%), cell (72.9%), organelle (62.9%), and membrane (28.5%). Under the molecular function group, binding (52.5%) and catalytic activity (49.2%) were found to be the two mainly distributed categories as described, which are in agreement with the active metabolic functions in the examined tissues. With the help of GO functional classification, a large number of the unigenes were assigned to a diverse range of experimentally derived annotation. Our annotations provide a great foundation and a valuable resource for gene expression profile analysis, gene location, and gene isolation experiment in *Miscanthus* species. In addition, the main GO classifications identified through de novo transcriptome analyses in fundamental biological processes, cellar component, and molecular function were similar to previous reported studies in *M. sinensis*, *Sorghum bicolor*, [52] and *Hemarthria* [53], suggesting that our transcripts are the representative of a comprehensive *Miscanthus* transcriptome within the Andropogoneae tribe.

The networks of gene interactions in cells could be well understood by the KEGG pathway analysis. In this study, all the unigenes were analyzed in the KEGG pathway database and 55,557 unigenes were mapped to twenty main

categories including 128 different KEGG pathways (Figure 4). Most of the assigned genes were involved in a metabolism process (50,235, 90.4%), such as amino acid metabolism, carbohydrate metabolism, nucleotide metabolism, energy metabolism, lipid metabolism, and glycan biosynthesis and metabolism, suggesting a large number of genes induced by various metabolic activities under drought stress. Furthermore, a significant portion of unigenes was involved in the genetic information processing pathways (26,942, 48.5%), including transcription, translation, folding, sorting and degradation, replication, and repair. In addition, some of the unigenes were classified into organismal systems (4377, 7.9%), cellular process (4522, 8.1%), and environmental information processing (3625, 6.5%). The annotation of unigenes provides a large information base involved in drought tolerance process and plant drought response pathways, which serve an efficient guidance for future gene expression, gene network analyses, and regulatory metabolic network identification.

In order to conduct the prediction of protein coding region (CDs), all unigenes are firstly aligned by BLASTx (e value < 0.00001) to protein databases in the priority order of NR, Swiss-Prot, KEGG, and COG. Proteins with the highest ranks in BLAST results are taken to decide the coding region sequences of unigenes, and the coding region sequences are translated into amino sequences with the standard codon table. Unigenes that cannot be aligned to any database are scanned by EST-Scan, producing nucleotide sequence (5′→3′) direction and amino sequence of the predicted coding region. In this study, a total of 84,633 (73.8%)

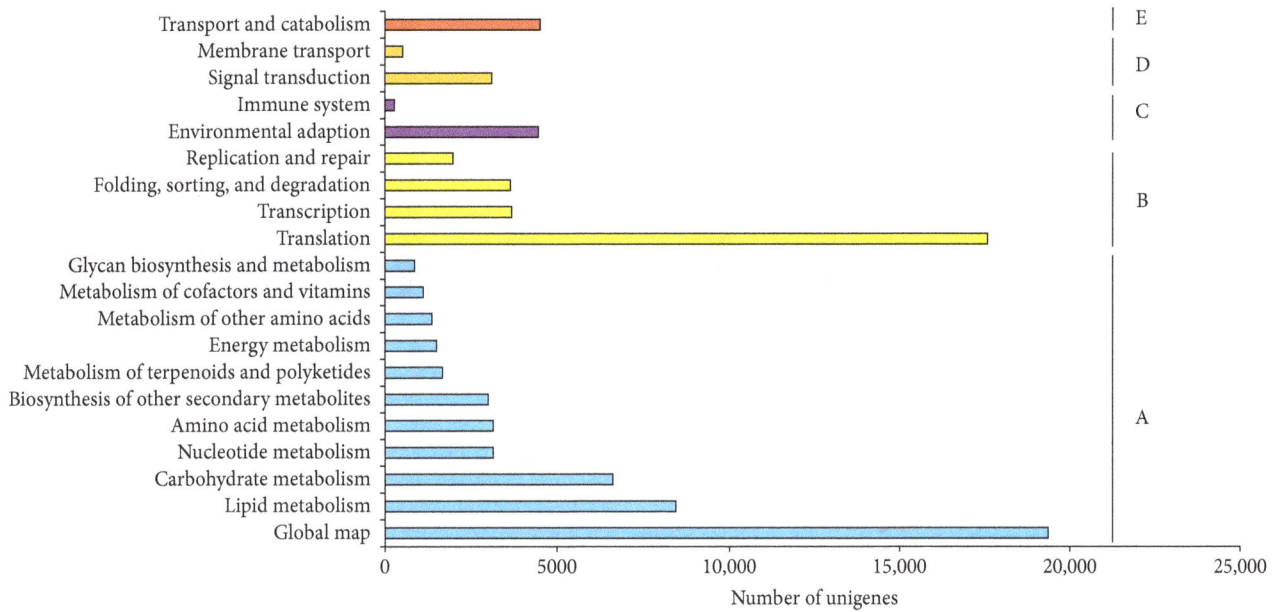

FIGURE 4: The KEGG pathway classification of all unigenes identified from six *M. sinensis* samples. A refers to metabolism category, B refers to genetic information processing category, C refers to organismal system category, D refers to environment information processing category, and E refers to cellular processes category.

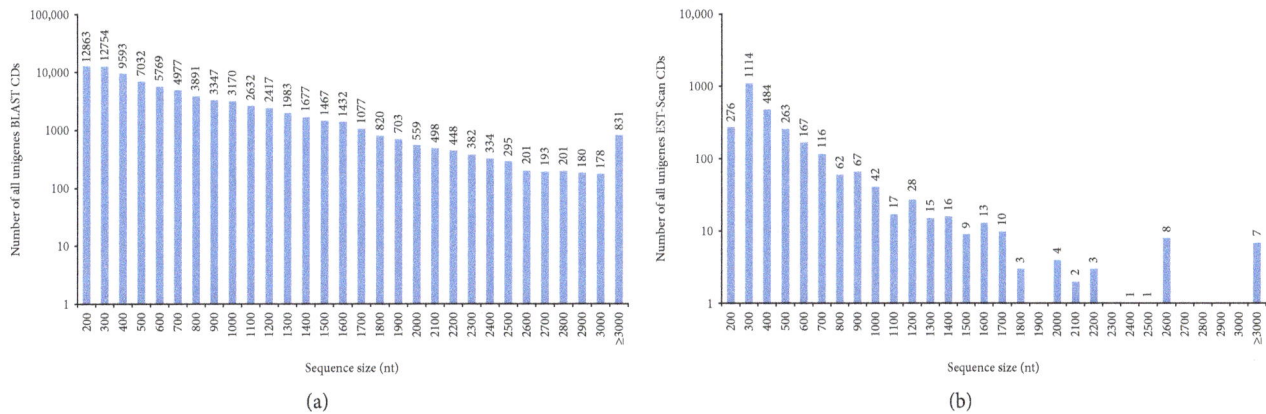

FIGURE 5: The length distribution of predicted CDs. (a) The length distribution of all unigenes BLAST CDs. (b) The length distribution of CDs scanned by EST-Scan.

unigenes were predicted to be CDs among the 114,747 assembled (Figure 5), of which 81,904 (96.8%) were aligned to the four previously discussed databases and another 2729 (3.3%) without BLAST hits were predicted by EST-Scan.

2.3. Differential Expression Gene Analysis. A total number of 5324 upregulated and 3276 downregulated differentially expressed genes (DEGs) were detected through the transcriptome comparison of well-watered (M0) versus drought stress 5 days (M1) in genotype "M20102208." The numbers of DEGs found in M0 versus M2 were 3467 upregulated and 4175 downregulated; in M0 versus M3, 829 upregulated and 4683 downregulated; in M0 versus M4, 4370 upregulated and 3792 downregulated; and in M0 versus M5, 7536 upregulated and 5297 downregulated (Figure 6(a)).

The results showed that the number of upregulated DEGs was decreasing with the drought stress treatment time prolonging, and when the drought stress treatment continued 15 days, the upregulated DEGs just have 829. However, the number of upregulated DEGs was increased when the plant was exposed to drought tress 20 days and 30 days.

For getting the dynamic change profiles of DEGs of *M. sinensis* under drought stress, another group of comparison was conducted (Figure 6(b)). The results showed that drought stress treatments from 0 day (M0) to 5 days (M1), 5 days (M1) to 10 days (M2), and 15 days (M3) to 20 days (M4) were three critical steps with the largest change of the number of DEGs, while from 10 days (M2) to 15 days (M3) and 20 days (M4) to 30 days (M5) were relatively stable stages where the number of DEG is no obvious change. Especially

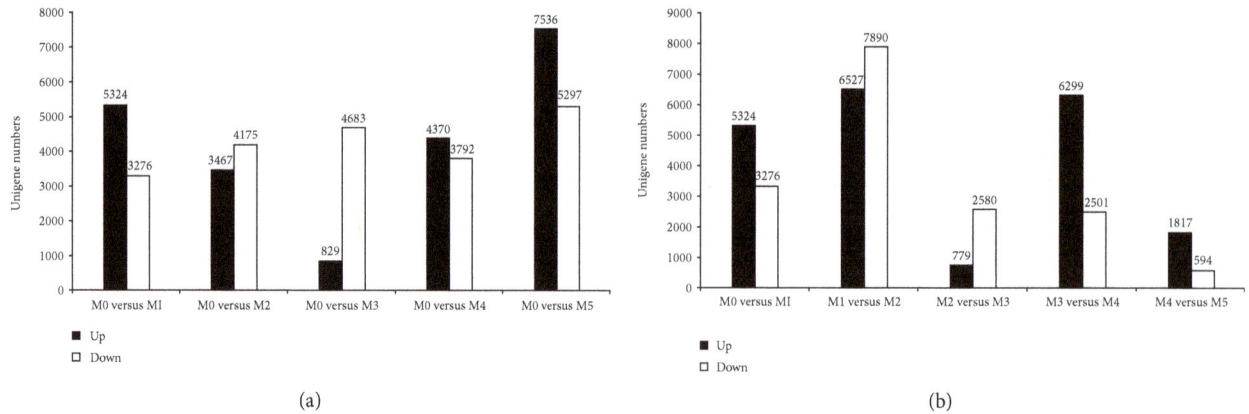

FIGURE 6: The number of DEGs at different time points under drought stress. (a) DEGs identified by the comparisons between well-watered (M0) and different drought stress treatments (M1, M2, M3, M4, and M5). (b) DEGs identified by the comparisons between two successive treatments.

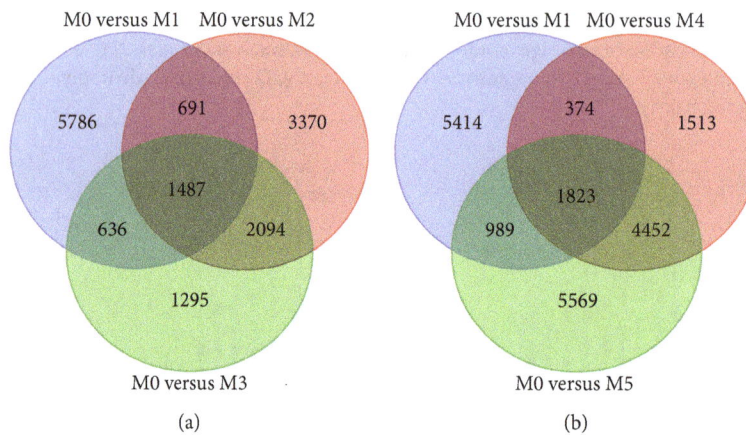

FIGURE 7: The similarity comparison of the DEGs between different time points under drought stress. (a) The DEGs identified from M0 versus M1, M0 versus M2, and M0 versus M3. (b) The DEGs identified from M0 versus M1, M0 versus M4, and M0 versus M5.

from 20 to 30 days, there just a total of 2411 DEGs were found indicating that *M. sinensis* exposed to drought stress 30 days almost had no molecular regulation or growth response and plant leaves showed obvious senescence phenotype.

The expression variation observed from RNA-seq provides a good representation of change in transcript profiles among samples [49]. Understanding the temporal expression change of regulated genes under drought stress may give insights into the gene related with drought stress adaptation in *M. sinensis*. After the similarity comparison of the DEGs identified from various stages of drought stress treatment, the DEGs from M0 versus M1 were highly dissimilar to other comparisons (just about 25% similar), while DEGs from M0 versus M3 have 65.0% matching to M0 versus M2, and the DEGs from M0 versus M4 have 76.9% similar to those from M0 versus M5 (Figure 7). The results were highly consistent with the previous results of dynamic change comparison in which the plants exposed to the 30-day drought stress treatment were divided into three main periods, that is, 0-day to 5-day slight stress, 5-day to 15-day medium stress, and 15-day to 30-day heavy stress.

At each defined phase, the differentially expressed genes are expected to have a specific function. More significantly, there were a substantial number of genes that exhibited modulated expression under drought stress. Interestingly, for the function analysis of the upregulated DEGs among medium stress periods, we found that after the 15-day drought stress treatment, the significantly high level expressed genes were mainly functions to cytochrome c oxidase, while stag-green related gene SORBIDRAFT [54] and were commonly highly expressed after the 10-day treatment. This indicates that a larger number of genes may participate in drought tolerance regulatory mechanisms and may be responsible for the plant phenotype under drought stress. In addition, various functions related to growth and development including putative ubiquitin carboxyl-terminal hydrolase superfamily, putative AP2/EREBP transcription factor superfamily, methyltransferase ZRP4, germin-like protein subfamily, lipoxygenase, and ATP synthase were also highly expressed within the initial slight stress and heavy stress period. These results indicate that drought stress 15 days could be a critical period for *M. sinensis* drought resistant molecular mechanism

TABLE 2: Identified assembled genes involved in ABA biosynthetic and catabolic pathway under drought stress.

Gene ID	Length	Swiss-Prot ID	Description	KO definition
Unigene18465	1664	sp|O24592|NCED1_MAIZE	9-cis-epoxycarotenoid dioxygenase 1	EC: 1.13.11.51
Unigene2508	615	sp|O24592|NCED1_MAIZE	9-cis-epoxycarotenoid dioxygenase 1	EC: 1.13.11.51
CL9892.Contig1	1319	sp|Q05JG2|ABAH1_ORYSJ	Abscisic acid 8'-hydroxylase 1	EC: 1.14.13.93
CL8871.Contig2	1441	sp|Q05JG2|ABAH1_ORYSJ	Abscisic acid 8'-hydroxylase 1	EC: 1.14.13.93
CL8871.Contig1	1635	sp|Q05JG2|ABAH1_ORYSJ	Abscisic acid 8'-hydroxylase 1	EC: 1.14.13.93

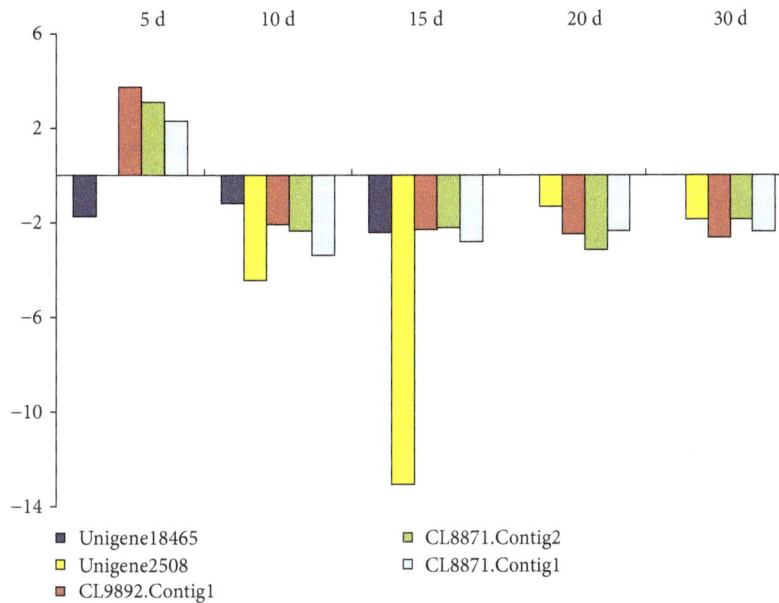

FIGURE 8: The differential expression level of key enzyme genes involved in ABA biosynthetic and catabolic pathway under drought stress. Unigene18465 and Unigene2508 functions were described as 9-cis-epoxycarotenoid dioxygenase 1. CL9892.Contig1, CL8871.Contig2, and CL8871.Contig1 functions were described as abscisic acid 8'-hydroxylase 1.

regulation, and when the relative soil water content reached 51.61% at 15 days, the chlorophyll of plant leaves begins to degrade to response to the stress.

In addition, the endogenous ABA level in plants is essential for various ABA-dependent stress responses, especially drought and salt stresses. Recently, with the molecular basis of ABA biosynthesis and catabolism established, the increasing concern about ABA biosynthetic and catabolic enzyme gene-expressed profiles is essential for the identification of drought stress response regulatory networks. Involved in carotenoid biosynthesis pathway, 9-cis-epoxycarotenoid dioxygenase (NCED) is a key enzyme for ABA biosynthesis, which was originally identified from maize viviparous 14 mutants playing a crucial role in drought stress-inducible genes [55]. On the other hand, abscisic acid 8'-hydroxylase (CYP707A) is an important enzyme gene for the ABA catabolism oxidative pathway in the drought stress response, which belongs to a class of cytochrome P450 monooxygenases [24]. In this study, we focused on the dynamic expression profile of these two DEG groups described above to identify the drought stress response mechanism of M. sinensis (Table 2). The results showed that the NCED gene downregulated differential expression under drought stress in M. sinensis and the highest expressed at 15 days. The NCED transcripts are rapidly induced by drought stress to promote the ABA accumulation and to enhance drought tolerance of the plant [23]. However, among all the periods, we did not find the upregulated differential expression of NCED, indicating that the 5-day drought stress treatment in this study could be late for discovering the NCED gene in time, since the short-chain dehydrogenase/reductase (ABA2) was detected upregulated at 5 days. For CYP707A, the DEGs upregulated at 5 days and downregulated at the following periods represent that when the plant accepts the stress signal molecular, the ABA catabolism process was slow down for ABA accumulation to forcing the drought tolerance, and this could be an important mechanism for M. sinensis drought tolerance (Figure 8). Various crops could regulate drought stress response in different ways, although we indicate that M. sinensis drought tolerance was enhanced by the downregulated ABA catabolism pathway through the differential expression patterns; further studies are needed for function validation in knockout mutants or transgenic plants.

3. Experimental Design

3.1. Drought Treatment and Sample Collection. A wild drought tolerance genotype "M20102208" was collected from Sichuan Province (highway side, N 30° 08′, E 103° 14′). All the plants were propagated through rhizome division from a single individual. The plants were planted in plastic pots (20 cm in diameter and 25 cm in height) with soil mixture (50% loam with 50% fine sandy). *M. sinensis* plants were grown in a growth chamber at 30°C/25°C, 16 h/8 h (day/night), 70% relative humidity, and 500 μmol photons m^{-2}·s^{-1}. After three-month establishment, all the replications were subjected to nature drought stress treatment. Prior to treatment, all the pots were well-watered and the soil water content (SWC) was measured by the Soil Moisture Equipment TDR 300 (Santa Barbara, CA, USA). Naturally, water stress was applied by stopping water for 30 days. Leaf samples were collected and immediately frozen in liquid nitrogen from three replications at 0- (SWC = 91.57%), 5- (SWC = 85.55%), 10- (SWC = 73.12%), 15- (SWC = 51.61%), 20- (SWC = 39.89%), and 30-day (SWC = 26.41%) drought stress treatments for RNA extraction. Total RNA was extracted using the RNeasy Plant Mini Kit (Qiagen, USA) according to the manufacturer's instructions. RNA purity, concentration, and integrity were assessed using the RNA Nano 6000 Kit for the Agilent 2100 Bioanalyzer 2100 System (Agilent Technologies, USA). After RNA isolation and quality assessment, samples were stored at −80°C until the cDNA library construction and transcriptomic assay were completed.

3.2. Library Construction and Sequencing. A total of 5 μg of total RNA per sample was used to construct the cDNA libraries. In all, six cDNA sequencing libraries of *M. sinensis* were constructed using the NEB Next® Ultra™ RNA Library Prep Kit for Illumina (New England Biolabs, USA). Initially, the total RNA was treated with RNase-free DNase I (NEB) for 30 min at 37°C and poly(A) mRNA was isolated from total RNA using poly-T oligo-linked magnetic beads. Following purification, the poly(A)-containing mRNA was fragmented into 200–250 bp pieces using fragment buffer (Ambion), and the first-strand cDNA was synthesized using random hexamer primers and the short fragments as templates. The products were then treated with RNase H, and second-strand cDNA was synthesized by DNA polymerase I (16°C for 2 h). Finally, NEBNext Adaptor with hairpin loop structure was ligated to the cDNA and the 3′ ends of the DNA fragments were adenylated in preparation for hybridization. Subsequently, the cDNA fragments were purified and the quality of the library was evaluated using the Agilent Bioanalyzer 2100 system. The index-coded samples were clustered on a cBot System using the TruSeq PE Cluster kit v3-cBot-HS (Illumina). After generating the clusters, Illumina sequencing (paired-end technology in the Illumina HiSeq 2000 platform) of the six libraries was performed for RNA-seq analysis.

3.3. RNA Sequence Analysis and Drought-Induced Transcriptomic Changes. Raw reads (accession numbers SRP095822 in the NCBI SRA database) from six libraries were transformed from sequencing-received image data. Raw reads were filtered to remove those with only adapters, low quality reads, and unknown or reads with less than 20 bp in length. Following the calculation of sequence duplication, Q20, and the GC content of the clean reads, the de novo assembly of RNA-seq was conducted using Trinity (http://trinityrnaseq.github.io), which is specific for high-throughput transcript assembly of RNA-Seq data without a reference genome [35]. The unigenes were generated by Trinity modules, and the processes of sequence splicing and redundancy removing were employed. The unigenes were annotated against the following protein databases: NR (nonredundant NCBI protein sequences), KOG/COGs (Clusters of Orthologous Groups of proteins), Swiss-Prot (a manually annotated and reviewed protein sequence database), and KEGG Ortholog database using BLASTx searches (e value < 1e−5). Protein function information can be predicted from the annotation of the most similar protein in those databases. If the results of the databases conflicted with each other, a priority order of NR, Swiss-Prot, KEGG, and COG was followed when deciding the sequence direction of unigenes. The KEGG pathway database records networks of molecular interactions in the cells and variants of them specific to particular organisms. GO functional annotation was conducted with NR annotation which offers a dynamic updated controlled vocabulary and a strictly defined concept to comprehensively describe properties of genes using BLAST2GO program.

3.4. Differential Expression Genes and Pathway Analysis. The gene expression level was calculated by the number of uniquely mapped reads per kilobase of exon fragments per million mappable reads (FPKM) using Cufflinks (http://cufflinks.cbcb.umd.edu/). For genes with more than one alternative transcript, the longest transcript was selected to calculate the FPKM. With the expression level of each gene calculated, the differential expression analysis was conducted. The false discovery rate (FDR) as a statistical method was used to determine the threshold of the p value in multiple hypothesis testing, and for the analysis, a threshold of the FDR ≤ 0.001 and an absolute value of log2 ratio ≥ 1 were used to judge the significance of the gene expression differences. Tool edgeR23 was used to identify significantly up- and downregulated genes on the read count values of genes. The differentially expressed genes (DEGs) were used for GO and KEGG enrichment analyses. First, all of the DEGs were blasted in the GO database (http://www.geneontology.org/) and the gene numbers were calculated for each GO term with GO-Term Finder version 0.86 (http://search.cpan.org/dist/GO-TermFinder/). GO terms were defined as significantly enriched GO terms in DEGs, if the corrected p value was ≤0.05. Pathway enrichment analysis identifies significantly enriched metabolic pathways or signal transduction pathways in DEGs when compared with the whole genome background. Both GO terms and KEGG pathways with a Q value ≤ 0.05 are significantly enriched in DEGs.

4. Conclusions

High-throughput RNA sequencing technology was proved a powerful tool for gene discovering, gene expression, and physiological and biochemical metabolism realization under abiotic stress. By using the Illumina HiSeq 2000 platform to sequence *M. sinensis* under drought stress, approximately 316 million high-quality trimmed reads were generated from 349 million raw reads and a total of 114,747 unigenes were obtained after de novo assembly of the trimmed reads. Furthermore, 95,897 (83.57%) unigenes were annotated to at least one database including NR, Swiss-Prot, KEGG, COG, GO, and NT, and most of the annotations (59.5%) were aligned to the sorghum database, which is consistent with previous studies with the high utility of sorghum as a reference genome sequence for genus *Miscanthus*, supporting that the sequences obtained in our study were annotated properly. Differentially expressed gene analysis under different stress periods indicates that drought stress 15 days (soil water content reached 51.61%) could be a critical period for *M. sinensis* response to drought stress. *M. sinensis* plays an important role in improving the genetic base, abiotic stress tolerance, and climatic adaptation for this genus as a nonfood bioenergy crop. Hence, the transcriptome sequencing of *M. sinensis* reported here provides useful information for gene identification and greatly enriches the genomic available resources. The comparison of DEGs under different periods of drought stress allowed us to identify a wealth of candidate genes involved in drought tolerance regulatory networks, which will facilitate further advancements in genetic and molecular mechanisms with desired traits in further *M. sinensis* breeding programs.

Conflicts of Interest

The authors declare no conflict of interest.

Authors' Contributions

Xinquan Zhang, Linkai Huang, and Xiao Ma conceived the project and designed the experiments. Gang Nie, Yajie Zhang, and Lu Tang performed the experiment. Gang Nie, Zhongjie Ji, and Linkai Huang wrote the paper. All authors discussed the results and commented on the manuscript.

Acknowledgments

This work was supported by the Earmarked Fund for the Modern Agro-Industry Technology Research System (no. CARS-34) and the Grassland Basic Resources Investigation Research in China (2017FY100602).

References

[1] I. Lewandowski, J. M. O. Scurlock, E. Lindvall, and M. Christou, "The development and current status of perennial rhizomatous grasses as energy crops in the US and Europe," *Biomass and Bioenergy*, vol. 25, no. 4, pp. 335–361, 2003.

[2] J. M. Greef, M. Deuter, C. Jung, and J. Schondelmaier, "Genetic diversity of European Miscanthus species revealed by AFLP fingerprinting," *Genetic Resources and Crop Evolution*, vol. 44, no. 2, pp. 185–195, 1997.

[3] T. R. Hodkinson, M. W. Chase, M. D. Lledo, N. Salamin, and S. A. Renvoize, "Phylogenetics of *Miscanthus, Saccharum* and related genera (Saccharinae, Andropogoneae, Poaceae) based on DNA sequences from ITS nuclear ribosomal DNA and plastid *trnL* intron and *trnL-F* intergenic spacers," *Journal of Plant Research*, vol. 115, no. 5, pp. 381–392, 2002.

[4] T. R. Hodkinson, M. W. Chase, C. Takahashi, I. J. Leitch, M. D. Bennett, and S. A. Renvoize, "The use of DNA sequencing (ITS and *trnL-F*), AFLP, and fluorescent in situ hybridization to study allopolyploid *Miscanthus* (Poaceae)," *American Journal of Botany*, vol. 89, no. 2, pp. 279–286, 2002.

[5] E. A. Heaton, F. G. Dohleman, and S. P. Long, "Meeting US biofuel goals with less land: the potential of Miscanthus," *Global Change Biology*, vol. 14, no. 9, pp. 2000–2014, 2008.

[6] E. A. Heaton, F. G. Dohleman, and S. P. Long, "Seasonal nitrogen dynamics of *Miscanthus × giganteus* and *Panicum virgatum*," *Global Change Biology Bioenergy*, vol. 1, no. 4, pp. 297–307, 2009.

[7] A. Hastings, J. Clifton-Brown, M. Wattenbach, P. Mitchell, and P. Smith, "The development of MISCANFOR, a new *Miscanthus* crop growth model: towards more robust yield predictions under different climatic and soil conditions," *Global Change Biology Bioenergy*, vol. 1, no. 2, pp. 154–170, 2009.

[8] A. Nishiwaki, A. Mizuguti, S. Kuwabara et al., "Discovery of natural *Miscanthus* (Poaceae) triploid plants in sympatric populations of *Miscanthus sacchariflorus* and *Miscanthus sinensis* in southern Japan," *American Journal of Botany*, vol. 98, no. 1, pp. 154–159, 2011.

[9] M. S. Dwiyanti, J. R. Stewart, A. Nishiwaki, and T. Yamada, "Natural variation in *Miscanthus sinensis* seed germination under low temperatures," *Grassland Science*, vol. 60, pp. 194–198, 2014.

[10] I. Lewandowski, J. C. Clifton-Brown, B. Andersson et al., "Environment and harvest time affects the combustion qualities of *Miscanthus* genotypes," *Agronomy Journal*, vol. 95, no. 5, pp. 1274–1280, 2003.

[11] L. V. Clark, J. E. Brummer, K. Glowacka et al., "A footprint of past climate change on the diversity and population structure of *Miscanthus sinensis*," *Annals of Botany*, vol. 114, no. 1, pp. 97–107, 2014.

[12] K. G. Anzoua, K. Suzuki, S. Fujita, Y. Toma, and T. Yamada, "Evaluation of morphological traits, winter survival and biomass potential in wild Japanese *Miscanthus sinensis* Anderss. populations in Northern Japan," *Grassland Science*, vol. 61, no. 2, pp. 83–91, 2015.

[13] H. Zhao, B. Wang, J. He et al., "Genetic diversity and population structure of *Miscanthus sinensis* germplasm in China," *PLoS One*, vol. 8, no. 10, article e75672, 2013.

[14] G. Nie, X. Q. Zhang, L. K. Huang et al., "Genetic variability and population structure of the potential bioenergy crop *Miscanthus sinensis* (Poaceae) in Southwest China based on SRAP markers," *Molecules*, vol. 19, no. 8, pp. 12881–12897, 2014.

[15] G. Nie, L. Huang, X. Zhang et al., "Marker-trait association for biomass yield of potential bio-fuel feedstock *Miscanthus sinensis* from Southwest China," *Frontiers in Plant Science*, vol. 7, p. 802, 2016.

[16] J. M. Gifford, W. B. Chae, K. Swaminathan, S. P. Moose, and J. A. Juvik, "Mapping the genome of *Miscanthus sinensis* for QTL associated with biomass productivity," *Global Change Biology Bioenergy*, vol. 7, no. 4, pp. 797–810, 2015.

[17] M. Seki, T. Umezawa, K. Urano, and K. Shinozaki, "Regulatory metabolic networks in drought stress responses," *Current Opinion in Plant Biology*, vol. 10, no. 3, pp. 296–302, 2007.

[18] K. Shinozaki and K. Yamaguchi-Shinozaki, "Gene networks involved in drought stress response and tolerance," *Journal of Experimental Botany*, vol. 58, no. 2, pp. 221–227, 2007.

[19] X. Yu, G. Bai, S. Liu et al., "Association of candidate genes with drought tolerance traits in diverse perennial ryegrass accessions," *Journal of Experimental Botany*, vol. 64, no. 6, pp. 1537–1551, 2013.

[20] M. Seki, M. Narusaka, J. Ishida et al., "Monitoring the expression profiles of 7000 *Arabidopsis* genes under drought, cold and high-salinity stresses using a full-length cDNA microarray," *The Plant Journal*, vol. 31, no. 3, pp. 279–292, 2002.

[21] K. Shinozaki, K. Yamaguchi-Shinozaki, and M. Seki, "Regulatory network of gene expression in the drought and cold stress responses," *Current Opinion in Plant Biology*, vol. 6, no. 5, pp. 410–417, 2003.

[22] M. Rabbani, K. Maruyama, H. Abe et al., "Monitoring expression profiles of rice genes under cold, drought and high-salinity stresses, and abscisic acid application using cDNA microarray and RNA gel-blot analyses," *Plant Physiology*, vol. 133, no. 4, pp. 1755–1767, 2003.

[23] S. Iuchi, M. Kobayashi, T. Taji et al., "Regulation of drought tolerance by gene manipulation of 9-*cis*-epoxycarotenoid dioxygenase, a key enzyme in abscisic acid biosynthesis in *Arabidopsis*," *The Plant Journal*, vol. 27, no. 4, pp. 325–333, 2001.

[24] T. Kushiro, M. Okamoto, K. Nakabayashi et al., "The *Arabidopsis* cytochrome P450 CYP707A encodes ABA 8′-hydroxylases: key enzymes in ABA catabolism," *The EMBO Journal*, vol. 23, no. 7, pp. 1647–1656, 2004.

[25] S. Saito, N. Hirai, C. Matsumoto et al., "Arabidopsis CYP707As encode (+)-abscisic acid 8′-hydroxylase, a key enzyme in the oxidative catabolism of abscisic acid," *Plant Physiology*, vol. 134, no. 4, pp. 1439–1449, 2004.

[26] J. Zhang, R. Creelman, and J. Zhu, "From laboratory to field. Using information from Arabidopsis to engineer salt, cold, and drought tolerance in crops," *Plant Physiology*, vol. 135, no. 2, pp. 615–621, 2004.

[27] F. Z. Wang, Q. B. Wang, S. Y. Kwon, S. S. Kwak, and W. A. Su, "Enhanced drought tolerance of transgenic rice plants expressing a pea manganese superoxide dismutase," *Journal of Plant Physiology*, vol. 162, no. 4, pp. 465–472, 2005.

[28] T. Umezawa, M. Fujita, Y. Fujita, K. Yamaguchi-Shinozaki, and K. Shinozaki, "Engineering drought tolerance in plants: discovering and tailoring genes to unlock the future," *Current Opinion in Biotechnology*, vol. 17, no. 2, pp. 113–122, 2006.

[29] R. P. Ellis, B. P. Forster, D. Robinson et al., "Wild barley: a source of genes for crop improvement in the 21st century?," *Journal of Experimental Botany*, vol. 51, no. 342, pp. 9–17, 2000.

[30] D. Bartels, "Targeting detoxification pathways: an efficient approach to obtain plants with multiple stress tolerance?," *Trends in Plant Science*, vol. 6, no. 7, pp. 284–286, 2001.

[31] V. Mittova, M. Guy, M. Tal, and M. Volokita, "Salinity up-regulates the antioxidative system in root mitochondria and peroxisomes of the wild salt-tolerant tomato species *Lycopersicon pennellii*," *Journal of Experimental Botany*, vol. 55, no. 399, pp. 1105–1113, 2004.

[32] J. C. Clifton-Brown, I. Lewandowski, B. Andersson et al., "Performance of 15 *Miscanthus* genotypes at five sites in Europe," *Agronomy Journal*, vol. 93, no. 5, pp. 1013–1019, 2001.

[33] P. Lan, W. Li, and W. Schmidt, "Complementary proteome and transcriptome profiling in phosphate-deficient Arabidopsis roots reveals multiple levels of gene regulation," *Molecular & Cellular Proteomics*, vol. 11, no. 11, pp. 1156–1166, 2012.

[34] L. Pan, X. Zhang, J. Wang et al., "Transcriptional profiles of drought-related genes in modulating metabolic processes and antioxidant defenses in *Lolium multiflorum*," *Frontiers in Plant Science*, vol. 7, p. 519, 2016.

[35] M. G. Grabherr, B. J. Haas, M. Yassour et al., "Full-length transcriptome assembly from RNA-Seq data without a reference genome," *Nature Biotechnology*, vol. 29, no. 7, pp. 644–652, 2011.

[36] M. Trick, Y. Long, J. Meng, and I. Bancroft, "Single nucleotide polymorphism (SNP) discovery in the polyploid *Brassica napus* using Solexa transcriptome sequencing," *Plant Biotechnology Journal*, vol. 7, no. 4, pp. 334–346, 2009.

[37] B. Wang, G. Guo, C. Wang et al., "Survey of the transcriptome of *Aspergillus oryzae* via massively parallel mRNA sequencing," *Nucleic Acids Research*, vol. 38, no. 15, pp. 5075–5087, 2010.

[38] O. Morozova, M. Hirst, and M. A. Marra, "Applications of new sequencing technologies for transcriptome analysis," *Annual Review of Genomics and Human Genetics*, vol. 10, no. 1, pp. 135–151, 2009.

[39] E. Mizrachi, C. A. Hefer, M. Ranik, F. Joubert, and A. A. Myburg, "*De novo* assembled expressed gene catalog of a fast-growing *Eucalyptus* tree produced by Illumina mRNA-Seq," *BMC Genomics*, vol. 11, no. 1, p. 681, 2010.

[40] R. Garg, R. Patel, A. Tyagi, and M. Jain, "*De novo* assembly of chickpea transcriptome using short reads for gene discovery and marker identification," *DNA Research*, vol. 18, no. 1, pp. 53–63, 2011.

[41] D. Zou, X. Chen, and D. Zou, "Sequencing, de novo assembly, annotation and SSR and SNP detection of sabaigrass (*Eulaliopsis binata*) transcriptome," *Genomics*, vol. 102, no. 1, pp. 57–62, 2013.

[42] S. Yates, M. Swain, M. Hegarty et al., "*De novo* assembly of red clover transcriptome based on RNA-Seq data provides insight into drought response, gene discovery and marker identification," *BMC Genomics*, vol. 15, no. 1, p. 453, 2014.

[43] C. Kim, T. H. Lee, H. Guo et al., "Sequencing of transcriptomes from two *Miscanthus* species reveals functional specificity in rhizomes, and clarifies evolutionary relationships," *BMC Plant Biology*, vol. 14, no. 1, p. 134, 2014.

[44] K. Swaminathan, M. S. Alabady, K. Varala et al., "Genomic and small RNA sequencing of *Miscanthus* × *giganteus* shows the utility of sorghum as a reference genome sequence for Andropogoneae grasses," *Genome Biology*, vol. 11, no. 2, article R12, 2010.

[45] G. T. Slavov, R. Nipper, P. Robson et al., "Genome-wide association studies and prediction of 17 traits related to phenology, biomass and cell wall composition in the energy grass *Miscanthus sinensis*," *The New Phytologist*, vol. 201, no. 4, pp. 1227–1239, 2014.

[46] S. Liu, L. V. Clark, K. Swaminathan, J. M. Gifford, J. A. Juvik, and E. J. Sacks, "High-density genetic map of *Miscanthus*

sinensis reveals inheritance of zebra stripe," *Global Change Biology Bioenergy*, vol. 8, pp. 616–630, 2015.

[47] K. Swaminathan, W. B. Chae, T. Mitros et al., "A framework genetic map for *Miscanthus sinensis* from RNAseq-based markers shows recent tetraploidy," *BMC Genomics*, vol. 13, no. 1, p. 142, 2012.

[48] X. F. Ma, E. Jensen, N. Alexandrov et al., "High resolution genetic mapping by genome sequencing reveals genome duplication and tetraploid genetic structure of the diploid *Miscanthus sinensis*," *PLoS One*, vol. 7, no. 3, article e33821, 2012.

[49] A. Barling, K. Swaminathan, T. Mitros et al., "A detailed gene expression study of the *Miscanthus* genus reveals changes in the transcriptome associated with the rejuvenation of spring rhizomes," *BMC Genomics*, vol. 14, no. 1, p. 864, 2013.

[50] M. A. Harris, J. Clark, A. Ireland et al., "The Gene Ontology (GO) database and informatics resource," *Nucleic Acids Research*, vol. 32, Supplement 1, pp. D258–D261, 2004.

[51] X. Tang, Y. Xiao, T. Lv et al., "High-throughput sequencing and *de novo* assembly of the *Isatis indigotica* transcriptome," *PLoS One*, vol. 9, no. 9, article e102963, 2014.

[52] P. Chouvarine, A. M. Cooksey, F. M. McCarthy et al., "Transcriptome-based differentiation of closely-related *Miscanthus* lines," *PLoS One*, vol. 7, no. 1, article e29850, 2012.

[53] X. Huang, H. D. Yan, X. Q. Zhang et al., "*De novo* transcriptome analysis and molecular marker development of two *Hemarthria* species," *Frontiers in Plant Science*, vol. 7, p. 496, 2016.

[54] G. N. Chaudhari and B. Fakrudin, "Candidate gene prediction and expression profiling of near isogenic lines (NILs) carrying stay-green QTLs in rabi sorghum," *Journal of Plant Biochemistry and Biotechnology*, vol. 26, no. 1, pp. 64–72, 2017.

[55] S. H. Schwartz, B. C. Tan, D. A. Gage, J. A. Zeevaart, and D. R. McCarty, "Specific oxidative cleavage of carotenoids by VP14 of maize," *Science*, vol. 276, no. 5320, pp. 1872–1874, 1997.

Permissions

The contributors of this book come from diverse backgrounds, making this book a truly international effort. This book will bring forth new frontiers with its revolutionizing research information and detailed analysis of the nascent developments around the world.

We would like to thank all the contributing authors for lending their expertise to make the book truly unique. They have played a crucial role in the development of this book. Without their invaluable contributions this book wouldn't have been possible. They have made vital efforts to compile up to date information on the varied aspects of this subject to make this book a valuable addition to the collection of many professionals and students.

This book was conceptualized with the vision of imparting up-to-date information and advanced data in this field. To ensure the same, a matchless editorial board was set up. Every individual on the board went through rigorous rounds of assessment to prove their worth. After which they invested a large part of their time researching and compiling the most relevant data for our readers.

The editorial board has been involved in producing this book since its inception. They have spent rigorous hours researching and exploring the diverse topics which have resulted in the successful publishing of this book. They have passed on their knowledge of decades through this book. To expedite this challenging task, the publisher supported the team at every step. A small team of assistant editors was also appointed to further simplify the editing procedure and attain best results for the readers.

Apart from the editorial board, the designing team has also invested a significant amount of their time in understanding the subject and creating the most relevant covers. They scrutinized every image to scout for the most suitable representation of the subject and create an appropriate cover for the book.

The publishing team has been an ardent support to the editorial, designing and production team. Their endless efforts to recruit the best for this project, has resulted in the accomplishment of this book. They are a veteran in the field of academics and their pool of knowledge is as vast as their experience in printing. Their expertise and guidance has proved useful at every step. Their uncompromising quality standards have made this book an exceptional effort. Their encouragement from time to time has been an inspiration for everyone.

The publisher and the editorial board hope that this book will prove to be a valuable piece of knowledge for researchers, students, practitioners and scholars across the globe.

List of Contributors

Alexandre Bueno Santos, Patrícia Silva Costa, Anderson Oliveira do Carmo, Evanguedes Kalapothakis, Edmar Chartone-Souza and Andréa Maria Amaral Nascimento
Departamento de Biologia Geral, Instituto de Ciências Biológicas, Universidade Federal de Minas Gerais, Belo Horizonte, MG, Brazil

Gabriel da Rocha Fernandes, Larissa Lopes Silva Scholte and Jeronimo Ruiz
Centro de Pesquisas René Rachou, FIOCRUZ, Belo Horizonte, MG, Brazil

Adam P. Sage, Brenda C. Minatel, Erin A. Marshall, Victor D. Martinez, Greg L. Stewart, Katey S. S. Enfield and Wan L. Lam
Department of Integrative Oncology, British Columbia Cancer Research Centre, Vancouver, BC, Canada

Chongyang Wu, Chaoqin Lin, Xinyi Zhu, Hongmao Liu and Yunliang Hu
The Second Affiliated Hospital and Yuying Children's Hospital of Wenzhou Medical University, Wenzhou, Zhejiang 325000, China

Wangxiao Zhou, Chongyang Wu, Chaoqin Lin, Xinyi Zhu, Hongmao Liu and Yunliang Hu
School of Laboratory Medicine and Life Sciences/ Institute of Biomedical Informatics, Wenzhou Medical University, Wenzhou 325035, China

Junwan Lu, Licheng Zhu and Cong Cheng
College of Medicine and Health, Lishui University, Lishui 323000, China

Hongyi Nie, Haiyang Geng, Yan Lin, Shupeng Xu, Zhiguo Li, Yazhou Zhao and Songkun Su
College of Bee Science, Fujian Agriculture and Forestry University, Fuzhou 350002, China

Yazhou Zhao
Institute of Apiculture, Chinese Academy of Agricultural Sciences, Beijing 100093, China

Luka Bolha, Metka Ravnik-Glavač and Damjan Glavač
Department of Molecular Genetics, Institute of Pathology, Faculty of Medicine, University of Ljubljana, Ljubljana, Slovenia

Metka Ravnik-Glavač
Institute of Biochemistry, Faculty of Medicine, University of Ljubljana, Ljubljana, Slovenia

Fahad Al-Qurainy, Salim Khan, Mohammad Nadeem, Abdel-Rhman Z. Gaafar and Mohamed Tarroum
Department of Botany and Microbiology, College of Science, King Saud University, Riyadh 11451, Saudi Arabia

Abdulhafed A. Al-Ameri
Department of Biology, Faculty of Education and Science, Rada'a Al-Baydha University, Al-Baydha, Yemen

Zehra Omeroğlu Ulu, Salih Ulu and Nehir Ozdemir Ozgenturk
Faculty of Art and Science, Molecular Biology and Genetics, Yildiz Technical University, Istanbul, Turkey

Soner Dogan
Department of Medical Biology, School of Medicine, Yeditepe University, Istanbul, Turkey

Bilge Guvenc Tuna
Department of Medical Biophysic, School of Medicine, Yeditepe University, Istanbul, Turkey

Xing Huang, Jingen Xi, Chunping He, Jinlong Zheng, Weihuai Wu, Yanqiong Liang and Kexian Yi
Environment and Plant Protection Institute, Chinese Academy of Tropical Agricultural Sciences, Haikou 571101, China

Bo Wang
National Key Laboratory of Crop Genetic Improvement, Huazhong Agricultural University, Wuhan, Hubei 430070, China

Yajie Zhang
Hainan Climate Center, Haikou 570203, China

Jianming Gao, Helong Chen and Shiqing Zhang
Institute of Tropical Bioscience and Biotechnology, Chinese Academy of Tropical Agricultural Sciences, Haikou 571101, China

Jianli Wang, Guiqing Han and Changhong Guo
College of Life Science and Technology of Harbin Normal University, Harbin 150080, China

Jianli Wang, Zhenying Wu, Lichao Ma and Chunxiang Fu
Key Laboratory of Biofuels, Shandong Provincial Key Laboratory of Energy Genetics, Qingdao Institute of Bioenergy and Bioprocess Technology, Chinese Academy of Sciences, Qingdao 266101, China

Jianli Wang, Zhongbao Shen, Duofeng Pan, Ruibo Zhang, Daoming Li, Hailing Zhang and Guiqing Han
Grass and Science Institute of Heilongjiang Academy of Agricultural Sciences, Harbin, Heilongjiang 150086, China

Peng Zhong
Rural Energy Research Institute of Heilongjiang Academy of Agricultural Sciences, Harbin, Heilongjiang 150086, China

Corinne E. Sexton, Meganne Ferrel, Perry G. Ridge and John S. K. Kauwe
Department of Biology, Brigham Young University, Provo, UT 84602, USA

Mark T. W. Ebbert
Department of Neuroscience, Mayo Clinic, Jacksonville, FL 32224, USA

Ryan H. Miller
Department of Oncological Sciences, University of Utah, Salt Lake City, UT 84112, USA

Jo Ann T. Tschanz
Department of Psychology, Utah State University, Logan, UT, USA
Center for Epidemiologic Studies, Utah State University, Logan, UT, USA

Christopher D. Corcoran
Department of Mathematics and Statistics, Utah State University, Logan, UT, USA
Alzheimer's Disease Neuroimaging Initiative, University of Southern California, Los Angeles, CA 90089, USA

Mbaye Tine
UFR des Sciences Agronomiques, de l'Aquaculture et des Technologies Alimentaires (UFR S2ATA), Universite Gaston Berger (UGB), Route de Ngallele BP 234, Saint-Louis, Senegal

Kazuo Araki, Jun-ya Aokic, Junya Kawase, Akiyuki Ozaki, Hiroshi Fujimoto, Ikki Yamamoto and Hironori Usuki
Research Center for Aquatic Breeding, National Research Institute of Aquaculture, Fisheries Research Agency, 224 Hiruda, Tamaki-cho, Watarai, Mie 519-0423, Japan

Kazuo Araki and Junya Kawase
Marine Biological Science, Faculty of Bio-resources, Mie University Graduate School, 1577 Kurimamachiya-cho, Tsu City, Mie 514-8507, Japan

Kazuhisa Hamada
Marine Farm Laboratory Limited Company, 309 Takahiro Tachibaura Otsuki-cho, Hata-gun, Kochi 788-0352, Japan

Valentina Donà and Vincent Perreten
Institute of Veterinary Bacteriology, Vetsuisse Faculty, University of Bern, Bern, Switzerland

Chunsheng Gao, Chaohua Cheng, Lining Zhao, Yongting Yu, Qing Tang, Pengfei Xin, Touming Liu, Zhun Yan, Yuan Guo, and Gonggu Zang
Institute of Bast Fiber Crops, Chinese Academy of Agricultural Sciences/Key Laboratory of the Biology and Process of Bast Fiber Crops, Ministry of Agriculture, Changsha 410205, China

Lingjie Ma, Sheng-Wei Ma, Qingyan Deng, Yang Yuan, Zhaoyan Wei, Haiyan Jia and Zhengqiang Ma
The Applied Plant Genomics Laboratory of Crop Genomics and Bioinformatics Centre, Nanjing Agricultural University, Jiangsu 210095, China

Mehdi Dehghani
Hematology and Medical Oncology Department, Hematology Research Center, Shiraz University of Medical Sciences, Shiraz, Iran

Zahra Samani, Hassan Abidi, Reza Mahmoudi and Mohsen Nikseresht
Cellular and Molecular Research Center, Yasuj University of Medical Sciences, Yasuj, Iran

Leila Manzouri
Social Determinant of Health Research Center, Yasuj University of Medical Sciences, Yasuj, Iran

Saeed Hosseini Teshnizi
Biostatistican, Molecular Medicine Research Center, Hormozgan University of Medical Sciences, Bandar Abbas, Iran

Vivek Kumar Singh, Amit Kumar, Subbaiyan Gopala Krishnan, Ranjith Kumar Ellur, Prolay Kumar Bhowmick and Ashok Kumar Singh
ICAR-Indian Agricultural Research Institute, Division of Genetics, New Delhi 110012, India

Brahma Deo Singh
Banaras Hindu University, School of Biotechnology, Varanasi 221005, Uttar Pradesh, India

Sadhna Maurya
ICAR-Indian Agricultural Research Institute, Division of Plant Physiology, New Delhi 110012, India

Kunnummal Kurungara Vinod
Rice Breeding and Genetics Research Centre, ICAR-Indian Agricultural Research Institute, Aduthurai 612 101, India

Gang Nie, Linkai Huang, Xiao Ma, Yajie Zhang, Lu Tang and Xinquan Zhang
Department of Grassland Science, Animal Science and Technology College, Sichuan Agricultural University, Chengdu, Sichuan 611130, China

Zhongjie Ji
Department of Agronomy, Purdue University, West Lafayette, IN 47906, USA

Index

A

Actinobacillus Porcitonsillarum, 137, 144
Alzheimer's Disease, 58, 104-106, 108-109
Alzheimer's Disease Neuroimaging Initiative, 104
Ampc Genes, 21, 28, 31
Angiogenesis, 17, 53, 60
Autism Spectrum Disorder, 32

B

Band 3 Anion Transport Protein, 111-112, 121
Biotic Stressors, 111, 113, 121
Bootstrap Method, 80

C

Cephalosporins, 20-22, 28
Chromobacterium, 1-9
Chromosome Sequences, 24, 26, 29
Chromosome-level Genome Assemblies, 125
Chronic Calorie Restriction, 69, 74, 77
Ciliogenesis, 41
Circular Genome Map, 3, 5
Circular Rnas, 44, 58-62
Clear Cell Renal Cell Carcinoma, 10-11, 18-19
Coding Sequences, 3, 5, 27, 29, 71, 92
Colorectal Cancer, 32, 41, 50, 52-55, 60-62, 176-177
Crassulacean Acid, 79, 88
Cytoplasm, 45, 53, 58, 82

D

Differentially Expressed Genes, 42, 69-74, 145, 147-151, 159, 196, 198
Digital Dna-dna Hybridization, 7
Dna-binding Domain, 32, 34, 91, 99

E

Enterobacter Cloacae Complex, 20, 26, 30
Enzyme Adenosine Deaminase, 45
Exon Circularization, 45-46, 58
Exonic Circrnas, 45

F

Farnesyl-diphosphate, 172

G

Gene Expression Omnibus, 69-70
Gene Ontology, 69-71, 81-82, 132-133, 145, 147, 151, 193, 201

Genome Sequence, 1-3, 8-9, 22, 31, 92, 125-127, 129-131, 133-135, 137-139, 142-144, 158, 192-193, 199-200
Genome Structural Variation, 125
Genome-wide Expression Profile, 146
Genotype-phenotype Linkage Analysis, 125

H

Heavy Metal Resistance Gene Clusters, 29-30
Hemolytic Enterotoxins, 1
Hidden Markov Model, 92, 94

I

Intermittent Calorie Restriction, 69, 74, 76
Intron Pairing, 45-46

L

Lactate Dehydrogenase, 111, 121, 123
Lung Cancer, 18, 50, 52-55, 60-62, 171-177
Lytic Proteins, 1, 5

M

Marker-aided Selection, 179, 181
Micrornas, 10, 19, 44, 59, 155
Mirna Loci, 11-13
Mirna Transcriptome, 10
Miscanthus Sinensis, 190, 199-201
Mitogen-activated Protein Kinase, 111-112, 121
Multiple Sequence Alignments, 42, 92

N

Naphthylacetic Acid, 91-92
Neutral Markers, 111, 118
Next Generation Sequencing, 70, 125-126
Noncoding Rnas, 18, 44, 58, 77
Nonredundant Protein, 21, 81
Nonsmall Cell Lung Cancer, 171
Nudix Hydrolases, 1, 5

O

Oligonucleotides, 65, 115

P

Parkinson's Disease, 105
Phylogenetic Analysis by Maximum Likelihood, 81
Phylogenetics, 8, 122, 199
Phylogeny, 3-4, 41, 80-81, 91, 102, 139, 142
Phytoene Synthase, 1, 6

Polygenic Scores, 104, 108
Proteomics, 123, 200
Pulsed-field Gel Electrophoresis, 20, 24-25

R
Rna Sequencing, 19, 69-70, 77, 126, 161, 190-191, 199-200

S
Schizophrenia, 104, 109
Sequence Characterized Amplified Region (SCAR) Marker, 63
Single Nucleotide Polymorphisms, 125, 139
Small Rna-sequencing, 10-12

T
Tetrathionate, 20, 25-27, 29
Transcriptome Analysis, 42-43, 67, 69-70, 133, 155, 191, 200-201
Transcriptomics, 77, 158, 167

W
Whole-genome Sequencing, 9, 126